Volume 29

CRM PROCEEDINGS & LECTURE NOTES

Centre de Recherches Mathématiques
Université de Montréal

Bäcklund and Darboux Transformations. The Geometry of Solitons

AARMS-CRM Workshop
June 4–9, 1999
Halifax, N.S., Canada

Alan Coley
Decio Levi
Robert Milson
Colin Rogers
Pavel Winternitz
Editors

The Centre de Recherches Mathématiques (CRM) of the Université de Montréal was created in 1968 to promote research in pure and applied mathematics and related disciplines. Among its activities are special theme years, summer schools, workshops, postdoctoral programs, and publishing. The CRM is supported by the Université de Montréal, the Province of Québec (FCAR), and the Natural Sciences and Engineering Research Council of Canada. It is affiliated with the Institut des Sciences Mathématiques (ISM) of Montréal, whose constituent members are Concordia University, McGill University, the Université de Montréal, the Université du Québec à Montréal, and the Ecole Polytechnique. The CRM may be reached on the Web at www.crm.umontreal.ca.

American Mathematical Society
Providence, Rhode Island USA

The production of this volume was supported in part by the Fonds pour la Formation de Chercheurs et l'Aide à la Recherche (Fonds FCAR) and the Natural Sciences and Engineering Research Council of Canada (NSERC).

2000 *Mathematics Subject Classification.* Primary 37K35, 35Q51, 17B80, 35Qxx, 37Kxx.

Library of Congress Cataloging-in-Publication Data

AARMS-CRM Workshop (1999 : Halifax, N.S.)
Bäcklund and Darboux transformations : the geometry of solitons : AARMS-CRM Workshop, June 4–9, 1999, Halifax, N.S., Canada / Alan Coley . . . [et al.], editors.
p. cm. — (CRM proceedings & lecture notes, ISSN 1065-8580 ; v. 29)
Includes bibliographical references.
ISBN 0-8218-2803-7 (alk. paper)
1. Solitons—Congresses. 2. Bäcklund transformations—Congresses. 3. Darboux transformations—Congresses. 4. Geometry, Differential—Congresses. I. Title: Geometry of solitons. II. Coley, A. A. III. Title. IV. Series.

QC174.26.W28 A2 1999
530.12′4—dc21 2001053367

∞ The paper used in this book is acid-free and falls within the guidelines
established to ensure permanence and durability.
This volume was submitted to the American Mathematical Society
in camera ready form by the Centre de Recherches Mathématiques.
Visit the AMS home page at URL: http://www.ams.org/

10 9 8 7 6 5 4 3 2 1 06 05 04 03 02 01

Selected Titles in This Series

29 **Alan Coley, Decio Levi, Robert Milson, Colin Rogers, and Pavel Winternitz, Editors,** Bäcklund and Darboux transformations. The geometry of solitons, 2001

28 **J. C. Taylor, Editor,** Topics in probability and Lie groups: Boundary theory, 2001

27 **I. M. Sigal and C. Sulem, Editors,** Nonlinear dynamics and renormalization group, 2001

26 **J. Harnad, G. Sabidussi, and P. Winternitz, Editors,** Integrable systems: From classical to quantum, 2000

25 **Decio Levi and Orlando Ragnisco, Editors,** SIDE III—Symmetries and integrability of difference equations, 2000

24 **B. Brent Gordon, James D. Lewis, Stefan Müller-Stach, Shuji Saito, and Noriko Yui, Editors,** The arithmetic and geometry of algebraic cycles, 2000

23 **Pierre Hansen and Odile Marcotte, Editors,** Graph colouring and applications, 1999

22 **Jan Felipe van Diejen and Luc Vinet, Editors,** Algebraic methods and q-special functions, 1999

21 **Michel Fortin, Editor,** Plates and shells, 1999

20 **Katie Coughlin, Editor,** Semi-analytic methods for the Navier-Stokes equations, 1999

19 **Rajiv Gupta and Kenneth S. Williams, Editors,** Number theory, 1999

18 **Serge Dubuc and Gilles Deslauriers, Editors,** Spline functions and the theory of wavelets, 1999

17 **Olga Kharlampovich, Editor,** Summer school in group theory in Banff, 1996, 1998

16 **Alain Vincent, Editor,** Numerical methods in fluid mechanics, 1998

15 **François Lalonde, Editor,** Geometry, topology, and dynamics, 1998

14 **John Harnad and Alex Kasman, Editors,** The bispectral problem, 1998

13 **Michel C. Delfour, Editor,** Boundaries, interfaces, and transitions, 1998

12 **Peter C. Greiner, Victor Ivrii, Luis A. Seco, and Catherine Sulem, Editors,** Partial differential equations and their applications, 1997

11 **Luc Vinet, Editor,** Advances in mathematical sciences: CRM's 25 years, 1997

10 **Donald E. Knuth,** Stable marriage and its relation to other combinatorial problems: An introduction to the mathematical analysis of algorithms, 1997

9 **D. Levi, L. Vinet, and P. Winternitz, Editors,** Symmetries and integrability of difference equations, 1996

8 **J. Feldman, R. Froese, and L. M. Rosen, Editors,** Mathematical quantum theory II: Schrödinger operators, 1995

7 **J. Feldman, R. Froese, and L. M. Rosen, Editors,** Mathematical quantum theory I: Field theory and many-body theory, 1994

6 **Guido Mislin, Editor,** The Hilton Symposium 1993: Topics in topology and group theory, 1994

5 **D. A. Dawson, Editor,** Measure-valued processes, stochastic partial differential equations, and interacting systems, 1994

4 **Hershy Kisilevsky and M. Ram Murty, Editors,** Elliptic curves and related topics, 1994

3 **Rémi Vaillancourt and Andrei L. Smirnov, Editors,** Asymptotic methods in mechanics, 1993

2 **Philip D. Loewen,** Optimal control via nonsmooth analysis, 1993

1 **M. Ram Murty, Editor,** Theta functions: From the classical to the modern, 1993

Contents

Preface

This meeting took place in Halifax, Nova Scotia, Canada between June 4th and June 9th, 1999. It brought together some sixty scientists from twenty countries. The program of the Workshop consisted of two parts:

- Two preliminary days of invited lectures, the latter being each of two hours duration and intended as an introduction for non-specialists to the subject of Darboux and Bäcklund transformations.
- The main body of technical lectures for those specialists in soliton theory with a research interest in its geometric underpinnings with emphasis on the role of Bäcklund and Darboux transformations.

Bäcklund and Darboux transformations have a long history and both owe their origin to geometric investigations carried out in the 19th century. Thus, Bäcklund transformations initially arose in connection with the construction of pseudospherical surfaces. The Gauss–Weingarten system for the latter, in asymptotic coordinates immediately provides a contact with soliton theory in that it yields, via its compatibility conditions, the integrable sine-Gordon equation. The latter was obtained originally in this geometric context by Edmond Bour in 1862. The following year, Julius Weingarten set down another integrable equation, namely the hyperbolic sinh–Gordon equation in his treatise on surfaces of constant mean curvature.

Both pseudospherical surfaces and surfaces of constant mean curvature are included in a more general category of surfaces that had been introduced in 1861 by Weingarten and which are defined by the requirement that the principal curvatures are functionally dependent. Ribacour in 1872 studied such Weingarten or so-called W surfaces and obtained results which are akin to those derived later by Bianchi in 1879 in his habilitation thesis. Therein, what has become known as the Bianchi transformation for the construction of pseudospherical surfaces was presented. The celebrated Bäcklund transformation for pseudospherical surfaces was subsequently derived in 1883 and constitutes a conjugation of the parameter-independent Bianchi transformation and a Lie symmetry which serves to intrude a vital "Bäcklund" parameter. The latter proved essential to the next major development which took place in 1892. In that year, under the title *Sulla Transformatione di Bäcklund per le Superficie Pseudosferiche* the Italian geometer Luigi Bianchi, in a masterly breakthrough, established that the Bäcklund transformation at the level of the sine-Gordon equation admits a commutative property. A consequence of this is a nonlinear superposition principle embodied in what Bianchi termed a permutability theorem. The evidence that Bianchi's permutability theorem has important application in nonlinear physics had to await the work of Seeger et al. in 1953 on crystal dislocations. Therein, in the context of Frenkel and Kontorova's dislocation theory of 1938, the superposition of so-called "eigenmotions" was obtained via the permutability theorem. Indeed, the interaction of what today is called a breather with a kink-type dislocation was both described analytically by means of the permutability theorem and displayed graphically. The typical solitonic features to be later discovered numerically by Zabusky and Kruskal in 1965 for the Korteweg–de Vries equation, namely preservation of velocity and shape following interaction, as well as the concomitant phase shift were all derived via the permutability theorem for the sine-Gordon equation in this remarkable work.

In 1967, Lamb obtained the classical sine-Gordon equation in an analysis of the propagation of ultrashort light pulses. He had become aware of the earlier work of Seeger et al. and exploited the permutability theorem associated with the Bäcklund transformation to generate an analytic expression for pulse decomposition corresponding to the two-soliton solution. Later in 1971, he used the permutability theorem to analyse the decomposition of $2N\pi$ light pulses into N stable 2π pulses. The experimental evidence for such a decomposition behaviour had been provided by Gibbs and Slusher in 1970 who recorded the decomposition of a 6π pulse into there 2π pulsed in a *Rb* vapour. In the same year, Scott had noted how the permutability theorem may be exploited in the study of long Josephson junctions.

It was in 1895, that two Dutch mathematicians Korteweg and de Vries had set down the nonlinear evolution equation which now bears their name and which models long wave propagation in a rectangular channel. However, it is less well-known that what is now called the Korteweg–de Vries (KdV) equation had, in fact, been set down earlier by Boussinesq in his memoir of 1877 entitled *Essai sur la Théorie des Eaux Courantes*. Indeed, a pair of equations equivalent to the KdV equation appeared as early as 1871 in two papers by Boussinesq devoted to the study of wave propagation in rectangular channels. The KdV equation admits a simple wave solution which at last provided a theoretical confirmation of the existence of the "large wave of translation" observed in 1834 by the Scottish engineer John Scott Russell on the Union Canal near Edinburgh. In 1973, Wahlquist and Estabrook demonstrated that the KdV equation, like the sine-Gordon equation, admits invariance under a Bäcklund-type transformation and, moreover, possesses an associated permutability theorem which was used iteratively to construct multi-soliton solutions. Interestingly, this nonlinear superposition principle arises elsewhere "mutatis mutandis" as the ε-algorithm of numerical analysis.

In 1974, a Bäcklund transformation for the nonlinear Schrödinger (NLS) equation was constructed by Lamb via a classical method developed by Clairin in 1910. It may also be generated in a purely geometric manner via an intrinsic formalism used in kinematic studies in hydrodynamics. A permutability theorem can again be constructed by means of the Bäcklund transformation. The NLS equation has its roots in an analysis published in 1905 by da Rios, a student of Levi Civita, of the three-dimensional motion of an isolated non-stretching vortex filament in an unbounded, inviscid fluid. Thus, da Rios obtained a pair of coupled nonlinear equations for the temporal evolution of the curvature and torsion of the vortex filament subsequently to be rediscovered by Betchov in 1965. It is these nonlinear equations that were composed by Hasimoto in 1972 to produce the NLS equation. The latter, seems to have been first set down explicitly by Kelley and Talanov in 1965 in independent studies of the self-focusing of optical beams in nonlinear Kerr media. In geometric terms, the work of Hasimoto showed that the NLS equation is generated by the evolution of an inextensible curve moving in space with local normal velocity in the direction of the binormal and with velocity magnitude proportional to the local curvature.

In 1973, the celebrated AKNS spectral scheme was introduced by Ablowitz et al. A broad spectrum of $1+1$-dimensional nonlinear evolution equations amenable to the Inverse Scattering Transform were encapsulated thereby as compatibility conditions. The linear structure of the AKNS system permits the application in soliton theory of another important class of transformations with their origin in the 19th Century, namely Darboux transformations. The latter arose in a studies by Darboux in 1882 both with regard to Sturm–Liouville problems and in the theory of surfaces having spherical representation. In actuality Darboux's classical transformation is but a special case of a result due to Moutard as earlier reported by the geometers Bertrand and Bonnet in Comptes Rendus in 1870 but only published by Moutard in 1878.

Iterated Darboux transformations were subsequently later constructed by Crum in 1955 in connection with connected Sturm–Liouville problems. Twenty years later, the Crum transformation was taken up by Wadati et al. and used to generate multi-soliton solutions of integrable equations associated with the AKNS system. In geometric terms, these iterated versions of Darboux transformations arise in the classical theory of surfaces as Levy sequences and are described in Eisenhart's *Transformations of Surfaces.* In a similar manner, the binary Darboux transformation of soliton theory is nothing but the fundamental transformation (F) of conjugate nets again discussed extensively in the treatise of Eisenhart.

By 1974, the Bäcklund transformations for the canonical $1+1$-dimensional soliton equations, namely the sine-Gordon, KdV and NLS equations were all in place. Moreover the derivation of the AKNS scheme had opened the way to the application of the classical Darboux transformation and its variants to the linear representations generic to solitonic equations. Thus it was timely that in 1974 a National Science Foundation meeting was convened at Vanderbilt University in the USA to assess the status and potential role of Bäcklund and associated transformations in soliton theory.

In the quarter of a century that has passed since that Meeting it has become evident that the privileged classes of surfaces that admit Bäcklund–Darboux transformations are intrinsically tied to the privileged classes of nonlinear equations that admit solitonic phenomena. It is this remarkable fact that prompted the organisation of the present international meeting. The title and scope of the Workshop was motivated by the geometric origins of Bäcklund and Darboux transformations and their subsequent important applications in soliton theory. The main objective of the Workshop was to bring together in an active forum, leading international experts in the areas of Bäcklund and Darboux transformations and the underlying geometry of soliton theory to discuss modern developments but set in an historical context. The latter aspect was regarded as of particular importance since it has only recently become apparent that many other classes of surfaces that arise in classical studies, such as, the isothermic system and associated Calapso equation, the Demoulin system of projective geometry and the Lamé system are solitonic in nature. Indeed, Bianchi diagrams associated with classical permutability theorems for such systems can be interpreted as quadrilaterals on integrable discrete surfaces.

The broad scope of the lectures presented at the Workshop bears eloquent testament to the vigour of the subject. It supports the underlying conviction that

Bäcklund and Darboux transformations and the study of the surfaces, both continuous and discrete, that admit such transformations have a permanent place in that important area of nonlinear physics that has become known as soliton theory.

The proceedings follow the structure of the meeting itself. Part 1 contains four of the six lecture series presented during the introductory part of the workshop. Part 2 contains 36 of the 47 contributions presented at the AARMS-CRM workshop. All contributions were refereed. For multiauthored contributions the name of the author who presented the talk in printed in italics. I some cases two speakers combined their contributions into one article. Then both names appear in italics.

A. Coley, D. Levi
R. Milson, C. Rogers
P. Winternitz

Montréal, 2001

Albert Victor Bäcklund

Born: Väsby, Sweden 1845

Died: Lund, Sweden 1922

Bäcklund received his tertiary education at the University of Lund. In 1864, he was appointed as an assistant at the Astronomical Observatory where he became a student of Professor Axel Möller. In 1868, Bäcklund received his Doktor Philosophiae for a thesis concerning the measurement of latitude from astronomical observations. Bäcklund thereafter turned to geometry and, in particular, to the work of the Norwegian mathematician Sophus Lie.

In 1874, Bäcklund was awarded a government travel scholarship to pursue his studies on the Continent for six months. He spent most of his time in Leipzig and Erlangen where he met both Klein and Lindemann. Ideas Bäcklund gained in this period inspired his later work on geometry on what have come to be known as Bäcklund transformations.

In 1878, Bäcklund was appointed to the new Extraordinary Chair in Mechanics and Mathematical Physics at Lund. He was elected a Fellow of the Royal Swedish Academy of Sciences (KVA) in 1888. In 1897, Bäcklund was awarded the Chair in Physics at the University of Lund where he was subsequently Rector during the period 1907–1909. Bäcklund retired in 1910 but continued his scholarly research until his death.

Bäcklund is usually associated with the type of transformation of surfaces that bears his name, the extensions of which have had major impact in soliton theory. Bäcklund's geometric work in this area was originally motivated by an attempt to extend Lie's theory of contact transformations. Bäcklund also made a significant incursion into the theory of characteristics which originated in the work of Monge and Ampère. Indeed, Bäcklund was regarded by both Goursat and Hadamard as the founder of the modern theory of characteristics.

Selective Bibliography

1. C. W. Oseen and A. V. Bäcklund, Year book of the Royal Swedish Academy of Sciences, 1924 (Swedish); German transl., Jahresber. Deutsch. Math. Verein **38** (1929), 113–152.

Gaston Darboux

Born: Nîmes, France 1842

Died: Paris, France 1917

Darboux entered the École Normale Supérieure in 1861 and completed his doctoral thesis sur "Les surfaces orthogonales" (1866). Between 1873 and 1878 he was suppléant to Liouville in the chair of rational mechanics at the Sorbonne and in 1880 succeeded to a chair of higher geometry previously held there by Chasles and retained it until his death. In 1884, he became a member of the Académie des Sciences. In 1902 he was elected a Fellow of the Royal Society of London and was awarded its Sylvester Medal in 1916.

Darboux's primary contributions were to geometry although he also made important incursions into analysis. He produced extensive works on orthogonal systems of surfaces, notably his "Leçons sur la théorie générale des surfaces", vol. I–IV (1887–1896) and the "Leçons sur les systèmes orthogonaux et les coordonnées curvilignes" (1898). The latter contains results on cyclides relevant to modern soliton theory.

Selective Bibliography

1. L. P. Eisenhart, *Darboux's contribution to geometry*, Bull. Amer. Math. Soc. (N.S.) **24** (1918), 227–237.
2. D. Hilbert and G. Darboux, Acta Math. **42** (1919), 269–273.
3. E. Lebon and G. Darboux, Paris, 1910, 1913.

List of Participants

Aminov, Yuriy
Instytut Matematyki
Inst. of Low Temperature
Uniwersytet w Bialymstoku
NAN of Ukraine (Kharkov)
UKRAINE
aminov@cksr.ac.bialystok.pl

Anderson, Ian
Department of Mathematics
Utah State University
Logan, UT 84322-4125
USA
anderson@sunfs.math.usu.edu

Aratyn, Henrik
Dept. of Physics
University of Illinois
Chicago, IL 60680
USA
aratyn@uic.edu

Atakishiyev, Natig
Instituto de Matematicas
Univ. Nacional Autonoma de Mexico
Apdo Postal 273-3
C.P. 62210
Cuernavaca, Morelos
MEXICO
natig@aluxe.matcuer.unam.mx

Cieslinski, Janek
Inst. Physics
Warsaw Unive. Division of Bialystok
ul. Lipowa 41
15-424 Bialystok
POLAND
janek@alpha.uwb.edu.pl

Clarkson, Peter A.
Dept. of Mathematics and Statistics
Univ. of Kent at Canterbury
Cornwall Building
Canterbury, CT2 7NF
UNITED KINGDOM
p.a.clarkson@ukc.ac.uk

Clelland, Nielsen Jeanne
Department of Mathematics
Univ. of Colorado Campus
Box 395 Boulder, CO 80309-0395
USA
jeanne.clelland@colorado.edu

Conte, Robert
Service de phys. de l'état condensé
CEA–Saclay
F-91191 Gif-sur-Yvette Cedex
FRANCE
conte@amoco.saclay.cea.fr

Ercolani, Nicholas
Department of Mathematics
Universit of Arizona
Tucson, Arizona 85721
USA
ercolani@math.arizona.edu

Estabrook, Frank
Jet Propulsion Laboratory
CALTECH
Pasadena, CA 91125
USA
frank@bottom.jpl.nasa.gov

Fels, Mark
Department of Mathematics
Utah State University
Logan, UT 84322-4125
USA
fels@sunfs.math.usu.edu

Ferapontov, Eugene
Center for Nonlinar Studies
Landau Inst. of Theoret. Physics
Academy of Science of Russia
Kosygina 2117940 MoscowGSP-1
RUSSIA
fer@landau.ac.ru

Finley III, J. Daniel
Dept. of Physics & Astronomy
The University of New Mexico
Albuquerque, NM 87131-1156
USA
finley@tagore.phys.unm.edu

Fokas, Athanassios S.
Department of Mathematics
Imperial College of Science
Technology & Medicine
180 Queen's GateLondon
SW7 2B7
ENGLAND
a.fokas@ic.ac.uk

Gegenberg, Jack
Dept. of Math. & Stat.
University of New Brunswick
Fredericton (New Brunswick)
E3B 5A3, CANADA
lenin@math.unb.ca

Gromak, Valery L.
Dept. of Math. & Mech.
Bielorussian State Univ.
F. Skorina 4220050 Minsk
BELARUS
grom@mmf.bsu.unibel.by

Grundland, Michel
Centre de recherches mathématiques
Université de Montréal
C.P. 6128, Succ. Centre-ville
Montréal (Québec)
H3C 3J7, CANADA
grundlan@CRM.UMontreal.CA

Gurses, Metin
Department of Mathematics
Bilkent University
Faculty of Sciences
06533 Bilkent
ANKARA, TURKEY
gurses@fen.bilkent.edu.tr

Havlicek, Miloslav
Faculty of Nuclear Sci. & Phys. Engr.
115 19 Prague 1 Brehova 7
CZECH REPUBLIC
havlicek@km1.fjfi.cvut.cz

Hietarinta, Jarno
Dept. of Physics
University of Turku
FIN-20014
FINLAND
hietarin@utu.fi

Hlavaty, Ladislav
Dept. of Physics
Czech Technical University
Brehova 7115 19 Prague 1
CZECH REPUBLIC
hlavaty@br.fjfi.cvut.cz

Hoenselaers, Cornelius
Dept. of Mathematical Sciences
University of Technology
Loughborough, LE11 3TU
UNITED KINGDOM
c.a.hoenselaers@lboro.ac.uk

Irwin, Patrick
Groupe de physique des particules
Université de Montréal
C.P. 6128, Succ. Centre-ville
Montréal (Québec)
H3C 3J7, CANADA
irwin@lps.umontreal.ca

Ivey, Thomas
Dept. of Mathematical Sciences
Ball State University
Muncie, IN 47306
USA
tivey@math.bsu.edu

Konopelchenko, Boris
Dipartimento di Fisica
Universita di LecceVia
Arnesano73100 Lecce
ITALY
boris.konopeltchenko@le.infn.it

Kovalyov, Mikhail
Department of Mathematical Sciences
University of Alberta
632 Central Academic Building
Edmonton (Alberta)
T6G 2H1, CANADA
misha@altair.math.ualberta.ca

Kruskal, Martin D.
Department of Mathematics
Rutgers University
Hill Center, Busch Campus
New Brunswick, NJ 08903
USA
kruskal@math.rutgers.edu

Lafortune, Stéphane
Department of Mathematics
University of Arizona
Tucson, AZ 85721-0001
USA
lafortus@math.arizona.edu

LeBlanc, Jason
Department of Mathematics
University of New Brunswick
P.O. Box 4400
Fredericton (New Brunswick)
E3B 5A3, CANADA
jason@math.unb.ca

Levi, Decio
Dipto di Fisica
Universita di Roma 3
Dip. Fisica - via della Vasca Navale 84
00146 Roma
ITALIA
levi@roma1.infn.it

Lu, Kau U.
Department of Mathematics
California State University, Long Beach
Long Beach, CA 90840
USA
lunotwen@csulb.edu

Ma, Wen-Xiu
Department of Mathematics
City University of Hong Kong
83 Tat Chee Avenue
Kowloon, HONG KONG
mawx@math.cityu.edu.hk

Mathieu, Pierre
Dép. de physique
Université Laval
Pav. Alexandre-Vachon, local 2634
Sainte-Foy (Québec)
G1K 7P4, CANADA
pmathieu@phy.ulaval.ca

Musette, Micheline
Dienst TENA
Vrije Universiteit Brussel
Pleinlaan 2
1050 Brussel
BELGIQUE
mmusette@vub.ac.be

Olver, Peter J.
School of Mathematics
University of Minnesota
127, Vincent Hall
Minneapolis, MN 55455-0488
USA
olver@ima.umn.edu

Rauch, Stefan
Department of Mathematics
Linkping Univ.
SE-58183 Linkping
SWEDEN
strauc@mai.liu.se

Reyes, Enrique G.
Department of Mathematics
Utah State University
Logan, UT 84322-4125
USA
reyes@sunfs.math.usu.edu

Richter, David
Dept. of Math. & Stat.
McGill University
805, Sherbrooke W.,
Montréal (Québec)
H3A 2K6, CANADA
richter@math.mcgill.ca

Rogers, Colin
Dept. of Applied Mathematics
University of New South Wales
School of Mathematics
Sydney 2052
AUSTRALIA
colinr@alpha.maths.unsw.edu.au

Schief, Wolfgang K.
School of Mathematics
The University of South Wales
2052 Syndey
AUSTRALIA
schief@alpha.maths.unsw.edu.au

Schmid, Rudolf
Department of Mathematics
Emory University
Atlanta, GA 30329
USA
rudolf@mathcs.emory.edu

Segur, Harvey
Program in Applied Mathematics
University of Colorado
Campus Box 426
Boulder, CO 80309-0426
USA
segur@newton.colorado.edu

Steudel, Heinz
Phys. Inst. "Nonclassical Radiation" Group
Humboldt University
Invalidenstr. 10
1010115 Berlin
GERMANY
steudel@physik.hu-berlin.de

Tenenblat, Keti
Departamento de Matematica
Universidade de Brasilia
70910-900 Brasilia DF
BRAZIL
keti@mat.unb.br

Thomova, Zora
School of Arts and Sciences
SUNY-Institute of Technology
P.O. Box 3050
Utica, NY 13504-3050
USA
thomovz@sunyit.edu

Tremblay, Sébastien
Centre de recherches mathématiques
Université de Montréal
C.P. 6128, succ. Centre-ville
Montréal (Québec)
H3C 3J7, CANADA
tremblay@CRM.UMontreal.CA

Wiegmann, Paul
James Frank Institute
University of Chicago
5640 S. Ellis Ave., RI-223
Chicago, IL 60637
USA
wiegmann@rainbow.uchicago.edu

Winternitz, Pavel
Centre de recherches mathématiques
Université de Montréal
C.P. 6128, Succ. Centre-ville
Montréal (Québec)
H3C 3J7, CANADA
wintern@CRM.UMontreal.CA

Zakrzewski, Wojciech J.M.
Department of Mathematical Sciences
University of Durham
Science Site
Durham DH1 3LE
UNITED KINGDOM
w.j.Zakrzewski@durham.ac.uk

Zhedanov, Alexei
Institut for Physics & Tech.
Donetsk University
Donetsk, 340114
UKRAINE
zhedanov@host.dipt.donetsk.ua

Part 1

Introductory Lectures

Centre de Recherches Mathématiques
CRM Proceedings and Lecture Notes
Volume **29**, 2001

Bäcklund Transformations of the Higher Order Painlevé Equations

V. I. Gromak

Abstract. The six Painlevé equations (P_1)–(P_6) were first derived around the turn of the century in an investigation by Painlevé and his colleagues [**11, 16, 17**]. Such nonlinear ordinary differential equations have the property that the singularities other than poles of any of the solutions are independent of the particular solution and so are independent only upon the equation; this property is known as the Painlevé property. Although first discovered from strictly mathematical considerations, the Painlevé equations have appeared in various of physical applications as cited in [**1–5, 9, 12**]. There has been considerable interest in Painlevé equations over the last few years due the fact that they arise as reductions of solutions of soliton equations solvable by the inverse scattering method [**1, 19**]. The generalized symmetries of such PDEs then immediately give higher order ODEs that may define new transcendental functions. The question of whether some of the higher order ODEs obtained in this way define new functions remains to be answered. Answering this question is made more difficult by the fact that since the time of Painlevé only restricted progress has been made in the systematic classification in order of ODEs having Painlevé property.

The Painlevé equations may also be thought of as nonlinear analogues of the classical special functions. There is currently much interest in studying the Painlevé equations using the isomonodromy deformation method. In this approach, the Painlevé equation is written as an integrability condition of a linear system.

There have been many recent studies of integrable mappings and discrete systems, including higher order and discrete analogues of the Painlevé equations. Therefore, any success in the study of Painlevé equations provides new results in the associated problems.

In this course we shall discuss recurrence relations and their application to the construction of various kinds of solutions of the Painlevé equations and their higher order analogues. In the context of the Painlevé equations, recurrence relations are usually referred to as Bäcklund transformations. Informally, a Bäcklund transformation is defined as a system of equations relating one solution of a given equation either to another solution of the same equation, possibly with different values of the parameters, or to a solution of another equation. In this approach one can derive various properties of the Painlevé equations and their higher order analogues, including exact solutions (rational, algebraic, classical), fundamental domain of the parameter space, special integrals and transcendence. This is the purpose of the present course.

2000 *Mathematics Subject Classification.* Primary: 34A20; Secondary: 32G34.
This is the final form of the paper.

Our course is organized as follows. In the Introduction, we discuss the Painlevé property and Bäcklund transformations. In Section 2 we review several important general properties of the second Painlevé equation (P_2), such as the character of the singularities, the representation of the solutions as a ratio of entire functions, τ-functions and the number of solution poles. Then from the Hamiltonian system we obtain differential and integral forms of Bäcklund transformations of (P_2). Using this result we find algebraic nonlinear superposition of solutions (Bianchi formula or a discrete analogue of (P_2)) and the fundamental domain of the parameter space. Since one of the principal aims of our investigation is to obtain solution hierarchies for the Painlevé equations by making multiple applications of Bäcklund transformations, in Sections 3–4 we obtain the rational solutions and classical solutions which can be expressed in terms of Airy functions. Then in Section 5 we discuss the transcendence of (P_2).

In Sections 6–7 we consider the higher order analogues of (P_1) and (P_2) which arise as exact reductions of the Korteweg–de Vries hierarchy. We show that the higher order analogues have the same properties of solutions as (P_1) and (P_2).

1. Introduction

Singular points and their classification. Cauchy's theorem. Fixed and movable singularities, examples. Painlevé property. First order P-type equations. Painlevé and Fuchs theorems. Second order P-type equations. The Painlevé list. Painlevé equations (P_1)–(P_6). Bäcklund transformations: definition and some examples.

In our course, we shall consider nonlinear ODEs (and PDEs) in the complex plane

$$w' = f(z, w), \tag{1.1}$$

$f : D \subset C^{n+1} \to C^n$, i.e. equations (or systems of equations) which are regarded as analytic functions of complex variables. Therefore a solution of equation (1.1) is an analytic function, which is determined by its singular points.

It is necessary to say something about the classification of the singular points of an analytic function $w = w(z)$.

DEFINITION 1.1. A point at which $w(z)$ fails to be analytic is called a singular point or singularity of $w(z)$.

Singular points can belong to the following classes:

(a) Isolated singular points.
(b) Nonisolated singular points.
(c) Single-valued ones, for which the function does not change its value as z goes around a given initial point z_0.
(d) Multi-valued or branch (or critical) points.
(e) The points for which the function has a limit, whether finite or infinite, as $z \to z_0$.
(f) The points for which the function $w = w(z)$ has no limit as $z \to z_0$, i.e. z_0 is an essential singular point.

A singular point may have more than one of the above properties. For example, it may be isolated and single-valued (it is then either a pole or a removable singularity or an essential singular point). It may be multi-valued, for example, finite-valued and isolated (e.g. an algebraic critical point or algebraic pole). Algebraic branch points, poles and algebraic poles are called algebraic singularities.

PROPOSITION 1.1. *Consider the system*

$$\frac{dw_j}{dz} = f_j(z, w_1, \dots, w_n), \quad w_j(z_0) = w_j^0, \qquad j = 1, \dots, n,$$

and let f be analytic in the domain $|z - z_0| \le a$, $|w_j - w_j^0| \le b$ and M be the upper bound of the set f in this domain. Then the system admits a unique solution $w(z)$, which is analytic within the circle $|z - z_0| \le \rho$ and which reduces to w_0, when $z \to z_0$, $\rho = a(1 - e^{-b/M(n+1)})$. In the linear case ρ is equal to a.

If (z_0, w_0) is a singularity of f then the solution $w(z)$ can have a singularity at z_0.

Singularities of solutions of the above system are divided into two classes: fixed singularities and movable singularities.

EXAMPLE 1.1.

$$w' = 1/z, \quad w = \ln z + C.$$

In this example $z = 0$ and $z = \infty$ are singularities for every solution and the coefficients of the equation are not analytic at $z = 0, \infty$. Those singularities are fixed.

EXAMPLE 1.2.

$$w' = -w^2, \quad w = (z + C)^{-1}.$$

In this case the solution $w(z)$ defined by the initial data w_0, z_0 has a pole at the point $-z_0 + 1/w_0$. In this example $z = -C$ is a simple pole for the general solution. It is a movable singularity.

EXAMPLE 1.3.

$$w' = e^{-w}, \quad w = \ln(z + C).$$

In this example $z = -C$ is a transcendental movable singularity.

Singularities of ODEs are of two types: fixed and movable.

Consider the nth order equation in the complex domain

$$x^{(n)} + \sum_{j=1}^{n-j} p_j(t) x^{(n-j)} = 0.$$

If the coefficients $p_j(t)$ are all analytic in the neighborhood of a point t_0, $|t - t_0| < \rho$, then t_0 is a regular point of the ODE, and there exists a unique solution which is expressed as a Taylor series in $t - t_0$, and which is convergent at least within some neighbourhood of t_0, $|t - t_0| < \rho$. Therefore the singularities of solutions of linear equations can be located only at singularities of the coefficients. These singularities are said to be fixed, since their location does not vary when we are going from one particular solution to another. For example, solutions of the hypergeometric equation

$$t(t-1)\frac{d^2x}{dt^2} + \big(\gamma - (\alpha + \beta + 1)t\big)\frac{dx}{dt} - \alpha\beta x = 0,$$

can only have singularities at the points $0, 1, \infty$. It is a general property of linear ODEs that all singularities of solutions are fixed. Nonlinear ODEs lose this property. Thus, consider the simple nonlinear ODE

$$x' + x^2 = 0.$$

Its general solution is $x(t) = 1/(t - t_0)$, where t_0 is an integration constant, which also denotes the location of the singularity. This singularity, a pole, is said to be movable since its location depends on the constant of integration t_0.

DEFINITION 1.2. The movable singularities of the solution are the singularities whose location depends on the constant of integration. Fixed singularities occur at points where the coefficients of equation are singular.

Nonlinear ODEs can have both movable and fixed singularities. For example, the equation $w' = w^2/z$ has a general solution $w = -1/\ln Cz$. In this case $z = 0$ is a fixed singular point but $z = 1/C$ is a movable singularity. A solution of an ODE can have various kinds of singularities.

DEFINITION 1.3. The ODE is said to possess the Painlevé property when all movable singularities are single-valued (poles), i.e. when solutions are free from movable critical points but can have fixed multivalued singularities.

ODEs which possess the Painlevé property are said to be of Painlevé type (P-type).

Painlevé proved that for the first order equation of the form

$$P(w', w, z) = 0,$$

where P is polynomial in w', w, and analytic in t, the only movable singularities of the solutions are poles and/or algebraic branch points.

L. Fuchs showed that an equation of P-type of the form

$$w' = \frac{P(z, w)}{Q(z, w)},$$

where P and Q are polynomials in x with coefficients analytic in z, is necessarily the Riccati equation. Fuchs solved this problem in general case for the first order ODEs.

Picard posed the problem of determining the equations of the form

$$w'' = R(w', w, z),$$

where R is a rational function of w', w and an analytic function of z, which have no solutions with movable critical points. The problem was solved by Painlevé and his colleagues. The general method used to find equations whose solutions are free from such singularities is known as the α-method (or method of small parameter). Painlevé showed that there are only 50 canonical equations which have the property of having no movable critical points. He and his colleagues further showed that forty-four equations were integrable in terms of known functions (such as elliptic functions and functions solutions of linear equations) or were reducible to one of six new nonlinear differential equations, namely

$$(\mathrm{P}_1)\quad w'' = 6w^2 + z,$$

$$(\mathrm{P}_2)\quad w'' = 2w^3 + zw + \alpha,$$

$$(\mathrm{P}_3)\quad w'' = \frac{w'^2}{w} - \frac{w'}{z} + \frac{1}{z}(\alpha w^2 + \beta) + \gamma w^3 + \frac{\delta}{w},$$

$$(\mathrm{P}_4)\quad w'' = \frac{w'^2}{2w} + \frac{3}{2}w^3 + 4zw^2 + 2(z^2 - \alpha)w + \frac{\beta}{w},$$

$$(\mathrm{P}_5)\quad w'' = \frac{3w-1}{2w(w-1)}w'^2 - \frac{1}{z}w' + \frac{1}{z^2}(w-1)^2\left(\alpha w + \frac{\beta}{w}\right) + \frac{\gamma}{z} + \frac{\delta w(w+1)}{w-1},$$

$$(\mathrm{P}_6)\quad w'' = \frac{1}{2}\left(\frac{1}{w} + \frac{1}{w-1} + \frac{1}{w-z}\right)w'^2 - \left(\frac{1}{z} + \frac{1}{z-1} + \frac{1}{w-z}\right)w' + \frac{w(w-1)(w-z)}{z^2(z-1)^2}\left(\alpha + \frac{\beta z}{w^2} + \frac{\gamma(z-1)}{(w-1)^2} + \frac{\delta z(z-1)}{(w-z)^2}\right),$$

with α, β, γ, δ arbitrary constants.

These equations are known as the Painlevé equations. The first three equations (P_1), (P_2), (P_3) were discovered by Painlevé; the fourth and fifth, (P_5) and (P_6), were added by Gambier; and the general form of the sixth, (P_6), was given by R. Fuchs. The solutions of (P_1), (P_2), (P_4) are meromorphic functions of z. If the substitution $z = e^t$ is made in (P_3) and (P_5), then the solutions become meromorphic functions of t. However, for (P_6) the points $0, 1, \infty$ are fixed critical points, and hence the general solution of (P_6) is not meromorphic throughout the complex plane. In fact, equation (P_6) contains the other five, since these may be obtained from (P_6) by passing to the limit $(\mathrm{P}_6) \to (\mathrm{P}_5) \to \cdots \to (\mathrm{P}_1)$. For example if in (P_2) $w \to \varepsilon w + \varepsilon^{-5}$, $z \to \varepsilon^2 z - 6\varepsilon^{10}$, $\alpha \to 4\varepsilon^{-5}$ and $\varepsilon \to 0$, then $(\mathrm{P}_2) \to (\mathrm{P}_1)$. Since Painlevé proved that solutions of (P_1) define new transcendental functions, if then follows that so also do the solutions of (P_2)–(P_6), except for special values of the constants α, β, γ, δ. But what are these parameter values? In our course we investigate this problem and the properties of solutions for (P_2)–(P_6). The main approach to the investigation is the method of Bäcklund transformations.

We start with some examples of Bäcklund transformations.

(1) The solutions of the following linear and Riccati equations

$$x'' + p_1(t)x' + p_2(t)x = 0, \quad y' + y^2 + p_1(t)y + p_2(t) = 0,$$

are related by the formulae

$$x = \exp\int y\,dt, \quad y = x'/x.$$

(2) The Cole–Hopf transformation $u = (\ln v)_x$ reduces the Burgers's equation (it is the simplest and rather interesting nonlinear evolution, a one-dimensional reduction of Navier–Stokes equation)

$$u_t = u_{xx} + 2uu_x,$$

to the linear diffusion equation $v_t = v_{xx}$.

(3) The transformation between the Liouville equation $u_{xy} = \exp u$ and the wave equation $\bar{u}_{xy} = 0$ has the following form

$$u_x + \bar{u}_x = \sqrt{2}\exp(u - \bar{u})/2, \quad u_y - \bar{u}_y = \sqrt{2}\exp(u + \bar{u})/2.$$

These transformations are examples of Bäcklund transformations between two different equations. If the equations are the same, these transformations are called auto-Bäcklund transformations.

(4) It is known that there exist recurrence relations between solutions of the Bessel equation with different values of the parameters

$$z^2 w'' + zw' + (z^2 - \nu^2)w = 0,$$

$$J_{\nu+1} = \frac{2\nu}{z}J_\nu - J_{\nu-1}, \quad J_{\nu+1} = J_{\nu-1} - 2J'_\nu.$$

(5) Having a solution ϕ of the sine–Gordon equation $\phi_{xy} = \sin\phi$, it is possible to construct a new solution ψ by means of the following equations

$$\psi_x = \phi_x + 2a \sin\frac{\phi+\psi}{2}, \qquad \psi_t = -\phi_t + \frac{2}{a}\sin\frac{\phi+\psi}{2}.$$

(6) The KdV equation

$$u_t - 6uu_x + u_{xxx} = 0,$$

can be reduced to the equation

$$w_t - 3w_x^2 + w_{xxx} = 0,$$

by the transformation $u = w_x$ and one integration. Then

$$w_x + \bar{w}_x = -m + (w - \bar{w})^2/2,$$
$$w_t + \bar{w}_t = (w - \bar{w})(w_{xx} - \bar{w}_{xx}) + 2(u^2 + u\bar{u} + \bar{u}^2),$$

where w and $\bar{w}$ are the solutions and u, $\bar{u}$ are the corresponding solutions of KdV equation.

DEFINITION 1.4. A relation

$$R(u, v, u_x, v_x, \dots) = 0,$$

or a system of such relations, or the explicit formulae

$$u = R_1(v, v_x, v_t, \dots), \qquad v = R_2(u, u_x, u_t, \dots),$$

is called a Bäcklund transformation of equations

$$E(u) = 0, \quad D(v) = 0,$$

if v is a solution of $D(v) = 0$ whenever u is a solution of $E(u) = 0$ or vice versa.

2. The Second Painlevé Equation

2.1. General properties of the solutions of the equation (P₂). Character of singularities. Representation of the solutions as a ratio of entire functions. τ-functions. 1-solutions. Hamiltonian form of an equivalent system. Hamiltonian of (P_2). The number of solution poles.

The second Painlevé equation has the form

$$w'' = 2w^3 + zw + \alpha. \tag{$P2$}$$

It has one parameter α and the simplest form of the Painlevé equations. Painlevé proved that any solution of (P_2) is a meromorphic function of z. Therefore the singularities of any solution are poles. In the neighbourhood of each pole z_0, a solution has the following expansion

$$w = \frac{\varepsilon}{\tau} - \frac{\varepsilon}{6}z_0\tau - \frac{\alpha+\varepsilon}{4}\tau^2 + h\tau^3 - \frac{3\alpha+\varepsilon}{72}z_0\tau^4 + \dots, \quad \varepsilon^2 = 1, \tag{2.1}$$

where $z - z_0 = \tau$, h is an arbitrary constant.

The theorem on meromorphic functions implies that each solution of (P_2) can be expressed as the ratio of two entire functions, i.e.

$$w(z) = \frac{v(z)}{u(z)}. \tag{2.2}$$

From (2.1) there follow the relations

$$w^2 = \frac{1}{(z-z_0)^2} - z_0/3 + O(z-z_o), \tag{2.3}$$

and

$$\int_{z_1}^{z} w^2\, dz = -\frac{1}{z-z_0} + O(z-z_0).$$

Therefore the function

$$u(z) = \exp\left(-\int_{z_1}^{z} dz \int_{z_1}^{z} w^2\, dz\right), \tag{2.4}$$

where the path of integration does not pass through singularities of $w(z)$, is for an arbitrary solution an entire function, and it has a simple zero at z_0. Therefore the function $v(z)$ is entire too. From (2.3) and equation (P_2), it follows that the equations for $u(z)$ and $v(z)$ have the form

$$\begin{gathered} uu'' = u'^2 - v^2, \\ v''u^2 + vu'^2 - 2u'v'u = v^3 + zvu^2 + \alpha u^3. \end{gathered} \tag{2.5}$$

It can be proved that all the solutions of system (2.4) are entire and that, if (u, v), $u \neq 0$, is an arbitrary solution of (2.4), then (2.2) determines a solution of (P_2).

PROPOSITION 2.1. *Any nonrational solution of* (P_2) *has an infinite number of simple poles.*

This result follows from the Vittikh theorem stating that an algebraic differential equation $P(z, w, w', \ldots, w^{(n)}) = 0$ with one dominant term has no entire transcendental solutions.

Equation (P_2) has a one-parameter family of solutions (1-solutions), which is defined by the general solution of some Riccati equation

$$w' = a(z)w^2 + b(z)w + c(z). \tag{2.6}$$

Indeed, from (2.6) we have

$$w'' = a'w^2 + b'w + c' + (2aw + b)(aw^2 + bw + c).$$

Therefore the coefficients a, b, c are determined by the system

$$a^2 = 1,\ a' + 3ab = 0, \quad b' + 2ac + b^2 = z, \qquad c' + bc = \alpha.$$

Then $a^2 = 1$, $b = 0$, $c = z/2a$, $\alpha = a/2$.

PROPOSITION 2.2. *All the solutions of the equation*

$$w' = \varepsilon w^2 + \varepsilon z/2, \quad \alpha = \varepsilon/2, \qquad \varepsilon^2 = 1, \tag{2.7}$$

are solutions of (P_2).

If $w = -\varepsilon u'/u$ in (2.6) then $u'' + (z/2)u = 0$. The last equation is an Airy equation. Therefore 1-solutions of (P_2) for $\alpha = \varepsilon/2$ can be expressed by means of Airy function.

Let us consider the system

$$w' = \varepsilon w^2 + \varepsilon z/2 + v, \quad v' = \alpha - \varepsilon/2 - 2\varepsilon wv, \qquad \varepsilon^2 = 1. \tag{2.8}$$

Elimination of v from (2.8) yields (P_2). On the other hand (2.8) yields

$$w = \frac{\varepsilon}{2v}(\alpha - \varepsilon/2 - v'), \quad v \neq 0, \tag{2.9}$$

and

$$v = w' - \varepsilon w^2 - \varepsilon z/2, \tag{2.10}$$

where

$$v'' = \frac{v'^2}{2v} - zv - 2\varepsilon v^2 - \frac{(\alpha - \varepsilon/2)^2}{2v}. \tag{2.11}$$

The first equation of the system (2.8) is a Riccati equation and it may be considered as the Miura transformation of (P_2). In this case equation (2.5) is the result of the Miura transformation. Also the system (2.8) is a linearization of the equation (P_2). System (2.8) is a Hamiltonian system with Hamiltonian

$$H(z, w, v) = \frac{v^2}{2} + \varepsilon(w^2 + z/2)v - \left(\alpha - \frac{\varepsilon}{2}\right)w.$$

Now we can denote by $h(z) = H\big(z, w(z), v(z)\big)$ and $\tau(z) = \exp\int_a^z h(z)\,dz$. In this case $\tau(z)$ is an entire function and $h(z)$ satisfies the equation

$$h''^2 + 4h'^3 + 2h'(zh' - h) - \big(\alpha + (1/2)\big)^2/4 = 0,$$

and

$$w = \frac{2h'' + \alpha + 1/2}{4h'}, \qquad v = -2h', \qquad \varepsilon = -1.$$

Note that if in the system (2.8) v vanishes, then we have 1-solutions (2.7) for $\alpha = \varepsilon/2$.

As for equation (2.5), it has the P-property. This equation is the 34th in the Painlevé list and its general solution is

$$v = -\varepsilon(y' + y^2 + z/2), \tag{2.12}$$

where y is the general solution of equation

$$y'' = 2y^3 + zy + \alpha - \varepsilon/2 - 1/2. \tag{2.13}$$

We can verify this fact by differentiating and substituting (2.12) into equation (2.11). Equation (2.13) is the second Painlevé equation with parameter $\alpha - (\varepsilon+1)/2$.

2.2. Bäcklund transformations of (P_2). T, T^{-1}, S-transformations. Integral form of B-transformation. Group of B-transformations. Nonlinear superposition of solutions (Bianchi's formula). Discrete form of (P_2). Fundamental domain of parameter space (Weyl chamber).

Now we can construct the Bäcklund transformation of (P_2). First of all from (2.8) and (2.12) we have

$$\varepsilon w' - w^2 + y' + y^2 = 0. \tag{2.14}$$

This equation relates solutions of (P_2) with two different values of parameters. Differentiating equation (2.14), we obtain

$$y = -\varepsilon w + \frac{(\alpha - 1/2)(\varepsilon + 1)}{2(\varepsilon w' - w^2 - z/2)}.$$

Of course, we can find w by means of y. In this case we have

$$w = -\varepsilon y - \frac{\alpha(\varepsilon+1)-1}{2\varepsilon(y' + y^2 + z/2)}.$$

PROPOSITION 2.3 (Bäcklund transformations). *If $w_\alpha = w(z,\alpha)$ is a solution of* (P_2) *for a given value of the parameter α, then the transformations*

$$S : w(z,\alpha) \to w(z,-\alpha) = -w(z,\alpha), \tag{2.15}$$

$$T : w_{\alpha-1} \to w_\alpha = -w_{\alpha-1} - \frac{\alpha - 1/2}{w'_{\alpha-1} + w^2_{\alpha-1} + z/2}, \tag{2.16}$$

$$T^{-1} : w_\alpha \to w_{\alpha-1} = -w_\alpha + \frac{\alpha - 1/2}{w'_\alpha - w^2_\alpha - z/2}, \tag{2.17}$$

determine other solutions of (P_2).

In (2.15)–(2.17) $TT^{-1} = T^{-1}T = S^2 = \mathrm{id}$. We can regard S, T, T^{-1} as the transformations on the set $F = \{\alpha, w\}$, where $\alpha \in C$ and w is a solution of (P_2). In this case S and T generate a transformation group G_2 on F. In addition, we define the transformations

$$\tau \colon C \to C, \quad \alpha \to -\alpha, \qquad \tau_+ \colon C \to C, \quad \alpha \to \alpha + 1,$$

associated with the transformations S, T. This is a group $\widetilde{G_2}$.

PROPOSITION 2.4 ([**15**]). *$\widetilde{G_2} \simeq G_2 \simeq W(A_1)$, where $W(A_1)$ is the Affine Weyl group associated with the Lie algebra of type A_1.*

Let $\alpha = \varepsilon/2$, $v \neq 0$ in (2.1). *This condition is equivalent to*

$$R(z,\varepsilon/2) \equiv w'(z,\varepsilon/2) - \varepsilon w^2(z,\varepsilon/2) - \varepsilon/2 \neq 0,$$

and the equation for v is

$$v'' = \frac{v'^2}{2v} - zv - 2\varepsilon v^2.$$

Let v be $v = \lambda y^2$. Then

$$y'' = -\varepsilon\lambda y^3 - zy/2.$$

Let z be $z = -2^{1/3}\tau$, $\lambda = -\varepsilon 2^{1/3}$. Then

$$y'' = 2y^3 + \tau y.$$

Thus we have obtained the following result.

PROPOSITION 2.5. *The solutions $w(z,\varepsilon/2)$, $w(z,0)$ of equation* (P_2) *are linked by the formulae*

$$\begin{aligned} \lambda w^2(\tau,0) &= w'(z,\varepsilon/2) - \varepsilon w^2(z,\varepsilon/2) - \varepsilon z/2, \\ w(z,\varepsilon/2) &= -\frac{\varepsilon}{2v} v'_z = 2^{-1/3}\varepsilon\frac{w'(\tau,0)}{w(\tau,0)}. \end{aligned} \tag{2.18}$$

The following result is a consequence of the Proposition 2.3, 2.5.

PROPOSITION 2.6. *In order to find the general solution of* (P_2) *for arbitrary values of the parameter α, it is sufficient to consider the values of α satisfying*

$$0 \leq \Re\alpha \leq 1/2. \tag{2.19}$$

This is the fundamental domain of parameter space (or Weyl chamber). Note that if w is an unknown function in (2.14) then $w = -\varepsilon y$ is a particular solution. Therefore we can solve the Riccati equation (2.14). If $w(z_0)$, $w'(z_0)$ are the initial data of solution w, then

$$w_{\alpha\varepsilon-1} = -\varepsilon w_\alpha + \frac{(\alpha\varepsilon - 1/2)J(w_\alpha)}{w_0'\varepsilon - w_0^2 - z_0/2 + (\alpha - 1/2)\int_{z_0}^{z} J(w_\alpha)\,dz}, \tag{2.20}$$

where

$$J(w_\alpha) = \exp\biggl(2\int_{z_0}^{z} \varepsilon w_\alpha\,dz\biggr).$$

This is the integral form of the Bäcklund transformation.

There are some results for the system (2.16), (2.17).

- Let $\alpha \to \alpha + 1$ in (2.16). Then

$$w_{\alpha+1} = -w_\alpha - \frac{\alpha + 1/2}{w_\alpha' + w_\alpha^2 + z/2}.$$

From (2.16) we find

$$w_\alpha' = w_\alpha^2 + z/2 + \frac{\alpha - 1/2}{w_{\alpha-1} + w_\alpha}.$$

Then we have

$$w_{\alpha+1} = -w_\alpha - \frac{\alpha + 1/2}{2w_\alpha^2 + z + (\alpha - 1/2)/(w_\alpha + w_{\alpha-1})}. \tag{2.21}$$

This result can be regarded as a Bianchi's formula or a discrete analogue of (P_2). It is a consequence of the following

PROPOSITION 2.7. *Assume that the functions w_α, $w_{\alpha-1}$ satisfy the system* (2.16), (2.17). *Then they together with $w_{\alpha+1}$ from* (2.21) *are solutions of equation* (P_2) *for the parameters α, $\alpha - 1$, $\alpha + 1$.*

The transformations (2.16), (2.17) allow us to construct solutions for different values of the parameters. But if $w(z,\alpha)$ is a solution of (P_2), then

$$S_j : w(z,\alpha) = w_j(z,\alpha) = \mu w(\mu z, \alpha), \quad \mu = \exp(2\pi i j/3).$$

This is an example of auto-Bäcklund transformation.

PROPOSITION 2.8. *Let $w(z,\alpha)$ be a solution of* (P_2). *Then*

$$\widetilde{w}(z,\alpha) = T^{\alpha} S T^{-\alpha} w(z,\alpha),$$

is a solution of (P_2) *too.*

It is easy to show that $\widetilde{w}(z,\alpha) = w(z,\alpha)$ if and only if $w(z,\alpha)$ is a rational solution.

3. Rational Solutions of (P_2) (0-Solutions)

General form of 0-solutions. System for polynomials $P(z)$ and $Q(z)$. Yablonski–Vorob'ev polynomials. Necessary and sufficient conditions. List of 0-solutions.

PROPOSITION 3.1. *An arbitrary 0-solution has the form*

$$w(z) = P(z)/Q(z), \tag{3.1}$$

where $P(z)$, $Q(z)$ are polynomials with $\deg P = n - 1$, $\deg Q = n$.

Consequence. An arbitrary 0-solution has an expansion

$$w(z) = \sum_{j=1}^{n} \frac{\varepsilon_j}{z - z_j}, \tag{3.2}$$

where $\operatorname{res}_{z=z_j} w(z) = \varepsilon_j$, $\varepsilon_j^2 = 1$.

PROPOSITION 3.2. *The solution w from (3.1) is a 0-solution of (P_2) if and only if $P(z)$, $Q(z)$ are solutions of the system*

$$\begin{aligned} QQ'' &= Q'^2 - P^2, \\ P''Q^2 + PQ'^2 &= P^3 + 2P'Q'Q + zPQ^2 + \alpha Q^3. \end{aligned} \tag{3.3}$$

PROOF. To establish this proposition we can use the system (2.4) in the entire functions $u(z)$, $v(z)$. Indeed, if (3.1) is a solution of (P_2), then system (2.4) has a solution

$$u(z) = Q(z)\exp\big(g(z)\big), \quad v(z) = P(z)\exp\big(g(z)\big), \tag{3.4}$$

where $g(z)$ is an entire function. Substituting (3.4) into (3.3) we find $g''(z) = 0$ and $P(z)$, $Q(z)$ satisfy the system (3.3). Let P, Q have the expansions

$$P = p_{n-1}z^{n-1} + p_{n-2}z^{n-2} + \dots, \qquad Q = q_n z^n + q_{n-1}z^{n-2} + \dots, \qquad q_n \neq 0.$$

Then from system (3.3) we have

$$nq_n^2 = p_{n-1}^2, \quad p_{n-1} = -\alpha q_n \Longrightarrow n = \alpha^2.$$

Therefore we obtain the following results. □

PROPOSITION 3.3. *(P_2) has a unique rational solution if $\alpha \in [0, 1/2)$. It is $w = 0$.*

PROPOSITION 3.4. *(P_2) has a unique rational solution if and only if α is an integer.*

PROOF. This fact follows from the Bäcklund transformation (2.16), (2.17) and Proposition 3.3. Indeed, if we start from the rational solution $w = 0$ with $\alpha = 0$ and apply T, T^{-1}, we can construct the 0-solution for any integer α. Moreover, by the property $TT^{-1} = T^{-1}T = \mathrm{id}$ and the uniqueness of $w = 0$ for $\alpha = 0$, (P_2) cannot have more than one 0-solution for every $\alpha \in Z$. If we admit the existence of a 0-solution for $\alpha \neq n$, $n \in Z$, this contradicts Proposition 3.3 due to the transformations T, T^{-1}. □

REMARK. Yablonski and Vorob'ev showed that the 0-solution of (P_2) has the form

$$w(z) = \frac{P'_\alpha}{P_\alpha} - \frac{Q'_\alpha}{Q_\alpha},$$

and $P_{\alpha+1} = Q_\alpha$, $Q_{\alpha+1} = (zQ_\alpha^2 + 4Q'^2_\alpha - 4Q_\alpha Q''_\alpha)/P_\alpha$, $P_0 = Q_0 = 1$. If $\alpha = 1$, then $P_1 = 1$, $Q_1 = z$. If $\alpha = 2$, then $P_2 = z$, $Q_2 = z^3 + 4$.

Polynomials P_α, Q_α are called Yablonski–Vorob'ev polynomials.

Every rational solution has only simple poles with residue $+1$ or -1. Now we can find how many poles there are with residue $+1$ and how many poles there are with residue -1. Let l_+ and l_- be the numbers of poles with residue $+1$ and -1 respectively; then $l_+ + l_- = n = \alpha^2$. Now we can find the residue of $z = \infty$. Let

TABLE 1. A list of 0-solutions of (P_2).

α	$w(z,\alpha)$
0	0
±1	$\mp1/z$
±2	$\mp\frac{2(z^3-2)}{z(z^3+4)}$
±3	$\mp\frac{3z^2(z^6+8z^3+160)}{(z^3+4)(z^6+20z^3-80)}$
±4	$\mp\frac{4(z^{15}+50z^{12}+1000z^9-22400z^6-112000z^3-224000)}{z(z^6+20z^3-80)(z^9+60z^6+11200)}$

z be $z=1/t$. Then $w'_z=-t^2w'_t$, $w''_{z^2}=2t^3w'_t+t^4w''_{t^2}$. Therefore equation (P_2) in the neighborhood of infinity has the form

$$t^4w''+2t^3w'=2w^3+\frac{w}{t}+\alpha.$$

Let the 0-solution have an expansion $w=a_1t+a_2t^2+\dots$ (around $z=\infty$).

Then $a_1=-\alpha$, $a_2=0$, $a_3=0$ and $\operatorname{res}_{z=\infty}w(z)=\alpha$. Therefore,

$$l_+-l_-=-\alpha,$$

because the sum of the residues of the poles is l_+-l_-. Hence,

$$l_++l_-=\alpha^2,\quad l_+-l_-=-\alpha,$$

and

$$l_+=\frac{\alpha(\alpha-1)}{2},\quad l_-=\frac{\alpha(\alpha+1)}{2}.$$

4. One-Parameter Families of Classical Solutions (1-Solutions)

Family of solutions expressible by Airy's functions. List of 1-solutions.

In the previous lecture we showed that (P_2) has 1-solutions generated by the general solution of the Riccati equation

$$w'=\varepsilon w^2+\varepsilon z/2,\quad \alpha=\varepsilon/2,\ \varepsilon^2=1, \tag{4.1}$$

and which are expressed by means of the Airy function

$$w=-\varepsilon u'/u,\quad u''+\frac{z}{2}u=0.$$

Now we use the Bäcklund transformations for the construction of 1-solutions for other values of the parameter α. We start with a solution (4.1) for $\alpha=1/2$. Then from (2.17) with $\alpha=3/2$ we obtain

$$w_{1/2}=-y+(y'-y^2-z/2)^{-1}, \tag{4.2}$$

where $y=w(z,3/2)$. Substituting (4.2) into (4.1) with $\varepsilon=1$, we find

$$y'^3-\left(y^2+\frac{z}{2}\right)y'^2-\left(y^4+zy^2+4y+\frac{z^2}{4}\right)y'+y^6 +\frac{3}{2}zy^4+4y^3+\frac{3}{4}z^2y^2+2yz+2+\frac{z^3}{8}=0. \tag{4.3}$$

TABLE 2. Airy solutions of (P_2), where $U = u'/u$ and u is any solution of the Airy equation $u'' + (z/2)u = 0$.

α	$w(z,\alpha)$
$\pm 1/2$	$\mp U$
$\pm 3/2$	$\pm \frac{2U^3+zU-1}{2U^2+z}$
$\pm 5/2$	$\pm \frac{4zU^4+6U^3+4z^2U^2+3zU+z^3-1}{(2U^2+z)(4U^3+2zU-1)}$

All solutions of this equation are simultaneously solutions of (P_2) with $\alpha = 3/2$. From (2.6) with $\alpha = 3/2$ we have

$$y = w(z,3/2) = w_{1/2} - (2w_{1/2}^2 + z)^{-1}, \quad w_{1/2} = -\frac{u'}{u}, \quad u'' + \frac{z}{2}u = 0. \tag{4.4}$$

By analogy with the above we can construct 1-solutions for $\alpha = 5/2$ and for all $\alpha = n + 1/2$, $n \in N$.

Finally, using the property $w(z,-\alpha) = -w(z,\alpha)$ we obtain the following

PROPOSITION 4.1. (P_2) *admits a one-parameter family of particular solutions represented by rational functions of the solution of an Airy equation and its derivative for* $\alpha = n + 1/2$, $n \in Z$. *These* 1-*solutions satisfy the Fuchsian equation*

$$w'^n + \sum_{j=1}^{n} P_j(z,w)w'^{n-j} = 0, \quad n \in N, \tag{4.5}$$

where

$$P_j(z,w) \equiv \sum_{k=0}^{2j} p_{jk}(z)w^{2j-k}, \quad j = \overline{1,n}. \tag{4.6}$$

5. Algebraic Nonintegrability of (P_2)

Arbitrary constants of the general solution of (P_2). Transcendence of (P_2).

Now we shall prove that the condition $\alpha = n + 1/2$, $n \in Z$ is necessary, i.e. if we have the equation

$$P(w', w, z) = 0, \tag{5.1}$$

where P is a polynomial in w', w which determines 1-solutions of (P_2), then (5.1) coincides with some equation from (4.5). Simultaneously we shall prove that equation (P_2) has no first integral of the form (5.1). To prove these facts it is sufficient to show that equation (5.1) does not determine 1-solutions in the fundamental domain $0 \leq \mathrm{Re}(\alpha) \leq 1/2$.

If equation (5.1) generates 1-solutions of (P_2), then (5.1) is a first integral of (P_2) and (5.1) is an equation of P-type. In this case it has the form of equation (4.5). In the neighborhood of a pole we have

$$w' = a_2w^2 + a_1w + a_0 + \sum_{j=1}^{\infty} a_{-j}w^{-j}. \tag{5.2}$$

Substituting (5.2) in (P_2), we find that

$$a_2 = \varepsilon, \quad \varepsilon^2 = 1, \quad a_1 = 0, \quad a_0 = \varepsilon z/2, \quad a_{-1} = \alpha\varepsilon - 1/2,$$

$a_{-2} = h$, where h is an arbitrary constant and $a_j = 0, j \leq 3$ are determined by the recurrence relations

$$a_{-n-1} = \frac{\varepsilon}{n-1}\left(\sum_{j=1}^{n-1} j a_{-j} a_{1+j-n} - a'_{-n}\right), \quad n \geq 2. \tag{5.3}$$

It follows from (5.3) that $a_{-j} = 0$, $j \geq 3$, if $a_{-1} = 0$, $h = 0$. Each root w of (5.1) can be expressed in the form (5.2). Hence, if $G(w, z, \alpha, \varepsilon, h)$ denotes the right-hand side of (5.2), then

$$P(w', w, z) = \prod_{i=1}^{l_+} \prod_{j=1}^{l_-} \big(w' - S(w, z, \alpha, 1, h_i)\big)\big(w' - S(w, z, \alpha, -1, h_j)\big), \tag{5.4}$$

where l_+, $l_- \in Z_0$ are the numbers of roots of (5.1) with $\varepsilon = 1$ and $\varepsilon = -1$ ($l_+ + l_- = n, n \geq 1$) respectively and h_ν is a value of the arbitrary coefficient a_2 for the ν-th root. Expanding (5.4), we obtain

$$P(w', w, z) = w'^n - \left(S_2 w^2 + S_1 w + S_0 + \sum_{j=1}^{\infty} \frac{S_{-j}}{w^j}\right) w'^{n-1} + \dots, \tag{5.5}$$

where $S_2 = l_+ - l_-$, $S_1 = 0$, $S_0 = (z/2)(l_+ - l_-)$,

$$S_{-1} = \alpha(l_+ - l_-) - (n/2), \quad S_{-j} = \sum_{k=0}^{n} p_{-j}^{(k)}, \qquad j \geq 2.$$

The left-hand side of (5.5) is a polynomial in w', w. Hence, $S_{-j} = 0$, $j \geq 1$. In particular, from condition $S_{-1} = 0$ we have

$$\alpha(l_+ - l_-) = n/2, \quad n = l_+ + l_-, \qquad n \in N. \tag{5.6}$$

From (5.6) it follows immediately that (P_2) with $\alpha = 0$ has no 1-solutions given by (5.1) and $l_+ - l_- \neq 0$. Consider the relation (5.6) in the fundamental domain $0 \leq \mathrm{Re}(\alpha) \leq 1/2$. Indeed, if either ($P_2$) has 1-solutions defined by (5.1) or (5.1) is a first integral, then there exist l_+, $l_- \in Z_0$ such that α defined by (5.6) belongs to the fundamental domain, i.e.

$$0 \leq \frac{l_+ + l_-}{l_+ - l_-} < 1. \tag{5.7}$$

Expanding (5.7), we see that this is not possible.

PROPOSITION 5.1. *Equation* (P_2) *has no algebraic first integral and no 1-solutions different from those obtained from the Riccati equation by application of the* T, T^{-1} *transformations.*

Thus all solutions of (P_2) can be divided into three classes:

(a) 0-solutions for $\alpha \in Z$, generated by $w(z, 0) = 0$.
(b) 1-solutions for $\alpha - 1/2 \in Z$, generated by the Riccati equation.
(c) Solutions which do not coincide with the solutions of classes a) and b). They define the second Painlevé transcendental functions.

6. Higher Analogue of $(\mathbf{P}_1)$

High analogue of the first Painlevé equation. Representation for the solutions of the first Painlevé equation and its higher analogues.

One of the important questions is nonlinear ordinary differential equations theory relates to the representation of the meromorphic solutions as the ratio of the entire functions similar to the Weierstrass function $\rho(z)$ which is a solution of an equation $\rho'^2 = 4\rho^3 + g_2\rho + g_3$. This has a representation through the entire function $\sigma(z)$ namely:

$$\rho(z) = (\sigma'^2 - \sigma\sigma'')/\sigma^2 = -\zeta'(z), \quad \zeta = \big(\ln(\sigma(z))\big)'.$$

It is well known, that in general, solutions of the Painlevé equations are meromorphic functions. This cannot be stated definitely about the higher order analogues of the Painlevé equations, which arise as exact reductions of completely integrable PDEs. In [**8**] representation of the Painlevé equations solutions as the ratio of the entire functions, one-to-one correspondence between the solutions of these equations and systems constructed with entire solutions were obtained. Now we continue the investigation of this problem and offer a unified method for representation of the meromorphic solution as the ratio of two entire functions for the Painlevé equations of the high order.

Lax [**14**] showed that the Korteveg–de Vries equation

$$u_t = 6uu_x - u_{xxx},$$

KdV(3) can be interpreted as the compatibility condition for two linear operators L and M. If L is Schrödinger operator $-d^2/dx^2 + u$ and $M = 4(d^3/dx^3) - 3u(d/dx) - 3(d/dx)u$, then L and M satisfy

$$L_t = [M, L] = ML - LM. \tag{6.1}$$

For L and M as chosen here, (1) implies that $u(x,t)$ satisfies the (KdV) equation. Commutator representation (6.1) is called Lax representation, and the operators L, M a Lax pair. We can consider A_m of the order $2m+1$ with coefficients depending of $u(x)$ and its derivatives. This operator equals $(1/2)P\big((1/i)(\partial/\partial x)\big)$ if $u \equiv 0$ and satisfies condition

$$\left[\frac{d^2}{dx^2} - u(x),\, A_m\right] = F(u, u_x, \ldots, u^{(2m+1)}),$$

where F is multiplicative operator. Such an operator A_m exists and it is possible to obtain recursion relations for its coefficients [**19**]. In this case, equation

$$u_t = F(u, u_x, \ldots, u^{(2m+1)}), \tag{6.2}$$

is called higher order analogues of (KdV).

Let us consider the (KdV) hierarchy in the form

$$\omega_t + \frac{\partial}{\partial x}B^n(\omega) = 0,$$

where $B^n(\omega)$ is determined by the Lenard recursion relations [**13, 18**] viz:

$$\frac{\partial}{\partial x}B^{n+1}(\omega) = (D^3 + 2\omega D + \omega_x)B^n(\omega),\ B^1(\omega) = \omega,\ D = \frac{\partial}{\partial x},\ n = 1, 2 \ldots.$$

We also consider the singular manifold equation for the (KdV) hierarchy, namely

$$z_t + z_x B^n\big(\{z; x\}\big) = 0.$$

Here $\{z;x\} = z_{xxx}/z_x - 3z_{xx}^2/2z_x^2$ is a Shwartz derivative. Let us seek solutions of this equation in the form

$$z = z(\xi), \quad \xi = x(\lambda t)^m. \tag{6.3}$$

Then

$$w(x,t) = \left\{z; \frac{\xi}{(\lambda t)^m}\right\} = (\lambda t)^{2m}\{z;x\} = (\lambda t)^{2m}F(\xi),$$

where $\{z(\xi);\xi\} \equiv F(\xi)$.

Let us introduce the operator d via

$$\frac{d}{d\xi}d^{n+1}(F) = (D^3 + 2FD + F_x)d^n(F), \quad d^1(F) = F, \quad D = \frac{d}{d\xi}, \qquad n = 1, 2, \ldots.$$

Then $B^n(\{z;x\}) = (\lambda t)^{2mn}d^n(F)$. Assuming $m = -1/(2n+1)$, $\lambda = -k/m$ equation (6.2) takes the form

$$d^n(\omega) - k\xi = 0.$$

In this case for $n = 1$ we have degenerate equation $F - k\xi = 0$. If $n = 2$, then we have the first Painlevé equation in the form $F'' + 3F^2/2 - k\xi = 0$. In order to write the first Painlevé equation in standard form we take $k = -4$ and scale transformation $F \to -4w$, $\xi \to z$, $n \to n+1$. Then we have

$$d^{n+1}(w) + 4z = 0, \tag{${}_{2n}\mathrm{P}_1$}$$

where

$$\frac{d}{dz}d^{n+1}(w) = (D^3 - 8wD - 4w_z)d^n(w), \quad d^1(w) = -4w, \quad D = \frac{d}{dz}, \qquad n = 1, 2, \ldots.$$

Then for $n = 1$, $n = 2$, $n = 3$ we have correspondingly

$$w'' = 6w^2 + z, \tag{${}_2\mathrm{P}_1$}$$

$$w^{(4)} = 20ww'' + 10(w')^2 - 40w^3 + z, \tag{${}_4\mathrm{P}_1$}$$

$$w^{(6)} = 42(w'')^2 + 56w'w''' + 28ww^{(4)} - 280w\big((w')^2 + ww'' - w^3\big) + z. \tag{${}_6\mathrm{P}_1$}$$

Equation (${}_2\mathrm{P}_1$) is the first Painlevé equation in standard form. We shall call (${}_{2n}\mathrm{P}_1$) with $n \geq 2$ the higher order analogue of the first Painlevé equation. The solution properties of (${}_2\mathrm{P}_1$) were studied in detail. For example, all solutions of (${}_2\mathrm{P}_1$) are meromorphic functions with movable double poles. In the neighborhood of such pole z_0 solution has the following expansion

$$w = t^{-2} - \frac{z_0}{10}t^2 - \frac{1}{6}t^3 + ht^4 + \frac{z_0^2}{300}t^6 + \sum_{k=7}^{\infty} a_k t^k, \quad (z - z_0) = t, \tag{6.4}$$

where h is an arbitrary constant. Proof of convergence of series (6.4) in the domain $0 < |z - z_0| < \rho$, $\rho > 0$ follows from the reduction of equation (${}_2\mathrm{P}_1$) into a Briot–Bouquet system. Specifically, rewrite (${}_2\mathrm{P}_1$) in the form

$$w' = x, \quad x' = 6w^2 + z. \tag{6.5}$$

In (6.2) we put

$$Y_1 = wt^2 - 1, \quad Y_2 = xt^3 + 2, \qquad (z - z_0) = t. \tag{6.6}$$

Hence, the functions Y_1, Y_2 satisfy the system

$$tY_1' = 2Y_1 + Y_2, \quad tY_2' = 12Y_1 + 3Y_2 + 6Y_1^2 + (t + z_0)t^4, \qquad z - z_0 = t, \tag{6.7}$$

where $\lambda_1 = -1$, $\lambda_2 = 6$ are the roots of characteristic equation of the linear part. Therefore [**10**], system (6.6) has solution with property $Y_1 \to 0$, $Y_2 \to 0$ as $t \to 0$. Due to the transformation (6.6) this solution generates the pole solution (6.4) of $({}_2\mathrm{P}_1)$.

We now consider solutions of $({}_4\mathrm{P}_1)$.

PROPOSITION 6.1. *Let any $z_0 \in C$. Then there exist solutions of $({}_4\mathrm{P}_1)$ which have at z_0 a double pole.*

PROOF. First of all we investigate the possible character of solution at point z_0. Let $w \sim c(z - z_0)^k$. Then substitution in $({}_4\mathrm{P}_1)$ gives $k = -2$ and the two possibilities $c = 1$ or $c = 3$.

Consider a case $c = 1$. Find the power which can give the arbitrary parameters of integrations, i.e. resonances. To do this, let us put $z - z_0 = t$ and substitute $w \sim t^{-2} + \beta t^r$ into the dominant monomials of the equation. Then we have

$$\beta\big[r(r-6)(r+3)(r-3)\big]t^{r-4} \sim 0,$$

which follows from mean order of β. Therefore the nonnegative resonances are $r = 0$, $r = 3$, $r = 6$, and the negative resonance which characterize the arbitrariness of z_0, is $r = -3$. Hence, a formal solution in the neighborhood of z_0 is

$$w = \frac{1}{t^2} + h_1 + a_2 t^2 + h_2 t^3 + a_4 t^4 + a_5 t^5 + h_3 t^6 + a_7 t^7 + \sum_{k=8}^{\infty} a_k t^k, \tag{6.8}$$

where $z - z_0 = t$, and h_1, h_2, h_3 are arbitrarily constants and

$$a_2 = -3h_1^2, \quad a_4 = -10h_1^3 - \frac{1}{56}z_0, \quad a_5 = \frac{3}{2}h_1 h_2 - \frac{1}{80}, \quad a_7 = -\frac{1}{980}h_1.$$

Let us prove the convergence of series (6.3) in domain $|z - z_0| < \rho$, $\rho > 0$ by reduction of initial equation into a system Briot–Bouquet. Rewrite $({}_4\mathrm{P}_1)$ in the equivalent system

$$y_1' = y_2,\ y_2' = y_3,\ y_3' = y_4,\ y_4' = 20y_1y_3 + 10y_2^2 - 40y_1^3 + z, \quad y \equiv y_1. \tag{6.9}$$

In this system we change variables according to

$$Y_1 = t^2 y_1 - 1, \quad Y_2 = t^3 y_2 + 2, \quad Y_3 = t^4 y_3 - 6, \quad Y_4 = t^5 y_4 + 24. \tag{6.10}$$

Then we have a Briot–Bouquet system

$$\begin{aligned} &tY_1' = 24Y_1, \quad tY_2' = 3Y_2 + Y_3, \quad tY_3' = 4Y_3 + Y_4,\\ &tY_4' = -40Y_2 + 20Y_3 + 5Y_4 + 20Y_1Y_3 + 10Y_2^2\\ &\qquad\qquad - 40Y_1^3 - 120Y_1^2 + (t + z_0)t^5, \end{aligned} \tag{6.11}$$

with characteristic roots $\lambda_1 = -1$, $\lambda_2 = 2$, $\lambda_3 = 5$, $\lambda_4 = 8$ of the linear part. From this there follows the existence of three-parametric family of holomorphic solutions of (6.4) with property $Y_j \to 0$, $j \in \{1, 2, 3, 4\}$ at $t \to 0$. Thus, from (6.6), (6.9), and (6.10) we obtain the local representation of general solution of $({}_4\mathrm{P}_1)$ in the neighborhood of arbitrary pole singularity.

Consider the case $c = 3$. The equation for resonances has a form

$$\beta(r+5)(r-6)(r+3)(r-8) = 0.$$

Therefore in this case we have two positive resonances $r = 6$, $r = 8$ and two negative resonances $r = -3$, $r = -5$. Formal pole expansion in the neighborhood of z_0 has the form

$$w = \frac{3}{t^2} + a_4 t^4 + a_5 t^5 + h_1 t^6 + h_2 t^8 + a_{10} t^{10} + a_{11} t^{11} + a_{12} t^{12} + \sum_{k=13}^{\infty} a_k t^k, \tag{6.12}$$

where

$$a_4 = z_0/504,\ a_5 = 1/240,\ a_{10} = a_4^2/39,\ a_{11} = 2/21,\ a_{12} = 5/51 h_1 a_4 + 29/612 a_6^2,$$

and h_1, h_2 are arbitrary parameters. Now we assume that in system (6.9)

$$Y_1 = t^2 y_1 - 3,\ Y_2 = t^3 y_2 + 6,\ Y_3 = t^4 y_3 - 18,\ Y_4 = t^5 y_4 + 72,\ z - z_0 = t.$$

Then we have again the system

$$\begin{aligned} &tY_1' = 24Y_1 + Y_2, \quad tY_2' = 3Y_2 + Y_3, \quad tY_3' = 4Y_3 + Y_4, \\ &tY_4' = -720Y_1 - 120Y_2 + 20Y_3 + 5Y_4 + 60Y_1Y_3 + 10Y_2^2 \\ &\qquad\qquad - 40Y_1^3 - 360Y_1^2 + (t + z_0)t^6, \end{aligned} \tag{6.13}$$

with roots $\lambda_1 = -1$, $\lambda_2 = -3$, $\lambda_3 = 8$, $\lambda_4 = 10$ of characteristic equation of the linear part. Therefore system (6.13) has a two-parametric family of holomorphic solutions with the property $Y_j \to 0$, $j \in \{1, 2, 3, 4\}$ as $t \to 0$. These solutions generate pole solutions of $({}_4\mathrm{P}_1)$ of form (6.12). Thus Proposition 6.1 holds. □

For the first Painlevé equation $({}_2\mathrm{P}_1)$ any solution is meromorphic, and it has the representation

$$w(z) = v(z)/u(z), \tag{6.14}$$

where $v(z), u(z)$ are entire functions.

Let us obtain the analogous representation for $({}_{2n}\mathrm{P}_1)$.

LEMMA 6.1. *For any solution $w(z)$ of $({}_{2n}\mathrm{P}_1)$ that has pole at $z_0 \in C$ the residues $c_j = \mathrm{res}_{z_0}(z - z_0)w(z)$ are entire positive numbers.*

PROOF. For equation $({}_2\mathrm{P}_1)$ $c_1 = 1$. For equation $({}_4\mathrm{P}_1)$ $c_1 = 1$, $c_2 = 3$. For equation $({}_6\mathrm{P}_1)$ $c_1 = 1$, $c_2 = 3$, $c_3 = 6$. Let $F \sim c/z^2$. Then from the recurrence relation determining d^n it follows $d^{n+1} \sim a/z^{2n+2}$, and c is determined by equation

$$\prod_{j=1}^{k} (c - j(j+1)/2) = 0. \tag{6.15}$$

The lemma follows from (6.15). □

PROPOSITION 6.2. *For any meromorphic solution $w(z)$ of equation $({}_{2n}\mathrm{P}_1)$ function*

$$u(z) = \exp\left(-2\int_a^z d\tau \int_a^\tau w(x)dx\right), \tag{6.16}$$

is entire.

PROOF. In the (6.8) $w(z)$ is holomorphic at a and the path of integration does not passes through the singularities of the function $w(z)$, which are simple poles.

At the holomorphic points of $w(z)$ the function $u(z)$ is also holomorphic. At the poles of $w(z)$ function $u(z)$ has zeros with expansions

$$u(z) = (z - z_0)^{2c}\big(u_1 + u_2(z - z_0) + \dots\big), \tag{6.17}$$

where $u_1 \neq 0$ and u_j are constants.

From (6.17) it follows that function $v(z) = u(z)w(z)$ is also entire. Thus one obtains the representation (6.14) of meromorphic solution of $({}_{2n}\mathrm{P}_1)$. A system determining $v(z)$ and $u(z)$ can be obtained by differentiating (6.16) and substituting (6.14) into equation $({}_{2n}\mathrm{P}_1)$. It has a form

$$uu'' = u'^2 - 2uv, \quad d^{n+1}(v/u) + 4z = 0. \tag{6.18}$$

Here derivatives $u^{(n)}$, $n \geq 2$ in the second equation of system (6.10) are replaced by corresponding expressions from the first equation. For $n = 2$ we find the system

$$uu'' = u'^2 - 2uv, \quad v''u^2 + v(u')^2 - 2uu'v' - 4uv^2 - zu^3 = 0. \tag{6.19}$$

For $n = 3$ the system (6.10) has a form

$$\begin{aligned} uu'' &= (u')^2 - 2uv,\\ v^{(4)}u^4 - v^{(3)}u^3u' &- 4u(u')^3v' + 6u^2v''(u')^2 + 16u'u^2vv' - 2u^3(v')^2\\ &-6u^3vv'' + v(u')^4 - 8u(u')^2v^2 + 16v^3u^2 - 3u'v^3u^3 - zu^5 = 0. \end{aligned} \tag{6.20}$$

□

Now we investigate the system (6.18). The choice of solution to system (6.18) is chosen with initial data

$$\begin{aligned} u(z_0) = u_0, \quad u'(z_0) &= u_0', \quad v(z_0) = v_0,\\ v'(z_0) = v_0', \dots, \quad v^{(2n-1)}&(z_0) = v_0^{(2n-1)}, \end{aligned} \tag{6.21}$$

where z_0, u_0, u_0', v_0, $v_0', \dots, v^{(2n-1)} \in C$. If $u(z_0)v(z_0) = 0$, then initial values (6.13) are singular.

LEMMA 6.2. *If* $(v \cdot u)$ *is a solution of system* (6.18) *then*

$$(\widetilde{v}, \widetilde{u}) = \big(\lambda(z)v, \quad \lambda(z)u\big), \lambda(z) \neq 0, \tag{6.22}$$

is a solution of system (6.18) *if and only if* $\lambda(z) = \exp(az + b)$, $a, b \in C$.

This lemma follows from direct substitution of (6.22) into (6.18).

PROPOSITION 6.3. *Any meromorphic solution* $w(z)$ *of equation* $({}_{2n}\mathrm{P}_1)$ *can be represented in the form* $w(z) = v(z)/u(z)$, *where* $\big(v(z), u(z)\big)$ *is the corresponding entire solution of system* (6.18) *determined up to the factor* $\exp(az + b)$, $a, b \in C$.

PROOF. Because $w(z) \not\equiv 0$, there exists $z_0 \in C$, with $w(z_0) \neq 0$. Let $w(z_0) = w_0 \neq 0$, $w'(z_0) = w_0', \dots, w_0^{(2n-1)} = w_0^{(2n-1)}$ be initial values of solution $w(z)$. We can construct a solution $\big(v(z), u(z)\big)$ of system (6.18) in accordance with (6.16). This solution is entire and has initial conditions

$$\begin{aligned} u(z_0) = 1, \quad u'(z_0) &= 0, \quad v(z_0) = w_0,\\ v^j(z_0) = \big(u(z)w(z)\big)^j(z_0), &\quad j \in \{2, \dots, 2n-1\}. \end{aligned} \tag{6.23}$$

Initial values (6.23) are not singular. Therefore the solution $\big(v(z), u(z)\big)$ with initial condition (6.23) is unique. However by virtue of Lemma 6.2 $\big(\widetilde{v}(z), \widetilde{u}(z)\big)$ is also solution and it gives the representation of $w(z)$ as (6.14). Therefore proposition is established. □

PROPOSITION 6.4. *Let (v,u), $u \neq 0$ is an arbitrary entire solution of system (6.10). Then v/u determines a meromorphic solution of* $({}_{2n}\mathrm{P}_1)$.

PROOF. From system (6.18) it follows that $v(z)$ is an arbitrary, if $u(z) \equiv 0$. If $v(z) \equiv 0$, then $u(z) \equiv 0$. Hence, if $u(z) \not\equiv 0$, then $v(z) \not\equiv 0$. Let z_0 be such that $u(z_0)v(z_0) \neq 0$, and $u_0 \neq 0$, u_0', $v_0 \neq 0$, $v_0', \dots, v_0^{(2n-1)}$ are initial values of solution $\big(v(z), u(z)\big)$. Let us take the solution $w(z)$ of $({}_{2n}\mathrm{P}_1)$ with initial values $w(z_0) = v_0/u_0$, $w'(z_0) = v_0'/u_0 - v_0u_0'/u_0^2, \dots, w^{(2n-1)}(z_0) = (v/u)^{(2n-1)}(z_0)$, where we assume $u''(z_0) = u_0'^2/u_0 - 2v_0$. Now we can construct a solution via the formula

$$u_1(z) = u_0 \exp\left(\frac{u_0'}{u_0}(z - z_0) - 2\int_{z_0}^{z} dz \int_{z_0}^{z} w(z)\,dz\right), \quad v_1(z) = u_1(z)w(z).$$

It is easy to verify that solution $\big(v_1(z), u_1(z)\big)$, satisfy the same initial values. Then by virtue of uniqueness it follows that $\big(v_1(z), u_1(z)\big) \equiv \big(v(z), u(z)\big)$. Therefore Proposition 6.3 holds.

The above results establish the one-to-one correspondence between meromorphic solutions of $({}_{2n}\mathrm{P}_1)$ and entire solutions of system (6.18). □

7. The Higher Order Analogue of (P_2)

Higher order analogue of the second Painlevé equation. Representation for the solutions of the second Painlevé equation and its higher order analogues. Rational and transcendental solutions.

Let us consider Korteweg–de Vries hierarchy in the form [**4, 7, 9**]

$$({}_m\mathrm{KdV}) \qquad (2m-1)u_t = X_m u,$$

where

$$X_1 = Du = u_x = D\frac{\partial H_1}{\partial u}, \quad X_m u = (2u + 2DuD^{-1} - D^2),$$
$$X_{m-1}u = D\frac{\partial H_m}{\partial u}, \quad m = 2, 3, \dots.$$

Under the change of variables

$$(7.1) \qquad z = xt^{-1/(2m-1)}, \quad u(x,t) = t^{-2/(2m-1)}\big(w'(z) + w^2(z)\big),$$

equation $({}_m\mathrm{KdV})$ is reduced into the ODE

$$({}_\nu\mathrm{P}_2) \qquad D^{-1}S_w^{m-1}(w') + zw + \alpha = 0,$$

where $S_w = 4w^2 + 4w'D^{-1}w - D^2$, $D = d/dz$, $v = 2m-2$. Thus, $v = 2, 4, \dots$, and the order of equation $({}_\nu\mathrm{P}_2)$ is v. For $m = 2$ equations $({}_m\mathrm{KdV})$ and $({}_\nu\mathrm{P}_2)$ become the Korteweg–de Vries equation

$$(\mathrm{KdV}) \qquad 3u_t = 6uu_x - u_{xxx},$$

and the second Painlevé equation

$$(\mathrm{P}_2) \qquad w'' = 2w^3 + zw + \alpha,$$

respectively with the connection $u(x,t) = t^{-2/3}\big(w'(z)+w^2(z)\big)$, $z = xt^{-1/3}$ [2]. For $m = 3$, $m = 4$, $m = 5$ in $({}_\nu\mathrm{P}_2)$ we have correspondingly

$$({}_4\mathrm{P}_2) \qquad w^{(4)} = 10w^2w'' + 10ww'^2 - 6w^5 - zw - \alpha,$$

$$({}_6\mathrm{P}_2) \qquad w^{(6)} = 14w^2w^{(4)} + 56ww'w^{(3)} + 42w(w'')^2 - 70\big(w^4 - (w')^2\big)w'' - 140w^3(w')^2 + 20w^7 + zw + \alpha,$$

$$({}_8\mathrm{P}_2) \qquad w^{(8)} = 18w^2w^{(6)} + 108ww'w^{(5)} - 6\big(21w^4 - 35(w')^2 - 38ww''\big)w^{(4)} - 138w(w^{(3)})^2 - 252w'(4w^3 - 3w'')w^{(3)} + 182(w'')^3 - 756w^3(w'')^2 + 84w^2\big(5w^4 - 37(w')^2\big)w'' - 798w(w')^4 + 1260w^5(w')^2 - 70w^9 - zw - \alpha.$$

Here we consider a case $m = 2$, i.e. $\upsilon = 4$ and equation $({}_4\mathrm{P}_2)$.

The solution properties of (P_2) have been studied in detail. For example, all solutions of (P_2) are meromorphic functions with movable double poles. In the neighborhood of such a pole z_0 the solution has the following expansion

$$(7.2) \quad w = \frac{\varepsilon}{t} - \frac{\varepsilon}{6}z_0t - \frac{\alpha+\varepsilon}{4}t^2 + ht^3 - \frac{3\alpha+\varepsilon}{72}z_0t^4 + \dots, \quad \varepsilon^2 = 1, \ (z - z_0) = t,$$

where h is an arbitrary constant. The convergence of series (7.2) in domain $0 < |z - z_0| < \rho$, $\rho > 0$ follows from reduction of equation (P_2) into Briot–Bouquet system [**10**]. Specifically, rewrite (P_2) in the form

$$w' = u, \quad u' = 2w^3 + zw + \alpha,$$

and let us put

$$Y_1 = wt - \varepsilon, \quad Y_2 = ut^2 + \varepsilon, \quad \varepsilon^2 = 1, \quad t = z - z_0.$$

Then we have a system

$$tY_1' = Y_1 + Y_2, \quad tY_2' = 6Y_1 + 2Y_2 + t^2(t + z_0)(Y_1 + \varepsilon) + 2Y_1^3 + \alpha t^3 + 6\varepsilon Y_1^2.$$

This is a Briot–Bouquet system with roots of the linear part: $\lambda_1 = -1$, $\lambda_2 = 4$. Therefore this is a case when the system has a one-parameter family of holomorphic solutions with the properties $Y_1 \to 0$, $Y_2 \to 0$, $t \to 0$. These solutions generate one-parameter family of local polar solutions at z_0.

The second Painlevé equation has only one dominant term $2w^3$. Therefore we have

PROPOSITION 7.1. *Any nonrational solution of* (P_2) *has an infinite number of simple poles.*

PROOF. This follows immediately by the standard Clunie reasoning. In fact, from $w(2w^2) = w'' - zw - \alpha$ we obtain that $m(r,w) = S(r,w)$. □

In the present paper we give a representation of the meromorphic solutions of equations $({}_\nu\mathrm{P}_2)$ as the ratio $w(z) = v(z)/u(z)$ of entire functions $v(z)$, $u(z)$. We establish one-to-one correspondence between the meromorphic solutions of the equation $({}_\nu\mathrm{P}_2)$ and entire solutions of the constructed system.

For the equation $({}_\nu\mathrm{P}_2)$ let us assume the following:

$$(7.3) \qquad u(z) = \exp\Big(-\int_{z_0}^{z} dz \int_{z_0}^{z} w^2\, dz\Big),$$

where the path of integration does not pass through the singularities of the function $w(z)$. In the neighborhood of the movable pole $z = \alpha$ the meromorphic solution $w(z)$ has expansion

$$w = a_{-1}(z-\alpha)^{-1} + a_1(z-\alpha) + \varphi_1(z,\alpha), \tag{7.4}$$

where a_{-1} takes any of the values $\pm 1, \pm 2, \dots \pm\nu/2$, and $\varphi_1(z,\alpha)$ is an analytic function in the neighborhood of α. Then from (7.3) it follows that $u(z)$ is an entire function for any meromorphic solution of the equation $({}_\nu\mathrm{P}_2)$, and at a pole α of the solution $w(z)$ the function $u(z)$ has zero of order a_{-1}^2. This fact is easily established by substitution (6.2) into the right part of (7.3). Hence, if we define the function $v(z)$ as $v(z) = w(z)u(z)$, then the function $v(z)$ is also entire for any meromorphic solution of the equation $({}_\nu\mathrm{P}_2)$. The system for finding the entire functions $u(z)$, $v(z)$ amerges by differentiation of (7.3) use of (6.14) and substitution of (7.3) into equation $({}_\nu\mathrm{P}_2)$. It has the form

$$uu'' - (u')^2 = -v^2, \quad D^{-1}S^{\nu/2}_{vu^{-1}}(v'u^{-1} - vu'u^{-2}) + zvu^{-1} + \alpha = 0. \tag{7.5}$$

For example, at $m = 3$ for the equation $({}_4\mathrm{P}_2)$ we have

$$uu'' - (u')^2 = -v^2,$$

$$v^{(4)}u^4 - 4v''v'(u')^3 + 6v''(u')^2u^2 - 2v''v^2u^2 - 4v'(u')^3u$$
$$+ 4v^2v'u' + v(u')^4 - 2v^3(u')^2 + v^5 + zvu^4 + \alpha u^5 = 0.$$

Let us consider the system (7.5).

LEMMA 7.1. *The system* (7.5) *has the solution*

$$v = 0, \quad u = \exp(az+b), \tag{7.6}$$

for any a, $b \in C$, $\alpha = 0$.

The choice of the solution of system (7.5) is defined by the initial conditions

$$\begin{gathered} u(z_0) = u_0, \quad u'(z_0) = u_0', \quad v(z_0) = v_0, \\ v'(z_0) = v_0', \dots, \quad v^{(\nu-1)}(z_0) = v_0^{(\nu-1)}, \end{gathered} \tag{7.7}$$

where z_0, u_0, u_0', v_0, $v_0', \dots,$ $v_0^{(\nu-1)} \in C$. Thus, if $u(z_0) = 0$, then the initial conditions $({}_{2n}\mathrm{P}_1)$ is singular.

LEMMA 7.2. *If* (v,u) *is a solution of system* (7.5), *then*

$$(\tilde v, \tilde u) = \big(\lambda(z)v, \lambda(z)u\big), \quad \lambda(z) \not\equiv 0, \tag{7.8}$$

is a solution of system (7.5) *if and only if* $\lambda = e^{az+b}$, $a, b \in C$.

The validity of Lemma 7.1 and Lemma 7.2 is confirmed by direct substitution (7.6) and (7.8) into system (7.5).

LEMMA 7.3. *Any solution of system* (7.5), *corresponding to the meromorphic solution of an equation* $({}_\nu\mathrm{P}_2)$ *and satisfying to the initial conditions*

$$\begin{gathered} u(z_0) = 1, \quad u'(z_0) = 0, \quad v(z_0) = v_0, \\ v'(z_0) = v_0', \dots, \quad v^{(\nu-1)}(z_0) = v_0^{(\nu-1)}, \end{gathered} \tag{7.9}$$

is entire.

PROOF. The initial conditions (7.9) are not singular. Let $w(z)$ be a meromorphic solution of an equation $({}_\nu P_2)$ with the initial conditions $w(z_0) = w_0$, $w'(z_0) = w_0', \dots, w^{(\nu-1)}(z_0) = w_0^{(\nu-1)}$. We shall consider the functions

$$u_1(z) = \exp\Big(-\int_{z_0}^{z} dz \int_{z_0}^{z} w^2\, dz\Big), \quad v_1(z) = w(z)u_1(z). \tag{7.10}$$

By construction, functions $u_1(z)$, $v_1(z)$ are entire and satisfy the system (7.5). By virtue of uniqueness, the statement is established if we take $w_0 = v_0$, $w_0' = v_0', \dots$, $w_0^{(\nu-1)} = v_0^{(\nu-1)}$. □

LEMMA 7.4. *Any solution of system (7.5), corresponding to the meromorphic solution of equation $({}_\nu P_2)$ and satisfying to the initial conditions*

$$u(z_0) = u_0 \neq 0,\ u'(z_0) = u_0',\ v(z_0) = v_0,\ v'(z_0) = v_0', \dots,\ v^{(\nu-1)}(z_0) = v_0^{(\nu-1)},$$

is entire.

PROOF. We take the solution $(\widetilde{v}, \widetilde{u})$ of system (7.5) with the initial conditions

$$\widetilde{u}(z_0) = 1,\ \widetilde{u}'(z_0) = 0,\ \widetilde{v}(z_0) = \widetilde{v}_0,\ \widetilde{v}'(z_0) = \widetilde{v}_0', \dots,\ \widetilde{v}^{(\nu-1)}(z_0) = \widetilde{v}_0^{(\nu-1)}.$$

By virtue of Lemma 7.3, this solution is entire, and by virtue of Lemma 7.2, the functions $(v, u) = \big(\widetilde{v}\exp(az+b), \widetilde{u}\exp(az+b)\big)$ will also be entire solutions of system (7.5). The proof of Lemma 7.4 follows from the choice a, b and $\widetilde{v}_0, \widetilde{v}_0', \dots, \widetilde{v}_0^{(\nu-1)}$ with

$$a = \frac{u_0'}{u_0}, \quad b = \ln(u_0) - z_0\frac{u_0'}{u_0}, \quad \widetilde{v}_0 = \frac{v_0}{u_0}, \dots, \quad \widetilde{v}_0^{(\nu-1)} = \Big(\frac{v}{u}\Big)^{(\nu-1)}(z_0),$$

where $u''(z_0) = u_0'^2/u_0$. □

PROPOSITION 7.2. *All solutions of system (7.5), corresponding to the meromorphic solutions of equation $({}_\nu P_2)$, are entire functions.*

PROOF. If $u \equiv 0$, then $v \equiv 0$ and this solution is entire. Let $u \not\equiv 0$. Then there exists domain D where function $u(z) \neq 0$ and it is analytic. Let $z_0 \in D$. Then by virtue of Lemma 7.4 we have the required result. □

PROPOSITION 7.3. *Let (v, u) be an arbitrary entire nonzero solution of system (7.5) (for some fixed value of the parameter α), which is different from the solutions (7.6). Then a ratio $v(z)/u(z)$ represents meromorphic solution of equation $({}_\nu P_2)$ for the same value of parameter α.*

PROOF. Let (v, u) be an entire nonzero solution of system (7.5). Then there exists z_0 such that $u(z_0) = u_0 \neq 0$. Let us take the meromorphic solution $w(z)$ of equation $({}_\nu P_2)$ with the initial conditions $w(z_0) = v_0/u_0$, $w'(z_0) = v_0'/u_0 - v_0u_0'/u_0^2, \dots$, $w^{(\nu-1)}(z_0) = (v/u)^{(\nu-1)}(z_0)$, where we assume that $u''(z_0) = u_0'^2/u_0$. We shall construct a solution of system (7.5) applying the formula

$$u_1(z) = u_0\exp\Big((z - z_0)\frac{u_0'}{u_0} - \int_{z_0}^{z} dz \int_{z_0}^{z} w^2(z)\, dz\Big), \quad v_1(z) = u_1(z)w(z).$$

It is not difficult to see that the solution $\big(v_1(z), u_1(z)\big)$ satisfies the same initial conditions, as $\big(v(z), u(z)\big)$. By virtue of uniqueness $\big(v_1(z), u_1(z)\big) \equiv \big(v(z), u(z)\big)$ and from the second parity of (7.10) statement of the Proposition 7.3 follows. □

Proposition 7.4. *Any meromorphic solution $w(z)$ of equation $({}_\nu\mathrm{P}_2)$ is represented in the form $w(z) = v(z)/u(z)$, where $\big(v(z), u(z)\big)$ is the corresponding entire solution of system* (7.5)*, determined up to the factor* $\exp(az+b)$.

Proof. If $w(z) \equiv 0$ at $\alpha = 0$, we shall take the entire solution of system (7.5) $\big(v(z), u(z)\big) = \big(0, \exp(az+b)\big)$. Let $w(z) \not\equiv 0$. Then there exists such z_0, that $w(z_0) \neq 0$, $w(z_0) \neq \infty$. Let us put

$$u(z) = \exp\Big(-\int_{z_0}^{z} dz \int_{z_0}^{z} w^2\, dz\Big), \quad v(z) = w(z)u(z).$$

These functions are entire solutions of system (7.5), and $w(z) = v(z)/u(z)$. But by virtue of (7.8) $w(z)$ is also expressed through the solution

$$\big(\widetilde{v}(z), \widetilde{u}(z)\big) = \big(v(z)\exp(az+b), u(z)\exp(az+b)\big).$$

The Propositions 7.2 and 7.3 establish one-to-one correspondence between the meromorphic solutions of equation $({}_\nu\mathrm{P}_2)$ and the entire solutions of system (7.5).

Let

$$w(z) = P(z)/Q(z), \tag{7.11}$$

be a rational solution of $({}_\nu\mathrm{P}_2)$, where $P(z)$ and $Q(z)$ are polynomials. By direct substitution (7.11) into $({}_\nu\mathrm{P}_2)$ follows that $\dim P(z) = \dim Q(z) - 1$. Therefore any rational solution can be represented in form

$$w(z) = \sum_{k=1}^{\nu/2} k\Bigg[\sum_{j_k=1}^{l_k^+} \frac{1}{z - z_{j_k}} - \sum_{j_{k+1}=1}^{l_k^-} \frac{1}{z - z_{j_{k+1}}}\Bigg] = \sum_{k=1}^{\nu/2} k\bigg(\frac{P_k'(z)}{P_k(z)} - \frac{Q_k'(z)}{Q_k(z)}\bigg),$$

where $l_k^\pm$ is a number of poles with residue k and $-k$ respectively, and

$$P_k(z) = \prod_{j_k=1}^{l_k^+} (z - z_{j_k}), \quad Q_k(z) = \prod_{i_k=1}^{l_k^-} (z - z_{i_k}).$$

Therefore,

$$Q(z) = \prod_{s=1}^{\nu/2} P_s Q_s, \quad P(z) = \sum_{k=1}^{\nu/2} k\widetilde{P}_k (P_k' Q_k - P_k Q_k'),$$

where

$$\widetilde{P}_k = \begin{cases} \prod_{s=1}^{\nu/2} P_s Q_s & \text{for } s \neq k, \\ 1 & \text{for } s = k. \end{cases}$$

Then system (7.5) has solution

$$u(z) = N(z)\exp\big(g(z)\big), \quad v(z) = M(z)\exp\big(g(z)\big), \tag{7.12}$$

where $g(z)$ is a polynomial and

$$N(z) = \prod_{k=1}^{\nu/2} (P_k Q_k)^{k^2}, \quad M(z) = P(z)\prod_{k=1}^{\nu/2} (P_k Q_k)^{k^2-1}. \tag{7.13}$$

Substituting (7.12) into system (7.5), we find that $g''(z) = 0$ and obtain a system for $M(z)$ and $N(z)$, which has form (7.5), where $u \to N$, $v \to M$. As the second equation of system (7.5) is result of substitution of (7.2) into $({}_\nu\mathrm{P}_2)$ the following result holds: □

PROPOSITION 7.5. *For the existence of rational solution of* $({}_\nu\mathrm{P}_2)$ *it is necessary and sufficient that system* (7.5), *where* $u \to N$, $v \to M$, *has polynomial solution* $\big(M(z), N(z)\big)$, $N(z) \neq 0$ *under the condition that at poles of* $w(z)$ *polynomials* $M(z)$ *and* $N(z)$ *have zeros of orders* $k^2 - 1$ *and* k^2 *respectively.*

Let

$$M(z) = \sum_{j=0}^{n-1} p_j z^{n-1-j}, \quad N(z) = \sum_{j=0}^{n} q_j z^{n-j}, \qquad q_0 = 1, \tag{7.14}$$

where

$$n = \sum_{k=1}^{\nu/2} (l_k^+ + l_k^-)k^2. \tag{7.15}$$

Substituting (7.14) into the first equation of system (7.5) we looking for $p_0^2 = n$. From the second equation it follows $p_0 = -\alpha$. Therefore $n = \alpha^2$. For a rational solution $z = \infty$ is a holomorphic point. Substituting $w(z) = \sum_{j=0}^{\infty} a_j z^{-j}$ into $({}_\nu\mathrm{P}_2)$ we obtain $a_0 = 0$, $a_1 = -\alpha$. Now from the total sum of residues of single-valued function in the complex plane we obtain

$$\sum_{k=0}^{\nu/2} (l_k^+ - l_k^-)k = -\alpha, \quad \sum_{k=0}^{\nu/2} (l_k^+ + l_k^-)k^2 = \alpha^2, \tag{7.16}$$

and our proposition follows.

PROPOSITION 7.6. *For the existence of a rational solution of* $({}_\nu\mathrm{P}_2)$ *a necessary and sufficient is that* $\alpha \in Z$. *For every integer* α, *equation* $({}_\nu\mathrm{P}_2)$ *has a unique rational solution.*

PROOF. Necessity follows from the first equation of system (7.16), because the left side of this equation is integer. Sufficiency follows from the Bäcklund transformations of $({}_\nu\mathrm{P}_2)$, which allows us to construct the rational solutions for all integer α from the seed solution $w = 0$ for $\alpha = 0$.

For example, for the equation $({}_4\mathrm{P}_2)$ the Bäcklund transformation has the form

$$\begin{aligned} S\colon w(z,\alpha) &\to -w(z,-\alpha), \\ T\colon w \to y &= -w - (2\alpha - 1)R^{-1}(r), \\ T^{-1}\colon y \to w &= -y - (2\alpha - 1)R^{-1}(q), \end{aligned}$$

where $R(r) = z + 6r^2 - 2r''$, $r = w^2 - w'$, $q = y' + y^2$, $w = w(z,\alpha)$, $y = w(z, \alpha - 1)$.

Applying the Bäcklund transformations to seed solution $w = 0$ for $\alpha = 0$ we obtain rational solutions of $({}_4\mathrm{P}_2)$:

$$\begin{array}{cc}
\alpha & w(z,\alpha) \\
0 & 0 \\
\pm 1 & \mp 1/z \\
\pm 2 & \mp 2/z \\
\pm 3 & \pm\big(2/z - 5z^4/(144+z^5)\big) \\
\pm 4 & \pm\frac{\left(-6967296-870912z^5-288z^{10}-4z^{15}\right)}{z(144+z^5)(-48384+1008z^5+z^{10})} \\
 & \cdots
\end{array}$$

□

References

1. M. Ablowitz and P. A. Clarkson, *Solitons, nonlinear evolution equations and inverse scattering*, London Mathematical Society Lecture Note Series, vol. 149, Cambridge University Press, Cambridge, 1991.
2. M. Ablowitz, A. Ramani, and H. Segur, *Nonlinear evolution equations and ordinary differential equations of Painlevé type*, Lett. Nuovo Cim. **23** (1978), 333–338.
3. ———, *A connection between nonlinear evolution equations and ordinary differential equations of P-type.* I; II, J. Math. Phys. **21** (1980), 1006–1015.
4. H. Airault, *Rational solutions of Painlevé equations*, Stud. Appl. Math. **61** (1979), 31–53.
5. A. S. Fokas and M. J. Ablowitz, *A unified approach to transformations and elementary solutions of Painlevé equations*, J. Math. Phys. **23** (1982), 2033–2042.
6. V. V. Golubev, *To the theory of Painlevé equations*, Math. Sb. **28** (1912), no. 2, 323–349.
7. V. I. Gromak, *To the theory of higher order Korteweg–de Vries equations*, Dep. in Differential Equations **3641–B86** (1986), 1–10.
8. ———, *On general representation of the solutions of Painlevé equations*, Differential Equations **31** (1997), 614–617.
9. V. I. Gromak and N. A. Lukashevich, *Analytical properties of the solutions of Painlevé equations*, University Press, Minsk, 1990.
10. J. Horn, *Gevőhnliche Differentiagleichungen Belibinger Ordiung*, G. I. Gőschensche Verlagchandlung, Leipzig, 1905.
11. E. L. Ince, *Ordinary differential equations*, Dover, New York, 1956.
12. M. D. Kruskal and P. A. Clarkson, *The Painlevé–Kowalevski and poly-Painlevé tests for integrability*, Stud. Appl. Math. **86** (1992), 87–165.
13. N. A. Kudryashov, *The first and second Painlevé equations of higher order and some relations between them*, Phys. Lett. A **224** (1997), 353–360.
14. P. D. Lax, *Integrals of nonlinear equations of evolution and solitary waves*, Comm. Pure Appl. Math. **21** (1968), 467–490.
15. K. Okamoto, *Studies of the Painlevé equations.* II, Japan J. Math. **13** (1987), 47–76.
16. P. Painlevé, *Leçons sur la théorie analytique des équations differentielles*, Leçons de Stockholm, 1895, Hermann, Paris, 1897; Reprinted in Oeuvres de Paul Painlevé, vol. 1, Éditions du CNRS, Paris, 1973.
17. ———, *Mémoire sur les équations différentielles dont l'intégrale générale est uniforme*, Bull. Soc. Math. France **28** (1900), 201–261.
18. J. Weiss, M. Tabor, and G. Carnevale, *The Painlevé property for partial differential equations*, J. Math. Phys. **24** (1983), 522–526.
19. V. E. Zakharov, S. V. Manakov, C. P. Novikov, and L. P. Pitaevsky, *Theory of solitons: Method of inverse problem*, Nauka, Moscow, 1980.

Department of mathematics and Mechanics, Bielorussian State University, F. Skorina 4, 220050 Minsk, Belarus.

E-mail address: grom@mmf.bsu.unibel.by

Centre de Recherches Mathématiques
CRM Proceedings and Lecture Notes
Volume 29, 2001

Dressing Method and Bäcklund and Darboux Transformations

D. Levi and O. Ragnisco

ABSTRACT. Here we present a set of results on various algebraic techniques that one can apply to get solutions of integrable Nonlinear Partial Differential Equations (NPDE's). These results are due to the work of many people starting from last century geometers.

1. Introduction

By the end of the nineteenth century several mathematicians, among them Korteweg, De Vries, Boussinesq, Bianchi, Darboux and Bäcklund considered certain NPDE's (among them the Korteweg–de Vries (KdV) equation, the Boussinesq equation, the sine–Gordon equation) which had spectacular properties:

- Infinite number of local conserved quantities.
- Infinite number of exact solutions.
- Simple algebraic superposition formulae relating explicit solutions among themselves.

After the numerical work of Fermi, Pasta and Ulam in 1955 on the recurrence properties of a nonlinear anharmonic oscillator on a lattice, Gardner, Green, Kruskal and Miura in 1967 introduced the Inverse Scattering Transform to integrate the KdV equation:

$$q_{,t}(x,t) = q_{,xxx}(x,t) - 6q(x,t)q_{,x}(x,t). \tag{1.1}$$

In the formalism introduced by Lax, an integrable NPDE is described by two operators, usually called the Lax pair, which in the case of the KdV equation (1.1) read:

$$L = -\partial_{xx} + q(x,t), \tag{1.2}$$

$$M = -4\partial_{xxx} + 3\big(q(x,t)\partial_x + \partial_x q(x,t)\big). \tag{1.3}$$

2000 *Mathematics Subject Classification.* Primary: 35G25, 35P05, 35Q53; Secondary: 35R30, 35R10, 35Q58, 35Q15, 35Q51.

We acknowledge support from NATO (07/17/96). One of the authors (O.R) would like to warmly thank W. Oevel, who has given an essential contribution to the Section "Bäcklund and Darboux transformations in Multidimensions": indeed, most of this part has been taken from an unpublished joint paper of O.R. with W. Oevel, dating back to 1993.

This is the final form of the paper.

The operator L characterizes the Spectral Problem:

$$L\psi(x,t;\lambda) = \lambda\psi(x,t;\lambda), \tag{1.4}$$

while the operator M characterizes the t-evolution of the wavefunction $\psi(x,t;\lambda)$:

$$\psi_{,t}(x,t;\lambda) = -M\psi(x,t;\lambda). \tag{1.5}$$

The compatibility of (1.4), (1.5) when λ is not t-dependent, i.e. when we are considering isospectral deformations of (1.4), implies:

$$L_{,t} = [L, M], \tag{1.6}$$

the *Lax Equation*.

Let us briefly present all the algebraic properties that we can derive from the existence of a Spectral Problem given by (1.4):

- The existence of a recursion operator $\mathcal{L}$ for constructing the hierarchy of NPDE's associated to L.
- The extension of the class of NPDE's by considering nonisospectral deformations of the Spectral Problem (1.4).
- The existence of Bäcklund and Darboux transformations.

1.1. Construction of the recursion operator. Starting from the Spectral Problem (1.4) we can look for operators M_j such that

$$L_{,t} = [L, M_j], \tag{1.7}$$

is satisfied. To do so let us consider the following set of operator equations:

$$[L, M] = V, \quad [L, \widetilde{M}] = \widetilde{V}. \tag{1.8}$$

Both Eqs. (1.8) are valid when V and $\widetilde{V}$ are scalar functions for some given operators M and $\widetilde{M}$. Taking into account the structure of the operator L, we can require that

$$\widetilde{M} = LM + F\partial_x + G, \tag{1.9}$$

where F and G are some scalar functions of $q(x,t)$ and its derivatives and of V, fixed by the requirement that Eqs. (1.8) are satisfied. From Eqs. (1.8), (1.9) we get that $\widetilde{V}$ can be expressed as an operator $\mathcal{L}$, the recursion operator, on the function V and some boundary conditions. The recursion operator reads:

$$\mathcal{L}V(x,t) = -\frac{1}{4}V_{,xx}(x,t) + q(x,t)V(x,t) - \frac{1}{2}q_{,x}(x,t)\int_x^\infty dy\, V(y,t). \tag{1.10}$$

Consequently in term of the recursion operator (1.10) the class of equations associated to the Spectral Problem (1.4) characterized by the operator L given by (1.2) reads:

$$q_{,t}(x,t) = f(\mathcal{L},t)q_{,x}(x,t), \tag{1.11}$$

where $f(\mathcal{L},t)$ is an entire function of its first argument. In correspondence with each particular function $f(\mathcal{L},t)$ we can compute from Eq. (1.9) the corresponding operator M which completes the Lax pair associated to Eq. (1.11).

The KdV equation (1.1) is obtained by choosing $f(\mathcal{L},t) = \mathcal{L}$.

1.2. Extension to the nonisospectral case. When $\lambda = \lambda(t)$ we have to modify our Lax equation. In this case we have:

$$L_{,t} = [L, M] + N, \tag{1.12}$$

where N is an operator which has to be a function of L, i.e. $N = f_1(L)$. The associated evolution of the spectral parameter is given by $\lambda_{,t} = f_1(\lambda)$.

The hierarchy of compatible NPDE's is obtained paraphrasing the procedure presented in Section 1.1 and is given by:

$$q_{,t}(x,t) = f_0(\mathcal{L}, t)q_{,x}(x,t) + f_1(\mathcal{L})\big[xq_{,x}(x,t) + 2q(x,t)\big]. \tag{1.13}$$

1.3. Bäcklund and Darboux transformations. Let us assume that we have two different Spectral Problems of the same form (1.4):

$$L\psi = \lambda\psi, \quad L = -\partial_{xx} + q(x,t), \tag{1.14}$$

$$\tilde{L}\widetilde{\psi} = \lambda\widetilde{\psi}, \quad \tilde{L} = -\partial_{xx} + \tilde{q}(x,t). \tag{1.15}$$

If we are looking for a relation between $q(x,t)$ and $\tilde{q}(x,t)$, we have to assume that the wavefunctions $\psi(x,t;\lambda)$ and $\widetilde{\psi}(x,t;\lambda)$ are related by an operator D, which we call Darboux operator, i.e. we must have:

$$\widetilde{\psi}(x,t;\lambda) = D\psi(x,t;\lambda). \tag{1.16}$$

From (1.15), (1.16) it follows the operator equation:

$$\tilde{L}D = DL. \tag{1.17}$$

A hierarchy of Darboux operators D can be constructed through a recursive procedure, as we did in Eq. (1.8) for the evolution operators M. Namely we can put:

$$\tilde{L}D - DL = W, \quad \tilde{L}\widetilde{D} - \widetilde{D}L = \widetilde{W}, \tag{1.18}$$

where

$$\widetilde{D} = \tilde{L}D + F\partial_x + G. \tag{1.19}$$

We can define a recursion operator Λ which transform a W into a $\widetilde{W}$ to which, taking into account Eq. (1.19), it will correspond a transformation from a Darboux operator D to a new Darboux operator $\widetilde{D}$. The operator Λ is given by:

$$\begin{aligned} \Lambda W(x,t) &= W_{,xx}(x,t) - 2\big[\tilde{q}(x,t) + q(x,t)\big]W(x,t) + \Gamma\int_x^\infty dy\, W(y,t), \\ \Gamma W(x,t) &= \big[\tilde{q}_{,x}(x,t) + q_{,x}(x,t)\big]W(x,t), \\ &\quad + [\tilde{q}(x,t) + q(x,t)]\int_x^\infty dy\big[\tilde{q}(y,t) - q(y,t)\big]W(y,t). \end{aligned} \tag{1.20}$$

Taking into account the boundary conditions for F and G we get the following Bäcklund transformations:

$$g_0(\Lambda,t)\big[q(x,t) - \tilde{q}(x,t)\big] + g_1(\Lambda,t)\Gamma 1 = 0. \tag{1.21}$$

g_0 and g_1 being entire function of their first argument. The simplest Bäcklund transformation corresponds to both g_0 and g_1 constant, say α and β. In this case Eq. (1.21) reads:

$$\alpha[q - \tilde{q}] + \beta\left\{\tilde{q}_x + q_x + (\tilde{q} - q)\int_x^\infty dy[\tilde{q} - q]\right\} = 0. \tag{1.22}$$

1.4. Solution technique for integrable NPDE's. The solution of the Cauchy problem for the KdV equation was presented by Gardner, Green, Kruskal and Miura in 1967 and goes under the name of Spectral Transform. It is based on the solution of the Spectral Problem (1.4).

Let us assume that, given an "initial value" $q(x,t_0)$ for the dependent variable of the NPDE, we can solve the Spectral Problem (1.4). This is certainly the case if we are able to solve the one-parameter family of differential equations

$$-\psi_{,xx}(x,t_0;k) + q(x,t_0)\psi(x,t_0;k) = k^2\psi(x,t_0;k), \quad k \in \mathbb{C}. \tag{1.23}$$

In particular we can identify the "scattering solution" defined by the asymptotic behavior ($k \in \mathbb{R}$):

$$\begin{aligned} \psi(x,t_0;k) &\xrightarrow[x\to+\infty]{} e^{-ikx} + R(k,t_0)e^{ikx}, \\ \psi(x,t_0;k) &\xrightarrow[x\to-\infty]{} T(k,t_0)e^{-ikx}, \end{aligned} \tag{1.24}$$

where $R(k,t_0)$ is the *reflection coefficient* and $T(k,t_0)$ is the *transmission coefficient.* The bound state solutions are obtained by requiring that $\psi_b(x,t_0;k) \to 0$ at both infinities. In this case k is purely imaginary, i.e. $k = ip$. For potentials that decay fast enough as $|x| \to \infty$, the reflection coefficient can be analytically continued outside the real k-axis, and has (generically simple) poles at the points $k = ip_j$: the corresponding residues give us the normalization constants $C_j(t_0)$. So we can associate to a given Spectral Problem (1.23) its spectral data $\mathcal{S}\big[q(x,t_0)\big]$, given by:

$$\mathcal{S}\big[q(x,t_0)\big] = \big\{R(k,t_0), k \in \mathbb{R}; p_j, C_j(t_0), j = 1,2,\ldots,N\big\}. \tag{1.25}$$

From the knowledge of the operator M we get the time evolution of the function $\psi(x,t;k)$ and by its asymptotic expansion the time evolution of the spectral data $\mathcal{S}\big[q(x,t)\big]$. Specifically, for $q(x,t)$ evolving according to (1.1) we find that p_j do not evolve in time while the reflection coefficient $R(k,t)$ and the normalization constants $C_j(t)$ obey the equations:

$$R_{,t} = -4ik^3R, \tag{1.26}$$

$$C_{j,t} = -4p_j^3C_j. \tag{1.27}$$

So we are able to get the spectral data in closed form at any time t and then by solving the Inverse Spectral Transform [**3**] (IST) we obtain the function $q(x,t)$ at any time t. Schematically we have:

$$\begin{array}{ccc} q(x,t_0) & \xrightarrow[x\to\infty]{} & \mathcal{S}\big[q(x,t_0)\big] \\ & & \downarrow \\ q(x,t) \leftarrow \text{IST} & \leftarrow & \mathcal{S}\big[q(x,t)\big]. \end{array}$$

Hence, by a finite number of linear steps (direct spectral problem, evolution of spectral data, inverse spectral problem) we solve the initial value problem for the NPDE.

This procedure has been extended in 1979 by Zakharov and Shabat [**30, 31**] to a much larger class of systems than those covered by the Lax Technique. This extension has been called the *Dressing Method* and will be described in Section 2.

Here we tersely recall the so-called zero-curvature approach introduced by Zakharov and Shabat. Let us consider the following overdetermined system of linear matrix equations:

$$\begin{aligned} \Psi_x(x,t;\lambda) &= U\big(q(x,t);\lambda\big)\Psi(x,t;\lambda), \\ \Psi_t(x,t;\lambda) &= V\big(q(x,t);\lambda\big)\Psi(x,t;\lambda), \end{aligned} \tag{1.28}$$

where U and V have a preassigned dependence on λ and $q(x,t)$. Here usually $\lambda \in \mathbb{C}$ while $q(x,t)$ may be a vector and by the symbols $U(q(x,t);\lambda)$, $V(q(x,t);\lambda)$ we mean that U and V are functions of $q(x,t)$, where the functional dependence is to be taken in the most general sense as possible, even through integral or differential operators.

As an example we can show that from the Schrödinger Spectral Problem (1.23) one can get, by rewriting it in matrix form, a matrix $U(q(x,t);\lambda)$. In fact, defining a vector function Ψ as:

$$\Psi(x,t;k) = \begin{pmatrix} \psi_x(x,t;k) + ik\psi(x,t;k) \\ \psi(x,t;k) \end{pmatrix}, \qquad \lambda = k^2, \tag{1.29}$$

we get for U:

$$U\big(q(x,t);k\big) = \begin{pmatrix} ik & q(x,t) \\ 1 & -ik \end{pmatrix}. \tag{1.30}$$

Let us notice that with a different choice of the representation of Ψ in terms of ψ we can associate to the same Spectral Problem (1.23) a different matrix U. For example, by choosing

$$\Psi(x,t;\lambda) = \begin{pmatrix} \psi_x(x,t;\lambda) \\ \psi(x,t;\lambda) \end{pmatrix}, \tag{1.31}$$

we get

$$U\big(q(x,t);\lambda\big) = \begin{pmatrix} 0 & q(x,t) - \lambda \\ 1 & 0 \end{pmatrix}. \tag{1.32}$$

The request that Eq. (1.28) have a nonsingular solution for Ψ implies that U and V have to satisfy the so called *zero curvature condition*:

$$U_{,t} - V_{,x} + [U,V] = 0. \tag{1.33}$$

The NPDE is then obtained requiring that Eq. (1.33) be fulfilled for any λ.

Other NPDEs are obtained by choosing more general forms of matrices U. For example U given by

$$U\big(u(x,t),v(x,t);k\big) = \begin{pmatrix} ik & u(x,t) \\ v(x,t) & -ik \end{pmatrix}, \tag{1.34}$$

provides, with a proper choice of the constraint between v and u, the Nonlinear Schrödinger equation:

$$iu_{,t} = u_{,xx} - u \mid u \mid^2, \tag{1.35}$$

or the Modified Korteweg–de Vries equation:

$$u_{,t} = u_{,xxx} - u^2 u_{,x}, \tag{1.36}$$

or the sine–Gordon equation:

$$u_{,tx} = \sin(u). \tag{1.37}$$

As a solution-generating technique the Dressing Method is based on the Hilbert–Riemann Problem, as we will see later.

The Dressing Method has been discretized with success and has proven to be an efficient way to integrate systems on the lattice. In fact by considering the following system of linear matrix equations

$$\begin{aligned}\Psi_{n+1}(t;\lambda) &= U_n\big(q_n(t),\lambda\big)\Psi_n(t;\lambda),\\ \Psi_{n,t}(t;\lambda) &= V_n\big(q_n(t),\lambda\big)\Psi_n(t;\lambda),\end{aligned} \tag{1.38}$$

the corresponding "zero curvature" equation reads:

$$U_{n,t} + U_n V_n = V_{n+1} U_n, \tag{1.39}$$

i.e. a differential-difference equation. Instead, by enforcing compatibility of the following discrete linear matrix equations

$$\begin{aligned}\Psi_{n+1,m}(\lambda) &= U_{n,m}(q_{n,m},\lambda)\Psi_{n,m}(\lambda),\\ \Psi_{n,m+1}(\lambda) &= V_{n,m}(q_{n,m},\lambda)\Psi_{n,m}(\lambda),\end{aligned} \tag{1.40}$$

we obtain:

$$U_{n,m+1} V_{n,m} = V_{n+1,m} U_{n,m}, \tag{1.41}$$

i.e. a difference-difference equation. Later on we will discuss the relationship of these results with those connected with the construction of Bäcklund and Darboux transformations for NPDE and for nonlinear differential-difference equations. It is sufficient now to stress the fact that for all equations which can be represented in this way by Eqs. (1.31), (1.33) one can construct algebraic techniques for finding classes of explicit exact solutions.

In Sections 2, 3, 4 we will deal with Dressing Method, Bäcklund Transformations, Darboux Transformations for NPDEs in $1+1$ dimensions. In Section 5 we will briefly outline how these notions and techniques can be extended to encompass NPDEs in $2+1$ dimensions.

2. Dressing Method

Let us now describe the Dressing Method as a solution technique. We refer to the literature on the subject [**14, 30, 31**] for more details.

Let us assume that we are able to compute a solution, *the bare solution* of Eqs. (1.28), (1.33), i.e. we have a $U_0(\lambda)$, $V_0(\lambda)$ and a $\Psi_0(x,t;\lambda)$ which solve Equations (1.28), (1.33), where $U_0(\lambda)$ and $V_0(\lambda)$ have the same dependence from λ as U and V. The function $\Psi_0(x,t;\lambda)$ is nonsingular for any λ.

To the linear problems $\Psi_{0,x} = U_0\Psi_0$ and $\Psi_{0,t} = V_0\Psi_0$ we can associate a matrix function $G_0(\lambda)$, defined on a simply connected and closed contour Γ in the complex λ plane.

We can construct a *new solution* U, V, Ψ, if we can solve the following Hilbert–Riemann Problem: recover the two matrices $\Psi_+(x,t;\lambda)$ and $\Psi_-(x,t;\lambda)$ from

$$G(x,t;\lambda) = \Psi_+(x,t;\lambda)\Psi_-(x,t;\lambda), \tag{2.1}$$

where

$$G(x,t;\lambda) = \Psi_0(x,t;\lambda)G_0(\lambda)\Psi_0^{-1}(x,t;\lambda). \tag{2.2}$$

If the matrix $G(x,t;\lambda)$ fulfills some technical requirements (related to the so-called "index theorem") the Hilbert–Riemann Problem (2.1) is completely defined whenever we give a boundary condition at one point in the λ plane, say $\lambda = \infty$ for $\Psi_+(x,t;\lambda)$ and we look for

- $\Psi_+(x,t;\lambda)$ which is analytic and nonsingular outside Γ excepts for a finite number of isolated poles;
- $\Psi_-(x,t;\lambda)$ which is analytic and nonsingular inside Γ excepts for a finite number of isolated poles.

Once the Hilbert–Riemann Problem is solved, i.e. we have found the functions $\Psi_-(x,t;\lambda)$ and $\Psi_+(x,t;\lambda)$, the new solution of the linear problem (1.28) is given by:

$$
\begin{aligned}
U(x,t;\lambda) &= \Psi_+^{-1}(x,t;\lambda)\{-\Psi_{+,x}(x,t;\lambda)+U_0(\lambda)\Psi_+(x,t;\lambda)\}\\
&= \{\Psi_{-,x}(x,t;\lambda)+\Psi_-(x,t;\lambda)U_0(\lambda)\}\Psi_-^{-1}(x,t;\lambda),\\
V(x,t;\lambda) &= \Psi_+^{-1}(x,t;\lambda)\{-\Psi_{+,t}(x,t;\lambda)+V_0(\lambda)\Psi_+(x,t;\lambda)\}\\
&= \{\Psi_{-,t}(x,t;\lambda)+\Psi_-(x,t;\lambda)V_0(\lambda)\}\Psi_-^{-1}(x,t;\lambda),\\
\Psi(x,t;\lambda) &= \Psi_+^{-1}(x,t;\lambda)\Psi_0(x,t;\lambda).
\end{aligned}
\tag{2.3}
$$

Explicit solutions are obtained by solving the Hilbert–Riemann Problem with zeroes, i.e. when $G_0(\lambda)=I$, where I is the identity matrix, and consequently also $G(x,t;\lambda)=I$. In such a case the function $\Psi_-(x,t;\lambda)$ (resp. $\Psi_+(x,t;\lambda)$) can have only zeroes inside (resp. outside) Γ, and we can assume

- $\Psi_+(x,t;\lambda)$ has N simples zeroes outside Γ at $\lambda=\lambda_j$ $(j=1,\dots,N)$;
- $\Psi_-(x,t;\lambda)$ has N simples poles at $\lambda=\lambda_j$ $(j=1,\dots,N)$;
- $\Psi_-(x,t;\lambda)$ has N simples zeroes inside Γ at $\lambda=\mu_j$ $(j=1,\dots,N)$;
- $\Psi_+(x,t;\lambda)$ has N simples poles at $\lambda=\mu_j$ $(j=1,\dots,N)$.

As $\Psi_-(x,t;\lambda)\Psi_+(x,t;\lambda)=I$ we can write, in full generality,

$$
\Psi_+(x,t;\lambda)=\prod_{j=1}^{N}\Psi_+^j(x,t;\lambda),\quad \Psi_-(x,t;\lambda)=\prod_{j=N}^{1}\Psi_-^j(x,t;\lambda),
\tag{2.4}
$$

where the product in Eq. (2.4) is to be considered as an ordered product, from the lower index to the upper index, and the single factors $\Psi_-^j(x,t;\lambda)$ and $\Psi_+^j(x,t;\lambda)$ are given by:

$$
\begin{aligned}
\Psi_+^j(x,t;\lambda) &= \left[I-\frac{\lambda_j-\mu_j}{\lambda-\mu_j}P_j(x,t)\right]\chi_j(x,t),\\
\Psi_-^j(x,t;\lambda) &= \chi_j^{-1}(x,t)\left[I-\frac{\lambda_j-\mu_j}{\lambda-\lambda_j}P_j(x,t)\right],
\end{aligned}
\tag{2.5}
$$

where $\chi_j(x,t)$ is a normalization matrix and $P_j(x,t)$ is a projection matrix, namely

$$
P_j^2(x,t)=P_j(x,t).
\tag{2.6}
$$

Taking into account the fact that $\Psi_-(x,t;\lambda)\Psi_+(x,t;\lambda)=I$ we can rewrite Eqs. (2.3) in a simpler form:

$$
\begin{aligned}
U(x,t;\lambda) &= \Psi_-(x,t;\lambda)\{-\partial_x+U_0(\lambda)\}\Psi_+(x,t;\lambda),\\
V(x,t;\lambda) &= \Psi_-(x,t;\lambda)\{-\partial_t+V_0(\lambda)\}\Psi_+(x,t;\lambda).
\end{aligned}
\tag{2.7}
$$

By requiring that $U(q,\lambda)$ and $U_0(\lambda)$ ($V(q,\lambda)$ and $V_0(\lambda)$) have the same rational dependence on λ and have neither poles nor zeroes at the points λ_j and μ_j, we get the following set of determining equations for P_j in terms of U_0:

$$
\begin{aligned}
&P_j\chi_j\left(\prod_{k=1}^{j-1}\Psi_+^k(\lambda_j)\right)^{-1}\left[-\partial_x+U_0(\lambda_j)\right]\prod_{k=1}^{j-1}\Psi_+^k(\lambda_j)\chi_j^{-1}(I-P_j)=0,\\
&(I-P_j)\chi_j\left(\prod_{k=1}^{j-1}\Psi_+^k(\mu_j)\right)^{-1}\left[-\partial_x+U_0(\mu_j)\right]\prod_{k=1}^{j-1}\Psi_+^k(\mu_j)\chi_j^{-1}P_j=0,
\end{aligned}
\tag{2.8}
$$

where j can take any value between 1 and N. Equations similar to (2.8) define the time dependence of P_j; of course, they are written in terms of V_0 and involve t derivatives. From Eqs. (2.8) and their counterpart with respect to t we get the complete definition of the projection matrices $P_j(x,t)$, in both their x and t dependence, in terms of the function $\Psi_0(x,t;\lambda)$ for $\lambda=\lambda_j$ and $\lambda=\mu_j$.

Now from Eq. (2.7) we can get an expression for $U\big(q(x,t),\lambda\big)$ as a function of all $P_j(x,t)$ and χ_j. For the sake of simplicity we present here the formulas corresponding to just one pole and one zero in the complex λ plane, entailing that we have only one function χ and one projection matrix $P(x,t)$

$$
\begin{aligned}
U\big(q(x,t),\lambda\big)&=\chi_{,x}(x,t)\chi^{-1}(x,t)+\chi(x,t)U_0(\lambda)\chi^{-1}(x,t)\\
&\quad+\frac{\lambda_1-\mu_1}{\lambda-\lambda_1}P(x,t)\chi(x,t)\big[U_0(\lambda)-U_0(\lambda_1)\big]\chi^{-1}\big[I-P(x,t)\big]\\
&\quad-\frac{\lambda_1-\mu_1}{\lambda-\mu_1}\big[I-P(x,t)\big]\chi(x,t)\big[U_0(\lambda)-U_0(\mu_1)\big]\chi^{-1}P(x,t),
\end{aligned}
\tag{2.9}
$$

and similarly for $V\big(q(x,t),\lambda\big)$:

$$
\begin{aligned}
V\big(q(x,t),\lambda\big)&=\chi_{,t}(x,t)\chi^{-1}(x,t)+\chi(x,t)V_0(\lambda)\chi^{-1}(x,t)\\
&\quad+\frac{\lambda_1-\mu_1}{\lambda-\lambda_1}P(x,t)\chi(x,t)\big[V_0(\lambda)-V_0(\lambda_1)\big]\chi^{-1}\big[I-P(x,t)\big]\\
&\quad-\frac{\lambda_1-\mu_1}{\lambda-\mu_1}\big[I-P(x,t)\big]\chi(x,t)\big[V_0(\lambda)-V_0(\mu_1)\big]\chi^{-1}P(x,t).
\end{aligned}
\tag{2.10}
$$

Let us now study the solution of Eqs. (2.8) in more details. To do so we decompose the vector space on which the matrix Ψ is acting into the direct sum of two subspaces $\mathcal{M}$ and $\mathcal{N}$, where $\mathcal{M}$ is the image of P, such that $P\mathcal{M}=\mathcal{M}$ and $\mathcal{N}$ is the kernel of P such that $P\mathcal{N}=0$. For simplicity, let us again consider the case of just one pole and one zero in the complex λ plane. In such a case Eq. (2.8) can be written as:

$$
\begin{aligned}
P_1\chi_1\big[-\partial_x+U_0(\lambda_1)\big]\chi_1^{-1}(I-P_1)&=0,\\
(I-P_1)\chi_1\big[-\partial_x+U_0(\mu_1)\big]\chi_1^{-1}P_1&=0.
\end{aligned}
\tag{2.11}
$$

Then, if by $|v>$ we mean a generic vector belonging to $\mathcal{N}$ and by $|u>$ a generic vector belonging to $\mathcal{M}$, we have:

$$
\begin{aligned}
\chi_1\big[-\partial_x+U_0(\lambda_1)\big]\chi_1^{-1}(I-P_1)|v>&=|v>,\\
\chi_1\big[-\partial_x+U_0(\mu_1)\big]\chi_1^{-1}P_1|u>&=|u>,
\end{aligned}
\tag{2.12}
$$

and similarly for the t derivatives. Eqs. (2.12) imply that

$$\begin{aligned} |u> &= \Psi_0(x,t;\mu_1)|u^0>, \\ |v> &= \Psi_0(x,t;\lambda_1)|v^0>, \\ P(x,t) &= \chi_1 \frac{|v><u|}{<u|v>}, \end{aligned} \tag{2.13}$$

where $|u^0>$ and $|v^0>$ are constant arbitrary vectors belonging to $\mathcal{M}$ and $\mathcal{N}$ respectively. The N soliton solutions, corresponding to N poles and N zeroes in the λ complex plane can be obtained recursively from the knowledge of the $N-1$ soliton solution. Let us end this section by pointing out that:

- The Dressing Method provide solutions independently for the x part and the t part of the linear problem. So we can just solve one of them, say the x part, and get a solution with t dependent coefficients. The t dependence is then determined by introducing the resulting solution into the NPDE.
- An extremely important role is played by the normalization matrix χ. An Hilbert–Riemann Problem is called canonical when $\chi = I$. It has been shown that one can always reduce oneself to canonical normalization by a gauge transformation which change the linear problem but leaves invariant the NPDE [**17**].

3. Bäcklund Transformations

The Bäcklund Transformations (BT's) reduce the integration of a NPDE to the solution of an Ordinary Differential Equation (ODE), in general, of low order. If the ODE relates two different solutions of the same NPDE, then we speak of an *auto-Bäcklund* transformation, otherwise we speak of a *Miura* transformation.

Given a solution $q(x,t)$ of a NPDE, the solution of the auto-Bäcklund transformation provide a new solution, say $\tilde{q}(x,t)$ of the same NPDE. The solution $q(x,t)$ of the NPDE is characterized by its matrix linear problems:

$$\Psi_x = U(q,\lambda)\Psi, \tag{3.1}$$

$$\Psi_t = V(q,\lambda)\Psi, \tag{3.2}$$

while $\tilde{q}(x,t)$ is characterized by

$$\begin{aligned} \widetilde{\Psi}_x &= U(\tilde{q},\lambda)\widetilde{\Psi}, \\ \widetilde{\Psi}_t &= V(\tilde{q},\lambda)\widetilde{\Psi}. \end{aligned} \tag{3.3}$$

Both linear problems (3.1), (3.2), and (3.3) give rise to the same NPDE

$$U_{,t} - V_{,x} + [U,V]. \tag{3.4}$$

The existence of a BT implies that there is a relation between the two solutions, i.e. between the two linear problems (3.1), (3.2), and (3.3), which is given by the fact that one can derive one from the other by *dressing it up.* There must exist a nonsingular matrix D, often called Darboux Matrix, such that:

$$\widetilde{\Psi}(x,t;\lambda) = D\big(q(x,t),\tilde{q}(x,t);\lambda\big)\Psi(x,t;\lambda), \tag{3.5}$$

whence it follows:

$$D_{,x}(q,\tilde{q};\lambda) = U(\tilde{q},\lambda)D(q,\tilde{q};\lambda) - D(q,\tilde{q};\lambda)U(q,\lambda), \tag{3.6}$$

$$D_{,t}(q,\tilde{q};\lambda) = V(\tilde{q},\lambda)D(q,\tilde{q};\lambda) - D(q,\tilde{q};\lambda)V(q,\lambda). \tag{3.7}$$

The compatibility of Eqs. (3.6), (3.7) implies:

$$U_{,t}(\tilde{q}) - V_{,x}(\tilde{q}) + \big[U(\tilde{q}), V(\tilde{q})\big] = D\big\{U_{,t}(q) - V_{,x}(q) + \big[U(q), V(q)\big]\big\}D^{-1}, \tag{3.8}$$

entailing that, if $q(x,t)$ is a solution of the NPDE then also $\tilde{q}(x,t)$ will be a solution of the same equation.

If we define $q(x) = q_n(x)$, $\Psi = \Psi_n$ and $\tilde{q}(x) = q_{n+1}(x)$, $\widetilde{\Psi} = \Psi_{n+1}$ then Eqs. (3.1), (3.5), (3.6) are equivalent to Eqs. (1.38), (1.39), having defined $U\big(q(x), \lambda\big) = U_n\big(q_n(x), \lambda\big)$ and $D\big(q(x), \tilde{q}(x); \lambda\big) = V_n\big(q_n(x), \lambda\big)$. More details on this can be found in [**9, 10**].

Denoting by $\mathcal{D}$ the set of all possible Darboux Matrices associated to a given Spectral Problem then we can easily proof that:

(1) In $\mathcal{D}$ there exists the identity BT and it is given by $D = I$.
(2) In $\mathcal{D}$ there exists the inverse of a BT and it is given by its inverse Darboux Matrix: $D' = D^{-1}$.
(3) In $\mathcal{D}$ is defined the product of two Darboux Matrices $D_1(q, \tilde{q}; \lambda)$ and $D_2(\tilde{q}, \hat{q}; \lambda)$ and it is given by: $D = D_2 D_1$.
(4) In $\mathcal{D}$ is defined the sum of two Darboux Matrices $D_1(q, \tilde{q}; \lambda)$ and $D_2(q, \tilde{q}; \lambda)$ and it is given by: $D = D_1 + D_2$.

These properties, linear at the level of the Darboux Matrix, reflects nonlinearly on the fields $\big(q(x,t), \tilde{q}(x,t)\big)$ in the BT's. For example, from the third property, taking into account that the order in which we do the BT's should not be important as the result should be the same if one does first D_1 and then D_2 or vice-versa, we can get the nonlinear superposition formulas.

Let us assume that U and V are such that there exists a nontrivial solution to the Eqs. (3.6), (3.7) for

$$D = D_0(x,t) + \lambda D_1(x,t). \tag{3.9}$$

This is the standard case when the one soliton solution exists. We will say that the Darboux Matrix corresponds to an *adding one soliton* BT; it adds one pole in the complex λ plane.

As an example we consider the case of

$$U\big(q(x,t), r(x,t); \lambda\big) = \begin{pmatrix} 0 & -q(x,t) \\ -1 & \lambda - r(x,t) \end{pmatrix}. \tag{3.10}$$

To this Spectral Problem we can associate, as it has been shown in [**17**], a class of NPDE's among which we can find the KdV–Burgers system of equations:

$$\begin{aligned} q_{,t} &= q_{,xxx} - 6qq_{,x} - 3(q_{,x}r - r^2 q)_{,x}, \\ r_{,t} &= r_{,xxx} + (3rr_{,x} + r^3 + 6rq)_{,x}. \end{aligned} \tag{3.11}$$

For $r = 0$, the above system reduces to the KdV equation while for $q = 0$ reduces to a higher order equation of the Burgers hierarchy [**15**]. If we solve Eq. (3.6) with U given by Eq. (3.10) and D by Eq. (3.9) we get:

$$\begin{aligned} D_0 &= \begin{pmatrix} p_1(p_2 + H) & p_0 K(r_{,x} + q) + p_1\big[\tilde{q} - \tilde{r}_{,x} - (\tilde{r} - r)(p_2 + r + H)\big] \\ p_1 + p_0 K & p_0 K r + p_1(\tilde{r} + p_2 + H) \end{pmatrix}, \\ D_1 &= \begin{pmatrix} p_1 & 0 \\ 0 & -p_1 \end{pmatrix}, \end{aligned} \tag{3.12}$$

where

$$K = \exp\Big[\int_x^{\infty} dy\big(\tilde{r}(y,t) - r(y,t)\big)\Big],$$
$$H = \int_x^{\infty} dy\big(\tilde{q}(y,t) - q(y,t)\big), \tag{3.13}$$

p_0, p_1 and p_2 are arbitrary x independent parameters and the BT reads:

$$p_0 K(\tilde{q} - q + r_{,x}) + p_1\big[\tilde{r}_{,x} - \tilde{q} + q + (\tilde{r} - r)(\tilde{r} + p_2 + H)\big] = 0,$$
$$p_0 K\big[\tilde{q}(\tilde{r} - r) - \tilde{q}_{,x}\big] + p_1\big[qr - q_{,x} - \tilde{r}\tilde{q} + (\tilde{q} - q)(p_2 - \tilde{r} - H)\big] = 0. \tag{3.14}$$

When $r = \tilde{r} = 0$ then $K = 1$ and Eqs. (3.14) are compatible only if $p_0 = p_1$; in this case we get the BT for KdV equation:

$$(\tilde{q} + q)_{,x} = (\tilde{q} - q)(p_2 - H). \tag{3.15}$$

The matrix D_1 provide us with the normalization matrix of the corresponding Hilbert–Riemann Problem χ as when $\lambda \longrightarrow \infty$, D behaves as D_1. In the case of the KdV equation we get that $\chi = \sigma_3$ the third Pauli matrix. In [**17**] it has been shown that under the following quite general conditions:

(1) $D_1(q, q)$ is a constant matrix,
(2) there exists the adding two soliton Darboux Matrix, which is a second degree polynomial in λ,

the matrix $D_1(\tilde{q}, q)$ can be decomposed in the following way

$$D_1(\tilde{q}, q) = A^{-1}(\tilde{q})\widehat{D}A(q), \tag{3.16}$$

where A is an invertible matrix of the same dimension as D and $\widehat{D}$ is a constant matrix. Introducing the new wave function $\Phi = A\Psi$ we get a linear problem for Φ whose Hilbert–Riemann Problem has constant asymptotic normalization $\chi = \widehat{D}$. In our case we have:

$$\widehat{D} = \begin{pmatrix} p_1 & 0 \\ 0 & -p_0 \end{pmatrix}, \tag{3.17}$$

$$A(q, r) = \begin{pmatrix} \frac{1}{p_1} & 0 \\ 0 & -\frac{1}{p_0}\exp\big(-\int_x^{\infty} dy\, r(y,t)\big) \end{pmatrix}, \tag{3.18}$$

and the linear problem read:

$$\Phi_{,x} = \begin{pmatrix} 0 & u \\ v & \lambda \end{pmatrix}\Phi, \tag{3.19}$$

where

$$u(x,t) = \frac{p_0}{p_1} q(x,t) \exp\Big[\int_x^{\infty} dy\, r(y,t)\Big], \tag{3.20}$$

$$v(x,t) = \frac{p_1}{p_0} \exp\Big[-\int_x^{\infty} dy\, r(y,t)\Big]. \tag{3.21}$$

Starting from $q = r = 0$ we get

$$\begin{pmatrix} 0 & u \\ v+1 & 0 \end{pmatrix} = (\lambda_1 - \mu_1)\Big[P_1, \begin{pmatrix} 0 & 0 \\ 0 & 1 \end{pmatrix}\Big], \tag{3.22}$$

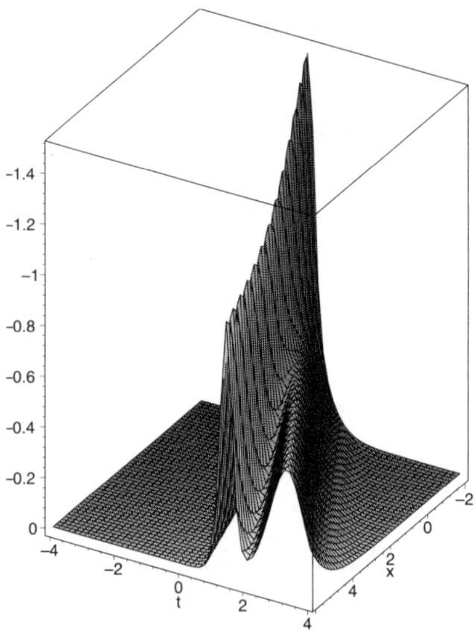

FIGURE 1. The function $q(x,t)$ for $\lambda_1 = 2$, $\mu_1 = -1.4$, $x_{1,0} = x_{2,0} = 0$.

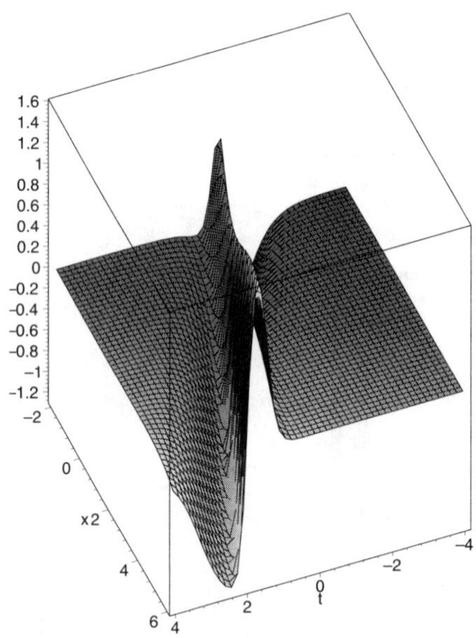

FIGURE 2. The function $r(x,t)$ for $\lambda_1 = 2$, $\mu_1 = -1.4$, $x_{1,0} = x_{2,0} = 0$.

and consequently

$$\begin{aligned} r(x,t) &\\ = (\lambda_1 - \mu_1) &\frac{\lambda_1^2 e^{\eta_1}(1+e^{\eta_2})^2 - \mu_1^2 e^{\eta_2}(1+e^{\eta_1})^2}{[\lambda_1 e^{\eta_1}(1+e^{\eta_2}) - \mu_1 e^{\eta_2}(1+e^{\eta_1})][\lambda_1(1+e^{\eta_2}) - \mu_1(1+e^{\eta_1})]}, \\ q(x,t) &\\ = (\lambda_1 - \mu_1)\lambda_1\mu_1 &\frac{\lambda_1 e^{\eta_1}(1+e^{\eta_2}) - \mu_1 e^{\eta_2}(1+e^{\eta_1})}{[\lambda_1 - \mu_1 + \lambda_1 e^{\eta_2} - \mu_1 e^{\eta_1}]^2}, \end{aligned} \tag{3.23}$$

where

$$\begin{aligned}\eta_1 &= \lambda_1(x - \lambda_1^2 t - x_{1,0}),\\ \eta_2 &= \mu_1(x - \mu_1^2 t - x_{2,0}).\end{aligned} \tag{3.24}$$

From the plotting of q and r (see Fig. 1 and Fig. 2) as function of x and t we see that, as t passes from $-\infty$ to ∞, q undergoes a splitting from one soliton to two solitons while r goes from two kinks and one antikink to one kink and two antikinks.

4. Classical Darboux Transformation

In 1882 [**4**] Gaston Darboux introduced the *Darboux Transformation* to produce new potentials for what is now called the stationary Schrödinger equation:

$$-\psi_{,xx}(x;k) + q(x)\psi(x;k) = k^2\psi(x;k). \tag{4.1}$$

This transformation is a mapping in the pair $\{\, q(x), \psi(x;k)\}$ to a new pair $\{\tilde{q}(x), \widetilde{\psi}(x;k)\}$. Given two *independent* solutions of Eq. (4.1), $\psi_1(x;k)$ and $\psi_2(x;k)$, choosen in such a way that they satisfy the following boundary conditions:

$$\psi_1(x,k) \xrightarrow[x\to-\infty]{} e^{-ikx}, \quad \psi_2(x,k) \xrightarrow[x\to-\infty]{} e^{ikx}, \tag{4.2}$$

we construct the so called *intermediate wave function*

$$f(x;k_0,a) = \psi_1(x;k_0) + a\psi_2(x;k_0), \tag{4.3}$$

and then the Darboux operator

$$\widehat{\mathcal{D}}(k_0,a) = \partial_x - \frac{f_x}{f}. \tag{4.4}$$

Taking into account the definitions (4.3), (4.4) we can define the new pair $\{\tilde{q}(x), \widetilde{\psi}(x;k)\}$ in the following way:

$$\tilde{q}(x) = q(x) - 2\partial_{xx}\log\big(f(x;k_0,a)\big), \tag{4.5}$$

$$\widetilde{\psi}(x) = \widehat{\mathcal{D}}(k_0,a)\psi(x;k). \tag{4.6}$$

As an example of applications let us consider the case when $q(x) = 0$ and $k_0 = 0$. Then $f = \alpha + \beta x$ and we get the new solution $\tilde{q} = -(f_{,x}/f)^2 = -\big(\beta^2/(\alpha+\beta x)^2\big)$.

The corresponding BT can be obtained starting from Eq. (4.5), substituting $q(x) = -w(x)_{,x}$ and using Eq. (4.1). We have:

$$\widetilde{w}(x) = w(x) + 2\left(\frac{f_{,x}}{f}\right) + 2p \Longrightarrow \frac{f_{,x}}{f} = \frac{1}{2}\big(\widetilde{w}(x) - w(x) - 2p\big). \tag{4.7}$$

By differentiating Eq. (4.7) and substituting into it Eq. (4.1) with $k_0^2 = -p^2$, in the case of bound states, we get:

$$\widetilde{w}_{,x}(x) + w_{,x}(x) = -\frac{1}{2}\big(\widetilde{w}(x) - w(x)\big)\big(\widetilde{w}(x) - w(x) + 4p\big), \tag{4.8}$$

i.e. the standard BT for KdV equation. A different form of the same BT can be obtained if we solve Eq. (4.5) for $(f_{,x}/f)$ in terms of q and $\tilde{q}$. We get:

$$\frac{f_{,x}}{f} = \big(\tilde{q}(x) + q(x) - 2k_0^2\big)^{1/2}, \tag{4.9}$$

and the BT reads:

$$\tilde{q}_{,x}(x) + q_{,x}(x) + 2\big(\tilde{q}(x) - k_0^2\big)\big(\tilde{q}(x) + q(x) - 2k_0^2\big)^{1/2} = 0. \tag{4.10}$$

The same result can be obtained using the Dressing Method and the definition of Darboux Matrix written out in terms of the wavefunction Ψ. In the case of the Schrödinger Spectral Problem we have:

$$\Xi_{,x}(x,k) = \begin{pmatrix} -ik & q(x) \\ 1 & ik \end{pmatrix} \Xi(x,k), \tag{4.11}$$

where by Ξ we denote the matrix solution

$$\Xi(x,k) \equiv \begin{pmatrix} \Xi_{11}(x,k) & \Xi_{12}(x,k) \\ \Xi_{21}(x,k) & \Xi_{22}(x,k) \end{pmatrix}. \tag{4.12}$$

The reduction given by the form of the linear problem (4.11) implies that we have to choose $\lambda_1 = -\mu_1 = ik_0$ and take Γ to be the real axis in the complex λ plane. Then

$$\begin{gathered} D = \Psi_+^{-1} = \left[I + \frac{2ik_0}{\lambda - ik_0} P(x) \right] \chi, \quad P = \frac{1}{1+\xi(x)\tilde{\xi}(x)} \begin{pmatrix} \xi\tilde{\xi} & \xi \\ \tilde{\xi} & 1 \end{pmatrix}, \\ \xi(x) = -\frac{\Xi_{11}(x,k_0) + a\Xi_{12}(x,k_0)}{\Xi_{21}(x,k_0) + a\Xi_{22}(x,k_0)}, \quad \tilde{\xi}(x) = -\frac{1}{2k_0 + \xi(x)}, \quad \chi = \sigma_3, \end{gathered} \tag{4.13}$$

and from Eq. (2.9) we get:

$$\tilde{q}(x) = q(x) + 2\xi_{,x}(x). \tag{4.14}$$

Taking into account the definition of the intermediate wave function f and of Ξ in terms of the wavefunction ψ of the Schrödinger Spectral problem we get:

$$\xi(x) = ik_0 - \frac{f_x}{f}. \tag{4.15}$$

Introducing (4.15) into Eq. (4.14) we reobtain the classical Darboux Transformation.

Let us now use the machinery of the Dressing Method to construct a New Darboux Transformation for the Schrödinger Spectral Problem. Let us start from the representation of the Schrödinger Spectral Problem (1.32) which depends on λ. The corresponding adding one soliton Darboux Matrix, being linear in λ will be quadratic in k and will thus directly provide two soliton solutions. From the determining equation (3.6) for the Darboux Matrix we get that

$$D_1\big(\widetilde{w}(x,t), w(x,t)\big) = \begin{pmatrix} 1 & \frac{1}{2}\big(w(x,t) - \widetilde{w}(x,t)\big) \\ 0 & 1 \end{pmatrix}, \tag{4.16}$$

having defined, as before $q(x,t) = w_{,x}(x,t)$. Taking into account the decomposition theorem on the normalization condition for the Hilbert–Riemann Problem mentioned in the previous section, we introduce the new wavefunction Θ by the gauge transformation $\Theta = D_1\big(0, w(x,t)\big)\Phi$. Then Θ will satisfy the following linear problem:

$$\Theta_{,x}(x,t) = \begin{pmatrix} \frac{1}{2}w(x,t) & \lambda + \frac{1}{2}w_{,x}(x,t) - \frac{1}{4}w(x,t)^2 \\ 1 & -\frac{1}{2}w(x,t) \end{pmatrix} \Theta(x,t). \tag{4.17}$$

By applying the Dressing Method to this linear problem (4.17) we get the Darboux Transformation:

$$\begin{gathered} \tilde{q} = q - 2\partial_{xx}\log\big(F(x)\big), \\ \widetilde{\psi} = f(\lambda)\left\{ \psi + \frac{\lambda_1 - \mu_1}{\lambda - \lambda_1} \frac{W\big(\psi, \psi^{(2)}(\lambda_1) + b\psi^{(1)}(\lambda_1)\big)}{F(x)} \right\}, \end{gathered} \tag{4.18}$$

where

$$W(A,B) = A_x B - AB_x,$$
$$F(x) = \beta + W\big(\psi^{(1)}(\mu_1) + a\psi^{(2)}(\mu_1), \psi^{(2)}(\lambda_1) + b\psi^{(1)}(\lambda_1)\big). \tag{4.19}$$

In the above formulas (4.18), (4.19) $\psi^{(1)}(\lambda)$ and $\psi^{(2)}(\lambda)$ are two independent solutions of the Schrödinger Spectral Problem, λ_1 and μ_1 are two complex numbers and β is a constant such that

$$\beta(\lambda_1 - \mu_1) = 0. \tag{4.20}$$

Eq. (4.18) provide a two complex parameter Darboux Transformation for the Schrödinger equation. There are two ways we can solve Eq. (4.20) and reduce Eqs. (4.18) to a one parameter Darboux Transformations:

- We set $\beta = 0$; then Eq. (4.20) is identically satisfied and there is no constraint between λ_1 and μ_1. To reduce the number of parameters we set $\lambda_1 = p_1^2$, $\mu_1 = p_2^2$, with p_1 and p_2 real parameters, and $f(\lambda) = (k - ip_1/k + ip_2)$. In the limit when p_1 goes to infinity the transformation (4.18) reduces to the classical Darboux transformation and gives the standard BT for KdV equation.
- We leave β arbitrary but set $\lambda_1 = \mu_1$ so as to solve Eq. (4.20). Choosing $\beta = (1/\rho)$, $b = (1/a)$ and $f(\lambda) = 1$ we get that

$$F(x) = 1 + \rho W(\psi_{,\lambda}, \psi)\big|_{\lambda=\mu_1}. \tag{4.21}$$

In this case it is easy to see that $F(x)$ satisfies the differential equation $(dF/dx) = \rho\psi^2$. Taking Eq. (4.21) into account we find that $w(x,t)$ satisfies the following second order differential equation:

$$(w - \widetilde{w})_{,xx} = -\frac{1}{8}(w - \widetilde{w})^3 - (w - \widetilde{w})(w_{,x} + \widetilde{w}_{,x} + 2\mu_1) + \frac{1}{2}\frac{(w - \widetilde{w})_{,x}}{w - \widetilde{w}}. \tag{4.22}$$

Let us notice that the dependence of this BT upon the difference between two solutions, namely $w - \widetilde{w}$, is such that it can provide a new potential $\widetilde{w}(x)$ even if $w(x)$ does not vanish asymptotically, $w(x) \xrightarrow[x\to\infty]{} g(x) \neq 0$. In particular it can provide solutions when $g(x) = x$ which corresponds to the case of the cylindrical KdV equation [**3**]:

$$q_{,t} = q_{,xxx} - 6qq_{,x} + \frac{q}{2t}. \tag{4.23}$$

This New Darboux Transformation, also called Levi–Darboux Transformation [**24**] or Binary Darboux Transformation [**7**], can be obtained for almost all integrable NPDE's both for continuous and discrete variables [**12, 13**].

5. Bäcklund and Darboux Transformations in Multidimensions

In the present section we give an outlook of the main ideas and techniques underlying the construction of Darboux and Bäcklund transformations for multidimensional equations. We restrict considerations to the KP and modified KP hierarchy, that will be treated at an introductory level, with no mention of some of the most interesting recent developments, having to do with "adjoint" and "binary" Darboux transformations, for which we refer to the bibliography [**7, 18, 27**].

We follow the approach based on the algebra of "pseudo-" (or "micro-")differential operators, introduced by the Sato school [**25**], and later developed by various authors [**1, 2, 19, 21, 22, 29**].

5.1. General ideas. Let us start with two hierarchies of Lax equations

$$L_{t_j} = [M_j, L], \quad \tilde{L}_{t_j} = [\widetilde{M}_j, \tilde{L}], \tag{5.1}$$

where j labels different equations associated with different evolution parameters ("times") t_j. In equation (5.1) the involved operators L, $\tilde{L}$ have the form

$$L = L_+ + L_-, \tag{5.2}$$

where

$$L_+ = \partial^n + \sum_{i=0}^{n-1} u_i \partial^i, \quad L_- = \sum_{i=-\infty}^{-1} u_i \partial^i; \quad \partial^1 = \frac{\partial}{\partial x}, \quad \partial^{-1} = \int_{-\infty}^{x} dx^1, \tag{5.3}$$

and the scalar functions u_i depend on infinitely many "times" $t_1 = x, t_2, t_3, \ldots$. We denote by g the set of pseudo-differential operators of the form (5.2), (5.3), which is clearly an associative algebra, and by g_+, g_- the subalgebras of purely differential and purely integral operators; as a vector space, g is the direct sum of g_+ and g_-, $g = g_+ \oplus g_-$, and accordingly $L_\pm$ are the projections $\pi_\pm(L)$ of L onto $g_\pm$. One assumes compatibility of (5.1) for different j, which means that the zero-curvature equations

$$M_{j_1 t_{j_2}} - M_{j_2 t_{j_1}} = [M_{j_2}, M_{j_1}], \quad \widetilde{M}_{j_1 t_{j_2}} - \widetilde{M}_{j_2 t_{j_1}} = [\widetilde{M}_{j_2}, \widetilde{M}_{j_1}], \tag{5.4}$$

do hold for all pairs of labels j_1, j_2. Let D be a field of operators satisfying the hierarchy of evolution equations

$$D_{t_j} = \widetilde{M}_j D - D M_j. \tag{5.5}$$

It is easily verified that due to (5.4) these equations are compatible when different j are considered simultaneously. The operator D provides a way to introduce a constraint between the hitherto independent Lax operators.

LEMMA. *The relation*

$$\tilde{L} D = D L, \tag{5.6}$$

is left invariant by the dynamics (5.1) *and* (5.5).

This is verified by a direct calculation. In the applications we will use the following interpretation of this fact. Assuming D to be invertible one may regard (5.6) as a gauge transformation

$$L \longrightarrow \tilde{L} = D L D^{-1}, \tag{5.7}$$

of the Lax operator. Let $L = L(u_1, u_2, \ldots)$ be parametrized by some fields u_i satisfying a set of nonlinear solvable equations encoded in the Lax representation. One wishes to construct the gauge operator D in such a way that the transformation leaves L invariant in form, i.e. $\tilde{L} = L(\tilde{u}_1, \tilde{u}_2, \ldots)$ is to be given by the same parametrization involving new fields $\tilde{u}_i$ satisfying the same set of nonlinear equations. Once these requirements are fulfilled, we call the gauge operator D a *Dressing* or *Darboux Operator* and the relation (5.6) a *Bäcklund Transformation.* The present definitions generalize those introduced in the previous sections for the one-dimensional case. Concretely, to construct a *Bäcklund Transformation* we proceed as follows:

(i) We first look for solutions $D = D(w_1, w_2, \dots)$ of (5.5) parametrized by *finitely* many fields w_i.

(ii) Then (5.7) is used to establish relations between the fields u_i in L, their dressed versions $\tilde{u}_i$ and the gauge fields w_i. These may, in principle, be used to express u_i in (nonlocal) terms of w_i.

In the relevant examples the crucial step is to parametrize the Lax equations by *finitely* many fields. The essential tool is the interpretation of (5.5) as a dressing equation which admits solutions with a finite number of fields.

Following the standard r-matrix scheme [**28**] we consider

$$L_{t_j} = [(L^j)_+, L], \quad \tilde{L}_{t_j} = \big[(\tilde{L}^j)_+, \tilde{L}\big], \quad j = 1, 2, \dots,$$

where the subscript $+$ denotes the projection to g_+. In the examples considered in the following section the gauge operation leaves the subalgebra g_- invariant, i.e. $D\xi D^{-1} \in g_-$ for arbitrary $\xi \in g_-$. From this property one finds

$$\widetilde{M}_j = (\tilde{L}^j)_+ = (DL^jD^{-1})_+ = (DM_jD^{-1})_+ + \big(D(L^j)_-D^{-1}\big)_+ = (DM_jD^{-1})_+,$$

so that (5.5) can be rewritten as the dressing equation

$$D_{t_j} = (DM_jD^{-1})_+D - DM_j = -(DM_jD^{-1})_-D. \tag{5.8}$$

The first representation involving the projection to g_+ shows that the dressing operator can be constrained to $D \in g_+$, since g_+ is a subalgebra. The second representation involving the projection to g_- provides the argument that a finite "highest order" in D may be assumed, where in our case the order is given by the highest power in the differential operator. Consequently, D may be chosen as a *finite* differential operator, and the resulting dressing transformations may be viewed as a finite iteration of elementary Darboux transformations.

By construction a compatible hierarchy of such dressing systems is obtained, since all labels j in (5.1) may be considered simultaneously. The existence of these symmetries indicate that the dressing equations, which may be interpreted as discretized equations [**9, 10**] are themselves integrable. Indeed, they can be cast in Lax form, as the resulting nonlinear system is the compatibility condition of the linear problems

$$\Phi_{t_j} = M_j\Phi, \quad \widetilde{\Phi} = D\Phi. \tag{5.9}$$

Following this procedure one can interpret the auto-Bäcklund transformations of the KP hierarchy as integrable lattice systems.

5.2. Examples: KP and modified KP hierarchy. Following [**23, 25**] the KP hierarchy may be constructed from the micro-differential Lax operator

$$L = \partial + \sum_{i \geq 1} u_i \partial^{-i}, \tag{5.10}$$

where $\partial = \partial/\partial x$. The operators M_j of the Lax pairs are constructed as

$$M_j = (L^j)_{\geq 0}, \quad q = 1, 2, \dots, \tag{5.11}$$

where the subscript ≥ 0 refers to the projection to nonnegative differential orders. The Lax representations (5.1) imply a hierarchy of nonlinear equations in two variables t_j and x for the infinite vector of fields $(u_1, u_2, \dots)$. In order to obtain closed equations for finitely many fields one uses the structure of the system associated with $y = t_2$:

$$(5.12)\qquad \begin{aligned} u_{1y} &= u_{1xx} + 2u_{2x}, \\ u_{2y} &= u_{2xx} + 2u_{3x} + 2u_1u_{1x}, \\ u_{3y} &= u_{3xx} + 2u_{4x} - 2u_1u_{1xx} + 4u_2u_{1x}. \\ &\cdots\cdots \end{aligned}$$

These relations may be integrated for $u_2, u_3, \ldots$, so that they can be expressed in (nonlocal) terms of u_1, its x and y-derivatives. Then the Lax representations (5.1) with $j \geq 3$ yield equations in the 3 independent variables x, y and t_j for the field u_1. The simplest closed equation arising from $t_3 = t$ is the KP equation proper,

$$(5.13)\qquad 4u_{1tx} = (u_{1xxx} + 12u_1u_{1x})_x + 3u_{1yy},$$

which may be integrated to its potential form

$$(5.14)\qquad 4U_{tx} = (U_{xxx} + 6U_x^2)_x + 3U_{yy},$$

for the field U introduced by

$$U_x = u_1.$$

One considers gauge transformations

$$L = \partial + \sum_{i\geq 1} u_i\partial^{-i} \longrightarrow \tilde{L} = DLD^{-1} = \partial + \sum_{i\geq 1} \tilde{u}_i\partial^{-i},$$

keeping the underlying Lax equations invariant, i.e. we assume that the transformed Lax pairs are again constructed by taking the evolution operators $\widetilde{M}_j$ as

$$\widetilde{M}_j = (\tilde{L}^j)_{\geq 0}.$$

In this case, for invertible D, the dressing equation (5.8) is given by

$$(5.15)\qquad D_{t_j} = -(DM_jD^{-1})_{<0}D = (DM_jD^{-1})_{\geq 0}D - DM_j,$$

where the subscript < 0 refers to the projection onto negative differential orders. From the two alternative representations (5.15) it is clear that this equation admits solutions of the form

$$(5.16)\qquad D = \partial^N + w_{N-1}\partial^{N-1} + \cdots + w_1\partial + w_0,$$

which—despite having a nontrivial kernel as an ordinary differential operator—admits a well defined inverse

$$D^{-1} = \partial^{-N} - w_{N-1}\partial^{-N-1} + \ldots,$$

in the algebra of pseudo-differential operators. The order N of the dressing operator is an arbitrary integer.

REMARK 5.1. For the special case $N = 1$ the operator D may be parametrized by an eigenfunction of the Lax operator

$$D = \Phi\partial\Phi^{-1} = \partial - \Phi_x\Phi^{-1}, \quad \Phi_{t_j} = M_j\Phi,$$

which generates a Darboux transformation $L \to \tilde{L}$ [**21,22**]. The solution (5.16) with general $N > 1$ may be regarded as an N-fold iteration of Darboux transformations generated by eigenfunctions $\Phi_1, \ldots, \Phi_N$, which can be retrieved from D as the elements of its kernel. Consequently, the coefficients of the dressing operator can be parametrized by the eigenfunctions via

$$w_i = (-1)^{N-i} \frac{\mathcal{W}_i(\Phi_1, \ldots, \Phi_N)}{\mathcal{W}(\Phi_1, \ldots, \Phi_N)}, \qquad i = 0, \ldots, N-1,$$

where $\mathcal{W}$ is the usual Wronskian determinant and

$$\mathcal{W}_i(\Phi_1, \ldots, \Phi_N) = \begin{vmatrix} \Phi_1 & \ldots & \frac{d^{i-1}\Phi_1}{dx^{i-1}} & \frac{d^{i+1}\Phi_1}{dx^{i+1}} & \ldots & \frac{d^N\Phi_1}{dx^N} \\ \Phi_2 & \ldots & \frac{d^{i-1}\Phi_2}{dx^{i-1}} & \frac{d^{i+1}\Phi_2}{dx^{i+1}} & \ldots & \frac{d^N\Phi_2}{dx^N} \\ \vdots & \vdots & \vdots & \vdots & \vdots & \vdots \\ \Phi_N & \ldots & \frac{d^{i-1}\Phi_N}{dx^{i-1}} & \frac{d^{i+1}\Phi_N}{dx^{i+1}} & \ldots & \frac{d^N\Phi_N}{dx^N} \end{vmatrix}.$$

In particular, we may consider an elementary Darboux transformation

$$D = \partial - w,$$

which gives rise to the relations

$$\tilde{u}_1 - u_1 = w_x, \quad \tilde{u}_2 - u_2 = w(\tilde{u}_1 - u_1) + u_{1x},$$

from (5.6). After integration one obtains

$$w = \widetilde{U} - U + \mu, \quad \tilde{u}_2 - u_2 = w(\widetilde{U}_x - U_x) + U_{xx}, \tag{5.17}$$

where μ is an integration constant. Elimination of w together with integration of (5.12) yields a closed set of differential-difference equations for U and u_2:

$$U_y = U_{xx} + 2u_2, \quad u_2^{(1)} - u_2 = \left(U_x + \frac{1}{2}(U^{(1)} - U + \mu)^2\right)_x. \tag{5.18}$$

Finally, elimination of the auxiliary field u_2 leads to

$$U_y^{(1)} - U_y = \left(U_x^{(1)} + U_x + (U^{(1)} - U + \mu)^2\right)_x. \tag{5.19}$$

This equation represents the spatial part of the standard auto-Bäcklund transformation for the potential form of the KP hierarchy.

Alternatively one may look at the corresponding equation for the field w generating the dressing transformation, where U and u_2 are given in (nonlocal) terms of w via (5.17). The evolution equation (5.5) directly yields the following system (which can be viewed as a differential-difference one):

$$w_y = (w_x + w^2 + 2U_x)_x, \quad w = \widetilde{U} - U + \mu, \tag{5.20}$$

which leads to (5.19) after elimination of w. On the other hand, elimination of U yields

$$\widetilde{w}_y - w_y = \left(\widetilde{w}_x + w_x + (\widetilde{w})^2 - w^2\right)_x. \tag{5.21}$$

By construction (5.20) is the compatibility condition of the Lax pair

$$\Phi_y = (L^2)_{\geq 0}\Phi = \Phi_{xx} + 2U_x\Phi, \quad \widetilde{\Phi} = D\Phi = \Phi_x - w\Phi. \tag{5.22}$$

We note that due to the homogeneous nature of the dressing equation (5.8) one can replace the dressing operator D by an arbitrary multiple, so that a spectral parameter may be introduced:

$$\lambda\widetilde{\Phi} = \Phi_x - w\Phi.$$

REMARK 5.2. We observe that one may consider a realisation of the dressing operation of the form

$$\widetilde{w}(x,y) = w(x+ih,y). \tag{5.23}$$

Subject to suitable analyticity assumptions one has

$$\frac{T+1}{T-1} = i\mathcal{T}, \quad (\mathcal{T}f)(z) = \frac{1}{h} P.V. \int_{-\infty}^{\infty} \coth\left(\frac{\pi(z'-z)}{h}\right) f(z')\, dz', \tag{5.24}$$

for the corresponding shift operator

$$T : f(z) \to f(z+ih).$$

With $w = iu$ "integration" of (5.21) yields

$$iu_y + \mathcal{T}u_{xx} + 2uu_x = 0,$$

which is the intermediate long wave (ILW) equation [**5**] with an imaginary evolution parameter y. It arises from the auto-Bäcklund transformation of the KP hierarchy subject to the interpretation (5.23). Its hierarchy is the compatibility condition of the linear problems

$$\lambda\widetilde{\Phi} = \Phi_x - iu\Phi, \quad \Phi_{t_j} = (L^j)_{\geq 0}\Phi,$$

where L is the Lax operator of the KP hierarchy.

5.3. The modified hierarchy. The modified KP hierarchy may be constructed from the pseudo-differential Lax operator [**1, 2, 21, 22**]

$$\hat{L} = \partial + \sum_{i\geq 0} \hat{u}_i \partial^{-i}, \tag{5.25}$$

where the operators $\widehat{M}_j$ of the Lax pairs are constructed as

$$\widehat{M}_j = (\hat{L}^j)_{\geq 1}, \quad j = 1, 2, \dots .$$

Here, in contrast to the KP hierarchy, zero-order terms are not involved in the projection ≥ 1 to strictly positive differential orders. Following [**21, 22**] solutions of $\hat{L}_{t_j} = [\widehat{M}_j, \hat{L}]$ are obtained from a gauge transformation of the Lax operator (5.10) for the KP hierarchy. The transformation involves an eigenfunction Φ satisfying $\Phi_{t_j} = (L^j)_{\geq 0}\Phi$ and is given by

$$\begin{aligned}\hat{L} = \frac{1}{\Phi} \circ L \circ \Phi &= \partial + \frac{\Phi_x}{\Phi} + u_1\partial^{-1} + \left(u_2 - \frac{\Phi_x}{\Phi}\right)\partial^{-2} + \dots \\ &= \partial + w + U_x\partial^{-1} + \sum_{i\geq 2} \hat{u}_i \partial^{-i}.\end{aligned}$$

With Remark 5.1 we have identified the gauge field $w = \Phi_x/\Phi$ of the KP hierarchy with the highest coefficient $\hat{u}_0$ in (5.25), and $\hat{u}_1$ coincides with the field u_1 satisfying the KP equation (5.13). The equation for w arising from $t_3 = t$ is the modified KP equation [**21, 22**]

$$4w_{tx} = (w_{xxx} - 6w^2 w_x)_x + 3w_{yy} + 6w_x w_y + 6w_{xx} \int^x w_y, \tag{5.26}$$

which may be integrated to its potential form

$$4V_{tx} = (V_{xxx} - 2V_x^3)_x + 3V_{yy} + 6V_{xx}V_y,$$

for the field V introduced by

$$w = V_x.$$

The equation (5.21) for w obtained from a discretization of the KP hierarchy also represents an auto-Bäcklund transformation of the modified KP hierarchy. Applying the general procedure of Section 1 we derive an extended version of (5.21) from a dressing transformation of (5.25).

The modified dressing equation (5.8) is given by

$$\widehat{D}_{t_j} = -(\hat{d}\widehat{M}_j\widehat{D}^{-1})_{<1}\widehat{D} = (\widehat{D}\widehat{M}_j\widehat{D}^{-1})_{\geq 1}\widehat{D} - \widehat{D}\widehat{M}_j, \tag{5.27}$$

where the subscript < 1 represents the projection onto nonpositive differential orders. This equation admits solutions of the form

$$\widehat{D} = g_N\partial^N + g_{N-1}\partial^{N-1} + \cdots + g_1\partial + g_0 + \partial^{-1}g_{-1}. \tag{5.28}$$

One may assume $g_{-1} = 0$ and $g_0 = \text{const}$.

REMARK 5.3. Special solutions of (5.27) are parametrized by an eigenfunction $\widehat{\Phi}$ via

$$\widehat{D} = -\frac{\widehat{\Phi}^2}{\widehat{\Phi}_x}\partial\frac{1}{\widehat{\Phi}} = -\frac{\widehat{\Phi}}{\widehat{\Phi}_x}\partial + 1, \quad \widehat{\Phi}_{t_j} = \widehat{M}_j\widehat{\Phi}, \tag{5.29}$$

which corresponds to a Darboux transformation $\hat{L} \to \hat{L}'$ [**21, 22**]. The solution (5.28) with general $N > 1$, $g_{-1} = 0$ and $g_0 = 1$ may be regarded as an N-fold iteration of Darboux transformations generated by eigenfunctions $\widehat{\Phi}_1, \ldots, \widehat{\Phi}_N$, which can be retrieved from $\widehat{D}$ as the elements of its kernel.

We consider a dressing operator of the simplest form

$$\widehat{D} = g\partial + \mu,$$

with some constant μ.

It gives rise to

$$g_x = g(w - w'), \tag{5.30}$$

from (5.6). The evolution equation (5.5) for $t_2 = y$ yields

$$g_y = g_{xx} + 2w'g_x - 2w_x g + 2\mu(w' - w). \tag{5.31}$$

By construction (5.30)/(5.31) is the compatibility condition of the Lax pair

$$\widehat{\Phi}_y = (\hat{L}^2)_{\geq 1}\widehat{\Phi} = \widehat{\Phi}_{xx} + 2w\widehat{\Phi}_x, \quad \widehat{\Phi}^{(1)} = \widehat{D}\widehat{\Phi} = g\widehat{\Phi}_x + \mu\widehat{\Phi}.$$

For $\mu = 0$ this leads to (5.21) upon elimination of g.

REMARK 5.4. Elimination of w from (5.30), (5.31) yields

$$if_y + \mathcal{T}f_{xx} - f_x\mathcal{T}f_x + 2\mu\big(\exp(f)\big)_x = 0, \tag{5.32}$$

where $f = \ln(g)$, the dressing operation is interpreted as in (5.23) and hence $\mathcal{T}$ is the integral operator (5.24). Equation (5.32) essentially is the modified ILW equation [**26**].

We conclude this subsection with a few comments on the interpretation of Bäcklund transformations. As it has been already pointed out, we can think of Bäcklund transformations of a given nonlinear PDE as an integrable *first discretization.* To get further integrable discretizations one could then consider Bäcklund transformations of the first discretization, and proceed until all independent variables have been discretized. However, one can follow a different path, that we are going to describe briefly in the sequel, and be sure that one obtains exactly the same result.

To stay on a solid ground, we will again consider the example of the KP hierarchy. We can consider the three gauge operators:

$$D_i = \partial - w_i, \quad i = 1, 2, 3,$$

which generate 3 different transformations $T_i : U \to U^{(i)}$ of the KP hierarchy. They give rise to 3 different discretizations with the first nontrivial equations (auto-Bäcklund transformations) given by

$$A_i : U_y^{(i)} - U_y = \left(U_x^{(i)} + U_x + (U^{(i)} - U + \mu_i)^2\right)_x, \quad i = 1, 2, 3.$$

Here

$$w_i = U^{(i)} - U + \mu_i,$$

with integration constants μ_i. For a second discretization one may eliminate the y-derivatives by choosing a pair of distinct indices $i, j \in \{1, 2, 3\}$ and computing $(T_i - 1)A_j - (T_j - 1)A_i$. This results in

$$B_{ij} : (U^{(i)} - U^{(j)})_x = (U^{(i)} - U^{(j)} + \mu_i - \mu_j)(U^{(i,j)} - U^{(i)} - U^{(j)} + U).$$

For a final discretization one eliminates the x-derivatives by adding $B_{12}+B_{23}+B_{31}$, where one obtains

$$\begin{aligned} U^{(1,2)}(U^{(1)} - U^{(2)} + \mu_1 - \mu_2) + U^{(2,3)}(U^{(2)} - U^{(3)} + \mu_2 - \mu_3) \\ + U^{(3,1)}(U^{(3)} - U^{(1)} + \mu_3 - \mu_1), \\ = (\mu_3 - \mu_2)U^{(1)} + (\mu_1 - \mu_3)U^{(2)} + (\mu_2 - \mu_1)U^{(3)}. \end{aligned} \tag{5.33}$$

Elimination of Φ_x from the gauge transformations

$$\Phi^{(i)} = \Phi_x - w_i \Phi, \quad i = 1, 2, 3,$$

of the eigenfunctions leads to the linear representation of (5.33):

$$\Phi^{(i)} - \Phi^{(j)} + v_{ij}\Phi = 0, \quad v_{ij} = w_i - w_j = U^{(i)} - U^{(j)} + \mu_i - \mu_j.$$

References

1. H. Aratyn, L. A. Ferreira, J. F. Gomes, and A. H. Zimerman, *On two-current realization of* KP *hierarchy*, Nuclear Phys. B **402** (1993), 85–117.
2. H. Aratyn, E. Nissimov, S. Pacheva, and I. Vaysburd, *R-matrix formulation of* KP *hierarchies and their gauge equivalence*, Phys. Lett. B **294** (1992), 167–176.
3. F. Calogero and A. Degasperis, *Spectral transform and solitons: Tools to solve and investigate nonlinear evolution equations*, North-Holland, Amsterdam, 1982.
4. M. G. Darboux, *Sur une proposition relative aux équations linéaires*, C. R. Acad. Sci. **94** (1882), 1456–1459.
5. A. Degasperis, D. Lebedev, M. Olshanetsky, S. Pakuliak, A. Perelomov, and P. M. Santini, *Generalized intermediate long-wave hierarchy in zero-curvature representation with noncommutative spectral parameter*, J. Math. Phys. **33** (1992), 3783–3793.
6. R. Hirota, *Discrete analogue of a generalized Toda equation*, J. Phys. Soc. Japan **50** (1981), 3785–3791.
7. B. G. Konopelchenko and C. Rogers, *Bäcklund and reciprocal transformations: gauge connections*, Nonlinear Equations in Applied Science, Academic Press, Boston, 1992, pp. 317–362.
8. B. A. Kupershmidt, *Discrete Lax equations and differential-difference calculus*, Astérisque **123** (1985).
9. D. Levi, *Nonlinear differential-difference equations as Bäcklund transformations*, J. Phys. A **14** (1981), 1083–1098.

10. D. Levi and R. Benguria, *Bäcklund transformations and nonlinear differential-difference equations*, Proc. Nat. Acad. Sci. USA **77** (1980), 5025–5027.
11. D. Levi, L. Pilloni, and P. M. Santini, *Integrable three-dimensional lattices*, J. Phys. A **14** (1981), 1567–1575.
12. D. Levi and O. Ragnisco, *Nonisospectral deformations and Darboux transformations for the third order spectral problem*, Inverse Problems **4** (1988), 815–828.
13. ———, *The inhomogeneous Toda lattice: Its hierarchy and Darboux–Bäcklund transformations*, J. Phys. A **24** (1991), 1729–1739.
14. D. Levi, O. Ragnisco, and M. Bruschi, *Extension of the Zakharov–Shabat generalized inverse problem to solve differential-difference and difference-difference equations*, Nuovo Cimento A **58** (1980), 56–66.
15. ———, *Continuous and discrete matrix Burgers hierarchy*, Nuovo Cimento B **74** (1983), 33–51.
16. D. Levi, O. Ragnisco, and A. B. Shabat, *Construction of higher local* $2+1$*-dimensional exponential lattice equations*, Canad. J. Phys. **72** (1994), 439–441.
17. D. Levi, A. Sym, and S. Wojciechowski, *A hierarchy of coupled Korteweg–de Vries equations and the normalization conditions of the Hilbert–Riemann problem*, J. Phys. A **16** (1983), 2423–2432.
18. Q. P. Liu and M. Manas, *Vectorial Darboux transformations for the Kadomtsev–Petviashvili hierarchy*, J. Nonlinear Sci. **9** (1999), 213–232.
19. F. Magri, M. Pedroni, and J. P. Zubelli, *On the geometry of Darboux transformations for the* KP *hierarchy and its connection with the discrete* KP *hierarchy*, Comm. Math. Phys. **188** (1997), 305–325.
20. T. Nakajima and S. Saito, *Lax formalism of discrete time integrable systems*, Phys. Lett. A **181** (1993), 17–23.
21. W. Oevel and C. Rogers, *Gauge transformations and reciprocal links in* $2+1$*-dimensions*, Rev. Math. Phys. **5** (1993), 299–330.
22. W. Oevel and W. Schief, *Darboux theorems and* KP *hierarchy*, Applications of Analytic and Geometric Methods to Nonlinear Differential Equations, Kluwer Acad. Publ. Dordrecht, 1993, pp. 193–206.
23. Y. Ohta, J. Satsuma, D. Takasaki, and T. Tokihiro, *An elementary introduction to Sato theory*, Recent Developments in Soliton Theory, Progr. Theoret. Phys. Suppl. **94** (1988), 210–241.
24. C. Rogers, W. K. Schief, and M. P. Stallybrass, *Initial/boundary value problem and Darboux–Levi-type transformations associated with a* $2+1$*-dimensional eigenfunction equation*, Internat. J. Non-Linear Mech. **30** (1995), 223–233.
25. M. Sato, *The* KP *hierarchy and infinite-dimensional Grassmann manifolds*, Proc. Sympos. Pure Math. **49**, Part I, Amer. Math. Soc., Providence, RI, 1989.
26. J. Satsuma and M. J. Ablowitz, *Solution of an internal wave equation describing a stratified fluid with finite depth*, Nonlinear Partial Differential Equations in Engineering and Applied Science (R. L. Stenberg, A. J. Kalinoski, and J. S. Papadakis, eds.), Dekker, New York, 1980, pp. 397–414.
27. W. K. Schief and C. Rogers, *Loewner transformations: Adjoint and binary Darboux connections*, Stud. Appl. Math. **100** (1998), 391–422.
28. M. A. Semenov–Tian–Shanski, *What is a classical* r*-matrix?*, Functional Anal. Appl. **17** (1983), 259–272.
29. K. Ueno and K. Takasaki, *Group representation and systems of differential equations*, Adv. Stud. Pure Math. **4** (1984), 1–95.
30. V. E. Zakharov and A. V. Mikhailov, *Relativistically invariant two-dimensional models of field theory which are integrable by means of the inverse scattering problem method*, Soviet Phys. JETP **47** (1978), 1071–1077.
31. V. E. Zakharov and A. B. Shabat, *Integration of the nonlinear equations of mathematical physics by the method of the inverse scattering problem* II, Functional Anal. Appl. **13** (1979), 166–174.

Dipartimento di Fisica, Università di Roma Tre and INFN-Sezione di Roma Tre, Via della Vasca Navale 84, I-00146 Roma, Italy.

E-mail address, D. Levi: `levi@amaldi.fis.uniroma3.it`

E-mail address, O. Ragnisco: `ragnisco@amaldi.fis.uniroma3.it`

Centre de Recherches Mathématiques
CRM Proceedings and Lecture Notes
Volume **29**, 2001

The Classical Geometry of Bäcklund Transformations. Introduction to Applications in Soliton Theory

C. Rogers and *W. K. Schief*

ABSTRACT. Here, the classical geometric origins of Bäcklund transformations are introduced "ab initio" and their fundamental role in modern soliton theory is described. Contact is made with recent research on discretisation of characteristic systems via classical Bäcklund transformations.

1. Background

The foundations of the differential geometry of curves and surfaces were set down in the early part of the XIXth century with the pioneering works of Monge (1746–1818) and Gauß (1777–1855). The major contributions of Monge are to be found in his Applications de l'Analyse à la Géométrie originally published in 1807. The major treatise on the geometry of surfaces of Gauß, instigated by a geodetic study sponsored by the Elector of Hannover, was the Disquitisiones Generales Circa Superficies Curvas published in 1827. Therein, Gauß laid down the system that bears his name and which is fundamental to the analysis of surfaces. It is this Gauß system and the symmetries that it admits for privileged classes of surfaces that underpin the remarkable connections between classical differential geometry and the study of the propagation and interaction of the important class of nonlinear waves that have become known as solitons and which arise in a diversity of physical applications.

The origins of modern soliton theory are also to be found in the XIXth century. Thus, it was in 1834 that the Scottish engineer John Scott Russell recorded the first sighting, along a canal near Edinburgh, of the solitary hump-shaped wave to be rediscovered in 1965 in the context of the celebrated Fermi–Pasta–Ulam problem by Kruskal and Zabusky and termed a soliton. Scott Russell observed that his so-called great wave of translation proceeded with a speed proportional to its height. He was later to reproduce this large amplitude surface phenomenon in water tank experiments described in a report to the British Association in 1844 wherein there is also depicted the creation of two such waves. However, the limited duration of Scott Russell's experiments apparently did not allow him to observe the dramatic interaction properties of these waves in their entirety. Moreover, at that time,

2000 *Mathematics Subject Classification.* Primary: 35Q51; Secondary: 35Q58.
This is the final form of the paper.

neither the nonlinear evolution equation descriptive of their propagation nor the means to predict their interaction properties were to hand.

It was in 1895 that two Dutch mathematicians, Korteweg and de Vries, derived the nonlinear wave equation which now bears their name and adopts the canonical form

$$\frac{\partial u}{\partial t} + 6u\frac{\partial u}{\partial x} + \frac{\partial^3 u}{\partial x^3} = 0.$$

This models long wave propagation in a rectangular canal and provides, through a simple travelling wave solution, a theoretical confirmation of the existence of the solitary wave observed some sixty years earlier by Scott Russell on the Union canal. However, it is less well-known that what is now called the Korteweg–de Vries (KdV) equation had, in fact, been set down earlier by Boussinesq in his memoir of 1877 entitled *Essai sur la Théorie des Eaux Courantes.* Indeed, a pair of equations equivalent to the KdV equation appeared as early as 1871 in two papers by Boussinesq devoted to wave propagation in rectangular channels.

In the pioneering study of Kruskal and Zabusky of 1965, the KdV equation was retrieved as a continuum limit of a lattice model with cubic nonlinearity. The existence in this nonlinear model of "solitonic" waves which possess the remarkable property that they preserve both their amplitude and speed subsequent upon interaction was uncovered via a numerical study. However, the problem of obtaining an analytical expression descriptive of the interaction of solitons remained.

It turns out, remarkably, that a generic method for the description of soliton interaction has its roots in a type of transformation originally introduced by Bäcklund in the XIXth century to construct pseudo-spherical surfaces [**11, 13**]. In our present introduction to the subject of Bäcklund transformations and the geometry of solitons it proves convenient to work in the context of a more general system due to Bianchi which describes hyperbolic surfaces. A Bäcklund transformation associated with a solitonic Bianchi system is derived. Normal propagation of pseudo-spherical surfaces is shown to lead to another solitonic system with its origin in classical differential geometry and the study of triply orthogonal coordinate systems. To conclude, it is shown that the classical Bäcklund transformation generating pseudo-spherical surfaces provides an integrable discretisation for a characteristic system associated with their Monge form. This provides a point of contact with the subject of discretisation and solitonic equations.

2. Geometric Preliminaries. The Gauß–Weingarten Equations for Hyperbolic Surfaces

The basic geometry of curves and surfaces required for an understanding of the classical geometry of solitons may be found in the classical work of Bianchi [**2**] and the monographs of Struick [**18**] or do Carmo [**4**]. Here, only the details required for our immediate purpose are set down.

Thus, let $\boldsymbol{r} = \boldsymbol{r}(u, v)$ denote the position vector of a generic point P on a surface Σ in $\mathbb{R}^3$. Then, the vectors $\boldsymbol{r}_u$ and $\boldsymbol{r}_v$ are tangential to Σ at P and at such points at which they are linearly independent $\boldsymbol{N} = \boldsymbol{r}_u \times \boldsymbol{r}_v / |\boldsymbol{r}_u \times \boldsymbol{r}_v|$ determines the unit normal to Σ.

The 1st and 2nd fundamental forms of Σ are defined by

$$\begin{aligned} \mathrm{I} &= d\boldsymbol{r} \cdot d\boldsymbol{r} = E\,du^2 + 2F\,du\,dv + G\,dv^2, \\ \mathrm{II} &= -d\boldsymbol{r} \cdot d\boldsymbol{N} = e\,du^2 + 2f\,du\,dv + g\,dv^2, \end{aligned} \tag{2.1}$$

where

$$
\begin{aligned}
E &= \boldsymbol{r}_u \cdot \boldsymbol{r}_u, \quad F = \boldsymbol{r}_u \cdot \boldsymbol{r}_v, \quad G = \boldsymbol{r}_v \cdot \boldsymbol{r}_v,\\
e &= -\boldsymbol{r}_u \cdot \boldsymbol{N}_u = \boldsymbol{r}_{uu} \cdot \boldsymbol{N}, \quad g = -\boldsymbol{r}_v \boldsymbol{N}_v = \boldsymbol{r}_{vv} \cdot \boldsymbol{N},\\
f &= -\boldsymbol{r}_u \cdot \boldsymbol{N}_v = -\boldsymbol{r}_v \cdot \boldsymbol{N}_u = \boldsymbol{r}_{uv} \cdot \boldsymbol{N}.
\end{aligned}
\tag{2.2}
$$

An important classical result due to Bonnet states that the sextuplet $\{E, F, G; e, f, g\}$ determines the surface Σ up to its position in space.

The celebrated Gauß equations associated with Σ are

$$
\begin{aligned}
\boldsymbol{r}_{uu} &= \Gamma^1_{11}\boldsymbol{r}_u + \Gamma^2_{11}\boldsymbol{r}_v + e\boldsymbol{N},\\
\boldsymbol{r}_{uv} &= \Gamma^1_{12}\boldsymbol{r}_u + \Gamma^2_{12}\boldsymbol{r}_v + f\boldsymbol{N},\\
\boldsymbol{r}_{vv} &= \Gamma^1_{22}\boldsymbol{r}_u + \Gamma^2_{22}\boldsymbol{r}_v + g\boldsymbol{N},
\end{aligned}
\tag{2.3}
$$

while the Weingarten equations comprise

$$
\boldsymbol{N}_u = \frac{fF - eG}{H^2}\boldsymbol{r}_u + \frac{eF - fE}{H^2}\boldsymbol{r}_v, \quad \boldsymbol{N}_v = \frac{gF - fG}{H^2}\boldsymbol{r}_u + \frac{fF - gE}{H^2}\boldsymbol{r}_v,
\tag{2.4}
$$

where $H^2 = |\boldsymbol{r}_u \times \boldsymbol{r}_v|^2 = EG - F^2$. The Γ^i_{jk} in (2.3) are the usual Christoffel symbols given by the relations

$$
\Gamma^i_{jk} = \frac{g^{il}}{2}(g_{jl,k} + g_{kl,j} - g_{jk,l}),
\tag{2.5}
$$

where, with $x^1 = u$, $x^2 = v$,

$$
\mathrm{I} = g_{jk}\, dx^j\, dx^k,
\tag{2.6}
$$

and

$$
g^{jk} g_{kl} = \delta^j_l.
\tag{2.7}
$$

In the above, the Einstein convention of summation over repeated indices has been adopted.

The compatibility conditions $(\boldsymbol{r}_{uu})_v = (\boldsymbol{r}_{uv})_u$ and $(\boldsymbol{r}_{uv})_v = (\boldsymbol{r}_{vv})_u$ applied to the linear system (2.3) produce the celebrated nonlinear Mainardi–Codazzi–Gauß system

$$
\begin{aligned}
&\partial_v(e/H) - \partial_u(f/H) + \Gamma^2_{22}(e/H) - 2\Gamma^2_{12}(f/H) + \Gamma^2_{11}(g/H) = 0,\\
&\partial_u(g/H) - \partial_v(f/H) + \Gamma^1_{22}(e/H) - 2\Gamma^1_{12}(f/H) + \Gamma^1_{11}(g/H) = 0,\\
&\mathcal{K} = \big[\partial_v(H\Gamma^2_{11}/E) - \partial_u(H\Gamma^2_{12}/E)\big]/H.
\end{aligned}
\tag{2.8}
$$

The last relation incorporates the "Theorema egregium" of Gauß which gives the total (Gaußian) curvature $\mathcal{K} = (eg - f^2)/H^2$ in terms of E, F, G alone. In physical terms, this implies that the total curvature of a surface Σ is invariant under bending without stretching.

If the total curvature of Σ is negative, that is, if Σ is a hyperbolic surface, then the asymptotic lines on Σ may be taken as parametric curves. Then $e = g = 0$ and the Mainardi–Codazzi equations $(2.8)_{1,2}$ reduce to

$$
\partial_u(f/H) + 2\Gamma^2_{12}(f/H) = 0, \quad \partial_v(f/H) + 2\Gamma^1_{12}(f/H) = 0,
\tag{2.9}
$$

while

$$
\mathcal{K} = -f^2/H^2 =: -\frac{1}{\rho^2},
\tag{2.10}
$$

and

$$\Gamma^1_{12} = (GE_v - FG_u)/2H^2, \quad \Gamma^2_{12} = (EG_u - FE_v)/2H^2. \tag{2.11}$$

If ω is the angle between the parametric lines then

$$\cos\omega = F/\sqrt{EG}, \tag{2.12}$$

and since $E, G > 0$, we may take, without loss of generality

$$E = \rho^2 a^2, \ G = \rho^2 b^2, \tag{2.13}$$

whence the fundamental forms I and II reduce to

$$\mathrm{I} = \rho^2(a^2 du^2 + 2ab\cos\omega\, du\, dv + b^2\, dv^2), \quad \mathrm{II} = 2\rho ab \sin\omega\, du\, dv. \tag{2.14}$$

The Mainardi–Codazzi equations (2.9) together with the Gauß representation $(2.8)_3$ for the total curvature now produce a classical system due to Bianchi [**2**], namely

$$\begin{aligned} a_v + \frac{1}{2}\frac{\rho_v}{\rho}a - \frac{1}{2}\frac{\rho_u}{\rho}b\cos\omega = 0, \quad b_u + \frac{1}{2}\frac{\rho_u}{\rho}b - \frac{1}{2}\frac{\rho_v}{\rho}a\cos\omega = 0,\\ \omega_{uv} + \frac{1}{2}\left(\frac{\rho_u}{\rho}\frac{b}{a}\sin\omega\right)_u + \frac{1}{2}\left(\frac{\rho_v}{\rho}\frac{a}{b}\sin\omega\right)_v - ab\sin\omega = 0.\end{aligned} \tag{2.15}$$

The importance of this nonlinear system in soliton theory has been noted by Cenkl [**3**] and subsequently by Levi and Sym [**8**]. In that connection it is subject to an additional constraint, namely $\rho_{uv} = 0$. The system then becomes integrable.

3. Pseudo-Spherical Surfaces. The Classical Bäcklund Transformation

If $\mathcal{K} = -1/\rho^2 < 0$ is a constant, then Σ is termed a pseudo-spherical surface. The Mainardi–Codazzi equations $(2.15)_{1,2}$ then yield $a = a(u)$, $b = b(v)$. If Σ is parametrised by arc length along asymptotic lines (corresponding to the transformation $du \to du' = \sqrt{E(u)}\, du$, $dv \to dv' = \sqrt{G(v)}\, dv$) then the fundamental forms become, on dropping the primes,

$$\begin{aligned} \mathrm{I} &= du^2 + 2\cos\omega\, du\, dv + dv^2,\\ \mathrm{II} &= \frac{2}{\rho}\sin\omega\, du\, dv,\end{aligned} \tag{3.1}$$

while $(2.15)_3$ reduces to the celebrated sine–Gordon equation

$$\omega_{uv} = \frac{1}{\rho^2}\sin\omega. \tag{3.2}$$

The associated Gauß equations yield

$$\begin{aligned} \boldsymbol{r}_{uu} &= \omega_u \cot\omega\, \boldsymbol{r}_u - \omega_u \operatorname{cosec}\omega\, \boldsymbol{r}_v,\\ \boldsymbol{r}_{uv} &= \frac{1}{\rho}\sin\omega\, \boldsymbol{N},\\ \boldsymbol{r}_{vv} &= -\omega_v \operatorname{cosec}\omega\, \boldsymbol{r}_u + \omega_v\cot\omega\, \boldsymbol{r}_v,\end{aligned} \tag{3.3}$$

while the Weingarten equations become

$$\boldsymbol{N}_u = 1/\rho[\cot\omega\, \boldsymbol{r}_u - \operatorname{cosec}\omega\, \boldsymbol{r}_v], \quad \boldsymbol{N}_v = 1/\rho[-\operatorname{cosec}\omega\, \boldsymbol{r}_u + \cot\omega\, \boldsymbol{r}_v]. \tag{3.4}$$

In the XXth century, the sine–Gordon equation has been shown, remarkably, to arise in a diversity of areas of physical interest (Rogers and Shadwick [**13**]). It was the work of Seeger et al. [**17**] that first demonstrated how the classical Bäcklund transformation for the sine–Gordon equation has important applications in the theory of crystal dislocations. Indeed, in [**17**], within the context of Frenkel's and Kontorova's dislocation theory, the superposition of so-called "eigen-motions" was obtained by means of the classical Bäcklund transformation. The interaction of what today are called breathers with kink-type dislocations was both described analytically and displayed graphically. The typical solitonic features to be subsequently discovered in [**19**] by Zabusky and Kruskal in 1965 for the KdV equation, namely preservation of velocity and shape following interaction as well as the concomitant phase shift were all recorded for the sine–Gordon equation in this remarkable paper of 1953.[1]

Later, Lamb [**7**] and Barnard [**1**] showed that the nonlinear superposition principle associated with the Bäcklund transformation for the sine–Gordon equation has application in the theory of ultra-short optical pulse propagation. In particular, solitonic decomposition phenomena observed experimentally in Rb vapour were thereby reproduced theoretically. In addition, the classical Bäcklund transformation has found application in the theory of long Josephson junctions (Scott [**16**]).

The preceding provides a historical motivation for introducing the study of Bäcklund transformations with the classical result for the sine–Gordon equation. It will be seen that this Bäcklund transformation, in fact, corresponds to a conjugation of invariant transformations due to Bianchi and Lie. The Lie symmetry serves to intrude a key Bäcklund parameter into the Bianchi transformation which enables its iteration and the generation of what are, in physical terms, multi-soliton solutions. Therein, the Bäcklund parameters turn out to have an important physical interpretation.

Underlying the original Bäcklund transformation for the sine–Gordon equation is a simple geometric construction for pseudo-spherical surfaces. Thus, if a point P is taken on an initial pseudo-spherical surface Σ and a line segment PP' of constant length and tangential to Σ at P is constructed in a manner dictated by a Bäcklund transformation as described below then the locus of the points P' as P traces out Σ is another pseudo-spherical surface Σ' with the same total curvature as Σ. The procedure may be iterated to generate a sequence of pseudo-spherical surfaces all with the same total curvature as the original seed surface Σ.

Let Σ be a pseudo-spherical surface with total curvature $\mathcal{K} = -1/\rho^2$ and with generic position vector $\boldsymbol{r} = \boldsymbol{r}(u, v)$ where u, v correspond to the parametrisation by arc length along asymptotic lines. In this parametrisation, $\boldsymbol{r}_u, \boldsymbol{r}_v$ and $\boldsymbol{N}$ are all unit vectors, but $\boldsymbol{r}_u$ and $\boldsymbol{r}_v$ are not orthogonal. Accordingly, it proves convenient to introduce an orthonormal triad $\{\boldsymbol{A}, \boldsymbol{B}, \boldsymbol{C}\}$ where

$$\boldsymbol{A} = \boldsymbol{r}_u, \quad \boldsymbol{B} = -\boldsymbol{r}_u \times \boldsymbol{N} = \operatorname{cosec}\omega\,\boldsymbol{r}_v - \cot\omega\,\boldsymbol{r}_u, \quad \boldsymbol{C} = \boldsymbol{N}. \tag{3.5}$$

The Gauß–Weingarten equations (3.3)–(3.4) now yield

$$\begin{aligned}
\begin{pmatrix} \boldsymbol{A} \\ \boldsymbol{B} \\ \boldsymbol{C} \end{pmatrix}_u &= \begin{pmatrix} 0 & -\omega_u & 0 \\ \omega_u & 0 & 1/\rho \\ 0 & -1/\rho & 0 \end{pmatrix} \begin{pmatrix} \boldsymbol{A} \\ \boldsymbol{B} \\ \boldsymbol{C} \end{pmatrix}, \\
\begin{pmatrix} \boldsymbol{A} \\ \boldsymbol{B} \\ \boldsymbol{C} \end{pmatrix}_v &= \begin{pmatrix} 0 & 0 & (1/\rho)\sin\omega \\ 0 & 0 & -(1/\rho)\cos\omega \\ -(1/\rho)\sin\omega & (1/\rho)\cos\omega & 0 \end{pmatrix} \begin{pmatrix} \boldsymbol{A} \\ \boldsymbol{B} \\ \boldsymbol{C} \end{pmatrix}.
\end{aligned} \tag{3.6}$$

---

[1] "Man sieht ..., daß beim Durchdringen von Wellengruppe und Versetzung weder die Energie noch die Geschwindigkeit beider geändert wird. Es tritt lediglich eine Verschiebung des Versetzungsmittelpunktes ... und des Schwerpunktes der Wellengruppe ... auf". [7, p. 189]

This linear system is compatible if and only if ω satisfies the sine–Gordon equation (3.2).

A new pseudo-spherical surface Σ' is now sought with position vector

$$\boldsymbol{r}' = \boldsymbol{r} + L\cos\phi\,\boldsymbol{A} + L\sin\phi\,\boldsymbol{B}, \tag{3.7}$$

where $L = |\boldsymbol{r}' - \boldsymbol{r}|$ is constant. Here, $\phi(u,v)$ is to be constrained by the requirement that on Σ', as on Σ, the coordinates u, v correspond to parametrisation along asymptotic lines. A necessary condition for this is that Σ' have a 1st fundamental form a primed analogue of $(3.1)_1$. In particular, this requires that

$$\boldsymbol{r}'_u \cdot \boldsymbol{r}'_u = 1, \quad \boldsymbol{r}'_v \cdot \boldsymbol{r}'_v = 1,$$

whence, we obtain

$$\phi_u = \omega_u + \frac{\beta}{\rho}\sin\phi, \quad \phi_v = \frac{1}{\beta\rho}\sin(\phi - \omega), \tag{3.8}$$

where

$$\beta = \frac{\rho}{L}\left(1 \pm \sqrt{1 - \frac{L^2}{\rho^2}}\right), \tag{3.9}$$

and

$$\boldsymbol{r}'_u \cdot \boldsymbol{r}'_v = \cos(2\phi - \omega). \tag{3.10}$$

Thus,

$$\mathrm{I}' = du^2 + 2\cos(2\phi - \omega)du\,dv + dv^2, \tag{3.11}$$

and it is readily shown that

$$\boldsymbol{r}'_u \cdot \boldsymbol{N}'_u = 0, \quad \boldsymbol{r}'_u \cdot \boldsymbol{N}'_v = \boldsymbol{r}'_v \cdot \boldsymbol{N}'_u = -\frac{1}{\rho}\sin(2\phi - \omega), \quad \boldsymbol{r}'_v \cdot \boldsymbol{N}'_v = 0, \tag{3.12}$$

so that

$$\mathrm{II}' = \frac{2}{\rho}\sin(2\phi - \omega)du\,dv. \tag{3.13}$$

These expressions for I' and II' show that Σ' is indeed a pseudo-spherical surface parametrised by arc length along asymptotic lines. The angle between the asymptotic lines on Σ' is given by

$$\omega' = 2\phi - \omega, \tag{3.14}$$

where

$$\omega'_{uv} = \frac{1}{\rho^2}\sin\omega'. \tag{3.15}$$

Use of the relation (3.14) to eliminate ϕ in (3.8) now produces the classical Bäcklund transformation $\mathbb{B}_\beta$ linking the sine–Gordon equation (3.2) and its primed analogue (3.15), namely

$$\begin{cases} \left(\frac{\omega' - \omega}{2}\right)_u &= \frac{\beta}{\rho}\sin\left(\frac{\omega' + \omega}{2}\right), \\ \left(\frac{\omega' + \omega}{2}\right)_v &= \frac{1}{\beta\rho}\sin\left(\frac{\omega' - \omega}{2}\right). \end{cases} \tag{$\mathbb{B}_\beta$}$$

It is noted that, under $\mathbb{B}_\beta$,

$$\boldsymbol{N}' \cdot \boldsymbol{N} = 1 - \frac{L\beta}{\rho} = \text{constant}, \tag{3.16}$$

that is, the tangent planes at corresponding points on Σ and Σ' meet at a constant angle ζ where $\beta = \tan(\zeta/2)$. In Bianchi's original geometric construction, of which the Bäcklund result is an extension,

$$L = \rho, \quad \beta = 1, \tag{3.17}$$

so that these tangent planes are orthogonal. Bäcklund's relaxation of the orthogonality requirement allows the key parameter β to be inserted into the Bianchi transformation. In fact, the Bäcklund transformation $\mathbb{B}_\beta$ may be viewed as a Bianchi transformation conjugated with a simple Lie group invariance. Thus, the sine–Gordon equation (3.2) is invariant under the scaling $u^* = \beta u$, $v^* = v/\beta$, $\beta \neq 0$ so that, any solution $\omega = \omega(u, v)$ generates a one-parameter class of solutions $\omega^*(u, v) = \omega(\beta u, v/\beta)$. Lie observed that composition of this invariance with the original Bianchi transformation produces $\mathbb{B}_\beta$.

In terms of the construction of pseudo-spherical surfaces, the Bäcklund transformation corresponds to the following result:

THEOREM 3.1. *Let $\boldsymbol{r}$ be the coordinate vector of the pseudo-spherical surface Σ corresponding to a solution ω of the sine–Gordon equation* (3.2). *Let ω' denote the Bäcklund transformation of ω via $\mathbb{B}_\beta$. Then, the coordinate vector $\boldsymbol{r}'$ of the pseudo-spherical surface Σ' corresponding to ω' is given by*

$$\boldsymbol{r}' = \boldsymbol{r} + \frac{L}{\sin\omega}\left[\sin\left(\frac{\omega - \omega'}{2}\right)\boldsymbol{r}_u + \sin\left(\frac{\omega + \omega'}{2}\right)\boldsymbol{r}_v\right], \tag{3.18}$$

where $L = \rho \sin\zeta$.

4. Bianchi's Permutability Theorem

It turns out, remarkably, that analytic expressions for multi-soliton solutions of the sine–Gordon equation which encapsulate their nonlinear interaction may be generated algorithmically by an elegant nonlinear superposition principle due to Bianchi (1890) and which is a consequence of the auto-Bäcklund transformation $\mathbb{B}_\beta$.

Thus, suppose ω is a seed solution of the sine–Gordon equation (3.2) and that ω_1 and ω_2 are the Bäcklund transforms of ω via $\mathbb{B}_{\beta_1}$ and $\mathbb{B}_{\beta_2}$, that is, $\omega_1 = \mathbb{B}_{\beta_1}\omega$, $\omega_2 = \mathbb{B}_{\beta_2}\omega$. Let $\omega_{12} = \mathbb{B}_{\beta_2}\omega_1$ and $\omega_{21} = \mathbb{B}_{\beta_1}\omega_2$. It is natural to enquire if there are any circumstances under which the commutative condition $\omega_{12} = \omega_{21}$ applies. To investigate this matter we set

$$\omega_{12} = \omega_{21} = \Omega, \tag{4.1}$$

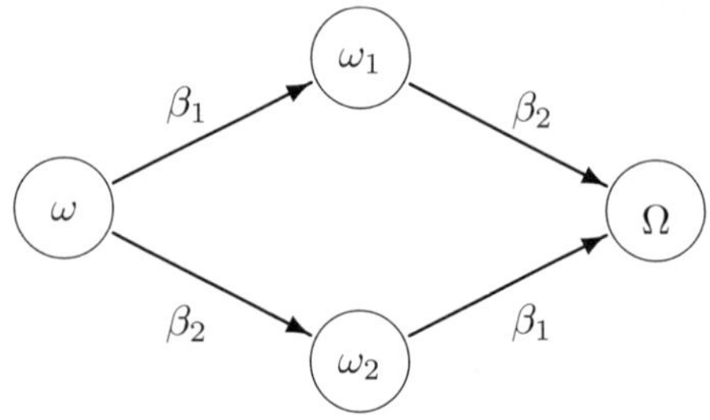

FIGURE 1. A commutative Bianchi diagram.

in the Bäcklund relations

$$\omega_{1,u} = \omega_u + \frac{2\beta_1}{\rho}\sin\left(\frac{\omega_1 + \omega}{2}\right), \tag{4.2}$$

$$\omega_{2,u} = \omega_u + \frac{2\beta_2}{\rho}\sin\left(\frac{\omega_2 + \omega}{2}\right), \tag{4.3}$$

$$\omega_{12,u} = \omega_{1,u} + \frac{2\beta_2}{\rho}\sin\left(\frac{\omega_{12} + \omega_1}{2}\right), \tag{4.4}$$

$$\omega_{21,u} = \omega_{2,u} + \frac{2\beta_1}{\rho}\sin\left(\frac{\omega_{21} + \omega_2}{2}\right). \tag{4.5}$$

The operations (4.2)–(4.3) + (4.4)–(4.6) yield

$$0 = \frac{2}{\rho}\left[\beta_1\left\{\sin\left(\frac{\omega_1+\omega}{2}\right) - \sin\left(\frac{\Omega+\omega_2}{2}\right)\right\} + \beta_2\left\{\sin\left(\frac{\Omega+\omega_1}{2}\right) - \sin\left(\frac{\omega_2+\omega}{2}\right)\right\}\right],$$

whence

$$\Omega = \omega + 4\tan^{-1}\left[\frac{\beta_2+\beta_1}{\beta_2-\beta_1}\tan\left(\frac{\omega_2-\omega_1}{4}\right)\right]. \tag{4.6}$$

This relation embodies the so-called Permutability Theorem with associated Bianchi diagram as in Figure 1.

The commutative property now allows a Bianchi lattice to be constructed corresponding to iterated application of the Permutability Theorem [**2**]. N-soliton solutions of the sine–Gordon equation may be thereby generated via purely algebraic procedures. These represent a nonlinear superposition of N single soliton solutions $\omega = 4\tan^{-1}\left[\exp\left((\beta_i/\rho)u + (1/\beta_i\rho)v + \alpha_i\right)\right]$ with Bäcklund parameters β_i, $i = 1\ldots N$. Thus, at each application of the Bäcklund transformation, a new Bäcklund parameter β_i is introduced and an ith order soliton generated. It is recalled that Seeger et al. [**17**] exploited the Permutability Theorem for the sine–Gordon equation to investigate interaction properties of kink and breather-type solutions in connection with a crystal dislocation model.

4.1. Pseudo-spherical surfaces associated with soliton solutions. If the pseudo-spherical surfaces are parametrised in terms of lines of curvature coordinates

$$x = u + v, \quad y = u - v, \tag{4.7}$$

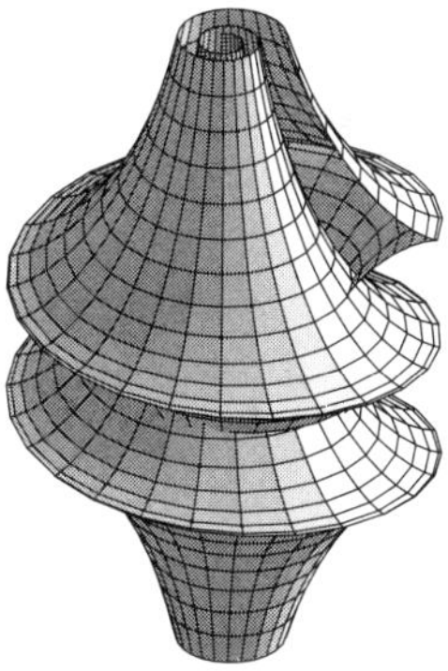

FIGURE 2. A surface of Dini.

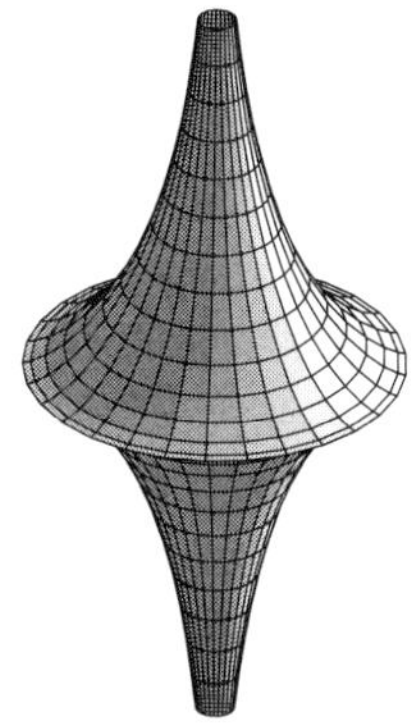

FIGURE 3. A pseudosphere ($\zeta = \pi/2$).

then the class of pseudo-spherical surfaces associated with the single soliton solution

$$\omega = 4\tan^{-1}\left[\exp\left(\frac{1}{2\rho}(\beta+\beta^{-1})x + \frac{1}{2\beta\rho}(\beta-\beta^{-1})y\right)\right], \tag{4.8}$$

consists of the Dini surfaces (c.f. Figure 2) with generic position vector

$$\boldsymbol{r}(x,y) = \begin{pmatrix} \rho\sin\zeta\,\operatorname{sech}\chi\cos(y/\rho) \\ \rho\sin\zeta\,\operatorname{sech}\chi\sin(y/\rho) \\ x-\rho\sin\zeta\tanh\chi \end{pmatrix}, \tag{4.9}$$

where $\chi = (x - y\cos\zeta)/\rho\sin\zeta$. In the case $\zeta = \pi/2$ corresponding to stationary single solitons these reduce to pseudospheres such as that displayed in Figure 3.

There is an important class of entrapped periodic two-soliton solutions known as breathers. An analytic expression for the breather solutions is readily obtained via the Permutability Theorem [**12**]. A gallery of pseudo-spherical surfaces associated with a particular class of solutions is reproduced below in Figure 4.

5. The Motion of Pseudo-Spherical Surfaces. A Weingarten System and its Bäcklund Transformation

The motion of a pseudo-spherical surface $\Sigma : \boldsymbol{r} = \boldsymbol{r}(u,v,t)$ parametrised in asymptotic coordinates u, v is now considered [**6**]. At each instant t, the total curvature $\mathcal{K} = \mathcal{K}(t) < 0$ is constant and negative on Σ. The orthonormal basis $\{\boldsymbol{A},\boldsymbol{B},\boldsymbol{C}\}$ as introduced in (3.5) is employed so that the Gauß–Weingarten equations are given by (3.6) wherein now, however, $\rho = \rho(t)$ and $\omega = \omega(u,v,t)$. The most general time evolution which maintains the orthonormality of the triad $\{\boldsymbol{A},\boldsymbol{B},\boldsymbol{C}\}$ is adjoined, namely that with

$$\begin{pmatrix} \boldsymbol{A} \\ \boldsymbol{B} \\ \boldsymbol{C} \end{pmatrix}_t = \begin{pmatrix} 0 & \alpha & \beta \\ -\alpha & 0 & \gamma \\ -\beta & -\gamma & 0 \end{pmatrix} \begin{pmatrix} \boldsymbol{A} \\ \boldsymbol{B} \\ \boldsymbol{C} \end{pmatrix}. \tag{5.1}$$

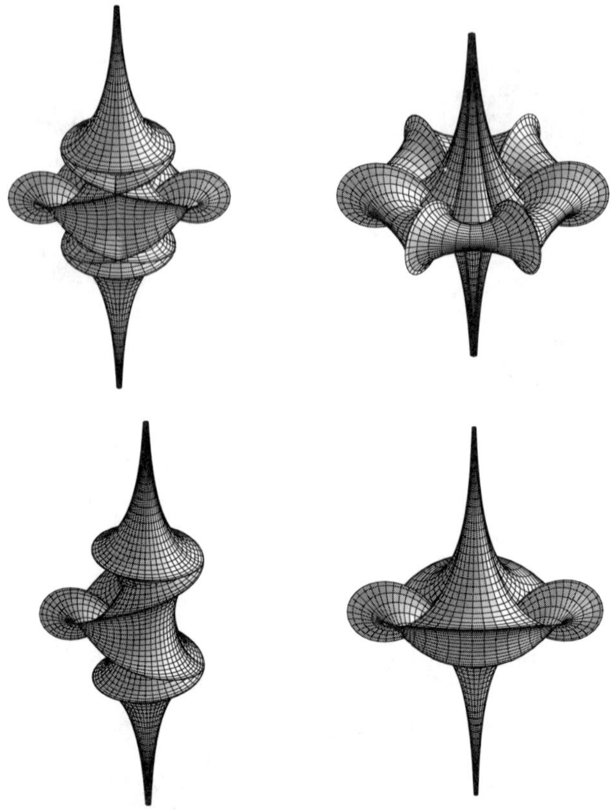

FIGURE 4. Stationary breather pseudospherical surfaces: $p/q = 1/4, 3/4, 1/5, 1/2$.

In particular, it may be shown that the normal motion with

$$r_t = \rho\theta_t \boldsymbol{N}, \quad (\theta = \omega/2), \tag{5.2}$$

is compatible provided that

$$\begin{gathered}
\theta_{xyt} - \theta_x\theta_{yt}\cot\theta + \theta_y\theta_{xt}\tan\theta = 0,\\
\left(\frac{\theta_{xt}}{\cos\theta}\right)_x - \frac{1}{\rho}\left(\frac{1}{\rho}\sin\theta\right)_t - \frac{1}{\sin\theta}\theta_y\theta_{yt} = 0,\\
\left(\frac{\theta_{yt}}{\sin\theta}\right)_y + \frac{1}{\rho}\left(\frac{1}{\rho}\cos\theta\right)_t + \frac{1}{\cos\theta}\theta_x\theta_{xt} = 0,\\
\theta_{xx} - \theta_{yy} = \frac{1}{\rho^2}\sin\theta\cos\theta.
\end{gathered} \tag{5.3}$$

This nonlinear system, originally introduced by Weingarten, appears in Eisenhart [**5**] and was studied extensively both by Darboux and Bianchi.

In the above motion, since Σ undergoes a purely normal propagation, it follows that the surfaces swept out by the parametric lines on Σ as t evolves will be orthogonal to Σ at all times. However, lines of curvature on a surface are mutually orthogonal, so that the family of surfaces swept out by the x-parametric lines will be orthogonal to the surfaces swept out by the y-parametric lines. Thus, the system consisting of these two families, augmented by the family whose members are the pseudo-spherical surfaces formed at each time t constitute a triply orthogonal system of surfaces. Triply orthogonal systems associated with multi-soliton solutions of the Weingarten system (5.3) may be readily constructed. These can be generated via an auto-Bäcklund transformation for the Weingarten system embodied in:

THEOREM 5.1. *The Weingarten system* (5.3) *is invariant under the Bäcklund transformation*

$$\omega'_u = \omega_u + \frac{2\beta}{\rho}\sin\left(\frac{\omega'+\omega}{2}\right), \quad \omega'_v = -\omega_v + \frac{2}{\beta\rho}\sin\left(\frac{\omega'-\omega}{2}\right),$$

$$\omega'_t = \left(\frac{1+\beta^2}{1-\beta^2}\right)\omega_t + \frac{2\rho\beta}{1-\beta^2}\operatorname{cosec}\omega\left[\omega_{vt}\sin\left(\frac{\omega'+\omega}{2}\right) - \omega_{ut}\sin\left(\frac{\omega'-\omega}{2}\right)\right], \tag{5.4}$$

where $\beta(t) = (\rho/L)[1 \pm \sqrt{1 - L^2/\rho^2}]$ *and* u, v *are asymptotic coordinates.*

A triply orthogonal system associated with a single-soliton solution of the Weingarten system is displayed in Figure 5. The system consists of two families of Dini surfaces and a family of spheres. It is determined by the position vector

$$\boldsymbol{r} = \big(\cos(y-t)/\cosh(x+y+t),\ \sin(y-t)/\cosh(x+y+t),\ x - \tanh(x+y+t)\big)^T.$$

In fact, the general Lamé system descriptive of triply orthogonal systems has been shown to be integrable in, e.g., [**14**]. The integrable nature of the Lamé system was also subsequently recorded in the work of Zakharov [**20**].

6. The Lelieuvre Formulae. An Auto-Bäcklund Transformation for the Bianchi System

The Lelieuvre Formulae (1888) were originally introduced in connection with the spherical representation of surfaces. They play a key role in the construction of a Bäcklund transformation for hyperbolic surfaces.

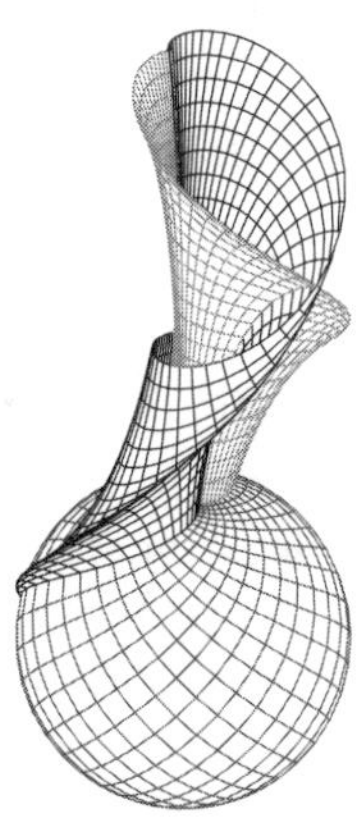

FIGURE 5. A triply orthogonal system of Dini surfaces and spheres.

Given a surface $\Sigma : \boldsymbol{r} = \boldsymbol{r}(u,v)$, its unit normal $\boldsymbol{N} = \big(X(u,v), Y(u,v), Z(u,v)\big)$ generates, as u and v vary, a coordinate system on the unit sphere $|\boldsymbol{N}|^2 = 1$. This mapping to the sphere $\boldsymbol{N} = \boldsymbol{N}(u,v)$ is called the spherical or Gauß representation of the surface $\Sigma : \boldsymbol{r} = \boldsymbol{r}(u,v)$. The corresponding 1st fundamental form of the sphere reads

$$d\boldsymbol{N}^2 = \mathcal{E}\, du^2 + 2\mathcal{F}\, du\, dv + \mathcal{G}\, dv^2, \tag{6.1}$$

where

$$\mathcal{E} = \boldsymbol{N}_u^2, \quad \mathcal{F} = \boldsymbol{N}_u \cdot \boldsymbol{N}_v, \quad \mathcal{G} = \boldsymbol{N}_v^2. \tag{6.2}$$

If Σ is now taken to be a hyperbolic surface with total curvature $\mathcal{K} = -1/\rho^2$ and is parametrised in terms of asymptotic coordinates then $e = g = 0$ so that

$$\boldsymbol{r}_u \cdot \boldsymbol{N}_u = 0, \quad \boldsymbol{r}_v \cdot \boldsymbol{N}_v = 0, \tag{6.3}$$

whence, since $\boldsymbol{r}_u \cdot \boldsymbol{N} = 0$, $\boldsymbol{r}_v \cdot \boldsymbol{N} = 0$,

$$\boldsymbol{r}_u = \bar{\lambda} \boldsymbol{N}_u \times \boldsymbol{N}, \quad \boldsymbol{r}_v = \bar{\mu} \boldsymbol{N}_v \times \boldsymbol{N}, \tag{6.4}$$

where the compatibility condition $\boldsymbol{r}_{uv} = \boldsymbol{r}_{vu}$ implies that $\bar{\lambda} + \bar{\mu} = 0$. Moreover,

$$E = \rho^2 \mathcal{E}, \quad F = -\rho^2 \mathcal{F}, \quad G = \rho^2 \mathcal{G}, \tag{6.5}$$

and

$$\mathcal{K} = -1/\bar{\lambda}^2, \tag{6.6}$$

whence $\bar{\lambda} = \pm\rho$. Accordingly, (6.4) yields

$$\boldsymbol{r}_u = \rho \boldsymbol{N}_u \times \boldsymbol{N}, \quad \boldsymbol{r}_v = -\rho \boldsymbol{N}_v \times \boldsymbol{N}. \tag{6.7}$$

These are the Lelieuvre formulae. They have integrability condition

$$(\rho \boldsymbol{N} \times \boldsymbol{N}_u)_v + (\rho \boldsymbol{N} \times \boldsymbol{N}_v)_u = \boldsymbol{0}. \tag{6.8}$$

Any solution of the latter vector equation with $\boldsymbol{N}^2 = 1$ guarantees the existence of an $\boldsymbol{r}(u,v)$ satisfying the Lelieuvre formulae. The latter, in turn, imply that $\boldsymbol{r}$ may be regarded as the position vector of a surface Σ parametrised in terms of asymptotic coordinates u, v since (6.8) shows that $e = \boldsymbol{r}_u \cdot \boldsymbol{N}_u = 0$, $g = -\boldsymbol{r}_v \cdot \boldsymbol{N}_v = 0$.

6.1. The Bäcklund transformation. Here, we restrict attention to the construction of a Bäcklund transformation that obeys the tangency condition

$$\boldsymbol{r}' = \boldsymbol{r} + \lambda \boldsymbol{r}_u + \mu \boldsymbol{r}_v. \tag{6.9}$$

If it is also assumed that Σ' like Σ is parametrised in terms of the asymptotic coordinates u, v then

$$\boldsymbol{r}'_{uu} \cdot \boldsymbol{N}' = 0, \quad \boldsymbol{r}'_{vv} \cdot \boldsymbol{N}' = 0. \tag{6.10}$$

These provide two second-order nonlinear equations for $\lambda(u,v)$ and $\mu(u,v)$. Any solution gives rise to a Bäcklund transformation between the hyperbolic surfaces Σ and Σ'. However, if in addition, it is demanded that $\boldsymbol{r}' - \boldsymbol{r}$ be tangential to Σ' then a Bäcklund transformation with such a property can be constructed via the Lelieuvre formulae and use of an orthonormal triad originally introduced by Bianchi, viz

$$\left\{ \boldsymbol{V} = \frac{1}{2\sin(\Omega/2)} \left(\frac{\boldsymbol{N}_u}{a} - \frac{\boldsymbol{N}_v}{b} \right),\ \boldsymbol{W} = \frac{1}{2\cos(\Omega/2)} \left(\frac{\boldsymbol{N}_u}{a} + \frac{\boldsymbol{N}_v}{b} \right),\ \boldsymbol{N} \right\}, \tag{6.11}$$

where a, b are as introduced in (2.13) and $\Omega = \omega + \pi$. It is noted that

$$dN^2 = a^2 du^2 + 2ab\cos\Omega\, du\, dv + b^2\, dv^2, \tag{6.12}$$

so that Ω is the angle between the coordinate lines on the sphere swept out by $\boldsymbol{N} = \boldsymbol{N}(u, v)$.

The Bäcklund transform $\boldsymbol{N}'$ of $\boldsymbol{N}$ may now be written as

$$\boldsymbol{N}' = \cos\sigma \boldsymbol{N} + \sin\sigma(\cos\theta \boldsymbol{V} + \sin\theta \boldsymbol{W}), \tag{6.13}$$

where σ is the angle between $\boldsymbol{N}$ and $\boldsymbol{N}'$. The total curvatures of Σ and Σ' are related by (Eisenhart [**5**])

$$\mathcal{K}\mathcal{K}' = \left(\frac{\sin\sigma}{d}\right)^4, \tag{6.14}$$

where $d = |\boldsymbol{r}' - \boldsymbol{r}|$ is the distance between corresponding points on Σ and Σ'. Σ and Σ' are assumed to be parametrised in terms of asymptotic coordinates so that their generic position vectors $\boldsymbol{r}$ and $\boldsymbol{r}'$ may be recovered via the Lelieuvre formulae

$$\begin{aligned} \boldsymbol{r}_u &= \boldsymbol{\nu}_u \times \boldsymbol{\nu}\,, \quad & \boldsymbol{r}_v &= -\boldsymbol{\nu}_v \times \boldsymbol{\nu}, \\ \boldsymbol{r}'_u &= \boldsymbol{\nu}'_u \times \boldsymbol{\nu}', \quad & \boldsymbol{r}'_v &= -\boldsymbol{\nu}'_v \times \boldsymbol{\nu}', \end{aligned} \tag{6.15}$$

where

$$\boldsymbol{\nu} = \sqrt{\rho}\boldsymbol{N}, \quad \boldsymbol{\nu}' = \sqrt{\rho'}\boldsymbol{N}', \tag{6.16}$$

and the total curvatures read

$$\mathcal{K} = -1/|\boldsymbol{\nu}|^4, \quad \mathcal{K}' = -1/|\boldsymbol{\nu}'|^4. \tag{6.17}$$

The requirement that $\boldsymbol{r}' - \boldsymbol{r}$ be tangential to both Σ and Σ' implies that

$$\boldsymbol{r}' - \boldsymbol{r} = m\boldsymbol{\nu}' \times \boldsymbol{\nu}, \tag{6.18}$$

leading to

$$d = |m||\boldsymbol{\nu}||\boldsymbol{\nu}'|\sin\sigma. \tag{6.19}$$

Accordingly, $m = \pm 1$ and the Bäcklund transformation becomes

$$\boldsymbol{r}' - \boldsymbol{r} = \sqrt{\rho\rho'}\boldsymbol{N} \times \boldsymbol{N}'. \tag{6.20}$$

If it is now required that the total curvature $\mathcal{K}$ be invariant under the Bäcklund transformation so that $\rho' = \rho$ then insertion of the parametrisation (6.13) into $(6.15)_{3,4}$ shows that

$$\begin{aligned} \theta_u &= \frac{1}{2}\Omega_u + a\tan\frac{\sigma}{2}\cos\left(\theta + \frac{\Omega}{2}\right) - \frac{1}{2}\frac{a}{b}\frac{\rho_v}{\rho}\sin\Omega, \\ \theta_v &= -\frac{1}{2}\Omega_v - b\cot\frac{\sigma}{2}\cos\left(\theta - \frac{\Omega}{2}\right) + \frac{1}{2}\frac{b}{a}\frac{\rho_u}{\rho}\sin\Omega, \end{aligned} \tag{6.21}$$

and

$$\sigma_u = -\frac{\rho_u}{\rho}\tan\frac{\sigma}{2}, \quad \sigma_v = \frac{\rho_v}{\rho}\cot\frac{\sigma}{2}, \tag{6.22}$$

with compatibility condition

$$\rho_{uv} = 0. \tag{6.23}$$

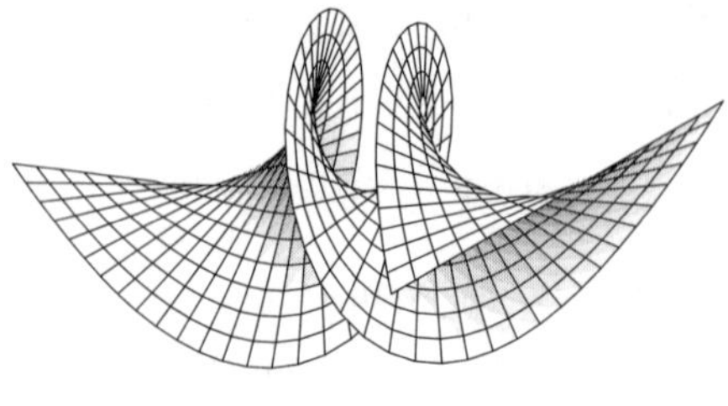

FIGURE 6. A Bianchi surface generated from a hyperbolic paraboloid.

Hyperbolic surfaces with this constraint on the total curvature are known as Bianchi surfaces. They admit the auto-Bäcklund transformation

$$\boldsymbol{r}' = \boldsymbol{r} - \left(\frac{2\mu}{1+\mu^2}\right)\frac{1}{\sin\Omega}\left[\cos\left(\theta - \frac{\Omega}{2}\right)\frac{\boldsymbol{r}_u}{a} + \cos\left(\theta + \frac{\Omega}{2}\right)\frac{\boldsymbol{r}_v}{b}\right], \tag{6.24}$$

where $\mu = \tan\sigma/2$ is a "nonisospectral" parameter. Bianchi applied this Bäcklund transformation to the seed hyperbolic paraboloid

$$x = \frac{1}{\sqrt{2}}(u+v), \quad y = \frac{1}{\sqrt{2}}(u-v), \qquad z = uv, \tag{6.25}$$

to generate the second surface Σ' given by

$$\begin{aligned} x' &= \frac{1}{\sqrt{2}}\left(3u - v\frac{1-3u^2}{1+u^2}\right), \quad y' = \frac{1}{\sqrt{2}}\left(3u + v\frac{1-3u^2}{1+u^2}\right), \\ z' &= \frac{uv}{\sqrt{2}}\left(\frac{3-u^2}{1+u^2}\right). \end{aligned} \tag{6.26}$$

A particular clipping of Σ' is illustrated in Figure 6. One-soliton Bianchi surfaces have been constructed and displayed graphically in [**9**].

7. Discretisation of Characteristic Systems via Bäcklund Transformations

The Bäcklund transformation (6.20) relating the hyperbolic surfaces Σ and Σ' delivers, on decomposition, the relations

$$\begin{aligned} x' - x &= \sqrt{\rho\rho'}\cos\sigma(X' - X)/(1 + TT' + XX'), \\ y' - y &= \sqrt{\rho\rho'}\cos\sigma(T' - T)/(1 + TT' + XX'), \\ z' - z &= \sqrt{\rho\rho'}\cos\sigma(T'X - TX')/(1 + TT' + XX'), \\ &= -T(x' - x) + X(y' - y), \\ &\frac{1 + TT' + XX'}{\sqrt{1+T^2+X^2}\sqrt{1+T'^2+X'^2}} = \cos\sigma, \end{aligned} \tag{7.1}$$

where

$$dz = -T\,dx + X\,dy, \quad dz' = -T'\,dx' + X'\,dy'.$$

The above Bäcklund transformation adopts the usual form

$$\mathbb{B}_i(x, y, z, z_x, z_y;\ x', y', z', z'_{x'}, z'_{y'}) = 0, \quad i = 1, \dots, 4, \tag{7.2}$$

as discussed in the classical literature.

If $\rho' = \rho =$ const then the system (7.1) has been shown by Rogers and Schief [**10**] to provide an integrable discretisation of the characteristic system associated with the Monge form of pseudo-spherical surfaces. The notion of a Bäcklund transformation as an integrable discretisation of a characteristic system has been extended to Tzitzeica surfaces in [**15**].

References

1. T. W. Barnard, *$2N\pi$ Ultrashort light pulses*, Phys. Rev. A **7** (1973), 373–376.
2. L. Bianchi, *Lezioni di geometria differenziale*, Spoerri, Pisa, 1903.
3. B. Cenkl, *Geometric deformations of the evolution equations and Bäcklund transformations*, Phys. D **18** (1986), 217–219.
4. M. P. do Carmo, *Differential geometry of curves and surfaces*, Prentice-Hall Inc., Englewood Cliffs, New Jersey, 1976.
5. L. P. Eisenhart, *A treatise on the differential geometry of curves and surfaces*, Dover, New York, 1960.
6. M. E. Johnston, C. Rogers, W. K. Schief, and M. L. Seiler, *On moving pseudospherical surfaces: A generalised Weingarten system*, Lie Groups Appl. **1** (1994), 124–136.
7. G. L. Lamb Jr., *Analytical descriptions of ultrashort optical pulse propagation in a resonant medium*, Rev. Modern. Phys. **43** (1971), 99–124.
8. D. Levi and A. Sym, *Integrable systems describing surfaces of nonconstant curvature*, Phys. Lett. A **149** (1990), 381–387.
9. R. Pruß, *Geometry of Bianchi surfaces in $\mathbb{E}^3$*, Master Thesis, Warsaw University, 1995.
10. C. Rogers and W. K. Schief, *The classical Bäcklund transformation and integrable discretisation of characteristic equations*, Phys. Lett A **232** (1997), 217–223.
11. ———, *Bäcklund and Darboux transformations. Geometry and modern applications in soliton theory*, Cambridge Texts Appl. Math., Cambridge University Press, 2001, (to be published).
12. C. Rogers, W. K. Schief, and M. Johnston, *Bäcklund and his works: Applications in soliton theory*, Geometric Approaches to Differential Equations (P. J. Vassiliou and I. G. Lisle, eds.), Australian Mathematical Society Lecture Series **15**, Cambridge University Press (2000), 16–55.
13. C. Rogers and W. F. Shadwick, *Bäcklund transformations and their applications*, Academic Press, New York, 1982.
14. W. K. Schief, *On a 2+1-dimensional Darboux system: Integrable reductions*, Inverse Problems **10** (1994), 1185–1198.
15. ———, *An integrable discretization of the characteristic equations associated with a gasdynamics system*, Phys. Lett A **238** (1998), 278–282.
16. A. C. Scott, *Propagation of magnetic flux on a long Josephson junction*, Nuovo Cimento B (11) **69** (1970), 241–261.
17. A. Seeger, H. Donth, and A. Kochendörfer, *Theorie der Versetzungen in eindimensionalen Atomreitien* III *Versetzungen*, Eigenbewegungen und ihre Wechselwirkung, Z. Phys. **134** (1953), 173–193.
18. D. J. Struick, *Lectures on classical differential geometry*, 2nd ed., Addison-Wesley Publishing Company, Inc. Reading, Mass, USA, 1961.
19. N. J. Zabusky and M. D. Kruskal, *Interaction of "solitons" in a collisionless plasma and recurrence of initial states*, Phys. Rev. Lett. **15** (1965), 240–243.
20. V. E. Zakharov, *Description of the n-orthogonal curvilinear coordinate systems and Hamiltonian integrable systems of hydrodynamic type.* I. *Integration of the Lamé equations*, Duke Math. J. **94** (1998), 103–139.

DEPARTMENT OF APPLIED MATHEMATICS, UNIVERSITY OF NEW SOUTH WALES, SYDNEY 2052, AUSTRALIA.

E-mail address, C. Rogers: `c.rogers@maths.unsw.edu.au`

E-mail address, W. K. Schief: `schief@maths.unsw.edu.au`

Centre de Recherches Mathématiques
CRM Proceedings and Lecture Notes
Volume **29**, 2001

An Introduction to Integrable Difference and Differential Geometries: Affine Spheres, Their Natural Generalization and Discretization

W. K. Schief

Abstract. We review the geometric and algebraic construction of discrete affine spheres and the underlying integrable discrete Tzitzeica equation.

1. Introduction

Difference ("discrete") geometries which are governed by integrable difference ("discrete") equations have recently been the subject of extensive studies. Due to the fact that a general theory of *discrete integrable geometry* is not yet available, one has so far been obliged to investigate the properties of particular continuous geometries and find their difference geometric analogues by ad hoc procedures. Alternatively, one may generate discrete geometries by means of Bäcklund transformations. However, a priori, it is by no means evident at which class of discrete surfaces one will arrive in this manner. Despite these theoretical shortcomings, an impressive body of integrable discrete geometries has already been established. A survey of the state of the art together with references to the origins of discrete (integrable) geometry are contained in the recent book *Discrete Integrable Geometry and Physics* [**9**]. References regarding the discretization of geometries using Bäcklund transformations may also be found in [**36**].

We here review work on the discretization of affine geometries which was undertaken in collaboration with A. I. Bobenko [**7, 8, 37**]. We show how classical affine spheres, which were introduced and shown to be "integrable" by Tzitzeica in 1907, may be discretized in a natural geometric manner, thereby preserving their integrable nature. This construction results in an integrable discrete version of the Tzitzeica equation which governs affine spheres. We also demonstrate how discrete surfaces may be generated by application of the Tzitzeica transformation to affine spheres. It is shown that these discrete surfaces contain the class of discrete affine spheres constructed in the above-mentioned geometric manner. As a by-product, a discrete version of the Tzitzeica transformation is obtained.

2000 *Mathematics Subject Classification.* Primary: 53A15; Secondary: 58F07.
This is the final form of the paper.

2. Affine Differential Geometry

For didactic reasons, we begin with a brief review of the classical differential geometry of surfaces in Euclidean space $\mathbb{R}^3$. A more elaborate account of these geometries in the context of soliton theory is given by C. Rogers and W. K. Schief in this volume [**30**].

2.1. Euclidean geometry. A classical result due to Bonnet [**14**] states that a surface $\Sigma\colon \boldsymbol{r} = \boldsymbol{r}(u,v) \in \mathbb{R}^3$ is determined up to its position in space by the *fundamental forms*

$$\begin{aligned}\mathcal{F}_1 &= d\boldsymbol{r}\cdot d\boldsymbol{r} = E\,du^2 + 2F\,du\,dv + G\,dv^2,\\ \mathcal{F}_2 &= -d\boldsymbol{r}\cdot d\boldsymbol{N} = e\,du^2 + 2f\,du\,dv + g\,dv^2,\end{aligned}\tag{2.1}$$

where $\boldsymbol{N} = \boldsymbol{r}_u \times \boldsymbol{r}_v/|\boldsymbol{r}_u \times \boldsymbol{r}_v|$ is the unit normal to Σ. The position vector $\boldsymbol{r}$ is governed by the linear *Gauss–Weingarten equations*

$$\begin{aligned}\boldsymbol{r}_{uu} &= \Gamma^1_{11}\boldsymbol{r}_u + \Gamma^2_{11}\boldsymbol{r}_v + e\boldsymbol{N}, \quad \boldsymbol{N}_u = \frac{fF - eG}{H^2}\boldsymbol{r}_u + \frac{eF - fE}{H^2}\boldsymbol{r}_v,\\ \boldsymbol{r}_{uv} &= \Gamma^1_{12}\boldsymbol{r}_u + \Gamma^2_{12}\boldsymbol{r}_v + f\boldsymbol{N},\\ \boldsymbol{r}_{vv} &= \Gamma^1_{22}\boldsymbol{r}_u + \Gamma^2_{22}\boldsymbol{r}_v + g\boldsymbol{N}, \quad \boldsymbol{N}_v = \frac{gF - fG}{H^2}\boldsymbol{r}_u + \frac{fF - gE}{H^2}\boldsymbol{r}_v,\end{aligned}\tag{2.2}$$

with $H^2 = EG - F^2$ and the *Christoffel symbols* Γ^k_{ij} are defined by

$$\Gamma^k_{ij} = \frac{1}{2}g^{kl}(g_{il,j} + g_{jl,i} - g_{ij,l}), \quad \mathcal{F}_1 = g_{ij}\,du^i\,du^j, \quad g_{ij}g^{jk} = \delta^k_i.\tag{2.3}$$

Here, Einstein's convention of summation over repeated indices has been adopted.

In the following, we focus on the class of *hyperbolic* surfaces, that is surfaces of negative *Gaussian curvature*

$$\mathcal{K} = \frac{\det(\mathcal{F}_2)}{\det(\mathcal{F}_1)} < 0.\tag{2.4}$$

In this case, one may choose *asymptotic coordinates* on the surfaces which are described by a purely off-diagonal second fundamental form, i.e.

$$e = g = 0, \quad \mathcal{K} = -\frac{f^2}{H^2}.\tag{2.5}$$

Accordingly, the compatibility conditions of the Gauss–Weingarten equations reduce to the *Mainardi–Codazzi equations*

$$\left(\frac{f}{H}\right)_u + 2\Gamma^2_{12}\frac{f}{H} = 0, \quad \left(\frac{f}{H}\right)_v + 2\Gamma^1_{12}\frac{f}{H} = 0,\tag{2.6}$$

supplemented by the *Theorema Egregium* of Gauss which provides an expression for the Gaussian curvature in terms of the first fundamental form alone. In fact, if one explicitly parametrizes the fundamental forms according to

$$\begin{aligned}\mathcal{F}_1 &= \bar{\rho}^2(\bar{a}^2du^2 + 2\bar{a}\bar{b}\cos\omega\,du\,dv + \bar{b}^2\,dv^2),\\ \mathcal{F}_2 &= 2\bar{\rho}\bar{a}\bar{b}\sin\omega\,du\,dv,\end{aligned}\tag{2.7}$$

then the Gaussian curvature takes the form

$$\mathcal{K} = -\frac{1}{\bar{\rho}^2},\tag{2.8}$$

and the Gauss equation becomes

$$\omega_{uv} + \frac{1}{2}\left(\frac{\bar{\rho}_u}{\bar{\rho}}\frac{\bar{b}}{\bar{a}}\sin\omega\right)_u + \frac{1}{2}\left(\frac{\bar{\rho}_v}{\bar{\rho}}\frac{\bar{a}}{\bar{b}}\sin\omega\right)_v = \bar{a}\bar{b}\sin\omega, \tag{2.9}$$

while the Mainardi–Codazzi equations read

$$\bar{a}_v + \frac{1}{2}\frac{\bar{\rho}_v}{\bar{\rho}}\bar{a} - \frac{1}{2}\frac{\bar{\rho}_u}{\bar{\rho}}\bar{b}\cos\omega = 0, \quad \bar{b}_u + \frac{1}{2}\frac{\bar{\rho}_u}{\bar{\rho}}\bar{b} - \frac{1}{2}\frac{\bar{\rho}_v}{\bar{\rho}}\bar{a}\cos\omega = 0. \tag{2.10}$$

The underdetermined system (2.9)–(2.10) is the standard form of the Gauss–Mainardi–Codazzi equations for hyperbolic surfaces in $\mathbb{R}^3$ which are parametrized in terms of asymptotic coordinates. In the case of pseudospherical surfaces, that is $\mathcal{K} = \text{const}$, the Gauss equation reduces to the sine–Gordon equation

$$\omega_{uv} = \sin\omega, \tag{2.11}$$

with $\bar{a} = \bar{b} = 1$ without loss of generality. However, from purely geometric considerations, Bianchi [**2**] derived the more general constraint

$$\bar{\rho}_{uv} = 0, \tag{2.12}$$

and showed that this condition is preserved by an extension of the classical *Bäcklund transformation* for pseudospherical surfaces [**14**] which is retrieved in the case $\bar{\rho} = \text{const}$. In the context of soliton theory, these *Bianchi surfaces* have been investigated in [**4, 11, 22**] (see also the contribution by C. Rogers and W. K. Schief in this volume).

2.2. Affine geometry. In the present form, the Gauss–Weingarten equations are invariant under *translations* and *rotations*

$$\boldsymbol{r} \to \mathcal{R}\boldsymbol{r} + \boldsymbol{t}, \quad \mathcal{R} \in \mathrm{SO}(3, \mathbb{R}), \quad \boldsymbol{t} \in \mathbb{R}^3. \tag{2.13}$$

An alternative formulation of the Gauss–Weingarten equations which allows for *centro-affine* transformations

$$\boldsymbol{r} \to \mathcal{A}\boldsymbol{r}, \quad \mathcal{A} \in \mathrm{GL}(3, \mathbb{R}), \tag{2.14}$$

is now obtained as follows. On solving the expressions (2.6) for Γ^1_{12} and Γ^2_{12}, the Gauss equation $(2.2)_3$ becomes

$$\boldsymbol{r}_{uv} = -\frac{1}{4}[\ln(-\mathcal{K})]_v\boldsymbol{r}_u - \frac{1}{4}[\ln(-\mathcal{K})]_u\boldsymbol{r}_v + f\boldsymbol{N}. \tag{2.15}$$

If we denote the distance from the origin of the ambient space $\mathbb{R}^3$ to the tangent plane to Σ at a generic point P by d (*algebraic distance*) then we deduce that

$$d = \boldsymbol{N}\cdot\boldsymbol{r}, \quad d_u = \boldsymbol{N}_u\cdot\boldsymbol{r}, \quad d_v = \boldsymbol{N}_v\cdot\boldsymbol{r}. \tag{2.16}$$

These relations may be regarded as projections of $\boldsymbol{r}$ onto the linearly independent triad $\{\boldsymbol{N}, \boldsymbol{N}_u, \boldsymbol{N}_v\}$ and, on use of the Weingarten equations, we obtain

$$\boldsymbol{r} = -\frac{d_v}{f}\boldsymbol{r}_u - \frac{d_u}{f}\boldsymbol{r}_v + d\boldsymbol{N}. \tag{2.17}$$

Thus, elimination of the unit normal from (2.15) and (2.17) results in the key equation

$$\boldsymbol{r}_{uv} = h\boldsymbol{r} + \rho_v\boldsymbol{r}_u + \rho_u\boldsymbol{r}_v, \tag{2.18}$$

with the definitions

$$h = \frac{f}{d}, \quad \frac{\mathcal{K}}{d^4} = -e^{-4\rho}. \tag{2.19}$$

This gives rise to the following

THEOREM 2.1 (The Gauss equations for hyperbolic surfaces in centro-affine geometry). *Surfaces of negative Gaussian curvature are governed by the* linear *system*

$$\boldsymbol{r}_{uu} = \left(\frac{h_u}{h} + \rho_u\right)\boldsymbol{r}_u + \frac{a}{h}\boldsymbol{r}_v, \quad \boldsymbol{r}_{uv} = h\boldsymbol{r} + \rho_v\boldsymbol{r}_u + \rho_u\boldsymbol{r}_v,$$
$$\boldsymbol{r}_{vv} = \left(\frac{h_v}{h} + \rho_v\right)\boldsymbol{r}_v + \frac{b}{h}\boldsymbol{r}_u, \tag{2.20}$$

which is compatible if and only if the functions h, a, b and ρ obey the nonlinear *system*

$$(\ln h)_{uv} = h - \frac{ab}{h^2} + \rho_u\rho_v, \quad a_v + \rho_u h_u = \rho_{uu}h, \quad b_u + \rho_v h_v = \rho_{vv}h. \tag{2.21}$$

The surface vector $\boldsymbol{r}$ is determined up to centro-affine transformations

$$\boldsymbol{r} \to \mathcal{A}\boldsymbol{r}, \quad \mathcal{A} = \text{const} \in \mathbb{R}^{3,3}. \tag{2.22}$$

2.3. Affine spheres. The particular parametrization (2.20) of the Gauss equations is a consequence of the compatibility conditions for $(2.2)_{1,5}$ and (2.18). Tzitzeica [**40, 41**] noticed that the quantity $\mathcal{K}/d^4$ is invariant (up to a constant multiplicative factor) under the group of centro-affine transformations (2.22), which led him to the ansatz

$$\mathcal{K} \sim d^4 \Leftrightarrow \rho = \text{const}\,. \tag{2.23}$$

In the generic case, this implies that we may set $a = \lambda$ and $b = 1/\lambda$ without loss of generality, where λ is an arbitrary constant. Consequently, the system (2.21) reduces to the *Tzitzeica equation*

$$(\ln h)_{uv} = h - \frac{1}{h^2}. \tag{2.24}$$

Tzitzeica's observation initiated a systematic study of properties of surfaces that are invariant under centro-affine transformations and translations. This work culminated in the treatise *Affine Differentialgeometrie* by Blaschke in 1923 [**3**]. The surfaces discovered by Tzitzeica have come to be known as (*proper*) *affine spheres* in affine differential geometry due to the fact that

$$D = e^{\rho} = \text{const}, \tag{2.25}$$

may be defined as the *affine distance* from the origin of $\mathbb{R}^3$ to a generic point on the surface Σ. As pointed out by Tzitzeica [**42**], they admit a Bäcklund transformation which, in analytical terms, is based on the Gauss equations

$$\boldsymbol{r}_{uu} = \frac{h_u}{h}\boldsymbol{r}_u + \frac{\lambda}{h}\boldsymbol{r}_v, \quad \boldsymbol{r}_{uv} = h\boldsymbol{r},$$
$$\boldsymbol{r}_{vv} = \frac{h_v}{h}\boldsymbol{r}_v + \frac{\lambda^{-1}}{h}\boldsymbol{r}_u. \tag{2.26}$$

These are compatible if and only if h is a solution of the Tzitzeica equation (2.24). It is important to note that even though the parameter λ does *not* appear in the Tzitzeica equation, it is a fundamental ingredient in the Gauss equations (2.26). In fact, the existence of the "spectral parameter" λ reflects the integrable nature of the Tzitzeica equation.

The *Tzitzeica transformation* constitutes a particular reduction of the *Moutard transformation* [**26**]. The latter constitutes a transformation which leaves invariant the *Moutard equation* $(2.26)_2$. The Moutard equation together with its invariance has been used extensively in differential geometry and soliton theory (see [**36**] and references therein). In particular, it has been exploited to construct integrable discretizations of a $2+1$-dimensional sine–Gordon system/equation [**27, 28**] introduced by Konopelchenko and Rogers in [**21**]. Since the Tzitzeica transformation is discussed in detail in Section 3, we here merely state Tzitzeica's observation.

THEOREM 2.2 (The Tzitzeica transformation for affine spheres). *Let Σ be an affine sphere and $\boldsymbol{r}, h$ the corresponding solution of the Gauss equations* (2.26) *and the Tzitzeica equation* (2.24). *Then, the position vector $\boldsymbol{r}'$ of another affine sphere Σ' is given by*

$$\boldsymbol{r}' = \boldsymbol{r} - \frac{2}{(\lambda-\mu)h}\left(\lambda\frac{\phi_u}{\phi}\boldsymbol{r}_v - \mu\frac{\phi_v}{\phi}\boldsymbol{r}_u\right), \tag{2.27}$$

where ϕ is a scalar solution of (2.26) *with parameter μ. The associated solution of the Tzitzeica equation reads*

$$h' = h - 2(\ln\phi)_{uv}. \tag{2.28}$$

The simplest seed solution of the Tzitzeica equation is given by $h = 1$. This choice corresponds to the affine sphere displayed in Figure 1(a). Application of the above Tzitzeica transformation to this seed produces the one-soliton solution of the Tzitzeica equation. The stationary one-soliton solution corresponds to a surface of revolution which is shown in Figure 1(b). The complete class of affine spheres of revolution has been investigated by Jonas [**20**]. In general, these surfaces of revolution are expressed in terms of elliptic functions. In the degenerate case of hyperbolic functions, the stationary one-soliton affine sphere is retrieved. An affine sphere corresponding to a moving one-soliton solution of the Tzitzeica equation is shown in Figure 1(c). A second application of the Tzitzeica transformation produces two-soliton affine spheres. An appropriate choice of the Bäcklund parameters gives rise to breather solutions of the Tzitzeica equation with associated affine spheres. A breather affine sphere is displayed in Figure 1(d). Details of the generation of these affine spheres may be found in [**31, 38**].

In 1977, the Tzitzeica equation was re-derived in soliton theory and subsequently analysed by several authors [**13**]-[**17**]. Later [**15**], a connection with $1+1$-dimensional anisentropic gas dynamics was recorded which turns out to be directly related to a new form of the affine sphere equation as set down by Jonas in 1953 [**20, 38**]. This observation has recently been exploited to show that the Tzitzeica transformation gives rise to an integrable discretization of the characteristic equations associated with Martin's formulation of the gasdynamics system in terms of a Monge–Ampère equation [**35**].

2.4. Generalized affine spheres. In Section 2.1, we have briefly mentioned Bianchi surfaces which are defined by the condition

$$\frac{\partial^2}{\partial u\partial v}\left(\frac{1}{\sqrt{-\mathcal{K}}}\right) = 0, \tag{2.29}$$

where u and v denote asymptotic coordinates. Since pseudospherical surfaces are obtained in the case $\mathcal{K} = \text{const}$, we may regard Bianchi surfaces as generalized pseudospherical surfaces. Similarly, it has been shown [**4**] that surfaces of *harmonic*

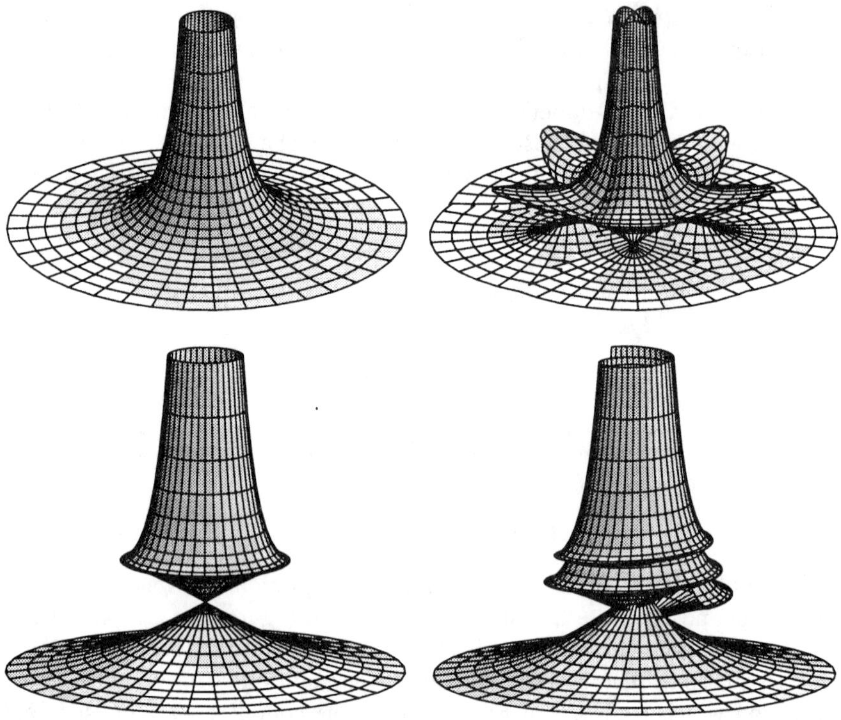

FIGURE 1. Affine spheres (anti-clockwise): (a) A seed affine sphere; (b) The stationary one-soliton affine sphere; (c) A moving one-soliton affine sphere; (d) A breather affine sphere.

inverse mean curvature, that is surfaces subjected to the constraint

$$\left(\frac{\partial^2}{\partial x^2} + \frac{\partial^2}{\partial y^2}\right)\left(\frac{1}{\mathcal{H}}\right) = 0, \tag{2.30}$$

with respect to *conformal* coordinates x and y, are also integrable. This class of surfaces contains *constant mean curvature* surfaces [**14**] which are represented by $\mathcal{H} = \text{const}$. It turns out that there indeed exist generalized affine spheres which are amenable to the techniques developed in soliton theory [**37**]:

Since affine distances (to the origin) are preserved by centro-affine transformations, it is reasonable from a geometric point of view to impose a constraint on the affine distance D. Thus, consider the condition

$$\frac{\partial^2}{\partial u \partial v}\left[\ln\left(-\frac{\mathcal{K}}{d^4}\right)\right] = 0, \tag{2.31}$$

and therefore

$$\rho_{uv} = (\ln D)_{uv} = 0. \tag{2.32}$$

In this case, the system (2.21), (2.32) is invariant under the Lie point symmetry

$$a \to \lambda a, \quad b \to b/\lambda, \quad \rho_u \to \lambda \rho_u, \quad \rho_v \to \rho_v/\lambda, \tag{2.33}$$

where λ is an arbitrary non-vanishing parameter. Surfaces which admit the constraint (2.32) may be termed *generalized affine spheres*. Since asymptotic coordinates are only defined up to an arbitrary re-parametrization of the coordinate lines, that is $u \to \tilde{u}(u)$, $v \to \tilde{v}(v)$, it is possible to bring ρ into canonical form:

THEOREM 2.3 (Generalized affine spheres). *Generalized affine spheres may be parametrized in such a way that ρ assumes the canonical form*

$$\rho = c_1 u + c_2 v + c_3, \tag{2.34}$$

and the Gauss equations read

$$\begin{aligned} \boldsymbol{r}_{uu} &= \left(\frac{h_u}{h} + \lambda c_1\right)\boldsymbol{r}_u + \lambda\frac{a}{h}\boldsymbol{r}_v, \quad \boldsymbol{r}_{uv} = h\boldsymbol{r} + \frac{1}{\lambda}c_2\boldsymbol{r}_u + \lambda c_1\boldsymbol{r}_v, \\ \boldsymbol{r}_{vv} &= \left(\frac{h_v}{h} + \frac{1}{\lambda}c_2\right)\boldsymbol{r}_v + \frac{1}{\lambda}\frac{b}{h}\boldsymbol{r}_u. \end{aligned} \tag{2.35}$$

The generalized Tzitzeica system (2.21), (2.32) *reduces to*

$$(\ln h)_{uv} = h - \frac{ab}{h^2} + c_1c_2, \quad a_v + c_1h_u = 0, \quad b_u + c_2h_v = 0. \tag{2.36}$$

In conclusion, it is noted that Tzitzeica's original Bäcklund transformation for affine spheres as formulated in Theorem 2.2 is also valid in the case of generalized affine spheres [**37**].

2.5. The affine normal. The condition $\rho = \text{const}$ defining affine spheres does not seem to admit a canonical discrete counterpart. In order to discretize affine spheres, it is therefore necessary to find an alternative formulation which allows for immediate discretization. It turns out that one should return to "first principles" and take into account the origin of the term "affine spheres", that is the fact that affine spheres may be regarded as the analogues of spheres in Euclidean geometry. To this end, we observe that the definition of the unit normal $\boldsymbol{N} = \boldsymbol{r}_u \times \boldsymbol{r}_v / |\boldsymbol{r}_u \times \boldsymbol{r}_v|$ is only meaningful in Euclidean geometry since it is invariant under rotations and translations. However, the connection between $\boldsymbol{N}$ and the tangent vectors $\boldsymbol{r}_u$, $\boldsymbol{r}_v$ is *not* affine-invariant. Thus, in affine geometry, the normal $\boldsymbol{N}$ must be replaced by an *affine normal*. Since we are interested in surfaces parametrized in terms of asymptotic coordinates, we here give a coordinate-dependent definition of the affine normal (direction). The invariant formulation of the affine normal may be found in [**3**].

Consider an infinitesimal quadrilateral composed of asymptotic line segments $(\boldsymbol{r}, \boldsymbol{r}_1, \boldsymbol{r}_2, \boldsymbol{r}_{12}) = (\boldsymbol{r}(u,v), \boldsymbol{r}(u+du,v), \boldsymbol{r}(u,v+dv), \boldsymbol{r}(u+du,v+dv))$ attached to a point P on a hyperbolic surface with associated position vector $\boldsymbol{r}$ as illustrated in Figure 2. In the generic case, let π_1 be the plane parallel to the edges $\boldsymbol{r}_1 - \boldsymbol{r}$ and $\boldsymbol{r}_{12} - \boldsymbol{r}_2$ and π_2 be the plane parallel to the edges $\boldsymbol{r}_2 - \boldsymbol{r}$ and $\boldsymbol{r}_{12} - \boldsymbol{r}_1$. We denote the straight line defined by the intersection of these two planes by $l(P) = \pi_1 \cap \pi_2$.

DEFINITION 2.4 (The affine normal). The direction of the affine normal $\boldsymbol{\xi}$ at a point P on a hyperbolic surface Σ is given by $l(P)$.

We are now in a position to give an alternative characterization of affine spheres.

THEOREM 2.5 (Or geometric definition of affine spheres). *A surface constitutes an affine sphere iff all affine normals intersect at a point $\mathcal{O}$.*

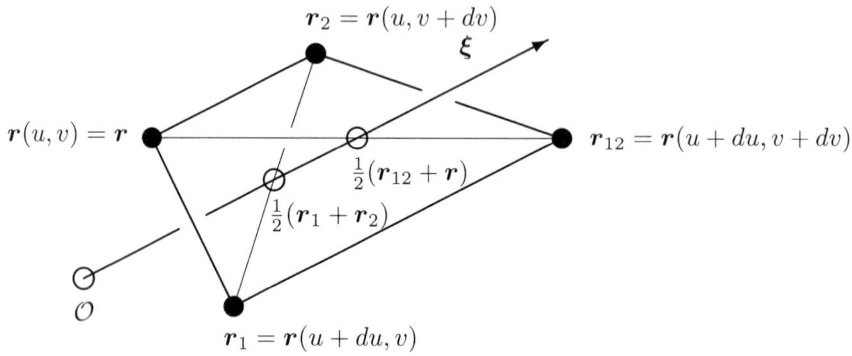

FIGURE 2. The affine normal.

PROOF. Without loss of generality, we may assume that the point $\mathcal{O}$ coincides with the origin of the ambient space. By definition, we have

$$r \sim r_{12} - r_1 - r_2 + r \approx r_{uv}\, du\, dv, \tag{2.37}$$

so that

$$r_{uv} \parallel r \Leftrightarrow \xi \parallel r. \tag{2.38}$$

Now, the left-hand side of the latter holds iff $\rho_u = \rho_v = 0$ by virtue of $(2.20)_2$ while the right-hand side is equivalent to the affine normals intersecting at the origin. This completes the proof. □

The above *geometric* characterization of affine spheres may also be used as an appropriate definition of discrete affine spheres if one regards (r, r_1, r_2, r_{12}) as a *finite* quadrilateral and applies the same construction of the (discrete) affine normal. However, since our definition of the affine normal is based on asymptotic coordinates, we first have to address the issue of canonical discrete asymptotic coordinates. This is done in Section 3.1.

2.6. Dual affine spheres. There exists yet another description of affine spheres which may be used in a natural *algebraic* discretization procedure. Remarkably, the discrete affine spheres constructed in this way have been shown to be identical with those alluded to in the previous section [**8**]. The first observation is that the position vector r of an affine sphere Σ may conveniently be normalized in such a way that

$$\frac{|r, r_u, r_v|}{h} = 1. \tag{2.39}$$

It is not difficult to verify that the left-hand side of (2.39) is indeed constant modulo the Gauss equations (2.26). Accordingly, the vector

$$\nu = \frac{1}{h} r_u \times r_v, \tag{2.40}$$

which is parallel to the Euclidean normal N, obeys the relation

$$r \cdot \nu = 1, \tag{2.41}$$

and hence constitutes a *co-normal*. Moreover, on use of the Gauss equations (2.26), it is readily shown that $\boldsymbol{\nu}$ is a solution of the *adjoint* linear system

$$\begin{aligned}\boldsymbol{\nu}_{uu} &= \frac{h_u}{h}\boldsymbol{\nu}_u - \frac{\lambda}{h}\boldsymbol{\nu}_v,\\ \boldsymbol{\nu}_{uv} = h\boldsymbol{\nu}, \quad \boldsymbol{\nu}_{vv} &= \frac{h_v}{h}\boldsymbol{\nu}_v - \frac{\lambda^{-1}}{h}\boldsymbol{\nu}_u.\end{aligned} \tag{2.42}$$

Thus, $\boldsymbol{\nu}$ defines a second affine sphere which may be regarded as *dual* to the affine sphere Σ. In fact, the definition (2.40) and the relation (2.41) give rise to the *Lelieuvre formulae* [**7, 14**] for affine spheres:

$$\begin{aligned}\boldsymbol{r}_u &= \boldsymbol{\nu}\times\boldsymbol{\nu}_u, \quad \boldsymbol{\nu}_u = \boldsymbol{r}_u\times\boldsymbol{r},\\ \boldsymbol{r}_v &= \boldsymbol{\nu}_v\times\boldsymbol{\nu}, \quad \boldsymbol{\nu}_v = \boldsymbol{r}\times\boldsymbol{r}_v.\end{aligned} \tag{2.43}$$

Conversely, the following theorem holds [**7**].

THEOREM 2.6 (Or algebraic definition of affine spheres). *If two vector-valued functions $\boldsymbol{r}$ and $\boldsymbol{\nu}$ obey the Lelieuvre formulae* (2.43) *then $\boldsymbol{r}$ and $\boldsymbol{\nu}$ are co-normals of each other, that is $\boldsymbol{r}\cdot\boldsymbol{\nu}=1$, and parametrize two affine spheres.*

PROOF. The Lelieuvre formulae imply that $\boldsymbol{r}_{uu}, \boldsymbol{r}_{vv}, \boldsymbol{r}_u, \boldsymbol{r}_v \perp \boldsymbol{\nu}$ so that $\boldsymbol{r}_{uu}$ and $\boldsymbol{r}_{vv}$ are linear combinations of $\boldsymbol{r}_u$ and $\boldsymbol{r}_v$. Moreover, the compatibility condition for $(2.43)_{2,4}$ leads to $\boldsymbol{r}_{uv} \parallel \boldsymbol{r}$. Thus, u and v constitute asymptotic coordinates and $\rho=$ const. The relation $\boldsymbol{r}\cdot\boldsymbol{\nu}=1$ is an algebraic consequence of (2.43). □

It is interesting to note that adjoint linear representations of integrable systems play a crucial role in connection with *binary Darboux transformations* [**24**]. These are particularly useful for generating coherent structure solutions (*dromions*) of 2+1-dimensional integrable equations such as the Davey–Stewartson [**10**] and Nizhnik–Veselov–Novikov [**1**] equations. Their connection with finite Loewner transformations [**23**], which are important in continuum mechanics, is discussed in [**39**] and by C. Rogers and W. K. Schief in this volume [**29**].

In conclusion, we observe that the above Lelieuvre formulae may indeed be discretized in a "naïve" manner by replacing the differential operators ∂_u and ∂_v by the standard difference operators. This substitution *preserves* integrability as demonstrated by Bobenko and Schief in [**7**].

3. Affine Difference Geometry

Discretizing surfaces usually means discretizing *parametrized surfaces*, that is one attempts to discretize natural *coordinate systems* on surfaces and additional geometric properties which define the particular class of surfaces under consideration [**9**]. Indeed, in the present case, it is necessary to discretize asymptotic coordinates and the affine normal and then propose a suitable definition of discrete affine spheres. To this end, it is assumed that "discrete coordinate lines" constitute coordinate polygons $\boldsymbol{r}=\boldsymbol{r}(n_i=\text{const})$, $i=1,2$ on a $\mathbb{Z}^2$ lattice

$$\boldsymbol{r}:\mathbb{Z}^2\to\mathbb{R}^3, \quad (n_1,n_2)\mapsto \boldsymbol{r}(n_1,n_2). \tag{3.1}$$

In the sequel, we refer to this lattice as a *discrete surface*. Moreover, the arguments of any vertex $\boldsymbol{r}=\boldsymbol{r}(n_1,n_2)$ will be suppressed and its four neighbours denoted by

$$\begin{aligned}\boldsymbol{r}_1 &= \boldsymbol{r}(n_1+1,n_2), \quad \boldsymbol{r}_2 = \boldsymbol{r}(n_1,n_2+1),\\ \boldsymbol{r}_{\bar{1}} &= \boldsymbol{r}(n_1-1,n_2), \quad \boldsymbol{r}_{\bar{2}} = \boldsymbol{r}(n_1,n_2-1).\end{aligned} \tag{3.2}$$

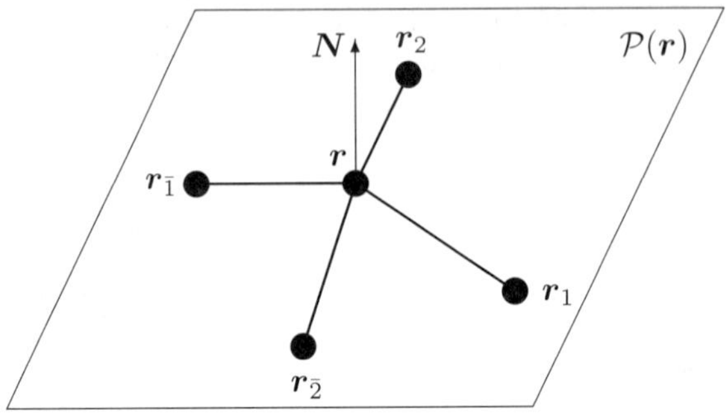

FIGURE 3. Discrete asymptotic nets.

Other vertices are represented by multiple indices. For instance, $(\boldsymbol{r}, \boldsymbol{r}_1, \boldsymbol{r}_2, \boldsymbol{r}_{12})$ designates an elementary quadrilateral. The symbols Δ_i denote the usual difference operators of first order, that is $\Delta_i f = f_i - f$.

3.1. Discrete asymptotic coordinates. As pointed out in Section 2.1, a surface $\Sigma : \boldsymbol{r} = \boldsymbol{r}(u, v)$ in Euclidean space $\mathbb{R}^3$ is parametrized in terms of asymptotic coordinates if the second fundamental form $\mathcal{F}_2 = -d\boldsymbol{r} \cdot d\boldsymbol{N}$ is off-diagonal, that is

$$\boldsymbol{N}_u \cdot \boldsymbol{r}_u = \boldsymbol{N}_v \cdot \boldsymbol{r}_v = 0. \tag{3.3}$$

Now, consider a point P on Σ and four infinitesimally close points $P_u, P_{\bar{u}}$ and $P_v, P_{\bar{v}}$ lying on the asymptotic lines $v = \text{const}$ and $u = \text{const}$ respectively which pass through P. The corresponding position vectors may be written as $\boldsymbol{r}(P) = \boldsymbol{r}(u, v)$, $\boldsymbol{r}(P_u) = \boldsymbol{r}(u + du, v)$, $\boldsymbol{r}(P_{\bar{u}}) = \boldsymbol{r}(u - du, v)$ and $\boldsymbol{r}(P_v) = \boldsymbol{r}(u, v + dv)$, $\boldsymbol{r}(P_{\bar{v}}) = \boldsymbol{r}(u, v - dv)$. If we introduce the difference vectors

$$\begin{aligned} \Delta_u \boldsymbol{r} &= \boldsymbol{r}(P_u) - \boldsymbol{r}(P), \quad \Delta_{\bar{u}} \boldsymbol{r} = \boldsymbol{r}(P_{\bar{u}}) - \boldsymbol{r}(P), \\ \Delta_v \boldsymbol{r} &= \boldsymbol{r}(P_v) - \boldsymbol{r}(P), \quad \Delta_{\bar{v}} \boldsymbol{r} = \boldsymbol{r}(P_{\bar{v}}) - \boldsymbol{r}(P), \end{aligned} \tag{3.4}$$

then Taylor's theorem delivers

$$\begin{aligned} |\Delta_{\bar{u}} \boldsymbol{r}, \Delta_u \boldsymbol{r}, \Delta_v \boldsymbol{r}| &\approx 4 |\boldsymbol{r}_u \times \boldsymbol{r}_v| (\boldsymbol{r}_{uu} \cdot \boldsymbol{N})\, du^3\, dv = 0, \\ |\Delta_{\bar{v}} \boldsymbol{r}, \Delta_u \boldsymbol{r}, \Delta_v \boldsymbol{r}| &\approx 4 |\boldsymbol{r}_u \times \boldsymbol{r}_v| (\boldsymbol{r}_{vv} \cdot \boldsymbol{N})\, du\, dv^3 = 0, \end{aligned} \tag{3.5}$$

by virtue of (3.3). This implies that, in the above approximation, the points P, $P_u, P_{\bar{u}}$ and $P_v, P_{\bar{v}}$ lie on a plane. The latter observation forms the basis of the following definition of discrete asymptotic nets.

DEFINITION 3.1. The coordinate polygons of a discrete surface $\boldsymbol{r} : \mathbb{Z}^2 \to \mathbb{R}^3$ form a *discrete asymptotic net* if any vertex $\boldsymbol{r}$ and its four neighbours $\boldsymbol{r}_1$, $\boldsymbol{r}_{\bar{1}}$, $\boldsymbol{r}_2$, $\boldsymbol{r}_{\bar{2}}$ lie on a plane $\mathcal{P}(\boldsymbol{r})$ (c.f. Figure 3).[1]

By construction, the unit normal $\boldsymbol{N}$ to the plane $\mathcal{P}(\boldsymbol{r})$, which may be regarded as "normal" to the discrete surface at $\boldsymbol{r}$, obeys the relations

$$\Delta_1 \boldsymbol{N} \cdot \Delta_1 \boldsymbol{r} = \Delta_2 \boldsymbol{N} \cdot \Delta_2 \boldsymbol{r} = 0, \tag{3.6}$$

which are nothing but the "naïve" discretization of the defining relations (3.3) for asymptotic coordinates.

[1]We usually commit the impropriety of referring to $\boldsymbol{r}$ as a vertex even though, strictly speaking, $\boldsymbol{r}$ is the position vector associated with a vertex.

The above definition of discrete asymptotic nets is due to Sauer [**32, 34**] and has been used to construct integrable discretizations of pseudospherical surfaces [**6, 33, 43**]. Remarkably, the corresponding underlying integrable discrete equation turns out to be Hirota's discrete sine–Gordon equation [**16**].

3.2. Discrete affine spheres. If a discrete surface is parametrized in terms of discrete asymptotic coordinates then we may associate with each quadrilateral $(\boldsymbol{r}, \boldsymbol{r}_1, \boldsymbol{r}_2, \boldsymbol{r}_{12})$ a *discrete affine normal (direction)* $\boldsymbol{r}$ given by

$$\boldsymbol{\xi} \sim \Delta_{12}\boldsymbol{r}, \tag{3.7}$$

and attach it to the centroid $\frac{1}{4}(\boldsymbol{r} + \boldsymbol{r}_1 + \boldsymbol{r}_2 + \boldsymbol{r}_{12})$. In other words:

DEFINITION 3.2 (The discrete affine normal). The discrete affine normal $\boldsymbol{\xi}$ associated with a quadrilateral $(\boldsymbol{r}, \boldsymbol{r}_1, \boldsymbol{r}_2, \boldsymbol{r}_{12})$ of a discrete asymptotic net passes through the centres of the diagonals $(\boldsymbol{r}, \boldsymbol{r}_{12})$ and $(\boldsymbol{r}_1, \boldsymbol{r}_2)$.

It is emphasized that this definition is analogous to that in the continuous case if one regards the quadrilateral in Figure 2 as *finite*. Once again, we are not concerned with the magnitude of the discrete affine normal. We now have all the necessary ingredients to define discrete affine spheres.

DEFINITION 3.3 (Discrete affine spheres). A discrete surface constitutes a *discrete affine sphere* if

(a) the coordinate polygons form discrete asymptotic nets and
(b) all affine normals intersect at a point $\mathcal{O}$.

In analytical terms, condition (a) in the definition of discrete affine spheres translates into

$$|\boldsymbol{r}_1 - \boldsymbol{r}, \boldsymbol{r}_2 - \boldsymbol{r}, \boldsymbol{r} - \boldsymbol{r}_{\bar{1}}| = 0, \quad |\boldsymbol{r}_1 - \boldsymbol{r}, \boldsymbol{r}_2 - \boldsymbol{r}, \boldsymbol{r} - \boldsymbol{r}_{\bar{2}}| = 0, \tag{3.8}$$

while the property (b) implies that

$$\boldsymbol{r}_{12} + \boldsymbol{r} \parallel \boldsymbol{r}_1 + \boldsymbol{r}_2, \tag{3.9}$$

if one chooses $\mathcal{O}$ to be the origin of $\mathbb{R}^3$ without loss of generality. Consequently, the position vector of discrete affine spheres obeys the *"discrete Gauss equations"*

$$\begin{aligned} \boldsymbol{r}_{11} - \boldsymbol{r}_1 &= \alpha(\boldsymbol{r}_1 - \boldsymbol{r}) + \beta(\boldsymbol{r}_{12} - \boldsymbol{r}_1), \\ \boldsymbol{r}_{12} + \boldsymbol{r} &= H(\boldsymbol{r}_1 + \boldsymbol{r}_2), \\ \boldsymbol{r}_{22} - \boldsymbol{r}_2 &= \gamma(\boldsymbol{r}_2 - \boldsymbol{r}) + \delta(\boldsymbol{r}_{12} - \boldsymbol{r}_2), \end{aligned} \tag{3.10}$$

where $\alpha, \beta, \gamma, \delta$ and H are as yet unspecified functions of n_i. However, the compatibility conditions for the above yield

$$\begin{gathered} \alpha = \frac{H_1 - 1}{H_1(H-1)}, \quad \gamma = \frac{H_2 - 1}{H_2(H-1)}, \\ \beta_2(H_2 - 1)H = \beta(H-1)H_1, \quad \delta_1(H_1 - 1)H = \delta(H-1)H_2, \\ H_{12} = \frac{H(H-1)}{H^2(H_1 + H_2 - H_1H_2) - H + \beta\delta H_1 H_2 (H-1)^2}. \end{gathered} \tag{3.11}$$

Hence, on setting

$$A = \beta(H-1), \quad B = \delta(H-1), \tag{3.12}$$

we obtain:

THEOREM 3.4 (The discrete Tzitzeica system). *Discrete affine spheres are governed by the discrete Gauss equations*

$$\begin{aligned} \boldsymbol{r}_{11}-\boldsymbol{r}_1 &= \frac{H_1-1}{H_1(H-1)}(\boldsymbol{r}_1-\boldsymbol{r})+\frac{A}{H-1}(\boldsymbol{r}_{12}-\boldsymbol{r}_1),\\ \boldsymbol{r}_{12}+\boldsymbol{r} &= H(\boldsymbol{r}_1+\boldsymbol{r}_2),\\ \boldsymbol{r}_{22}-\boldsymbol{r}_2 &= \frac{H_2-1}{H_2(H-1)}(\boldsymbol{r}_2-\boldsymbol{r})+\frac{B}{H-1}(\boldsymbol{r}_{12}-\boldsymbol{r}_2). \end{aligned} \tag{3.13}$$

They are compatible modulo

$$\begin{aligned} A_2 = \frac{H_1}{H}A, \quad B_1 = \frac{H_2}{H}B,\\ H_{12} = \frac{H(H-1)}{H^2(H_1+H_2-H_1H_2)-H+ABH_1H_2}, \end{aligned} \tag{3.14}$$

which is termed the discrete Tzitzeica system.

From a geometric point of view, it is evident that the term "discrete Tzitzeica system" is indeed justified. In fact, if one regards a discrete function $f:\mathbb{Z}^2\to\mathbb{R}$ as an approximation of a smooth function $\tilde{f}:\mathbb{R}^2\to\mathbb{R}$, that is

$$f(n_1,n_2)=\tilde{f}(x,y)\Big|_{(x,y)=(\varepsilon_1 n_1,\varepsilon_2 n_2)}, \tag{3.15}$$

for small ε_i, then, on dropping the tilde, the Taylor expansions

$$f_1 = f+\varepsilon_1 f_x+O(\varepsilon_1^2), \quad f_2 = f+\varepsilon_2 f_y+O(\varepsilon_2^2), \tag{3.16}$$

apply. Now, the form of the discrete Gauss equations (3.13) suggests the natural expansion

$$\mathrm{H}=1+\frac{1}{2}\varepsilon_1\varepsilon_2\mathrm{h}, \quad \mathrm{A}=\frac{1}{2}\varepsilon_1^3\mathrm{a}, \quad \mathrm{B}=\frac{1}{2}\varepsilon_2^3\mathrm{b}, \tag{3.17}$$

so that the discrete Tzitzeica system reduces to the gauge-invariant form of the Tzitzeica equation

$$(\ln h)_{xy}=h-abh^{-2}, \quad a=a(x), \quad b=b(y), \tag{3.18}$$

in the limit $\varepsilon_i\to 0$. Similarly, the discrete Gauss equations (3.13) become the continuous ones $(2.26)_{\lambda=1}$ in the corresponding gauge-invariant form.

The linear equations $(3.14)_{1,2}$ for A and B imply the existence of a potential τ defined according to

$$\begin{aligned} \tau_{11}=\tilde{c}\frac{\tau_1^2}{\tau A}\Rightarrow A=\tilde{c}\frac{\tau_1^2}{\tau\tau_{11}}, \quad \tau_{12}=\frac{\tau_1\tau_2}{\tau H}\Rightarrow H=\frac{\tau_1\tau_2}{\tau\tau_{12}},\\ \tau_{22}=\hat{c}\frac{\tau_2^2}{\tau B}\Rightarrow B=\hat{c}\frac{\tau_2^2}{\tau\tau_{22}}, \end{aligned} \tag{3.19}$$

where $\tilde{c}$ and $\hat{c}$ are arbitrary constants. Insertion of these relations into the nonlinear equation $(3.14)_3$ for H results in the *discrete Tzitzeica equation*

$$\begin{vmatrix} \tau & \tau_1 & \tau_{11}\\ \tau_2 & \tau_{12} & \tau_{112}\\ \tau_{22} & \tau_{122} & \tau_{1122} \end{vmatrix} + \tilde{c}\hat{c}\tau_{12}^3 = 0. \tag{3.20}$$

It is noted that the Tzitzeica equation

$$(\ln h)_{uv}=h-h^{-2}, \tag{3.21}$$

may be brought into the form

$$\begin{vmatrix} \tau & \tau_u & \tau_{uu} \\ \tau_v & \tau_{uv} & \tau_{uuv} \\ \tau_{vv} & \tau_{uvv} & \tau_{uuvv} \end{vmatrix} + \frac{1}{4}\tau^3 = 0, \tag{3.22}$$

in terms of a τ-function defined via

$$h = -2(\ln\tau)_{uv}. \tag{3.23}$$

This underlines the analogy between the classical continuous case and the discrete formalism presented here.

In conclusion, it is observed that the discrete Tzitzeica system is invariant under $A \to \lambda A$, $B \to \lambda^{-1}B$ where λ is an arbitrary constant. Thus, we have the following theorem.

THEOREM 3.5 (A Lax triad for the discrete Tzitzeica system). *Every solution of the discrete Tzitzeica system* (3.14) *corresponds to a one-parameter family of discrete affine spheres governed by*

$$\begin{aligned} \boldsymbol{r}_{11} - \boldsymbol{r}_1 &= \frac{H_1 - 1}{H_1(H-1)}(\boldsymbol{r}_1 - \boldsymbol{r}) + \lambda\frac{A}{H-1}(\boldsymbol{r}_{12} - \boldsymbol{r}_1), \\ \boldsymbol{r}_{12} + \boldsymbol{r} &= H(\boldsymbol{r}_1 + \boldsymbol{r}_2), \\ \boldsymbol{r}_{22} - \boldsymbol{r}_2 &= \frac{H_2 - 1}{H_2(H-1)}(\boldsymbol{r}_2 - \boldsymbol{r}) + \frac{1}{\lambda}\frac{B}{H-1}(\boldsymbol{r}_{12} - \boldsymbol{r}_2). \end{aligned} \tag{3.24}$$

As in the continuous case, the parameter λ is key to the integrability of the discrete Tzitzeica system/equation.

3.3. Generation of discrete surfaces via Bäcklund transformations. Discrete surfaces may also be generated by iterative application of Bäcklund transformations to continuous surfaces. This method of constructing discrete surfaces has been applied to pseudospherical [**6, 43**] and isothermic surfaces [**5, 12**]. A priori, it is not guaranteed that the discrete surfaces constitute discretizations of their continuous counterparts. In the two cases mentioned above, the discrete surfaces indeed reduce to the continuous ones in the natural continuum limit. However, we here show that the Tzitzeica transformation for affine spheres gives rise to discrete generalized affine spheres which may be specialized in such a way that the class of discrete affine spheres discussed in the preceding section is retrieved.

Let us assume that there exists a one-parameter (λ) family of transformations $\mathbb{B}_\lambda$ which acts pointwise within a given class of surfaces $\{\Sigma : \boldsymbol{r} = \boldsymbol{r}(u,v)\}$. Even though we have not yet specified any particular properties of these transformations, we refer to these transformations as *Bäcklund transformations* for simplicity. Thus, given a seed surface Σ in this class of surfaces, we may choose any *Bäcklund parameter* λ_1 to construct a new surface Σ_1 by means of the mapping

$$\mathbb{B}_{\lambda_1} : \Sigma \mapsto \Sigma_1. \tag{3.25}$$

A third surface Σ_2 is generated by another Bäcklund transformation $\mathbb{B}_{\lambda_2}$ with parameter λ_2. We now require that the family of Bäcklund transformations possess an additional property. It is assumed that the surface Σ_{12} obtained by applying the Bäcklund transformation $\mathbb{B}_{\lambda_2}$ to Σ_1 coincides with the surface Σ_{21} which is generated from Σ_2 by means of the Bäcklund transformation $\mathbb{B}_{\lambda_1}$, viz

$$\Sigma_{12} = \mathbb{B}_{\lambda_2}(\Sigma_1) = \mathbb{B}_{\lambda_1}(\Sigma_2) = \Sigma_{21}. \tag{3.26}$$

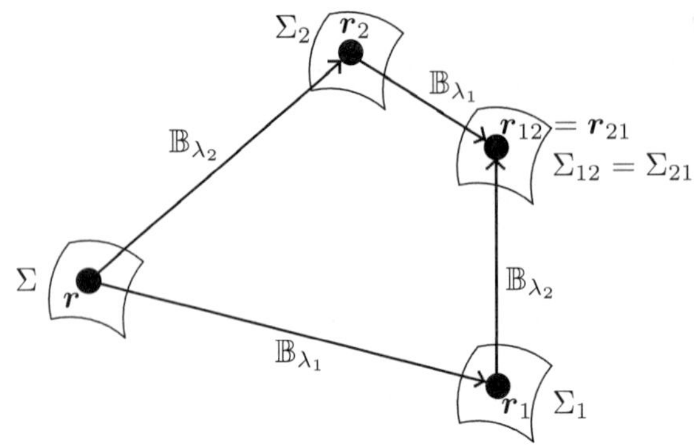

FIGURE 4. The closed Bianchi diagram.

This property has come to be known as *closure of the Bianchi diagram.* The situation is illustrated in Figure 4.

A discrete surface is now generated as follows. Let $\Sigma = \Sigma(0,0)$ be a surface in the above class. If $\mathbb{B}_1^{(i)}$, $i = 1, \ldots, n_1$ denote Bäcklund transformations associated with the parameters $\lambda_1^{(i)}$ then iterative application of the Bäcklund transformations generates the suite of surfaces

$$\Sigma(0,0) \xrightarrow{\lambda_1^{(1)}} \Sigma(1,0) \xrightarrow{\lambda_1^{(2)}} \Sigma(2,0) \xrightarrow{\lambda_1^{(3)}} \ldots \xrightarrow{\lambda_1^{(n_1)}} \Sigma(n_1,0). \tag{3.27}$$

Another set of Bäcklund transformations $\mathbb{B}_2^{(i)}$, $i = 1, \ldots, n_2$ corresponding to the Bäcklund parameters $\lambda_2^{(i)}$ gives rise to the sequence

$$\Sigma(0,0) \xrightarrow{\lambda_2^{(1)}} \Sigma(0,1) \xrightarrow{\lambda_2^{(2)}} \Sigma(0,2) \xrightarrow{\lambda_2^{(3)}} \ldots \xrightarrow{\lambda_2^{(n_2)}} \Sigma(0,n_2). \tag{3.28}$$

Since the Bianchi diagram is closed, application of $\mathbb{B}_1^{(1)}$ to $\Sigma(0,1)$ and $\mathbb{B}_2^{(1)}$ to $\Sigma(1,0)$ delivers the same surface which we denote by $\Sigma(1,1)$. Further iteration therefore generates the two-dimensional lattice of surfaces

$$(n_1, n_2) \mapsto \Sigma(n_1, n_2). \tag{3.29}$$

Accordingly, the collection of associated position vectors $\boldsymbol{r}(n_1, n_2)$ may be regarded as a function of two continuous and two discrete variables, that is

$$\boldsymbol{r} : \mathbb{R}^2 \times \mathbb{Z}^2 \to \mathbb{R}^3, \quad \boldsymbol{r} = \boldsymbol{r}(u, v; n_1, n_2). \tag{3.30}$$

Thus, for any fixed values of the original variables u and v, we may suppress the dependence on these variables and consider $\boldsymbol{r}$ as a two-dimensional lattice in $\mathbb{R}^3$:

$$\boldsymbol{r} : \mathbb{Z}^2 \to \mathbb{R}^3, \quad (n_1, n_2) \mapsto \boldsymbol{r}(n_1, n_2). \tag{3.31}$$

Evidently, the mapping (3.31) constitutes a discrete surface.

The following section is devoted to the properties of the discrete surfaces constructed via the classical Tzitzeica transformation.

3.4. Discrete (generalized) affine spheres. Consider two affine spheres Σ_1 and Σ_2 which are Tzitzeica transforms (c.f. Theorem 2.2) of an affine sphere Σ.

If ϕ^1 and ϕ^2 are the corresponding solutions of the Gauss equations (2.26) with parameters λ_1 and λ_2 respectively then the position vectors $\boldsymbol{r}_1$ and $\boldsymbol{r}_2$ read

$$\begin{aligned}\Sigma_1:\quad \boldsymbol{r}_1 &= \boldsymbol{r} - \frac{2}{(\lambda-\lambda_1)h}\left(\lambda\frac{\phi^1_u}{\phi^1}\boldsymbol{r}_v - \lambda_1\frac{\phi^1_v}{\phi^1}\boldsymbol{r}_u\right),\\ \Sigma_2:\quad \boldsymbol{r}_2 &= \boldsymbol{r} - \frac{2}{(\lambda-\lambda_2)h}\left(\lambda\frac{\phi^2_u}{\phi^2}\boldsymbol{r}_v - \lambda_2\frac{\phi^2_v}{\phi^2}\boldsymbol{r}_u\right).\end{aligned}\tag{3.32}$$

In order to iterate the Tzitzeica transformation, one needs scalar solutions of the transformed Gauss equations. Thus, by construction, the quantities

$$\begin{aligned}\phi^2_1 &= \phi^2 - \frac{2}{(\lambda_2-\lambda_1)h}\left(\lambda_2\frac{\phi^1_u}{\phi^1}\phi^2_v - \lambda_1\frac{\phi^1_v}{\phi^1}\phi^2_u\right),\\ \phi^1_2 &= \phi^1 - \frac{2}{(\lambda_1-\lambda_2)h}\left(\lambda_1\frac{\phi^2_u}{\phi^2}\phi^1_v - \lambda_2\frac{\phi^2_v}{\phi^2}\phi^1_u\right),\end{aligned}\tag{3.33}$$

satisfy the Gauss equations associated with Σ_1 and Σ_2 respectively. Consequently, application of the Tzitzeica transformation to Σ_1 and Σ_2 produces two affine spheres Σ_{12} and Σ_{21} whose position vectors are given by

$$\begin{aligned}\Sigma_{12}:\quad \boldsymbol{r}_{12} &= \boldsymbol{r}_1 - \frac{2}{(\lambda-\lambda_2)h_1}\left(\lambda\frac{\phi^2_{1u}}{\phi^2_1}\boldsymbol{r}_{1v} - \lambda_2\frac{\phi^2_{1v}}{\phi^2_1}\boldsymbol{r}_{1u}\right),\\ \Sigma_{21}:\quad \boldsymbol{r}_{21} &= \boldsymbol{r}_2 - \frac{2}{(\lambda-\lambda_1)h_2}\left(\lambda\frac{\phi^1_{2u}}{\phi^1_2}\boldsymbol{r}_{2v} - \lambda_1\frac{\phi^1_{2v}}{\phi^1_2}\boldsymbol{r}_{2u}\right).\end{aligned}\tag{3.34}$$

It turns out that these two surfaces coincide, that is

$$\boldsymbol{r}_{12} = \boldsymbol{r}_{21},\tag{3.35}$$

which implies that the Bianchi diagram associated with the Tzitzeica transformation is closed. This property is inherited from the Moutard transformation which likewise admits this *permutability theorem* [**19**].

3.5. The discrete Gauss equations. By construction, the discrete Gauss equations associated with discrete surfaces generated by a Bäcklund transformation may, in principle, be obtained through an elimination process from the continuous Gauss equations and the action of the Bäcklund transformation on the position vector $\boldsymbol{r}$. This procedure has been demonstrated for discrete generalized affine spheres in [**37**]. For brevity, we here focus on the geometric properties of the discrete Gauss equations and refer to [**37**] for the technical details. Thus, the transformation law $(3.32)_1$ implies that the difference vector $\boldsymbol{r}_1 - \boldsymbol{r}$ is tangent to the affine sphere Σ at the point $\boldsymbol{r}$. On the other hand, it is straight forward to show that "inversion" of $(3.32)_1$ yields

$$\boldsymbol{r}_1 = \boldsymbol{r} - \frac{2}{(\lambda+\lambda_1)h_1}\left(\lambda\frac{\phi^1_u}{\phi^1}\boldsymbol{r}_{1v} + \lambda_1\frac{\phi^1_v}{\phi^1}\boldsymbol{r}_{1u}\right),\tag{3.36}$$

which reveals that $\boldsymbol{r}_1 - \boldsymbol{r}$ is also tangent to the affine sphere Σ_1 at the point $\boldsymbol{r}_1$. Thus, we have retrieved the well-known *tangency property* of the classical Tzitzeica transformation which, in shorthand notation, may be formulated as

$$\boldsymbol{r}_1 - \boldsymbol{r} \parallel \Sigma, \Sigma_1.\tag{3.37}$$

It is known that any Bäcklund transformation with commuting Bianchi diagram and tangency property generates discrete asymptotic nets. This is evident since the line segments $(\boldsymbol{r}_1, \boldsymbol{r})$, $(\boldsymbol{r}_2, \boldsymbol{r})$, $(\boldsymbol{r}_{\bar{1}}, \boldsymbol{r})$, $(\boldsymbol{r}_{\bar{2}}, \boldsymbol{r})$ which connect the point $\boldsymbol{r}$ on the surface Σ to corresponding points $\boldsymbol{r}_1, \boldsymbol{r}_2, \boldsymbol{r}_{\bar{1}}, \boldsymbol{r}_{\bar{2}}$ on the neighbouring surfaces $\Sigma_1, \Sigma_2, \Sigma_{\bar{1}}, \Sigma_{\bar{2}}$ respectively are all tangent to Σ at $\boldsymbol{r}$ and hence the points $\boldsymbol{r}$ and $\boldsymbol{r}_1, \boldsymbol{r}_2, \boldsymbol{r}_{\bar{1}}, \boldsymbol{r}_{\bar{2}}$ lie on a plane. Thus, the following theorem holds:

THEOREM 3.6 (Discrete asymptotic nets generated by the Tzitzeica transformation). *Iterative application of the Tzitzeica transformation* (2.27) *to affine spheres generates a two-parameter* (u, v) *family of discrete asymptotic nets.*

The above tangency property for the Tzitzeica transformation implies that there exist functions $\tilde{\alpha}$ and $\tilde{\beta}$ such that

$$\boldsymbol{r}_{11} - \boldsymbol{r} = \tilde{\alpha}(\boldsymbol{r}_1 - \boldsymbol{r}) + \tilde{\beta}(\boldsymbol{r}_{12} - \boldsymbol{r}_1). \tag{3.38}$$

The explicit form of $\tilde{\alpha}$ and $\tilde{\beta}$ has been derived in [**37**]. It turns out that

$$\tilde{\alpha} = \frac{\lambda + \lambda_1}{\lambda - \lambda_1}\hat{\alpha}, \quad \tilde{\beta} = \frac{\lambda - \lambda_2}{\lambda - \lambda_1}\hat{\beta}, \tag{3.39}$$

where $\hat{\alpha}$, $\hat{\beta}$ are functions independent of λ and λ_1, λ_2 are assumed to be constants. Note that, in general, λ_1 and λ_2 may be regarded as functions of n_1 and n_2 respectively. In fact, as discrete functions, λ_1 and λ_2 are defined by $\lambda_1(n_1) = \lambda_1^{(n_1)}$ and $\lambda_2(n_2) = \lambda_2^{(n_2)}$, where $\lambda_1^{(n_1)}$ and $\lambda_2^{(n_2)}$ constitute the Bäcklund parameters in the sequences (3.27) and (3.28) respectively.

Expressions analogous to (3.38) and (3.39) are obtained in connection with the linear dependence of the difference vectors $\boldsymbol{r}_{22} - \boldsymbol{r}_2, \boldsymbol{r}_2 - \boldsymbol{r}$ and $\boldsymbol{r}_{12} - \boldsymbol{r}_2$. A third condition is obtained by solving $(3.32)_1$ for the tangent vectors $\boldsymbol{r}_u$ and $\boldsymbol{r}_v$ and substituting the solution into the expression $(3.34)_1$ for $\boldsymbol{r}_{12}$. This is summarized below [**8, 37**].

THEOREM 3.7 (The Gauss equations for discrete generalized affine spheres). *The Tzitzeica transformation applied to affine spheres generates discrete surfaces for which linear difference equations of the form*

$$\begin{aligned}
\boldsymbol{r}_{11} - \boldsymbol{r}_1 &= \frac{\lambda + \lambda_1}{\lambda - \lambda_1}\hat{\alpha}(\boldsymbol{r}_1 - \boldsymbol{r}) + \frac{\lambda - \lambda_2}{\lambda - \lambda_1}\hat{\beta}(\boldsymbol{r}_{12} - \boldsymbol{r}_1),\\
\boldsymbol{r}_{12} + c_0\boldsymbol{r} &= \widehat{H}(c_3\boldsymbol{r}_1 + c_4\boldsymbol{r}_2),\\
\boldsymbol{r}_{22} - \boldsymbol{r}_2 &= \frac{\lambda + \lambda_2}{\lambda - \lambda_2}\hat{\gamma}(\boldsymbol{r}_2 - \boldsymbol{r}) + \frac{\lambda - \lambda_1}{\lambda - \lambda_2}\hat{\delta}(\boldsymbol{r}_{12} - \boldsymbol{r}_2),
\end{aligned} \tag{3.40}$$

hold. The coefficients c_i are given by

$$c_0 = c_3c_4, \quad c_3 = -\frac{\lambda + \lambda_2}{\lambda - \lambda_2}, \quad c_4 = \frac{\lambda + \lambda_1}{\lambda - \lambda_1}. \tag{3.41}$$

The system (3.40) *may be regarded as the Gauss equations for* discrete generalized affine spheres *in that they reduce to the Gauss equations* (2.35) *in a natural continuum limit.*

PROOF. As usual, we regard n_1 and n_2 as discretizations of some continuous variables

$$x = \varepsilon_1 n_1, \quad y = \varepsilon_2 n_2. \tag{3.42}$$

The natural expansion of the coefficients in the discrete Gauss equations reads

$$\begin{aligned}\widehat{H} &= -1 - \frac{1}{2}\varepsilon_1\varepsilon_2(h + fg), \quad \lambda_1 = \frac{2}{\varepsilon_1 f}, \quad \lambda_2 = \frac{\varepsilon_2 g}{2},\\ \hat{\alpha} &= -(1+\varepsilon_1\alpha)\frac{f}{f_1}, \quad \hat{\beta} = -\frac{2\varepsilon_1}{\varepsilon_2 f_1}\beta, \quad \widehat{\gamma} = 1 + \varepsilon_2\gamma, \quad \hat{\delta} = -\frac{\varepsilon_2^2 f}{2}\delta,\end{aligned} \tag{3.43}$$

where we assume that the functions on the right-hand side are amenable to Taylor expansions, for instance

$$f_1 = f + \varepsilon_1 f_x + O(\varepsilon_1^2). \tag{3.44}$$

In the limit $\varepsilon_i \to 0$, the discrete Gauss equations become

$$\begin{aligned}\boldsymbol{r}_{xx} &= (\alpha + \lambda f)\boldsymbol{r}_x + \lambda\beta\boldsymbol{r}_y, \quad \boldsymbol{r}_{xy} = h\boldsymbol{r} + \frac{1}{\lambda}g\boldsymbol{r}_x + \lambda f\boldsymbol{r}_y,\\ \boldsymbol{r}_{yy} &= \left(\gamma + \frac{1}{\lambda}g\right)\boldsymbol{r}_y + \frac{1}{\lambda}\delta\boldsymbol{r}_x,\end{aligned} \tag{3.45}$$

and it is readily verified that the compatibility conditions lead to a parametrization of the form (2.35) with $\rho = \int f\,dx + \int g\,dy$. □

The compatibility conditions $\boldsymbol{r}_{112} = \boldsymbol{r}_{121}$ and $\boldsymbol{r}_{221} = \boldsymbol{r}_{122}$ for the discrete Gauss equations (3.40) give rise to a parametrization of the coefficients $\hat{\alpha}$, $\hat{\beta}$, $\widehat{\gamma}$, $\hat{\delta}$ and $\widehat{H}$ in terms of three functions A, B and H which obey a discrete version of the generalized Tzitzeica system (2.36). The explicit parametrization of the associated discrete Gauss equations has been given in [**8**].

In conclusion, we consider the limit $\lambda_1 \to \infty$, $\lambda_2 \to 0$. In this case, the constants c_i reduce to $c_0 = 1$ and $c_3 = c_4 = -1$ so that $(3.40)_2$ assumes the form of the Gauss equation $(3.24)_2$. Consequently, the discrete surfaces obtained in the degenerate case $\lambda_1 \to \infty$, $\lambda_2 \to 0$ constitute discrete affine spheres as defined in a purely geometric manner in Section 2.

3.6. The discrete Tzitzeica transformation. The discrete analogue of Tzitzeica's transformation for (generalized) affine spheres may be obtained in a purely algebraic manner. This is due to the fact that the Bäcklund transformation for discrete generalized affine spheres is inherited from that for affine spheres. Thus, application of a Tzitzeica transformation $\mathbb{B}_\mu$ with Bäcklund parameter μ to the two-dimensional lattice (3.29) of affine spheres produces another lattice of affine spheres

$$\widetilde{\Sigma}(n_1, n_2) = \mathbb{B}_\mu(\Sigma(n_1, n_2)). \tag{3.46}$$

Since the Bianchi diagram is closed, it is guaranteed that the surfaces $\widetilde{\Sigma}(n_1, n_2)$ are interrelated by Tzitzeica transformations, that is

$$\widetilde{\Sigma}(n_1, n_2) = \mathbb{B}_1^{(n_1)}(\widetilde{\Sigma}(n_1 - 1, n_2)) = \mathbb{B}_2^{(n_2)}(\widetilde{\Sigma}(n_1, n_2 - 1)). \tag{3.47}$$

The situation is illustrated in Figure 5 by way of a one-dimensional lattice of affine spheres. Consequently, for fixed parameters u and v, the discrete generalized affine sphere $(n_1, n_2) \mapsto \boldsymbol{r}(n_1, n_2)$ is mapped to another discrete generalized affine sphere $(n_1, n_2) \mapsto \tilde{\boldsymbol{r}}(n_1, n_2)$ by means of the Tzitzeica transformation $\mathbb{B}_\mu$. The action of this Bäcklund transformation on discrete generalized affine spheres may be obtained explicitly if one performs an elimination procedure analogous to that

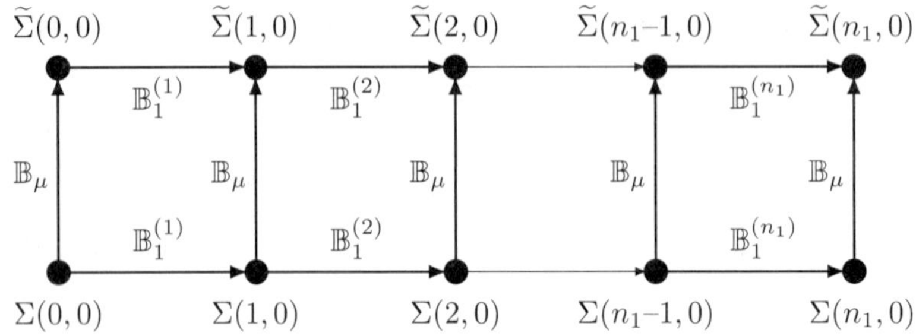

FIGURE 5. Two Bäcklund-related sequences of affine spheres.

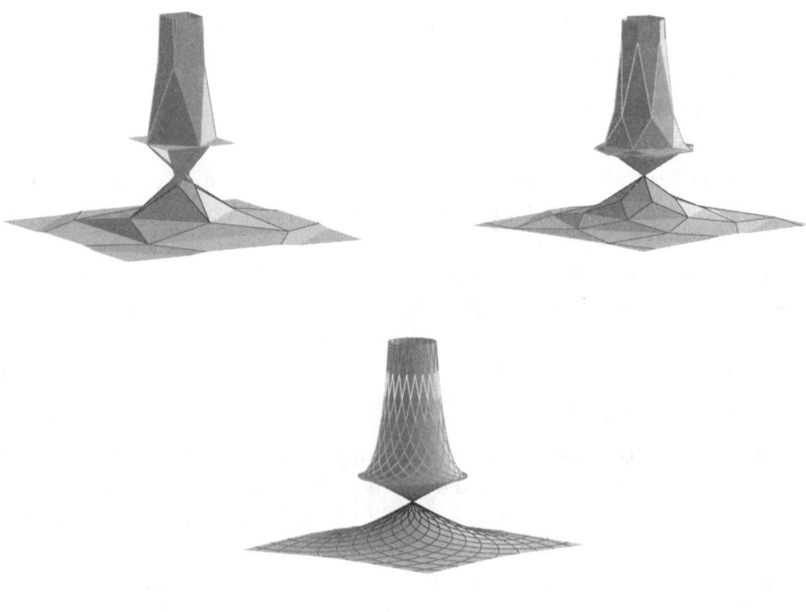

FIGURE 6. Discrete affine spheres of revolution.

applied in the derivation of the discrete Gauss equations for discrete generalized affine spheres. In the limit $\lambda_1 \to \infty$, $\lambda_2 \to 0$, one therefore obtains a *discrete Tzitzeica transformation* for discrete affine spheres [**8**].

Discrete affine spheres which have been generated by means of the discrete Tzitzeica transformation are displayed in Figure 6. These possess a $\mathbb{Z}_N$ rotational symmetry and exhibit either a closed planar polygon parallel to the (x, y) plane or a vertex which connects their upper and lower parts. It has been shown that these two properties exclude each other [**8**]. It is noted that for large N the discrete affine spheres resemble the stationary one-soliton affine sphere (Figure 1(b)). Indeed, the latter is obtained in the limit $N \to \infty$.

References

1. C. Athorne and J. J. C. Nimmo, *Moutard transformations for integrable partial differential equations*, Inverse Problems **7** (1991), 809–826.
2. L. Bianchi, *Lezioni di geometria differenziale*, Spoerri, Pisa, 1902.
3. W. Blaschke, *Vorlesungen über Differentialgeometrie* II, Springer, Berlin, 1923.
4. A. Bobenko, *Surfaces in terms of* 2 *by* 2 *matrices. Old and new integrable cases*, Harmonic Maps and Integrable Systems (A. Fordy and J. C. Wood, eds.), Vieweg, Braunschweig, 1994, 83–128.
5. A. Bobenko and U. Pinkall, *Discrete isothermic surfaces*, J. Reine Angew. Math. **475** (1996), 187–208.
6. ———, *Discrete surfaces with constant negative Gaussian curvature and the Hirota equation*, J. Differential Geom. **43** (1996), 527–611.
7. A. I. Bobenko and W. K. Schief, *Affine spheres: discretization via duality relations*, Experiment. Math. **8** (1999), 261–280.
8. ———, *Discrete indefinite affine spheres*, Discrete Integrable Geometry and Physics (A. I. Bobenko and R. Seiler, eds.), Oxford Lecture Ser. Math. Appl., vol. 16, Oxford Univ. Press, New York, 1999.
9. A. I. Bobenko and R. Seiler (eds.), *Discrete integrable geometry and physics*, Condensed Matter Physics and Discrete Geometry (Vienna, 1996), Oxford Lecture Ser. Math. Appl., vol. 16, Oxford Science Publications, The Clarendon Press, Oxford Univ. Press, New York, 1999.
10. M. Boiti, J. Leon, L. Martina, and F. Pempinelli, *Scattering of localized solitons in the plane*, Phys. Lett. A **132** (1988), 432–439.
11. B. Cenkl, *Geometric deformations of the evolution equations and Bäcklund transformations*, Phys. D **18** (1986), 217–219.
12. J. Cieśliński, *The Bäcklund transformation for discrete isothermic surfaces*, Symmetries and Integrability of Difference Equations (Canterbury, 1996) (P. A. Clarkson and F. Nijhoff, eds.), London Math. Soc. Lecture Note Ser., vol. 255, Cambridge University Press, 1999, pp. 109–121.
13. R. K. Dodd and R. K. Bullough, *Polynomial conserved densities for the sine–Gordon equations*, Proc. Roy. Soc. London Ser. A **352** (1977), 481–503.
14. L. P. Eisenhart, *A treatise on the differential geometry of curves and surfaces*, Dover, New York, 1960.
15. B. Gaffet, *A class of* $1-d$ *gas flows solvable by the inverse scattering transform*, Phys. D **26** (1987), 123–139.
16. R. Hirota, *Nonlinear partial difference equations.* III. *Discrete sine–Gordon equation*, J. Phys. Soc. Japan **43** (1977), 2079–2086.
17. A. R. Its, *Liouville theorem and inverse scattering problem*, J. Soviet Math. **31** (1985), 3330–3338.
18. A. V. Jiber and A. B. Shabat, *A Klein–Gordon equation with nontrivial group*, Dokl. Akad. Nauk SSSR **247** (1979), 1103–1105.
19. H. Jonas, *Sopra una classe di trasformazioni asintotiche, applicabili in particolare alle superficie la cui curvatura è proporzionale alla quarta potenza della distanza del piano tangente da un punto fisso*, Ann. of Math. (2) **30** (1921), 223–255.
20. ———, *Die Differentialgleichung der Affinsphären in einer neuen Gestalt*, Math. Nachr. **10** (1953), 331–352.
21. B. G. Konopelchenko and C. Rogers, *On* $2+1$*-dimensional nonlinear systems of Loewner type*, Phys. Lett. A **158** (1991), 391–397.
22. D. Levi and A. Sym, *Integrable systems describing surfaces of non-constant curvature*, Phys. Lett. A **149** (1990), 381–387.
23. C. Loewner, *A transformation theory of partial differential equations of gasdynamics*, Tech. Notes Nat. Adv. Comm. Aeronaut. **1950** (1950). no. 2065.
24. V. B. Matveev and M. A. Salle, *Darboux transformations and solitons*, Springer Ser. Nonlinear Dynam., Springer, Berlin, 1991.
25. A. V. Mikhailov, *The reduction problem and the inverse scattering method*, Phys. D **3** (1981), 73–117.

26. T. Moutard, *Sur la construction des équations de la forme* $1/z \quad \partial^2 z/\partial x \partial y = \lambda(x,y)$, *qui admettent une intégrale générale explicite*, J. École Polytechnique **45** (1878), 1–11.
27. J. J. C. Nimmo and W. K. Schief, *Superposition principles associated with the Moutard transformation: an integrable discretization of a* $(2+1)$*-dimensional sine–Gordon system*, Proc. Roy. Soc. London Ser. A **453** (1997), 255–279.
28. ———, *An integrable discretization of a* $2{+}1$*-dimensional sine–Gordon equation*, Stud. Appl. Math. **100** (1998), 295–309.
29. C. Rogers and W. K. Schief, *Infinitesimal Bäcklund transformations of* K*-nets. The* $2+1$*-dimensional sinh–Gordon system*, in this volume.
30. ———, *The classical geometry of Bäcklund transformations. Introduction to applications in soliton theory*, in this volume.
31. C. Rogers, W. K. Schief, and M. E. Johnston, *Bäcklund and his works: Applications in soliton theory*, Geometric Approaches to Differential Equations (Canberra, 1995) (P. Vassiliou, ed.,), Austral. Math. Soc. Lect. Ser., vol. 15, Cambridge Univ. Press, Cambridge, 2000.
32. R. Sauer, *Finite Analoga zur Differentialgeometrie der Asymptotennetze*, Ber. Math.-Tagung Tübingen **1946**, (1947), 127–129.
33. ———, *Parallelogrammgitter als Modelle pseudosphärischer Flächen*, Math. Z. **52** (1950), 611–622.
34. ———, *Differenzengeometrie*, Springer-Verlag, Berlin-New York, 1970.
35. W. K. Schief, *An integrable discretization of the characteristic equations associated with a gasdynamics system*, Phys. Lett. A **238** (1998), 278–282.
36. ———, *Isothermic surfaces in spaces of arbitrary dimension: Integrability, discretization and Bäcklund transformations. A discrete Calapso equation*, Stud. Appl. Math. (1999), (to appear).
37. ———, *Hyperbolic surfaces in centro-affine geometry, Integrability and discretization*, Chaos Solitons Fractals **11** (2000), 97–106.
38. W. K. Schief and C. Rogers, *The affinsphären equation. Moutard and Bäcklund transformations*, Inverse Problems **10** (1994), 711–731.
39. ———, *Loewner transformations. Adjoint and binary Darboux transformations*, Stud. Appl. Math. **100** (1998), 391–422.
40. G. Tzitzeica, *Sur une (nouvelle) classe de surfaces*, C. R. Acad. Sci. Paris **144** (1907), 1257–1259.
41. ———, *Sur une (nouvelle) classe de surfaces*, C. R. Acad. Sci. Paris **146** (1908), 165–166.
42. ———, *Sur une nouvelle classe de surfaces*, C. R. Acad. Sci. Paris **150** (1910), 955–956; 1227–1229.
43. W. Wunderlich, *Zur Differenzengeometrie der Flächen konstanter negativer Krümmung*, Österreich. Akad. Wiss. Math.-Nat. Kl. S.-B. II **160** (1951), 39–77.

School of Mathematics, The University of New South Wales, Sydney, NSW 2052, Australia.
E-mail address: schief@maths.unsw.edu.au

Part 2

Original Contributions

Centre de Recherches Mathématiques
CRM Proceedings and Lecture Notes
Volume **29**, 2001

On Bianchi and Bäcklund Transformations of Two Dimensional Surfaces in Four Dimensional Euclidean Space

Yu. Aminov and A. Sym

In 1879 L. Bianchi introduced the so called "complementary transformation" of pseudospherical surfaces $F^2 \subset E^3$ into pseudospherical surfaces $\overline{F}^2 \in E^3$ [**4**]. Let the Gauss curvature of F^2 be $K \equiv -1$ and x, y be the horocyclic system of coordinates on F^2. The metric of F^2 is

$$ds^2 = e^{2y}dx^2 + dy^2.$$

Let $r = r(x, y)$ be a position vector to F^2, then according to Bianchi the position vector to the new surface $\overline{F}^2$ is

$$\bar{r} = r - r_y.$$

By the remarkable Bianchi theorem, the Gauss curvature of $\overline{F}^2$ *is again* $\overline{K} \equiv -1$.

In 1883 Bäcklund generalized the Bianchi transformation in [**3**]. For the Bäcklund transformation we have

$$\bar{r} = r + a,$$

where a is a tangent vector field on F^2 of a constant length. *If* a *satisfies two well-known equations, then* $\overline{K} \equiv -1$.

In 1978 Y. Aminov generalized the Bianchi transformation for an immersed region F^n of n-dimensional Lobachevski space L^n into Euclidean space E^{2n-1} [**1**]. By the Cartan–Liber theorem the dimension $(2n-1)$ of the ambient space can not be lowered. If $x_1, \dots, x_{n-1}, y$ is the horocyclic system of coordinates on F^n with the metric

$$ds^2 = e^{2y}(dx_1^2 + \cdots + dx_{n-1}^2) + dy^2,$$

then the position vector of $\overline{F}^n$ is defined by

$$\bar{r} = r - r_y.$$

The theorem in [**1**] states that *the curvature of* $\overline{F}^n$ *for all tangent planes is* $\overline{K} \equiv -1$.

In 1979 K. Tenenblat and C.-L. Terng generalized the Bäcklund transformation for an immersed region of L^n into E^{2n-1} [**6**] and in 1985 K. Tenenblat constructed the Bäcklund transformation for submanifolds of space forms [**5**].

2000 *Mathematics Subject Classification.* Primary: 53A05; Secondary: 53A25, 37K35.

The article is supported by the Polish Committee for Science Research KBN Grant 2-PO3B-143-15.

This is the final form of the paper.

In all these generalizations the codimension of the immersed submanifolds was exactly $n - 1$. A natural question arises: can one extend the transformations discussed above to the case of codimension $\geq n$. In this paper we discuss the simplest case $F^2 \subset E^4$.

Let F^2 denote some isometric immersion of a domain of the Lobachevski plane L^2 of the curvature $K = -1$ into E^4 equipped with the horocyclic system of coordinates (x, y). Again $r = r(x, y)$ stands for a position vector to F^2. The Bianchi transformation is defined as above

$$\bar{r} = r - r_y.$$

THEOREM 1. *There exist pseudospherical surfaces F^2 in E^4 such that their Bianchi transform has K non equal to -1.*

For a pseudospherical surface in E^3 the vector r_y is tangent to both F^2 and $\overline{F}^2$. For surfaces in E^4 this is not true. If r_y is tangent to both F^2 and $\overline{F}^2$ we say that the vector field r_y on F^2 enjoys a bi-tangency property. For simplicity we say that F^2 enjoys a bi-tangency property as well.

THEOREM 2. *If F^2 enjoys the bi-tangency property, then $\overline{K} = -1$.*

THEOREM 3. *The set of all pseudospherical surfaces in E^4 with a bi-tangency property is of the arbitrariness of four functions of a single variable each.*

Finally, let us discuss briefly some results related to a still unsolved problem of Bäcklund transformation in the discussed case.

Let F^2 be an arbitrary surface in E^4 and its transform $\overline{F}^2$ be given by its position vector

$$\bar{r} = r + a,$$

where a is bi-tangent vector field on F^2.

Let τ be a tangent vector to F^2 at a point $x \in F^2$ and γ be a curve on F^2 passing through x in the direction τ. If $r = r(s)$ is a position vector of γ and s is an arc length, then the curvature vector of γ is r_{ss}. Projection of r_{ss} on the normal space at x gives the vector of normal curvature $k_n(\tau)$ of F^2 for the direction τ. Let us imagine the vector $k_n(\tau)$ as starting at x and ending at $Q(\tau)$. A set of all the points $Q(\tau)$ when τ rotates in the tangent plane T_xF^2 is called after E. Cartan the *ellipse of normal curvature.*

THEOREM 4. (1) *If the point x is inside the ellipse of normal curvature, then no vector field a tangent to both F^2 and $\overline{F}^2$ exists.*
(2) *If the point x is outside of the ellipse of normal curvature, then at every point $x \in F^2$ there exist at most two directions of the vector field a tangent to both F^2 and $\overline{F}^2$.*

We recall that the surface F^2 is of flat normal connection if the matrices of coefficients of the first and two second fundamental forms can be diagonalized simultaneously.

THEOREM 5. *Suppose $F^2 \subset E^4$ is of flat normal connection and its Gaussian curvature $K \neq 0$. If there exists a vector field a tangent to $\overline{F}^2$, then $F^2 \subset E^3$.*

A complete presentation of the results above including proofs appeared in [**2**].

References

1. Yu. Aminov, *Bianchi transform for the domain of multidimensional Lobachevski space*, Ukraïn. Geom. Sb. **21** (1978), 3–5 (in Russian).
2. Yu. Aminov and A. Sym, *On Bianchi and Bäcklund transformations of two-dimensional surfaces in E^4*, Math. Phys. Anal. Geom. **3** (2000), no. 1, 75–89.
3. A. V. Bäcklund, *Om ytor med konstant negativ krokning*, Lunds Univ. Arsskr. **19** (1883), 1–41.
4. L. Bianchi, *Ricerche sulle superficie a curvatura costante e sulle elicoidi*, Ann. Scuola Norm. Sup. Pisa Cl. Sci. (4) **2** (1879), no. 1, 285–340.
5. K. Tenenblat, *Bäcklund's theorem for submanifolds of space form and generalized wave equation*, Bol. Soc. Brasil Mat. (N.S.) **16** (1985), no. 2, 69–94.
6. K. Tenenblat and C. L. Terng, *A higher dimension generalization of the sine–Gordon equation and its Bäcklund transformation*, Bull. Amer. Math. Soc. (N.S.) **1** (1979), no. 3, 589–593.

VERKIN'S INST. FOR LOW TEMPERATURE, PHYSICS AND ENGINEERING, LENIN AV. 47, 310140, KHARKOV, UKRAINE AND INST. MATH. OF THE BIALYSTOK UNIVERSITY, AKADEMICKA 2, 15-267, BIALYSTOK, POLAND.

E-mail address: aminov@ilt.kharkov.ua; aminov@cksr.ac.bialystok.pl

INST. FOR THEOR. PHYSICS OF WARSAW UNIVERSITY, ST. HOZA 69, 00-681 WARSAW, POLAND.

E-mail address: Antoni.Sym@fuw.edu.pl

Centre de Recherches Mathématiques
CRM Proceedings and Lecture Notes
Volume **29**, 2001

Group Invariant Solutions without Transversality and the Principle of Symmetric Criticality

Ian M. Anderson, Mark E. Fels, and Charles G. Torre

ABSTRACT. We extend Lie's classical method for finding group invariant solutions to the case of nontransverse group actions. For this extension of Lie's method we identify a local obstruction to the principle of symmetric criticality. Two examples of nontransverse symmetry reductions for the potential form of Maxwell's equations are then examined.

Introduction

Let $\Delta = 0$ be a system of differential equations and let $\tilde{\Delta} = 0$ be a second system which is related to the first by a geometric transformation. For example, the second system could be derived from the first by the process of group reduction, a Bäcklund transformation, or a differential substitution. The question then arises whether formal geometric properties of the second system $\tilde{\Delta} = 0$ can be inferred from those of the original. For example, if $\Delta = 0$ is a system of Euler–Lagrange equations is $\tilde{\Delta} = 0$ also a system of Euler–Lagrange equations? To answer such questions a rigorous mathematical description of the process relating the two systems of equations is needed. In this article we will address this issue in the case where $\tilde{\Delta} = 0$ are the reduced equations for the group invariant solutions to $\Delta = 0$.

Lie's method of symmetry reduction for finding the group invariant solutions to partial differential equations iswell-known (see, for example, Bluman and Kumei [**5**], Olver [**8**], Winternitz [**11, 12**]). However, these references make the hypothesis of *transversality* of the group action, an assumption which is not valid for most problems in field theory and differential geometry. In the next section we show how one can dispense with this assumption and find the reduced differential equations for group actions which are not transverse. In Section 3 we extend the results in the article [**1**] and find another local obstruction to the principle of symmetric criticality [**10**] for nontransverse group actions. Finally in Section 4 we find the reduced equations for two nontransverse symmetry groups of Maxwell's equations and we check the principle of symmetric criticality on these.

2000 *Mathematics Subject Classification.* Primary: 58G35, 58E30; Secondary: 58D19, 58A20.

Research supported by NSF grants DMS–9804833 and PHY–9732636.

This is the final form of the paper.

1. Group Invariant Solutions Without Transversality

1.1 Preliminaries. Let M be an n-dimensional manifold and $\pi: E \to M$ a bundle over M. For this article it is sufficient to consider E as a trivial bundle $E = U \times \mathbb{R}^m$ where U is an open set in $\mathbb{R}^n$. The manifold M will serve as the space of independent variables and the bundle E plays the role of the total space of independent and dependent variables. Points of M will be labeled with local coordinates (x^i) and points of E with local coordinates (x^i, u^α). In terms of these coordinates the projection map π is given by $\pi(x^i, u^\alpha) = (x^i)$. We let $E_x = \pi^{-1}(x)$ for $x \in M$.

Let G be a finite dimensional Lie group which acts smoothly and projectably on E. That is, the action of each element of G is a fiber preserving transformation on E and, consequently, there is a smooth induced action of G on M such that the diagram

$$\begin{array}{ccc} E & \xrightarrow{\ g\ } & E \\ {\scriptstyle\pi}\downarrow & & \downarrow{\scriptstyle\pi} \\ M & \xrightarrow{\ g\ } & M \end{array}$$

commutes. For projectable group actions the action of G on the space of sections of E is given by

$$(g \cdot s)(x) = g \cdot s(g^{-1} \cdot x),$$

where $s: M \to E$ is a smooth section.

DEFINITION 1.1. *Let G be a smooth projectable group action on the bundle $\pi: E \to M$ and let $U \subset M$ be open. Then a smooth section $s: U \to E$ is G* **invariant**, *if for all $x \in U$ and $g \in G$ such that $g \cdot x \in U$,*

$$s(g \cdot x) = g \cdot s(x). \tag{1.1}$$

Let Γ be the Lie algebra of infinitesimal generators for the action of G on E. Since the action of G on E is assumed projectable any basis V_a, $a = 1, \ldots, p$, for Γ assumes the local coordinate form

$$V_a = \xi_a^i(x)\partial_{x^i} + \eta_a^\alpha(x, u)\partial_{u^\alpha}. \tag{1.2}$$

Suppose that $v_a \in \mathfrak{g}$ (the Lie algebra of G), and that V_a is the corresponding infinitesimal generator. Let s be an invariant section given in local coordinates by $U^\alpha = S^\alpha(x)$. If we substitute $g = \exp(tv_a)$ into equation (1.1), differentiate with respect to t and put $t = 0$ we obtain the **infinitesimal invariance equations**

$$\xi_a^i(x)\frac{\partial s^\alpha}{\partial x^i} = \eta_a^\alpha(x, s^\alpha), \tag{1.3}$$

for the invariant section s. If s is globally defined on all of M and if G is connected, then the infinitesimal invariance criterion (1.3) implies (1.1).

The condition that a group with infinitesimal generators V_a in (1.2) acts (infinitesimally) transversally is (see [**5, 8**]).

$$\operatorname{rank}\big[\xi_1(x), \ldots, \xi_p(x)\big] = \operatorname{rank}\big[\xi_1(x), \ldots, \xi_p(x), \eta_1(x, u), \ldots, \eta_p(x, u)\big].$$

Under the assumption of transversality the invariant section equation (1.3) is easily solved in terms of the infinitesimal invariants for Γ. In the next section we describe how to determine the invariant sections without this hypothesis.

1.2 Kinematic reduction. In order to study group invariant solutions to a differential equation we need to first characterize the G invariant sections of E and for this we make the following observation. Suppose that $p \in E$ and that there is a G invariant section $s: U \to E$ with $s(x) = p$, where $x \in U$. Let $G_x = \{g \in G \mid g \cdot x = x\}$ be the **isotropy subgroup of** G **at** x. Then for every $g \in G_x$, we compute from (1.1)

$$g \cdot p = g \cdot s(x) = s(g \cdot x) = s(x) = p. \tag{1.4}$$

Consequently, if $g \in G_x$ and $p \in E_x$ and $g \cdot p \neq p$, then there can be **no** invariant section through the point p. In view of this, we define the **kinematic bundle** $\kappa(E)$ **for the action of** G **on** E by

$$\kappa(E) = \bigcup_{x \in M} \kappa_x(E),$$

where

$$\kappa_x(E) = \{p \in E_x \mid g \cdot p = p \text{ for all } g \in G_x\}.$$

It is easy to check that $\kappa(E)$ is a G invariant subset of E and therefore the action of G restricts to an action on $\kappa(E)$. The set $\kappa(E)$ is a generalization of the construction in [**7**].

Now construct the quotient spaces $\widetilde{M} = M/G$ and $\tilde{\kappa}(E) = \kappa(E)/G$ for the actions of G on M and $\kappa(E)$ and define the **kinematic reduction diagram for the action of** G **on** E to be the commutative diagram

$$\begin{array}{ccccc} \tilde{\kappa}(E) & \xleftarrow{\mathbf{q}_\kappa} & \kappa(E) & \xrightarrow{\iota} & E \\ {\scriptstyle\tilde{\pi}}\downarrow & & {\scriptstyle\pi}\downarrow & & \downarrow{\scriptstyle\pi} \\ \widetilde{M} & \xleftarrow{\mathbf{q}_M} & M & \xrightarrow{id} & M. \end{array} \tag{1.5}$$

In this diagram ι is the inclusion map of the kinematic bundle $\kappa(E)$ into E, the maps $\mathbf{q}_M$ and $\mathbf{q}_\kappa$ are the projection maps to the quotient spaces and $\tilde{\pi}$ is the surjective map induced by π.

We mention a few simple technical facts about these constructions.

LEMMA 1.2.
(i) *For all* $p \in \kappa_x(E)$, $G_p \cong G_x$.
(ii) *If* $\tilde{p} \in \tilde{\kappa}(E)$ *and* $x \in M$ *satisfy* $\tilde{\pi}(\tilde{p}) = \mathbf{q}_M(x)$, *then there exists a unique point* $p \in \kappa_x(E)$ *such that* $\mathbf{q}_\kappa(p) = \tilde{p}$.

From (ii) it can be immediately inferred that for every local section $\tilde{s}: \widetilde{U} \to \tilde{\kappa}(E)$ there exists a uniquely determined G invariant section $s: \mathbf{q}_M^{-1}(\widetilde{U}) \to \kappa(E)$ such that

$$\mathbf{q}_\kappa\big(s(x)\big) = \tilde{s}\big(\mathbf{q}_M(x)\big). \tag{1.6}$$

The kinematic reduction diagram, along with (1.6), leads to the following existence theorem for invariant sections.

THEOREM 1.3. *Suppose that E admits a kinematic reduction diagram such that*
(i) $\kappa(E)$ *is an embedded subbundle of* E,
(ii) *the quotient spaces* $\widetilde{M}$ *and* $\tilde{\kappa}(E)$ *are smooth manifolds, and*
(iii) $\tilde{\pi}: \tilde{\kappa}(E) \to \widetilde{M}$ *is a bundle.*

Let $\widetilde{U}$ be any open set in $\widetilde{M}$ and let $U = \mathbf{q}_M^{-1}(\widetilde{U})$. Then (1.6) defines a one-to-one correspondence between the G invariant smooth sections $s\colon U \to E$ and the smooth sections $\tilde{s}\colon \widetilde{U} \to \tilde{\kappa}(E)$.

The proof of this theorem can be found in [**2**].

For generic group actions the transversality condition is replaced by the hypothesis that $\iota\colon \kappa(E) \to E$ is an embedding. When G is a compact Lie group acting by isometries on a Hermetian vector-bundle E this condition holds [**6**].

We now characterize infinitesimally invariant sections. If the rank of the coefficient matrix $\left[\xi_a^i(x)\right]$ is q, then there are locally defined functions $\phi_\varepsilon^a(x)$, where $\varepsilon = 1, \dots, p-q$, such that $\sum_{a=1}^p \phi_\varepsilon^a(x)\xi_a^i(x) = 0$. Consequently, using the infinitesimal invariant section equation (1.3), we find that

$$\sum_{a=1}^{p} \phi_\varepsilon^a(x)\Big(\xi_a^i(x)\frac{\partial s^\alpha}{\partial x^i} + \eta_a^\alpha\big(x, s^\beta(x)\big)\Big) = \sum_{a=1}^{p} \phi_\varepsilon^a(x^j)\eta_a^\alpha\big(x^j, s^\beta(x^j)\big) = 0, \tag{1.7}$$

are the algebraic equations constraining the invariant sections. Accordingly we define the **infinitesimal kinematic bundle** $\kappa^0(E) = \bigcup_{x\in M} \kappa_x^0(E)$ where

$$\begin{aligned} \kappa_x^0(E) &= \Big\{p = (x^j, u^\beta) \in E_x \;\Big|\; \sum_{a=1}^{p} \phi_\varepsilon^a(x^j)\eta_a^\alpha(x^j, u^\beta) = 0\Big\} \\ &= \{p \in E_x \mid Z(p) = 0 \text{ for all } Z \in \Gamma_x\}, \end{aligned} \tag{1.8}$$

and $\Gamma_x = \big\{Z \in \Gamma \mid \pi_*(Z)(x) = 0\big\}$. It is easy to see by using infinitesimal methods that $\kappa(E) \subset \kappa^0(E)$ and if, for each $x \in M$, G_x is connected, then $\kappa(E) = \kappa^0(E)$.

The (infinitesimal) kinematic reduction diagram (1.5) in local coordinates takes the following form

$$\begin{array}{ccccc} (\tilde{x}^r, v^a) & \xleftarrow{\;\mathbf{q}_\kappa\;} & (\tilde{x}^r, \hat{x}^k, v^a) & \xrightarrow{\;\iota\;} & \big(\tilde{x}^r, \hat{x}^k, \iota^\alpha(\tilde{x}^r, \hat{x}^k, v^a)\big) \\ \tilde{\pi}\big\downarrow & & \pi\big\downarrow & & \pi\big\downarrow \\ (\tilde{x}^r) & \xleftarrow{\;\mathbf{q}_M\;} & (\tilde{x}^r, \hat{x}^k) & \xrightarrow{\;id\;} & (\tilde{x}^r, \hat{x}^k). \end{array} \tag{1.9}$$

To derive this coordinate description of the kinematic reduction diagram we begin with the local coordinates $\tilde{\pi}\colon (\tilde{x}^r, v^a) \to (\tilde{x}^r)$ on the bundle $\tilde{\kappa}(E) \to \widetilde{M}$. Since $\mathbf{q}_M\colon M \to \widetilde{M}$ is a submersion, we can use the coordinates $\tilde{x}^r$ as part of a local coordinate system $(\tilde{x}^r, \hat{x}^k)$ on M where $k = 1 \dots n-q$, and q is the dimension of the orbits of G on M. As a consequence of Lemma 1.1 (ii) one can use $(\tilde{x}^r, \hat{x}^k, v^a)$ as a system of local coordinates on $\kappa(E)$. Let $(\tilde{x}^r, \hat{x}^k, u^\alpha) \to (\tilde{x}^r, \hat{x}^k)$ be a system of local coordinates on E. Since $\kappa(E)$ is an embedded subbundle of E, the inclusion map $\iota\colon \kappa(E) \to E$ assumes the form

$$\iota(\tilde{x}^r, \hat{x}^k, v^a) = \big(\tilde{x}^r, \hat{x}^k, \iota^\alpha(\tilde{x}^r, \hat{x}^k, v^a)\big). \tag{1.10}$$

If $v^a = \tilde{s}^a(\tilde{x}^r)$ is a local section of $\tilde{\kappa}(E)$, then the corresponding G invariant section of E is given by

$$s^\alpha(\tilde{x}^r, \hat{x}^k) = \iota^\alpha\big(\tilde{x}^r, \hat{x}^k, \tilde{s}^a(\tilde{x}^r)\big). \tag{1.11}$$

This is the general local form of an invariant section without the transversality assumption which we will use in the examples.

1.3 Dynamic reduction. Let $\pi^k: J^k(E) \to M$ be the kth order jet bundle of $\pi: E \to M$. A point $\sigma = j^k(s)(x)$ in $J^k(E)$ represents the values of a local section s and all its derivatives to order k at the point $x \in M$. Since G acts naturally on the space of sections of E by (1.3) the action of G on E prolongs to an action on $J^k(E)$ by setting

$$g \cdot \sigma = j^k(g \cdot s)(gx) \text{ where } \sigma = j^k(s)(x).$$

Now let $\pi: \mathcal{D} \to J^k(E)$ be a vector bundle over $J^k(E)$ and suppose that the Lie group G acts projectably on $\mathcal{D}$ in a manner which covers the action of G on $J^k(E)$. A **differential operator** is a section $\Delta: J^k(E) \to \mathcal{D}$. The differential operator Δ is G invariant if

$$g \cdot \Delta(\sigma) = \Delta(g \cdot \sigma),$$

for all $g \in G$ and all points $\sigma \in J^k(E)$. A section s of E defined on an open set $U \subset M$ is a solution to the differential equations $\Delta = 0$ if

$$\Delta\big(j^k(s)(x)\big) = 0 \quad \text{for all } x \in U.$$

Often the action of G on $\mathcal{D}$ is naturally defined in terms of the action of G on E as in the case of Euler–Lagrange operators (see Subsection 2.2).

We want to construct a bundle $\widetilde{\mathcal{D}} \to J^k\big(\tilde{\kappa}(E)\big)$ and a differential operator $\tilde{\Delta}: J^k(\tilde{\kappa}(E)) \to \widetilde{\mathcal{D}}$ such that the correspondence (1.6) defines a $1-1$ correspondence between the G invariant solutions of $\Delta = 0$ and the solutions of $\tilde{\Delta} = 0$. The required bundle $\widetilde{\mathcal{D}} \to J^k\big(\tilde{\kappa}(E)\big)$ can not be constructed by a direct application of the kinematic reduction diagram to $\mathcal{D} \to J^k(E)$. This difficulty is circumvented by introducing the **bundle of invariant** k**-jets**, (Olver [**8**])

$$\begin{aligned}\operatorname{Inv}^k(E) = \big\{\sigma \in J^k(E) \mid \sigma = j^k(s)(x_0),& \\ \text{where } s \text{ is a } G \text{ invariant section defined in a neighborhood of } x_0&\big\}.\end{aligned}$$

The fundamental property we need of $\operatorname{Inv}^k(E)$ is that the quotient space $\operatorname{Inv}^k(E)/G$ coincides with the jet space $J^k\big(\tilde{\kappa}(E)\big)$. We let $\mathcal{D}_{\text{Inv}} \to \operatorname{Inv}^k(E)$ be the restriction of $\mathcal{D}$ to the bundle of invariant k-jets and to this bundle we now apply our reduction procedure to arrive at the **dynamic reduction diagram**

$$\begin{array}{ccccccc}
\tilde{\kappa}(\mathcal{D}\,\text{Inv}) & \xleftarrow{\ \mathbf{q}\ } & \kappa(\mathcal{D}_{\text{Inv}}) & \xrightarrow{\ \iota\ } & \mathcal{D}_{\text{Inv}} & \xrightarrow{\ \iota^k\ } & \mathcal{D} \\
\downarrow{\scriptstyle\tilde{\pi}} & & \downarrow{\scriptstyle\pi} & & \downarrow{\scriptstyle\pi} & & \downarrow{\scriptstyle\pi} \\
J^k\big(\tilde{\kappa}(E)\big) & \xleftarrow{\ \mathbf{q}_{\text{Inv}}\ } & \operatorname{Inv}^k(E) & \xrightarrow{\ \text{Id}\ } & \operatorname{Inv}^k(E) & \xrightarrow{\ \iota^k\ } & J^k(E).
\end{array} \tag{1.12}$$

Any G invariant differential operator $\Delta: J^k(E) \to \mathcal{D}$ restricts to a G invariant differential operator $\Delta: \operatorname{Inv}^k(E) \to \mathcal{D}_{\text{Inv}}$ which determines a differential operator $\tilde{\Delta}: J^k\big(\tilde{\kappa}(E)\big) \to \tilde{\kappa}(\mathcal{D}_{\text{Inv}})$. We call the differential operator $\tilde{\Delta}$ the reduced operator and the equations $\tilde{\Delta} = 0$ the reduced equations. It is not difficult to prove the following theorem.

THEOREM 1.4. *The solutions to the reduced differential equations $\tilde{\Delta} = 0$ are in one-to-one correspondence with the G invariant solutions for the original equations $\Delta = 0$.*

To describe diagram (1.12) in local coordinates, we begin with the coordinate description (1.9) of the kinematic reduction diagram and we let

$$(\tilde{x}^r, \hat{x}^k, u^\alpha, u^\alpha_r, u^\alpha_k, u^\alpha_{rs}, u^\alpha_{rk}, u^\alpha_{kl}, \dots), \tag{1.13}$$

be coordinates on $J^k(E)$. Since the invariant sections are parameterized by functions $v^a = v^a(\tilde{x}^r)$ the coordinates for $\mathrm{Inv}^k(E)$ are

$$(\tilde{x}^r, \hat{x}^k, v^a, v^a_r, v^a_{rs}, \dots),$$

and the inclusion map

$$\iota^k \colon \mathrm{Inv}^k(E) \to J^k(E),$$

is given by

$$\iota^k(\tilde{x}^r, \hat{x}^k, v^a, v^a_r, v^a_{rs}, \dots) = (\tilde{x}^r, \hat{x}^k, u^\alpha, u^\alpha_r, u^\alpha_k, u^\alpha_{rs}, u^\alpha_{rk}, u^\alpha_{kl}, \dots), \tag{1.14}$$

where the derivative terms are obtained by taking derivatives of $u^\alpha = \iota^\alpha(\tilde{x}^r, \hat{x}^k, v^a)$, for example,

$$u^\alpha_r = \frac{\partial \iota^\alpha}{\partial \tilde{x}^r} + \frac{\partial \iota^\alpha}{\partial v^a} v^a_r, \qquad u^\alpha_k = \frac{\partial \iota^\alpha}{\partial \hat{x}^k}.$$

The coordinates on $J^k\big(\tilde{\kappa}(E)\big)$ are then

$$(\tilde{x}^r, v^a, v^a_r, v^a_{rs}, \dots).$$

Next let $\mathbf{f}^A$ be a local frame field for the vector bundle $\mathcal{D}$. The differential operator $\Delta \colon J^k(E) \to \mathcal{D}$ can be written in terms of the coordinates (1.13) on $J^k(E)$ and this local frame as

$$\Delta = \Delta_A(\tilde{x}^r, \hat{x}^k, u^\alpha, u^\alpha_r, u^\alpha_k, u^\alpha_{rs}, u^\alpha_{rk}, u^\alpha_{kl} \dots)\mathbf{f}^A. \tag{1.15}$$

The restriction of Δ to $\mathrm{Inv}^k(E)$ defines the section $\Delta_{\mathrm{Inv}} \colon \mathrm{Inv}^k(E) \to \mathcal{D}_{\mathrm{Inv}}$ by

$$\Delta_{\mathrm{Inv}} = \Delta_{\mathrm{Inv},A}(\tilde{x}^r, \hat{x}^k, v^a, v^a_r, v^a_{rs}, \dots)\mathbf{f}^A, \tag{1.16}$$

where the component functions are $\Delta_{\mathrm{Inv},A}(\tilde{x}^r, \hat{x}^k, v^a, v^a_r, v^a_{rs}, \dots) = \Delta_A \circ \iota^k$. Since Δ is a G invariant differential operator, Δ_{Inv} is a G invariant differential operator and hence necessarily factors through the kinematic bundle $\kappa(\mathcal{D}_{\mathrm{Inv}})$, and is a section

$$\Delta_{\mathrm{Inv}} \colon \mathrm{Inv}^k(E) \to \kappa(\mathcal{D}_{\mathrm{Inv}}).$$

The existence theorem for invariant sections implies that we can also find a locally defined G invariant frame $\mathbf{f}^Q_{\mathrm{Inv}}$ for $\kappa(\mathcal{D}_{\mathrm{Inv}}) \to \mathrm{Inv}^k(E)$ in terms of which the invariant operator Δ_{Inv} can be expressed as

$$\Delta_{\mathrm{Inv}} = \Delta_{\mathrm{Inv},Q}(\tilde{x}^r, \hat{x}^k, v^a, v^a_r, v^a_{rs}, \dots)\mathbf{f}^Q_{\mathrm{Inv}}.$$

The reduced operator is now easily determined.

2. The Principle of Symmetric Criticality

2.1 Lagrangian reduction. Let $\lambda = L\nu$ be a G invariant kth order Lagrangian, where L is a smooth function on $J^k(E)$ and ν is a volume form on M. For simplicity ν is assumed to be G invariant and L a differential invariant. In order to define the reduced Lagrangian $\tilde{\lambda}$ on $J^k(\tilde{\kappa}(E))$, we start with a G invariant q chain $\chi: M \to \wedge^q(TM)$ and define the equivariant bundle map

$$\rho_\chi: \wedge^*(T^*M) \to \wedge^{*-q}(T^*M),$$

by $\rho_\chi(\alpha) = \chi \lrcorner\, \alpha$. The map ρ_χ induces a map on G invariant sections

$$\rho_\chi: \Omega^*(M)^G \to \Omega^{*-q}(M)^G.$$

The image of a differential form under ρ_χ is a G basic form, from which we further obtain a map

$$\tilde{\rho}_\chi: \Omega^*(M)^G \to \Omega^{*-q}(\widetilde{M}).$$

For example, if $G = \mathrm{SO}(3)$ is acting in the standard way on $M = \mathbb{R}^3 - (0,0,0)$, then ([**1,** p. 615])

$$\chi = r(z\partial_x \wedge \partial_y - y\partial_x \wedge \partial_z + x\partial_y \wedge \partial_z), \tag{2.1}$$

where $r = \sqrt{x^2 + y^2 + z^2}$, and

$$\rho_\chi(dx \wedge dy \wedge dz) = r(x\,dx + y\,dy + z\,dz),$$

so

$$\tilde{\rho}_\chi(dx \wedge dy \wedge dz) = r^2\,dr.$$

Again see [**1**] or [**4**] for more details and examples.

Using the maps $\tilde{\rho}_\chi$ and $\iota^k: \mathrm{Inv}^k(E) \to J^k(E)$ we define the reduced Lagrangian as

$$\tilde{\lambda} = \tilde{\rho}_\chi\big((\iota^k)^*\lambda\big). \tag{2.2}$$

We point out that the role of the map $\tilde{\rho}_\chi$ is to reduce the number of independent variables in the volume form.

Let $\mathrm{Vert}(E)$ be the vertical bundle for $\pi: E \to M$. Then from the maps $\pi^{2k}_E: J^{2k}(E) \to E$ and $\pi^{2k}_M: J^{2k}(E) \to M$ we construct the **bundle of source forms** $\mathcal{D} \to J^{2k}(E)$ by

$$\mathcal{D} = (\pi^{2k}_E)^*\big(\mathrm{Vert}^*(E)\big) \wedge (\pi^{2k}_M)^*(\wedge^n T^*M),$$

where * denotes pullback. A point $\omega \in \mathcal{D}_\sigma$, $\sigma \in J^{2k}(E)$ is a differential form

$$\omega = A_\alpha\,du^\alpha \wedge \nu.$$

The Euler–Lagrange form $\mathcal{E}(\lambda)$ of a kth order Lagrangian is a section of $\mathcal{D}$ and may be written in local coordinates (x^i, u^α) as

$$\mathcal{E}(\lambda) = E_\alpha(L)\,du^\alpha \wedge \nu,$$

where $E_\alpha(L)$ are the Euler–Lagrange expressions for $\lambda = L\nu$.

At this point there are two ways to proceed in the reduction of the Euler–Lagrange equations. First, given a G invariant Lagrangian λ the Euler–Lagrange operator $\mathcal{E}(\lambda)$ is also G-invariant. Using the dynamic reduction diagram (1.12) in Subsection 1.3 we can then compute the reduced operator $\widetilde{\mathcal{E}(\lambda)}$ whose solutions determine the group invariant solutions. The reduced operator $\widetilde{\mathcal{E}(\lambda)}$ is a section of $\tilde{\kappa}(\mathcal{D}_{\mathrm{Inv}})$. Alternatively, if $\tilde{\lambda}$ is the reduced Lagrangian, then $\mathcal{E}(\tilde{\lambda})$ is a source form on $J^{2k}(\tilde{\kappa}(E))$ which, using the coordinates in (1.9), is a form

$$\mathcal{E}(\tilde{\lambda}) = E_a(\tilde{L})dv^a \wedge \tilde{\rho}_\chi(\nu),$$

where $\tilde{\rho}_\chi(\nu)$ is a volume form on $\widetilde{M}$. Therefore it is **not** possible to compare the sections $\widetilde{\mathcal{E}(\lambda)}$ and $\mathcal{E}(\tilde{\lambda})$ since they are sections of different bundles and consequently, the solutions to $\widetilde{\mathcal{E}(\lambda)} = 0$ and $\mathcal{E}(\tilde{\lambda}) = 0$ may be different. This leads to the following principle.

THE PRINCIPLE OF SYMMETRIC CRITICALITY (PSC). *Let G be a Lie group acting projectably on E and suppose there exists a G invariant chain χ. Let λ be a fixed but arbitrary G invariant Lagrangian with $\tilde{\lambda}$ the reduced Lagrangian, and let $\widetilde{\mathcal{E}(\lambda)}$ be the reduced Euler–Lagrange operator. If, for every choice of λ the submanifolds $\mathcal{E}(\tilde{\lambda})^{-1}(0) \subset J(\tilde{\kappa}(E))$ and $\widetilde{\mathcal{E}(\lambda)}^{-1}(0) \subset J(\tilde{\kappa}(E))$ coincide, then we say the principle of symmetric criticality holds for the action of G on E.*

Note that if PSC holds then every locally defined solution $\tilde{s}\colon \widetilde{U} \to \tilde{\kappa}(E)$ to $\mathcal{E}(\tilde{\lambda}) = 0$ determines though (1.6) a solution s to $\mathcal{E}(\lambda) = 0$. Of course the principle can often fail, see [**1, 10**].

Our objective is to determine the local obstruction to this principal which arises for nontransverse group actions.

A local obstruction to the commutation relation

$$\mathcal{E}\big(\tilde{\rho}_\chi(\iota^k)^*\lambda\big) = \tilde{\rho}_\chi\mathcal{E}\big((\iota^k)^*\lambda\big), \tag{2.3}$$

was found in [**1**] and is described in the following theorem.

THEOREM 2.1. *If $H^q(\Gamma, G_x) \neq 0$ for all $x \in M$, then there exists a locally defined map $\tilde{\rho}_\chi\colon \Omega^*(M)^G \to \Omega^{*-q}(\widetilde{M})$ such that (2.3) holds, and moreover $\tilde{\rho}_\chi$ is a co-chain map, i.e. $d\tilde{\rho}_\chi = \tilde{\rho}_\chi d$.*

For a description of the Lie algebra cohomology condition

$$H^q(\Gamma, G_x) \neq 0, \tag{2.4}$$

see [**1,** p. 650]. We now find

LEMMA 2.2. *Suppose that $H^q(\Gamma, G_x) \neq 0$ for all $x \in M$. Let $\widetilde{U} \subset \widetilde{M}$ and let $\tilde{s}\colon \widetilde{U} \to \tilde{\kappa}(E)$ be a solution to the Euler–Lagrange equations $\mathcal{E}(\tilde{\lambda}) = 0$ for the reduced Lagrangian. Then the corresponding invariant section $s\colon U \to E$ from (1.6) is a solution to $(\iota^{2k})^*\mathcal{E}(\lambda) = 0$.*

PROOF. It follows from the Lemma 2.1 that

$$\mathcal{E}(\tilde{\lambda}) = \mathcal{E}\big(\rho_\chi(\iota^k)^*\lambda\big) = \tilde{\rho}_\chi\mathcal{E}\big((\iota^k)^*\lambda\big),$$

which together with the basic fact about the Euler–Lagrange operators [**8**]

$$\mathcal{E}\big((\iota^k)^*\lambda\big) = (\iota^{2k})^*\mathcal{E}(\lambda), \tag{2.5}$$

proves the lemma.

Note however that this lemma does **not** guarantee that the invariant section $s: U \to E$ is a solution to the original Euler–Lagrange equations $\mathcal{E}(\lambda) = 0$. Let's see why.

2.2 The Palais condition. In this section we will assume that the cohomology condition (2.4) in Theorem 2.1 holds so that (2.3) is valid.

The obstruction to PSC we have hinted at arises from differences in the two types of reductions which are occurring. First, if λ is a G invariant Lagrangian, then the Euler–Lagrange operator $\Delta = \mathcal{E}(\lambda)$ is G invariant, and the right side of the dynamic reduction diagram gives

$$\Delta_{\mathrm{Inv}} = E_\alpha(L)\big|_{\mathrm{Inv}}\, du^\alpha \wedge \nu,$$

and a solution $\tilde{s}: \widetilde{U} \to \tilde{\kappa}(E)$ to the reduced equations $\widetilde{\mathcal{E}(\lambda)} = 0$ defines through (1.6) an invariant solution $s: U \to E$ to $E_\alpha(L)\big|_{j(s)(x)} = 0$. On the other hand if we reduce the Lagrangian λ as in (2.2), we find, using the formula in (2.5)

$$\mathcal{E}\big((\iota^k)^*\lambda\big) = (\iota^{2k})^*\mathcal{E}(\lambda) = \left(\frac{\partial \iota^\alpha}{\partial v^a} E_\alpha(L)\right)\bigg|_{\mathrm{Inv}} dv^a \wedge \nu.$$

Therefore if $\tilde{s}: \widetilde{U} \to \tilde{\kappa}(E)$ is a solution to the equations $\mathcal{E}(\tilde{\lambda}) = 0$ we have by Lemma 2.2 that the corresponding invariant section s in (1.6) is a solution to

$$\left(\frac{\partial \iota^\alpha}{\partial v^a} E_\alpha(L)\right)\bigg|_{j(s)(x)} = 0.$$

Clearly this does not guarantee that $E_\alpha(L)\big|_{j(s)(x)} = 0$, or that $\tilde{s}$ is a solution to the reduced equations.

Conditions under which solutions to the equations $\mathcal{E}(\tilde{\lambda}) = 0$ are solutions to $\widetilde{\mathcal{E}(\lambda)} = 0$ can be obtained by a detailed analysis of Δ_{Inv} and $(\iota^{2k})^*\mathcal{E}(\lambda)$.

Suppose $\Delta = \mathcal{E}(\lambda)$ is an invariant section of the bundle of source forms $\mathcal{D}$, and let $\sigma \in \mathrm{Inv}^{2k}(E)$ and $x_0 = \pi^M(\sigma) \in M$ and $Z \in \Gamma_{x_0}$. Then on account of the invariance of ν the Lie derivative of Δ is

$$(\mathcal{L}_Z\Delta)\big|_\sigma = \left(\mathrm{pr}^{2k} Z\big(E_\alpha(L)\big) + E_\beta(L)\frac{\partial \eta^\beta}{\partial u^\alpha}\right)\bigg|_\sigma du^\alpha \wedge \nu. \tag{2.6a}$$

This leads to the isotropy condition

$$0 = \left(\Delta_\beta \frac{\partial \eta^\beta}{\partial u^\alpha}\right)\bigg|_\sigma du^\alpha \wedge \nu, \tag{2.6b}$$

which determines the fibres of $\kappa^0(\mathcal{D}_{\mathrm{Inv}})$. A geometric description of (2.6) can be given in terms of the **vertical linear isotropy representation**. Suppose $\kappa(E) \subset E$ is an embedded subbundle and let $\mathrm{Vert}\big(\kappa(E)\big)$ be the restriction (or pullback) of the vertical bundle over E to $\kappa(E)$. Since G is fibre-preserving, the differential of the action of G on E induces an action of G on $\mathrm{Vert}(E)$ which, because $\kappa(E) \subset E$ is a G invariant set, restricts to an action on $\mathrm{Vert}\big(\kappa(E)\big)$. We then make the following definition.

DEFINITION 2.2. *Let $p \in \kappa(E)$, $g \in G_p$ and $Y_p \in \mathrm{Vert}_p\,\kappa(E)$. Then the* **vertical linear isotropy representation** *on $\kappa(E)$, $\rho_p\colon G_p \to GL\big(\mathrm{Vert}_p\,\kappa(E)\big)$ is defined by*

$$\rho_p(g)Y_p = g_*Y_p. \tag{2.7}$$

Note by Lemma 1.2 (i) this is also a representation of G_x, where $x = \pi(p)$.

The differential of ρ_p is a homomorphism $\rho_p\colon \Gamma_p \to gl\big(\mathrm{Vert}_p\,\kappa(E)\big)$ called the **infinitesimal vertical linear isotropy representation**. If $Z = \xi^i(x)\partial_{x^i} + \eta^\alpha(x,u)\partial_{u^\alpha} \in \Gamma_{(x_0,u_0)}$ then, in the standard basis ∂_{u^α}, we have

$$\rho_{(x_0,u_0)}(Z) = \left(\frac{\partial\eta^\beta}{\partial u^\alpha}\bigg|_{(x_0,u_0)}\right).$$

Equation (2.6) now proves the following Lemma.

LEMMA 2.4. *If $\sigma \in \mathrm{Inv}^{2k}(E)$ and $p = \pi_E(\sigma)$, then $\kappa_\sigma(\mathcal{D}_{\mathrm{Inv}}) \cong \big[\mathrm{Vert}^*_p\big(\kappa(E)\big)\big]^{G_x}$.*

The invariants $\big[\mathrm{Vert}^*_p\big(\kappa(E)\big)\big]^{G_x}$ are computed using the dual of the representation (2.7). It is not difficult to check

LEMMA 2.5. $\iota_*\,\mathrm{Vert}_p\big(\kappa(E)\big) = \big[\mathrm{Vert}_{\iota(p)}\,E\big]^{G_x}$.

Define the subspace

$$\big([\|_p(E)]^{G_x}\big)^\perp = \{\alpha \in \mathrm{Vert}^*_p(E) \mid \alpha(X) = 0 \text{ for all } X \in \big[\mathrm{Vert}_p(E)\big]^{G_x}\}.$$

The pullback construction for $\mathcal{D}$ and Lemma 2.5 allows us to identify $\big(\big[\mathrm{Vert}_p(E)\big]^{G_x}\big)^\perp$ with the subspace of $\mathcal{D}_{\mathrm{Inv},\sigma}$, where $\pi^{2k}_E(\sigma) = p$, given by

$$\big(\big[\mathrm{Vert}_p(E)\big]^{G_x}\big)^\perp \cong \left\{A_\alpha\, du^\alpha \wedge \nu \in \mathcal{D}_{\mathrm{Inv},\sigma} \;\middle|\; A_\alpha \frac{\partial \iota^\alpha}{\partial v^a} = 0\right\},$$

and the coordinates in (1.14) are being used. We now come to the key result.

THEOREM 2.5. *Suppose λ is a G invariant Lagrangian and that $H^q(\Gamma, G_x) \neq 0$ for all $x \in M$. If*

$$\big[\mathrm{Vert}^*_p(E)\big]^{G_x} \cap \big(\big[\mathrm{Vert}_p(E)\big]^{G_x}\big)^\perp = 0 \text{ for all } p \in \kappa(E), \tag{2.8}$$

then every solution $\tilde{s}\colon \widetilde{M} \to \tilde{\kappa}(E)$ to $\mathcal{E}(\tilde{\lambda}) = 0$, where $\tilde{\lambda}$ is reduced Lagrangian in (2.2), defines by (1.6), a solution to $\mathcal{E}(\lambda) = 0$. In other words, the PSC *holds.*

We will prove this under the hypothesis that the isotropy groups are connected so that we may use infinitesimal methods.

PROOF. Let $\tilde{s}\colon \widetilde{U} \to \tilde{\kappa}(E)$ be a solution to $\mathcal{E}(\tilde{\lambda}) = 0$ and let $s\colon U \to \kappa(E)$ be the corresponding invariant section from (1.6). Using the isomorphism in Lemma 2.4, we have

$$E_\alpha(L)\big|_{j(s)(x)} \in \big[\mathrm{Vert}^*_{(x,s(x))}(E)\big]^{G_x}.$$

We also have, by Lemma 2.2, that s satisfies

$$\left(\frac{\partial\iota^\alpha}{\partial v^a} E_\alpha(L)\right)\bigg|_{j(s)(x)} = 0,$$

therefore by Lemma 2.5

$$E_\alpha(L)\big|_{j(s)(x)} \in \left(\left[\mathrm{Vert}_p(E)\right]^{G_x}\right)^\perp,$$

where $p = \big(x, s(x)\big)$. Thus, for each $x \in U$,

$$E_\alpha(L)\big|_{j(s)(x)} \in \left[\mathrm{Vert}^*_{(x,s(x))}(E)\right]^{G_x} \cap \left(\left[\mathrm{Vert}_{(x,s(x))}(E)\right]^{G_x}\right)^\perp,$$

which vanishes by (2.8) and s is an invariant solution to $E_\alpha(L) = 0$.

It is also possible to prove that (2.8) is necessary for the PSC to hold. We call (2.8) *the Palais condition* due to its similarity to the condition given in Palais' original article [**10**] .

3. Example—Maxwell's Equations

The base manifold in this case is $M = \mathbb{R}^4$ with coordinates (t, x, y, z) and $E = T^*M$. The Minkowski metric is $\eta = \mathrm{diag}(+,-,-,-)$ and we use (x^a), $a = 0$, 1, 2, 3, and $(x, y, z) = (x^i)$, $i = 1$, 2, 3 when convenient. A section of E is a differential form $\alpha = u_a dx^a$. The form α is the potential in Maxwell's equations which are the Euler–Lagrange equations of the first-order Lagrangian

$$\lambda = |d\alpha|^2 \nu, \tag{3.1}$$

where $\nu = dx^0 \wedge dx^1 \wedge dx^2 \wedge dx^3$ and $|d\alpha|^2 = \eta^{ab}\eta^{cd} u_{[a,c]} u_{[b,d]}$. The Euler–Lagrange equations are $\Delta = \Delta^a \, du_a \wedge \nu$, where

$$\begin{aligned}
\Delta^0 &= u_{0,xx} + u_{0,yy} + u_{0,zz} - u_{1,tx} - u_{2,ty} - u_{3,tz}, \\
\Delta^1 &= -u_{0,tx} + u_{1,tt} - u_{1,yy} - u_{1,zz} + u_{2,xy} + u_{3,xz}, \\
\Delta^2 &= -u_{0,ty} + u_{1,xy} + u_{2,tt} - u_{2,xx} - u_{2,zz} + u_{3,yz}, \\
\Delta^3 &= -u_{0,tz} + u_{1,xz} + u_{3,yz} + u_{3,tt} - u_{3,xx} - u_{3,yy}.
\end{aligned}$$

We now consider two nontransverse groups actions, find the reduced equations and investigate the principle of symmetric criticality.

EXAMPLE 3.1. $\Gamma = \{\varepsilon_{kij} x^i \partial_{x^j} + \varepsilon^i_{kj} u_i \partial_{u_j}\}$.

The Lie algebra Γ is obtained from the infinitesimal generators of SO(3) acting in the standard way on E. At the point $\mathbf{x}_0 = (t_0, x_0^k)$ we have the isotropy vector-field

$$Z = x_0^k(\varepsilon_{kij} x^i \partial_{x^j} + \varepsilon^i_{kj} u_i \partial_{u_j}) \in \Gamma_{\mathbf{x}_0}. \tag{3.2}$$

The subbundle $\kappa(E) \subset E$ at the point $\mathbf{x}_0$ is given by the vanishing of $x_0^k \varepsilon^i_{kj} u_i$ (the cross product of (x_0^i) and (u_i)). Therefore $u_0 = v$, $u_i = w x_0^i$, with v, w being fibre coordinates on $\kappa_{\mathbf{x}_0}(E)$. The inclusion $\kappa(E) \to E$ is given by

$$\iota^k(t, x, y, z; v, w) \to (t, x, y, z; v, wx, wy, wz).$$

The functions t and $r = \sqrt{x^2+y^2+z^2}$ are invariants on M so that in local coordinates the kinematic reduction diagram is

$$\begin{array}{ccccc} (t,r;v,w) & \xleftarrow{\mathbf{q}_\kappa} & (t,x^i;v,w) & \xrightarrow{\iota} & (t,x^i;\,u_a) \\ \tilde\pi\downarrow & & \pi\downarrow & & \downarrow\pi \\ (t,r) & \xleftarrow{\mathbf{q}_M} & (t,x^i) & \xrightarrow[id]{} & (t,x^i). \end{array} \tag{3.3}$$

The invariant sections are now easily determined to be

$$u_0 = v(t,r), \quad u_i = w(t,r)x^i.$$

In order to determine the reduced equations we compute the fibres of $\mathcal{D}_{\mathrm{Inv}}$ by using Lemma 2.4. At the point $p_0 = (t_0, x_0, y_0, z_0; v_0, w_0x_0, w_0y_0, w_0z_0) \in \kappa(E)$ the vector-field Z in (3.2) lies in Γ_{p_0}. The (infinitesimal) vertical linear isotropy representation in the coordinate basis ∂_{u_a} is

$$\rho_{p_0}(Z) = \frac{\partial}{\partial u_l} x_0^k \varepsilon_{kj}^{\ i} u_i = \begin{pmatrix} 0 & 0 & 0 & 0 \\ 0 & 0 & -z_0 & y_0 \\ 0 & z_0 & 0 & -x_0 \\ 0 & -y_0 & x_0 & 0 \end{pmatrix}.$$

The fibres of $\kappa(\mathcal{D}_{\mathrm{Inv}})$ are two dimensional and the inclusion into $\mathcal{D}_{\mathrm{Inv}}$ is easily found by computing the null space of the transpose of the matrix above.

The sections

$$S_0 = du_0 \wedge \nu \quad \text{and} \quad S_1 = \sum_{i=1}^{3} x^i\, du_i \wedge \nu,$$

of $\mathcal{D}$ are invariant. The reduced operator is easily obtained from

$$\Delta_{\mathrm{Inv}} = \Big(v_{rr} + \frac{2}{r}v_r - rw_{tr} - 3w_t\Big) S_0 + \Big(w_{tt} - \frac{1}{r}v_{tr}\Big) S_1.$$

To compute the Palais condition (2.8) we have

$$\big[\mathrm{Vert}_p\,\kappa(E)\big]^{G_x} = \begin{pmatrix} a \\ bx_0 \\ by_0 \\ bz_0 \end{pmatrix}, \quad \big[\mathrm{Vert}_p^*\,\kappa(E)\big]^{G_x} = (a, bx_0, by_0, bz_0) \qquad a, b \in \mathbb{R},$$

and the Palais condition (2.8) is satisfied. The cohomology condition (2.4) is also easily checked, and the principle of symmetric criticality holds. (Note that this also follows from the symmetry group being compact.)

For completeness we compute the reduced Lagrangian. The Lagrangian in (3.1) restricted to $\mathrm{Inv}^k(E)$ is

$$\lambda\big|_{\mathrm{Inv}} = -\frac{1}{2}(rw_t - v_r)^2\nu,$$

and evaluating $\lambda\big|_{\mathrm{Inv}}$ on the chain in (2.1) we obtain the reduced Lagrangian

$$\tilde\lambda = \lambda\big|_{\mathrm{Inv}}(\chi) = \frac{1}{2}(rw_t - v_r)^2 r^2\, dt \wedge dr.$$

The reduced equations agree with the Euler–Lagrange equations of the reduced Lagrangian.

EXAMPLE 3.2. $\Gamma = \{V_1 = \partial_y, V_2 = \partial_t - \partial_z, V_3 = (t+z)\partial_y + y(\partial_t - \partial_z) + (u_3 - u_0)\partial_{u_2} - u_2(\partial_{u_0} + \partial_{u_3})\}$.

At the point $\mathbf{x}_0 = (t_0, x_0^k)$ the isotropy vector-field is

$$Z = V_3 - (t_0 + z_0)V_1 - y_0 V_2 = (u_3 - u_0)\partial_{u_2} - u_2(\partial_{u_0} + \partial_{u_3}) \in \Gamma_{\mathbf{x}_0}. \tag{3.4}$$

The subbundle $\kappa(E) \subset E$ at the point $\mathbf{x}_0$ is given by $u_3 = u_0$, $u_2 = 0$ and the inclusion $\kappa(E) \to E$ is

$$\iota(t, x, y, z; v, w) \to (t, x, y, z; v, w, 0, v).$$

The functions $r = t + z$ and x are invariants and the kinematic reduction diagram is

$$\begin{array}{ccccc} (r, x; v, w) & \xleftarrow{\mathbf{q}_\kappa} & (t, x^i; v, w) & \xrightarrow{\iota} & (t, x^i; u_a) \\ \tilde{\pi}\big\downarrow & & \pi\big\downarrow & & \big\downarrow\pi \\ (r, x) & \xleftarrow{\mathbf{q}_M} & (t, x^i) & \xrightarrow[id]{} & (t, x^i). \end{array} \tag{3.5}$$

The invariant sections are then

$$u_0 = v(r, x), \quad u_1 = w(r, x), \quad u_2 = 0, \quad u_3 = v(r, x),$$

and the reduced operator is

$$\tilde{\Delta} = (v_{xx} - w_{rx})(du_0 - du_3) \wedge \nu.$$

The infinitesimal linear isotropy representation is

$$\rho_{p_0}(Z) = \begin{pmatrix} 0 & 0 & -1 & 0 \\ 0 & 0 & 0 & 0 \\ -1 & 0 & 0 & 1 \\ 0 & 0 & -1 & 0 \end{pmatrix}.$$

To compute the Palais Condition (2.8) we have

$$\left[\operatorname{Vert}_p \kappa(E)\right]^{G_x} = \begin{pmatrix} a \\ b \\ 0 \\ a \end{pmatrix}, \quad \left[\operatorname{Vert}_p \kappa(E)^*\right]^{G_x} = (a, b, 0, -a) \qquad a, b \in \mathbb{R},$$

and

$$\left[\operatorname{Vert}_p^*(E)\right]^{G_x} \cap \left(\left[\operatorname{Vert}_p(E)\right]^{G_x}\right)^{\perp} = \{(a, 0, 0, -a), a \in \mathbb{R}\} \neq 0,$$

so the principle of symmetric criticality fails. In fact the Lagrangian (3.1) restricted to the invariant sections is easily computed to be

$$\lambda|_{\text{Inv}} = 0.$$

Note however that the Lie algebra cohomology condition (2.4) is satisfied so that the Palais condition is independent of the cohomology condition (2.4).

References

1. I. M. Anderson and M. E. Fels, *Symmetry reduction of variational bicomplexes and the principle of symmetry criticality*, Amer. J. Math. **112** (1997), 609–670.
2. ———, *Transverse group actions on bundles*, Topology Appl. (2001) (to appear).
3. I. M. Anderson, M. E. Fels, and C. G. Torre, *Group invariant solutions without transversality*, Comm. Math. Phys. **212**, 653–686.
4. ———, *Symmetry reduction of differential equations*, in progress.
5. G. W. Bluman and S. Kumei, *Symmetries and differential equations*, Appl. Math. Sci., vol. 81, Springer-Verlag, New York–Berlin, 1989.
6. J. Brúning and E. Heintze, *Representations of compact Lie groups and elliptic operators*, Invent. Math. **50** (1979), 169–203.
7. M. E. Fels and P. J. Olver, *On relative invariants*, Math. Ann. **308** (1997), 609–670.
8. P. J. Olver, *Applications of Lie groups to differential equations*, Graduate Texts in Mathematics, 2nd ed., vol. 107, Springer-Verlag, New York–Berlin, 1986.
9. L. V. Ovsiannikov, *Group analysis of differential equations*, Academic Press, New York–London, 1982.
10. R. S. Palais, *The principle of symmetric criticality*, Comm. Math. Phys. **69** (1979), 19–30.
11. P. Winternitz, *Group theory and exact solutions of partially integrable equations*, Partially Integrable Evolution Equations in Physics (Les Houches, 1989) (R. Conte and N. Boccara, eds.), NATO Adv. Sci. Inst. Ser. C: Math. Phys. Sci., vol. 310, Kluwer Acad. Publ., Dordrecht, 1990, pp. 515–567.
12. ———, *Lie groups and solutions of nonlinear partial differential equations*, Integrable Systems, Quantum Groups, and Quantum Field Theories (Salamanca, 1992) (A. Ibort and M. A. Rodriguez, eds.), NATO Adv. Sci. Inst. Ser. C: Math. Phys. Sci., vol. 409, Kluwer Acad. Publ., Dordrecht, 1993, pp. 429–495.

Department of Mathematics, Utah State University, Logan, Utah 84322, USA.
Email address, Ian M. Anderson: anderson@sunfs.math.usu.edu
Email address, Mark E. Fels: fels@sunfs.math.usu.edu
Email address, Charles G. Torre: torre@cc.usu.edu

Centre de Recherches Mathématiques
CRM Proceedings and Lecture Notes
Volume **29**, 2001

Multi-Component Matrix KP Hierarchies as Symmetry-Enhanced Scalar KP Hierarchies and Their Darboux–Bäcklund Solutions

H. Aratyn, E. Nissimov, and S. Pacheva

ABSTRACT. We show that any multi-component matrix KP hierarchy is equivalent to the standard one-component (scalar) KP hierarchy endowed with a special infinite set of Abelian additional symmetries, generated by squared eigenfunction potentials. This allows to employ a special version of the familiar Darboux–Bäcklund transformation techniques within the ordinary scalar KP hierarchy in the Sato formulation for a systematic derivation of explicit *multiple*-Wronskian tau-function solutions of all multi-component matrix KP hierarchies.

1. Introduction. Background on the KP Hierarchy and Ghost Symmetries

Multi-component generalizations of the Kadomtsev–Petviashvili (KP) hierarchy of integrable nonlinear soliton equations [**5, 18, 23**] attract a lot of interest both from physical and mathematical point of view. They are known to contain such physically relevant nonlinear integrable systems as the Davey–Stewartson, two-dimensional Toda lattice and three-wave resonant interaction ones [**6, 24**]. On the other hand, multi-component KP hierarchies turn out to be intimately connected to classical geometry of conjugate nets and the classification problem of Hamiltonian systems of hydrodynamical type [**9, 11**].

There exist several equivalent formulations of multi-component KP hierarchies: matrix pseudo-differential operator (Sato) formulation; tau-function approach via matrix Hirota bilinear identities; multi-component free fermion formulation. Here we will offer yet another approach. Namely, we will show that any multi-component $N \times N$ matrix KP hierarchy [**5,18,23**] is equivalent to the standard one-component (scalar) KP hierarchy endowed with $N-1$ copies of mutually commuting infinite-dimensional algebras of Abelian additional ("ghost") symmetries, generated by squared eigenfunction potentials of the initial hierarchy (see definition in (1.7) below). The latter "ghost" symmetry enhanced scalar KP hierarchy will be called

2000 *Mathematics Subject Classification*. Primary: 37K10; Secondary: 35Q55.

The article is supported by NSF grant INT-9724747.
The authors gratefully acknowledge support by the NSF grant INT-9724747.
This is the final form of the paper.

"multiple-KP hierarchy". The principal advantage of the multiple-KP hierarchy formulation over the standard Sato matrix pseudo-differential operator formulation, lies in the fact that the former allows to use a special version of the well-known Darboux–Bäcklund (DB) transformation techniques within the ordinary scalar KP hierarchy in the Sato formulation for a systematic derivation of soliton-like *multiple-*Wronskian tau-function solutions of the multi-component KP hierarchies (Section 5 below).

The starting point of our presentation is the pseudo-differential Lax operator $\mathcal{L}$ of the scalar KP hierarchy obeying KP evolution equations w.r.t. the multi-time $(t) \equiv (t_1 \equiv x, t_2, \ldots)$ (for notations and review, see [**7, 8**]):

$$\mathcal{L} = D + \sum_{i=1}^{\infty} u_i D^{-i}; \quad \frac{\partial \mathcal{L}}{\partial t_l} = \left[(\mathcal{L}^l)_+, \mathcal{L}\right], \quad l = 1, 2, \ldots \tag{1.1}$$

The symbol D stands for the differential operator $\partial/\partial x$, whereas $\partial \equiv \partial_x$ will denote derivative of a function. The subscripts $(\pm)$ indicate the purely differential/pseudo-differential part of the corresponding operator. Equivalently, one can represent Eq. (1.1) in terms of the dressing operator W whose pseudo-differential series are expressed in terms of the so called tau-function $\tau(t)$:

$$\mathcal{L} = W D W^{-1}, \quad \frac{\partial W}{\partial t_l} = -(\mathcal{L}^l)_- W, \quad W = \sum_{n=0}^{\infty} \frac{p_n(-[\partial])\tau(t)}{\tau(t)} D^{-n}, \tag{1.2}$$

with the following notation $[y] \equiv (y_1, y_2/2, y_3/3, \ldots)$ for any multi-variable $(y) \equiv (y_1, y_2, y_3, \ldots)$, in particular $(\partial) \equiv (\partial/\partial t_1, \partial/\partial t_2, \ldots)$, and with $p_k(y)$ being the Schur polynomials. The tau-function is related to the Lax operator as (below "Res" denotes the coefficient in front of D^{-1}):

$$\partial_x \frac{\partial}{\partial t_l} \ln \tau(t) = \operatorname{Res} \mathcal{L}^l, \quad \tau(t) \to \tau(t) e^{\sum_{l=1}^{\infty} c_l t_l}. \tag{1.3}$$

The second relation (1.3) tells that $\tau(t)$ is defined up to an exponential linear w.r.t. (t).

In the present approach a basic notion is that of (adjoint) eigenfunctions (EF's) $\Phi(t)$, $\Psi(t)$ of the scalar KP hierarchy satisfying:

$$\frac{\partial \Phi}{\partial t_k} = \mathcal{L}^k_+(\Phi); \quad \frac{\partial \Psi}{\partial t_k} = -(\mathcal{L}^*)^k_+(\Psi). \tag{1.4}$$

Throughout this paper we will rely on an important tool provided by the spectral representation of EF's [**2**]. The spectral representation is equivalent to the following statement: Φ and Ψ are (adjoint) EF's if and only if they obey the integral representation:

$$\Phi(t) = \int dz \frac{e^{\xi(t-t',z)}}{z} \frac{\tau(t-[z^{-1}])\tau(t'+[z^{-1}])}{\tau(t)\tau(t')} \Phi(t'+[z^{-1}]), \tag{1.5}$$

$$\Psi(t) = \int dz \frac{e^{\xi(t'-t,z)}}{z} \frac{\tau(t+[z^{-1}])\tau(t'-[z^{-1}])}{\tau(t)\tau(t')} \Psi(t'-[z^{-1}]), \tag{1.6}$$

where $\int dz$ denotes normalized contour integral around origin.

Further crucial notion to be employed in the present construction is the so called squared eigenfunction potential (SEP) $S(\Phi, \Psi)$ of arbitrary pair of an EF Φ

and an adjoint EF Ψ [**19, 20**] (c.f. also [**2**]):

$$S(\Phi,\Psi) = \partial^{-1}(\Phi,\Psi),$$

$$\frac{\partial}{\partial t_n} S\big(\Phi(t),\Psi(t)\big) = \mathrm{Res}\big(D^{-1}\Psi(\mathcal{L}^n)_+\Phi D^{-1}\big), \quad n = 1,2,\ldots \tag{1.7}$$

The flow equations above fix the ambiguity in applying the inverse derivative ∂^{-1} in the SEP definition up to an overall trivial constant. In what follows ∂^{-1} will appear always in the form of SEP's (first Eq. (1.7)).

Consider now an infinite system of independent (adjoint) EF's $\{\Phi_j,\Psi_j\}_{j=1}^{\infty}$ of the standard one-component KP hierarchy Lax operator $\mathcal{L}$ and define the following infinite set of the additional "ghost" symmetry flows [**1**]:

$$\frac{\partial}{\partial \bar{t}_s}\mathcal{L} = [\mathcal{M}_s,\mathcal{L}], \quad \mathcal{M}_s = \sum_{j=1}^{s}\Phi_{s-j+1}D^{-1}\Psi_j, \tag{1.8}$$

$$\frac{\partial}{\partial \bar{t}_s}\Phi_k = \sum_{j=1}^{s}\Phi_{s-j+1}\partial^{-1}(\Psi_j\Phi_k) - \Phi_{k+s},$$
$$\frac{\partial}{\partial \bar{t}_s}\Psi_k = \sum_{j=1}^{s}\Psi_j\partial^{-1}(\Phi_{s-j+1}\Psi_k) + \Psi_{k+s}, \tag{1.9}$$

$$\frac{\partial}{\partial \bar{t}_s}F = \sum_{j=1}^{s}\Phi_{s-j+1}\partial^{-1}(F\Psi_j); \quad \frac{\partial}{\partial \bar{t}_s}F^* = \sum_{j=1}^{s}\Psi_j\partial^{-1}(\Phi_{s-j+1}F^*), \tag{1.10}$$

where $s, k = 1, 2, \ldots$, and F, F^* denote generic (adjoint) EF's which do not belong to the "ghost" symmetry generating set $\{\Phi_j,\Psi_j\}_{j=1}^{\infty}$.

Transformations in (1.8)–(1.10) contain terms in $a\partial^{-1}b$, whose study was initiated in [**21, 22**] and exploited in [**10**] in which inhomogeneous terms in symmetry transformations appeared in the particular case of constrained KP hierarchies.

It is easy to show that the "ghost" symmetry flows $\partial/\partial\bar{t}_s$ from Eqs. (1.8)–(1.10) commute, namely, that the ∂-pseudo-differential operators $\mathcal{M}_s$ (1.8) satisfy the zero-curvature equations: $\partial\mathcal{M}_r/\partial\bar{t}_s - \partial\mathcal{M}_s/\partial\bar{t}_r - [\mathcal{M}_s,\mathcal{M}_r] = 0$.

Let us consider the (adjoint)-EF Eqs. (1.4) for $k = 2$ and $\Phi = \Phi_1$, $\Psi = \Psi_1$ (the pair generating the lowest $s = 1$ "ghost" symmetry flows (1.8)–(1.10)) together with their $s = 2$ "ghost" symmetry flow Eqs. (1.9) (i.e., $s = 2$, $k = 1$). The latter system can be written in the form:

$$\frac{\partial}{\partial t_2}\Phi_1 = (\partial^2 + 2u_1)\Phi_1, \qquad \frac{\partial}{\partial \bar{t}_2}\Phi_1 = \big[-\bar{\partial}^2 + 2\bar{\partial}\big(\partial^{-1}(\Phi_1\Psi_1)\big)\big]\Phi_1, \tag{1.11}$$

$$\frac{\partial}{\partial t_2}\Psi_1 = -(\partial^2 + 2u_1)\Psi_1, \quad \frac{\partial}{\partial \bar{t}_2}\Psi_1 = \big[\bar{\partial}^2 - 2\bar{\partial}\big(\partial^{-1}(\Phi_1\Psi_1)\big)\big]\Psi_1, \tag{1.12}$$

where $\bar{\partial} \equiv \partial/\partial\bar{t}_1$. Introducing new notations: $Q \equiv u_1 - 2(\Phi_1\Psi_1) - \bar{\partial}\big(\partial^{-1}(\Phi_1\Psi_1)\big)$; $-iT = t_2 - \bar{t}_2$, $X = t_1 + \bar{t}_1$, $Y = t_1 - \bar{t}_1$, the system (1.11)–(1.12) acquires the following form:

$$i\partial_T\Phi_1 = (\partial_X^2 + \partial_Y^2)\Phi_1 + 2(\Phi_1\Psi_1)\Phi_1 + Q\Phi_1,$$
$$(\partial_X^2 - \partial_Y^2)Q + 4\partial_X^2(\Phi_1\Psi_1) = 0. \tag{1.13}$$

$$-i\partial_T\Psi_1 = (\partial_X^2 + \partial_Y^2)\Psi_1 + 2(\Phi_1\Psi_1)\Psi_1 + Q\Psi_1, \tag{1.14}$$

which is nothing but the standard Davey–Stewartson (DS) system [**6, 24**][1]. Thus, we succeeded to express solutions of DS system in terms of a pair of (adjoint) EF's of ordinary one-component KP hierarchy supplemented with two additional "ghost" symmetry flows ($\partial/\partial\bar{t}_1$ and $\partial/\partial\bar{t}_2$ in the notations of (1.8)–(1.9)).

Consider now the τ-function of $\mathcal{L}$ (1.2) and let us act with $(\partial/\partial\bar{t}_s)$ on both sides of (1.3) obtaining $(\partial/\partial\bar{t}_s)\ln\tau = -\sum_{j=1}^{s}\partial^{-1}(\Phi_{s-j+1}\Psi_j)$ where we used (1.8) as well as the t_r-flow eqs. $(\partial/\partial t_r)\,\mathrm{der}\,\mathcal{M}_s = [\mathcal{L}_+^r, \mathcal{M}_s]_-$. With the help of the well-known recurrence relation for the Schur polynomials: $sp_s\big(-[\bar{\partial}]\big) = \sum_{k=1}^{s}(-\partial/\partial\bar{t}_k)$ $p_{s-k}\big(-[\bar{\partial}]\big)$, we find the following important identities [**1, 3**] relating the τ-function with the "ghost" symmetry generating (adjoint) EF's (here s, k, $j = 1, 2, \ldots$):

$$\frac{p_s(-[\bar{\partial}])(\Phi_1\tau)}{\tau} = \Phi_{s+1}, \quad \frac{p_s([\bar{\partial}])(\Psi_1\tau)}{\tau} = \Psi_{s+1},$$
$$\partial^{-1}(\Phi_k\Psi_j) = \sum_{l=0}^{j-1}\frac{p_{k+l}(-[\bar{\partial}])p_{j-l+1}([\bar{\partial}])\tau}{\tau}. \tag{1.15}$$

2. Double-KP System and Its Equivalence to Two-Component KP Hierarchy

In Refs. [**1, 3**] we have shown that the "ghost" symmetry flows from Eqs. (1.8)–(1.10) admit their own Lax representation in terms of a $\bar{D} \equiv \partial/\partial\bar{x} \equiv \partial/\partial\bar{t}_1$ pseudo-differential Lax operator $\bar{\mathcal{L}}$ w.r.t. multi-time $(\bar{t}) \equiv (\bar{t}_1 \equiv \bar{x}, \bar{t}_2, \ldots)$:

$$\bar{\mathcal{L}} \equiv \overline{W}\bar{D}\overline{W}^{-1} = \bar{D} + \sum_{i=1}^{\infty}\bar{u}_i\bar{D}^{-i}, \quad \frac{\partial\bar{\mathcal{L}}}{\partial\bar{t}_s} = [\bar{\mathcal{L}}_+^s, \bar{\mathcal{L}}], \tag{2.1}$$

$$\overline{W} = 1 + \sum_{j=1}^{\infty}\frac{p_j(-[\bar{\partial}])\tau(t,\bar{t})}{\tau(t,\bar{t})}\bar{D}^{-j}, \quad \frac{p_j(-[\bar{\partial}])\tau}{\tau} = \partial^{-1}(\Phi_j\Psi_1), \tag{2.2}$$

where Φ_j, Ψ_1 are the "ghost" symmetry generating (adjoint) EF's of the original KP hierarchy (1.1), which we denote as KP_1. Accordingly, we will denote the "ghost" KP system (2.1)–(2.2) as KP_2. Let us stress that the tau-function $\tau = \tau(t,\bar{t})$ of the KP_1 hierarchy (1.1)–(1.2), with time evolution parameters (t) and $(\bar{t})$ = fixed, is simultaneously the tau-function of KP_2 hierarchy (2.1)–(2.2) with time evolution parameters $(\bar{t})$ and (t) = fixed.

Next, we have found an infinite system of (adjoint) EF's $\{\bar{\Phi}_k, \bar{\Psi}_k\}_{k=1}^{\infty}$ of KP_2 system (i.e., obeying Eqs. (1.4) with all quantities replaced with the "barred" ones):

$$\bar{\Phi}_k = \frac{p_{k-1}(-[\partial])(\Psi_1\tau)}{\tau}, \quad \bar{\Psi}_k = \frac{p_{k-1}([\partial])(\Phi_1\tau)}{\tau}, \tag{2.3}$$

which generate "ghost" symmetry flows for KP_2 (2.1) analogous to (1.8)–(1.10) such that the corresponding "ghost" symmetry flow parameters coincide with the isospectral flow parameters (t) of the original KP_1 system (1.1). In particular, we note from Eqs. (2.3) that $\bar{\Phi}_1 = \Psi_1$ and $\bar{\Psi}_1 = \Phi_1$. Also, for any generic (adjoint) EF's F, F^* of KP_1 the SEP functions:

$$\overline{F} \equiv \partial^{-1}(F\Psi_1), \quad \overline{F}^* \equiv \partial^{-1}(\Phi_1 F^*), \tag{2.4}$$

are, respectively, an EF and adjoint EF of KP_2 [**1, 3**].

---

[1] The fact that DS system is contained within the "ghost" symmetry enhanced scalar KP hierarchy (1.1)–(1.2), (1.8)–(1.9) was first pointed out in [**17**].

Both KP_1 (original KP hierarchy) together with KP_2 ("ghost" symmetry flows' KP hierarchy) form a new larger hierarchy called *double*-KP [**1, 3**] possessing the property of "duality" symmetry, i.e., symmetry under interchanging the roles of KP_1 and KP_2. It is defined as follows:

$$(t) \equiv (\overset{(1)}{t}), \quad (\bar{t}) \equiv (\overset{(2)}{t}), \quad (\overset{(\alpha)}{t}) = (\overset{(\alpha)}{t_1}, \overset{(\alpha)}{t_2}, \ldots), \quad \bar{\mathcal{L}} \equiv \overset{(1)}{\mathcal{L}}, \quad \bar{\mathcal{L}} \equiv \overset{(2)}{\mathcal{L}}. \tag{2.5}$$

$$\overset{(\alpha)}{\mathcal{L}} = \overset{(\alpha)}{W}\overset{(\alpha)}{D}\overset{(\alpha)}{W}^{-1}, \quad \overset{(\alpha)}{W} = \sum_{j=0}^{\infty} \frac{p_j(-[\overset{(\alpha)}{\partial}])\tau}{\tau} \overset{(\alpha)}{D}^{-j}, \quad \overset{(\alpha)}{D} \equiv \partial/\partial \overset{(\alpha)}{t_1}. \tag{2.6}$$

$$\begin{aligned} \Phi_j(t,\bar{t}) &\equiv \overset{(12)}{\Phi_j}(\overset{(1)}{t}, \overset{(2)}{t}), & \Psi_j(t,\bar{t}) &\equiv \overset{(12)}{\Psi_j}(\overset{(1)}{t}, \overset{(2)}{t}), \\ \bar{\Phi}_j(t,\bar{t}) &\equiv \overset{(21)}{\Phi_j}(\overset{(1)}{t}, \overset{(2)}{t}), & \bar{\Psi}_j(t,\bar{t}) &\equiv \overset{(21)}{\Psi_j}(\overset{(1)}{t}, \overset{(2)}{t}). \end{aligned} \tag{2.7}$$

$$\overset{(\alpha\beta)}{\Phi_j} = \varepsilon_{\alpha\beta} \frac{p_{j-1}(-[\overset{(\beta)}{\partial}])\tau_{\alpha\beta}}{\tau}, \quad \overset{(\alpha\beta)}{\Psi_j} = \varepsilon_{\beta\alpha} \frac{p_{j-1}([\overset{(\beta)}{\partial}])\tau_{\beta\alpha}}{\tau}, \tag{2.8}$$
$$\varepsilon_{\alpha\beta} = \pm 1 \quad \text{for } \alpha \leq \beta, \ \alpha > \beta,$$

where in (2.8) we have introduced new tau-functions:

$$\tau_{\alpha\beta} = \varepsilon_{\alpha\beta}\tau \overset{(\alpha\beta)}{\Phi_1} = \varepsilon_{\alpha\beta}\tau \overset{(\beta\alpha)}{\Psi_1}. \tag{2.9}$$

Here the indices (α, β) are taking values $\alpha,\ \beta = 1,\ 2,\ \alpha \neq \beta$, and Eqs. (1.15), (2.3) have been taken into account.

Accordingly, Eqs. (1.1), (2.1), (1.4), (1.8)–(1.10) and their "duals" (for $\bar{\Phi}_k, \bar{\Psi}_k$) can be rewritten in a manifestly "duality"-symmetric form:

$$\begin{aligned} &\overset{(\alpha)}{\partial_s}\overset{(\alpha)}{\mathcal{L}} = \left[(\overset{(\alpha)}{\mathcal{L}})^s_+, \overset{(\alpha)}{\mathcal{L}}\right], \quad \overset{(\beta)}{\partial_s}\overset{(\alpha)}{\mathcal{L}} = \left[\overset{(\alpha\beta)}{\mathcal{M}_s}, \overset{(\alpha)}{\mathcal{L}}\right], \\ &\overset{(\alpha)}{\partial_s} \equiv \frac{\partial}{\partial \overset{(\alpha)}{t_s}}, \quad \overset{(\alpha\beta)}{\mathcal{M}_s} \equiv \sum_{j=1}^{s} \overset{(\alpha\beta)}{\Phi}_{s-j+1}\overset{(\alpha)}{D}^{-1}\overset{(\alpha\beta)}{\Psi_j}. \end{aligned} \tag{2.10}$$

$$\begin{aligned} &\overset{(\alpha)}{\partial_s}\overset{(\alpha\beta)}{\Phi_k} = (\overset{(\alpha)}{\mathcal{L}})^s_+(\overset{(\alpha\beta)}{\Phi_k}), \quad \overset{(\alpha)}{\partial_s}\overset{(\alpha)}{F} = (\overset{(\alpha)}{\mathcal{L}})^s_+(\overset{(\alpha)}{F}), \\ &\overset{(\alpha)}{\partial_s}\overset{(\alpha\beta)}{\Psi_k} = -(\overset{(\alpha)}{\mathcal{L}}{}^*)^s_+(\overset{(\alpha\beta)}{\Psi_k}), \quad \overset{(\alpha)}{\partial_s}\overset{(\alpha)}{F}{}^* = -(\overset{(\alpha)}{\mathcal{L}}{}^*)^s_+(\overset{(\alpha)}{F}{}^*). \end{aligned} \tag{2.11}$$

$$\begin{aligned} &\overset{(\beta)}{\partial_s}\overset{(\alpha\beta)}{\Phi_k} = \sum_{j=1}^{s} \overset{(\alpha\beta)}{\Phi}_{s-j+1}\overset{(\alpha)}{\partial}^{-1}(\overset{(\alpha\beta)}{\Psi_j}\overset{(\alpha\beta)}{\Phi_k}) - \overset{(\alpha\beta)}{\Phi}_{k+s}, \\ &\overset{(\beta)}{\partial_s}\overset{(\alpha\beta)}{\Psi_k} = \sum_{j=1}^{s} \overset{(\alpha\beta)}{\Psi_j}\overset{(\alpha)}{\partial}^{-1}(\overset{(\alpha\beta)}{\Phi}_{s-j+1}\overset{(\alpha\beta)}{\Psi_k}) + \overset{(\alpha\beta)}{\Psi}_{k+s}. \end{aligned} \tag{2.12}$$

$$\overset{(\beta)}{\partial_s}\overset{(\alpha)}{F} = \sum_{j=1}^{s} \overset{(\alpha\beta)}{\Phi}_{s-j+1}\overset{(\alpha)}{\partial}^{-1}(\overset{(\alpha\beta)}{\Psi_j}\overset{(\alpha)}{F}), \quad \overset{(\beta)}{\partial_s}\overset{(\alpha)}{F}{}^* = \sum_{j=1}^{s} \overset{(\alpha\beta)}{\Psi_j}\overset{(\alpha)}{\partial}^{-1}(\overset{(\alpha\beta)}{\Phi}_{s-j+1}\overset{(\alpha)}{F}{}^*), \tag{2.13}$$

where again $\alpha,\ \beta = 1,\ 2,\ \alpha \neq \beta$, $\overset{(\alpha)}{\partial}{}^{-1} \equiv (\overset{(\alpha)}{\partial_1})^{-1}$ and $s = 1, 2, \ldots$ In (2.13) and (2.11) $\overset{(\alpha)}{F}$, $\overset{(\alpha)}{F}{}^*$ denote generic (adjoint) EF's of $\overset{(\alpha)}{\mathcal{L}}$, i.e., such that they do not belong to the sets $\{\overset{(\alpha\beta)}{\Phi_k}, \overset{(\alpha\beta)}{\Psi_k}\}$.

In [**3**] we have shown the equivalence of double-KP hierarchy with the two-component (matrix) Sato KP hierarchy [**5, 18, 23**], originally formulated within the matrix pseudo-differential Lax formalism, which can be equivalently described by three tau-functions τ_{11}, τ_{12}, τ_{21} depending on two sets of multi-time variables (t), $(\bar{t})$ and obeying the 2×2 matrix Hirota bilinear identities (see Eq. (3.1) below). The proof proceeds by identifying two-component KP tau-functions with the tau-functions of double-KP hierarchy $\tau_{11}=\tau$, $\tau_{12}=\tau\Phi_1$, $\tau_{21}=-\tau\Psi_1$ as in (2.6)–(2.9).

3. Generalization to Multi-Component KP Hierarchies

Let us turn our attention to arbitrary N-component matrix KP hierarchies [**5, 18, 23**] $(N\geq 2)$. They can be equivalently characterized by the set of Hirota bilinear identities:

$$\sum_{\gamma=1}^{N}\varepsilon_{\alpha\gamma}\varepsilon_{\beta\gamma}\int dz\, z^{\delta_{\alpha\gamma}+\delta_{\beta\gamma}-2}e^{\xi(\overset{(\gamma)}{t}-\overset{(\gamma)}{t'},z)}\tau_{\alpha\gamma}(\ldots,\overset{(\gamma)}{t}-[z^{-1}],\ldots)\,\tau_{\gamma\beta}(\ldots,\overset{(\gamma)}{t'}+[z^{-1}],\ldots)=0, \tag{3.1}$$

for $N(N-1)+1$ tau-functions $\tau_{\alpha\alpha}\equiv\tau$ and $\tau_{\alpha\beta}(\alpha\neq\beta)$, where now the indices $\alpha,\,\beta,\,\gamma=1,\ldots,N$, $\delta_{\alpha\beta}$ are Kronecker symbols and $\varepsilon_{\alpha\beta}$ are the same as in (2.7). Also, we are using the standard notation $\xi(t,z)\equiv\sum_{l=1}^{\infty}t_l z^l$. Let us recall that (3.1) contain the following interesting systems of nonlinear equations:

$$\overset{(\alpha)}{\partial_1}\overset{(\beta)}{\partial_1}\ln\tau=\frac{\tau_{\alpha\beta}}{\tau}\frac{\tau_{\beta\alpha}}{\tau},\quad \overset{(\gamma)}{\partial_1}\left(\frac{\tau_{\alpha\beta}}{\tau}\right)=\varepsilon_{\alpha\beta}\varepsilon_{\alpha\gamma}\varepsilon_{\gamma\beta}\frac{\tau_{\alpha\gamma}}{\tau}\frac{\tau_{\gamma\beta}}{\tau},\qquad \alpha\neq\beta\neq\gamma, \tag{3.2}$$

the second one being the so called N'-wave system $(N'=N(N-1)/2)$.

Our main statement is that N-component KP hierarchy defined by (3.1) is equivalent to the *multiple*-KP hierarchy defined by Eqs. (2.6), (2.8)–(2.13) where now the indices α,β take values $\alpha,\,\beta=1,\ldots,N$. In other words, this multiple-KP hierarchy consists of N ordinary one-component KP hierarchies KP_α $(\alpha=1,\ldots,N)$ given by Lax operators $\overset{(\alpha)}{\mathcal{L}}$ (2.6) in different spaces and generating isospectral flows $\overset{(\alpha)}{\partial_s}$ w.r.t. different sets of evolution parameters $(\overset{(\alpha)}{t})$, such that the flows $\overset{(\alpha)}{\partial_s}$ act on the rest of KP subsystems KP_β $(\beta\neq\alpha)$ as "ghost" symmetry flows. In particular, Eqs. (2.13) apply also for $\overset{(\alpha)}{F}=\overset{(\alpha\gamma)}{\Phi_k}$ and $\overset{(\alpha)}{F^*}=\overset{(\alpha\gamma)}{\Psi_k}$ with $\gamma\neq\beta\neq\alpha$:

$$\overset{(\beta)}{\partial_s}\overset{(\alpha\gamma)}{\Phi_k}=\sum_{j=1}^{s}\overset{(\alpha\beta)}{\Phi}_{s-j+1}\overset{(\alpha)}{\partial}{}^{-1}(\overset{(\alpha\beta)}{\Psi_j}\overset{(\alpha\gamma)}{\Phi_k}),\qquad \overset{(\beta)}{\partial_s}\overset{(\alpha\gamma)}{\Psi_k}=\sum_{j=1}^{s}\overset{(\alpha\beta)}{\Psi_j}\overset{(\alpha)}{\partial}{}^{-1}(\overset{(\alpha\beta)}{\Phi}_{s-j+1}\overset{(\alpha\gamma)}{\Psi_k}), \tag{3.3}$$

since $\overset{(\alpha\gamma)}{\Phi_k}$, $\overset{(\alpha\gamma)}{\Psi_k}$ are generic (adjoint) EF's of $\overset{(\alpha)}{\mathcal{L}}$ w.r.t. "ghost" symmetry flows $\overset{(\beta)}{\partial_s}$ when $\gamma\neq\beta\neq\alpha$.

The detailed proof follows closely the pattern of the proof for the $N=2$ component KP case [**3**], making heavy use of the spectral representation identities (1.5)–(1.6), and will be given in a separate paper.

Here we will only present an additional important property of multiple-KP system (2.6), (2.8)–(2.13), (3.3), which appears only in the $N\geq 3$ cases. Namely, for any $\alpha\neq\beta\neq\gamma$ the following identities hold among (adjoint) EF's:

$$\overset{(\gamma)}{\partial_1}\overset{(\alpha\beta)}{\Phi_k}=\overset{(\alpha\gamma)}{\Phi_1}\overset{(\gamma\beta)}{\Phi_k},\quad \overset{(\gamma)}{\partial_1}\overset{(\alpha\beta)}{\Psi_k}=\overset{(\alpha\gamma)}{\Psi_1}\overset{(\gamma\beta)}{\Psi_k}. \tag{3.4}$$

In particular, taking $k = 1$ in the first Eq. (3.4) and taking into account (2.9), we see that the latter coincides with the N'-wave system (3.2). In other words, the N'-wave system is reformulated entirely in terms of EF's of an underlying "ghost" symmetry enhanced ordinary KP hierarchy. Moreover, it is easy to check that (3.4) are compatible with the "ghost" flow Eqs. (3.3), and in fact (3.4) are equivalent to (3.3).

4. Darboux–Bäcklund Orbits of Multi-Component KP Hierarchies

We will consider DB transformations in their form appropriate for the Sato formulation of KP hierarchies [**19, 20**]. Namely, DB transformations are pseudo-differential operator "gauge" transformations of the pertinent Lax operators given in terms of (adjoint) EF's (1.4).

Let us temporarily return to the simpler case of double-KP hierarchy. In our construction a very instrumental role will be played by the following *nonstandard* orbit of successive DB transformations for the original KP_1 system (1.1) ($\mathcal{L} \equiv \mathcal{L}(n)$, $\overset{(12)}{\Phi}_j \equiv \Phi_j \equiv \P_j^{(n)}$, etc.):

$$\mathcal{L}(n+1) = T(n)\mathcal{L}(n)T^{-1}(n), \quad T(n) = \Phi_1 D \Phi_1^{-1} \equiv \Phi_1^{(n)} D \Phi_1^{(n)^{-1}}. \tag{4.1}$$

$$\begin{aligned} \Phi_l^{(n+1)} &= \Phi_1^{(n)} \partial\Big(\frac{\Phi_{l+1}^{(n)}}{\Phi_1^{(n)}}\Big), \quad l \geq 1; \qquad \Psi_1^{(n+1)} = \frac{1}{\Phi_1^{(n)}}, \\ \Psi_j^{(n+1)} &= -\frac{1}{\Phi_1^{(n)}} \partial^{-1}(\Phi_1^{(n)} \Psi_{j-1}^{(n)}), \quad j \geq 2. \end{aligned} \tag{4.2}$$

$$F^{(n+1)} = \Phi_1^{(n)} \partial\Big(\frac{F^{(n)}}{\Phi_1^{(n)}}\Big), \qquad F^{*(n+1)} = -\frac{1}{\Phi_1^{(n)}} \partial^{-1}(\Phi_1^{(n)} F^{*(n)}), \tag{4.3}$$

for transformations in "positive" direction, as well as adjoint DB transformations, i.e., transformations in "negative" direction:

$$\mathcal{L}(n-1) = \widehat{T}^{*-1}(n)\mathcal{L}(n)\widehat{T}^{*}(n), \quad \widehat{T}(n) = \Psi_1 D \Psi_1^{-1} \equiv \Psi_1^{(n)} D \Psi_1^{(n)^{-1}}. \tag{4.4}$$

$$\begin{aligned} \Phi_1^{(n-1)} &= \frac{1}{\Psi_1^{(n)}}, \quad \Phi_l^{(n-1)} = \frac{1}{\Psi_1^{(n)}} \partial^{-1}(\Psi_1^{(n)} \Phi_{l-1}^{(n)}), \quad l \geq 2, \\ \Psi_j^{(n-1)} &= -\Psi_1^{(n)} \partial\Big(\frac{\Psi_{j+1}^{(n)}}{\Psi_1^{(n)}}\Big), \quad j \geq 1. \end{aligned} \tag{4.5}$$

$$F^{(n-1)} = \frac{1}{\Psi_1^{(n)}} \partial^{-1}(\Psi_1^{(n)} F^{(n)}), \qquad F^{*(n-1)} = -\Psi_1^{(n)} \partial\Big(\frac{F^{*(n)}}{\Psi_1^{(n)}}\Big). \tag{4.6}$$

In what follows, the DB "site" index (n) will be skipped for brevity whenever this would not lead to ambiguities. Note that under the above DB transformations, the tau-function $\tau \equiv \tau^{(n)}$ transforms as: $\tau^{(n+1)} = \Phi_1 \tau \equiv \Phi_1^{(n)} \tau^{(n)}$, $\tau^{(n-1)} = -\Psi_1 \tau \equiv -\Psi_1^{(n)} \tau^{(n)}$.

REMARK 1. Let us stress the noncanonical form of the (adjoint) DB transformations (4.2), (4.5) on the "ghost" symmetry generating (adjoint) EF's. On the other hand, for generic (adjoint) EF's F, F^* the (adjoint) DB transformations (4.3), (4.6) read as usual [**19, 20**].

The crucial property of the above DB orbit of the original one-component KP_1 hierarchy is that it induces an orbit of DB transformations for the whole double-KP system (2.6)–(2.13). More precisely, as shown in [**1, 3**], "ghost" symmetries (1.8) commute with DB transformations (4.1)–(4.6) of the KP_1 hierarchy, and induce the following DB transformations on the "ghost" KP_2 hierarchy (recall from Eqs. (2.3), that $\bar{\Phi}_1^{(n)} = \Psi_1^{(n)}$, $\bar{\Psi}_1^{(n)} = \Phi_1^{(n)}$):

$$\begin{aligned} \bar{\mathcal{L}}(n+1) &= \Big(\frac{1}{\bar{\Psi}_1^{(n)}} \bar{D}^{-1} \bar{\Psi}_1^{(n)}\Big) \bar{\mathcal{L}}(n) \Big(\frac{1}{\bar{\Psi}_1^{(n)}} \bar{D} \bar{\Psi}_1^{(n)}\Big), \\ \bar{\mathcal{L}}(n-1) &= \Big(\bar{\Phi}_1^{(n)} \bar{D} \frac{1}{\bar{\Phi}_1^{(n)}}\Big) \bar{\mathcal{L}}(n) \Big(\bar{\Phi}_1^{(n)} \bar{D}^{-1} \frac{1}{\bar{\Phi}_1^{(n)}}\Big). \end{aligned} \tag{4.7}$$

$$\begin{aligned} \bar{\Phi}_j^{(n-1)} &= \bar{\Phi}_1^{(n)} \bar{\partial}\Big(\frac{\bar{\Phi}_{j+1}^{(n)}}{\bar{\Phi}_1^{(n)}}\Big), \quad j \geq 1; \qquad \bar{\Psi}_1^{(n-1)} = \frac{1}{\bar{\Phi}_1^{(n)}}, \\ \bar{\Psi}_l^{(n-1)} &= -\frac{1}{\bar{\Phi}_1^{(n)}} \bar{\partial}^{-1}(\bar{\Phi}_1^{(n)} \bar{\Psi}_{l-1}^{(n)}), \quad l \geq 2. \end{aligned} \tag{4.8}$$

$$\overline{F}^{(n-1)} = \bar{\Phi}_1^{(n)} \bar{\partial}\Big(\frac{\overline{F}^{(n)}}{\bar{\Phi}_1^{(n)}}\Big), \quad \overline{F}^{*(n-1)} = -\frac{1}{\bar{\Phi}_1^{(n)}} \bar{\partial}^{-1}(\bar{\Phi}_1^{(n)} \overline{F}^{*(n)}). \tag{4.9}$$

$$\begin{aligned} \bar{\Phi}_1^{(n+1)} &= \frac{1}{\bar{\Psi}_1^{(n)}}, \quad \bar{\Phi}_l^{(n+1)} = \frac{1}{\bar{\Psi}_1^{(n)}} \bar{\partial}^{-1}(\bar{\Psi}_1^{(n)} \bar{\Phi}_{l-1}^{(n)}), \quad l \geq 2, \\ \bar{\Psi}_j^{(n+1)} &= -\bar{\Psi}_1^{(n)} \bar{\partial}\Big(\frac{\bar{\Psi}_{j+1}^{(n)}}{\bar{\Psi}_1^{(n)}}\Big), \quad j \geq 1. \end{aligned} \tag{4.10}$$

$$\overline{F}^{(n+1)} = \frac{1}{\Psi_1^{(n)}} \bar{\partial}^{-1}(\Psi_1^{(n)} \overline{F}^{(n)}), \quad \overline{F}^{*(n+1)} = -\Psi_1^{(n)} \bar{\partial}\Big(\frac{\overline{F}^{*(n)}}{\Psi_1^{(n)}}\Big), \tag{4.11}$$

where $\overline{F}$, $\overline{F}^*$ are generic (adjoint) EF's of $\bar{\mathcal{L}}$. For each DB "site" (n) the corresponding Lax operators and (adjoint) EF's from (4.1)–(4.11) define a double-KP system as in (2.6)–(2.13). Therefore, the DB orbit (4.1)–(4.11) defines an orbit of DB transformations for the associated two-component (matrix) KP hierarchy.

One can prove a similar statement also for generic DB transformations of KP_1 hierarchy:

$$\widetilde{\mathcal{L}} = (FDF^{-1})\mathcal{L}(FD^{-1}F^{-1}), \quad \widetilde{\Phi}_j = F\partial\Big(\frac{\Phi_j}{F}\Big), \quad \widetilde{\Psi}_j = -\frac{1}{F}\partial^{-1}(F\Psi_j). \tag{4.12}$$

$$\begin{aligned} \widehat{\mathcal{L}} &= (F^{*-1}D^{-1}F^*)\widehat{\mathcal{L}}(F^{*-1}DF^*), \\ \widehat{\Phi}_j &= \frac{1}{F^*}\partial^{-1}(F^*\Phi_j), \quad \widehat{\Psi}_j = -F^*\partial\Big(\frac{\Psi_j}{F^*}\Big), \end{aligned} \tag{4.13}$$

where F, F^* are generic (adjoint) EF's of $\mathcal{L}$, namely, the "ghost" symmetries (1.8) commute with generic DB transformations (4.12)–(4.13).

REMARK 2. Comparing (4.1)–(4.3) with (4.7), (4.10)–(4.11), and (4.4)–(4.6) with (4.7)–(4.9), we find that DB transformations of KP_1 hierarchy w.r.t. Φ_1 corresponds to adjoint-DB transformations of KP_2 hierarchy w.r.t. $\bar{\Psi}_1$, and *vice versa.*

Let us now go back to the general N-component KP hierarchy. Picking up any pair KP_α and KP_β (with $\alpha \neq \beta$) of one-component KP subsystems of multiple-KP hierarchy (2.10)–(2.13) and (3.3), we can construct a DB orbit w.r.t. $\overset{(\alpha\beta)}{\Phi}_1$ for this pair as in (4.1)–(4.3), (4.8), (4.10)–(4.11) by literally repeating the above construction with the identifications: $\mathcal{L} \equiv \overset{(\alpha)}{\mathcal{L}}$, $\bar{\mathcal{L}} \equiv \overset{(\beta)}{\mathcal{L}}$, $\Phi_j \equiv \overset{(\alpha\beta)}{\Phi}_j$, $\Psi_j \equiv \overset{(\alpha\beta)}{\Psi}_j$, $F \equiv \overset{(\alpha\gamma)}{\Phi}_k$, $F^* \equiv \overset{(\alpha\gamma)}{\Psi}_k$, $\bar{\Phi}_j \equiv \overset{(\beta\alpha)}{\Phi}_j$, $\bar{\Psi}_j \equiv \overset{(\beta\alpha)}{\Psi}_j$, $\bar{F} \equiv \overset{(\beta\gamma)}{\Phi}_k$, $\bar{F}^* \equiv \overset{(\beta\gamma)}{\Psi}_k$, where $\gamma \neq \alpha \neq \beta$. Hence such DB orbit, which we will call $\mathrm{DB}_{(\alpha\beta)}$ orbit, preserves the whole multiple-KP hierarchy. Also, according to the last Remark in Section 3, $\mathrm{DB}_{(\alpha\beta)}$ orbit is equivalent to adjoint-$\mathrm{DB}_{(\beta\alpha)}$ orbit, i.e., the DB orbit w.r.t. $\overset{(\beta\alpha)}{\Psi}_1$. Then, a general DB orbit preserving multiple-KP hierarchy, or equivalently, N-component matrix KP hierarchy, is obtained by combining the $\mathrm{DB}_{(\alpha\beta)}$ orbits for all pairs (α, β) with $\alpha \neq \beta$.

We are particularly interested in DB orbits passing through the "free" N-component KP hierarchy, i.e., with $\overset{(\alpha)}{\mathcal{L}} = \overset{(\alpha)}{D}$ and hence $\tau = \text{const}$ in (2.6). Then one can easily show that the general DB orbit $\mathrm{DB}_{(12,13,\ldots,1N)}$ for the N-component KP hierarchy passing through the "free" one is built up from a union of $\mathrm{DB}_{(1\beta)}$ orbits of the form (4.2)–(4.3), (4.10)–(4.11) with $\beta = 2, \ldots, N$. Accordingly, it is labelled by the set of integers $(n_{12}, n_{13}, \ldots, n_{1N}; m)$ indicating n_{12} DB iterations w.r.t. $\overset{(12)}{\Phi}_1$ or $|n_{12}|$ adjoint-DB iterations w.r.t. $\overset{(12)}{\Psi}_1$ for $n_{12} < 0$, n_{13} DB iterations w.r.t. $\overset{(13)}{\Phi}_1$ or $|n_{13}|$ adjoint-DB iterations w.r.t. $\overset{(13)}{\Psi}_1$ for $n_{13} < 0$, etc., whereas the last integer indicates m steps of DB transformations w.r.t. generic EF's $\overset{(1)}{F}_1, \ldots, \overset{(1)}{F}_m$ of KP_1 subsystem, i.e., such that they do not belong to any of the "ghost" symmetry generating sets of $\overset{(1\beta)}{\Phi}_k$.

5. Darboux–Bäcklund Solutions

Taking into account relations (2.9) and the structure of the $\mathrm{DB}_{(1\beta)}$ suborbits ($\beta = 2, \ldots, N$) of the form (4.2)–(4.3), (4.10)–(4.11) allows us to express all nondiagonal tau-functions of the N-component KP hierarchy on any given site $(n_{12}, n_{13}, \ldots, n_{1N}; m)$ of the full orbit $\mathrm{DB}_{(12,13,\ldots,1N)}$ in terms of DB shifts of the diagonal tau-function as follows (from now on we will use the shortened notations $n_{1\alpha} \equiv n_\alpha$):

$$\tau_{1\alpha}^{(\ldots,n_\alpha,\ldots)} = \eta_\alpha \tau^{(\ldots,n_\alpha+1,\ldots)}, \quad \tau_{\alpha 1}^{(\ldots,n_\alpha,\ldots)} = -\eta_\alpha \tau^{(\ldots,n_\alpha-1,\ldots)}. \tag{5.1}$$

$$\begin{aligned} \tau_{\alpha\beta}^{(\ldots,n_\alpha,\ldots,n_\beta,\ldots)} &= \eta_\alpha \eta_\beta \tau^{(\ldots,n_\alpha-1,\ldots,n_\beta+1,\ldots)}, \\ \tau_{\beta\alpha}^{(\ldots,n_\alpha,\ldots,n_\beta,\ldots)} &= \eta_\alpha \eta_\beta \tau^{(\ldots,n_\alpha+1,\ldots,n_\beta-1,\ldots)}, \end{aligned} \tag{5.2}$$

where $\eta_\alpha \equiv \mathrm{sign}(n_\alpha)$ and $1 < \alpha < \beta \leq N$. Since τ is the tau-function of scalar KP_1 sub-hierarchy, the problem of finding the complete DB solution of the full N-component KP hierarchy is reduced to applying the well-known techniques of (adjoint) DB iterations in ordinary one-component KP hierarchy within the Sato/tau-function formulation. Using the above techniques and taking into account the specific form of DB orbits (4.2)–(4.3) and (4.5)–(4.6), we obtain (assuming explicitly that part of the DB iterations are adjoint-DB ones) the following Wronskian-type expression for the diagonal tau-function:

$$\tau^{(-n_2,\ldots,-n_k,n_{k+1},\ldots,n_N;m)}/\tau^{(0,\ldots,0;0)} = (-1)^{\sum_{j=2}^{k} n_j -1} \tag{5.3}$$
$$\times \widetilde{W}[\overset{(1,k+1)}{\Phi}_1,\ldots,\overset{(1,k+1)}{\Phi}_{n_{k+1}},\ldots,\overset{(1N)}{\Phi}_1,\ldots,\overset{(1N)}{\Phi}_{n_N},\overset{(1)}{F}_1,\ldots,\overset{(1)}{F}_m;$$
$$\overset{(12)}{\Psi}_1,\ldots,\overset{(12)}{\Psi}_{n_2},\ldots,\overset{(1k)}{\Psi}_1,\ldots,\overset{(1k)}{\Psi}_{n_k}],$$

with the short-hand notation for Wronskian-type determinants:

$$\widetilde{W}[f_1,\ldots,f_k;f_1^*,\ldots,f_l^*] \equiv \det \begin{Vmatrix} \overset{(1)}{\partial}{}^{a-1} f_b & \overset{(1)}{\partial}{}^{a-1} f_{k-l+d} \\ \overset{(1)}{\partial}{}^{-1}(f_b f_c^*) & \overset{(1)}{\partial}{}^{-1}(f_{k-l+d} f_c^*), \end{Vmatrix}, \tag{5.4}$$
$$a,b = 1,\ldots,k-l;\; c,d = 1,\ldots,l.$$

Recalling property (2.4) and the "ghost" symmetry flow Eqs. (2.12), (1.10) and (3.4), we find that we can replace the SEP's in (5.3) (c.f. (5.4)) as follows:

$$\begin{gathered} \overset{(1)}{\partial}{}^{-1}(\overset{(1\beta)}{\Phi}_{i_\beta}\overset{(1\alpha)}{\Psi}_1) = \overset{(\alpha\beta)}{\Phi}_{i_\beta}, \quad \overset{(1)}{\partial}{}^{-1}(\overset{(1)}{F}_b\overset{(1\alpha)}{\Psi}_1) = \overset{(\alpha)}{F}_b, \\ \overset{(1)}{\partial}{}^{-1}(\overset{(1)}{F}_b\overset{(1\alpha)}{\Psi}_{j_\alpha}) = \overset{(\alpha)}{\partial}{}^{j_\alpha-1}\overset{(\alpha)}{F}_b + \ldots, \end{gathered} \tag{5.5}$$

where $\alpha = 2,\ldots,k$, $j_\alpha = 1,\ldots,n_\alpha$, $\beta = k+1,\ldots,N$, $i_\beta = 1,\ldots,n_\beta$ and $b = 1,\ldots,m$, and where the dots in second Eqs. (5.5) indicate terms with lower derivatives on the corresponding EF's which cancel in the determinant. Therefore, the Wronskian-type determinant on the r.h.s. of Eq. (5.3) together with (5.5) acquires the form of a *multiple* Wronskian, generalizing the double Wronskians of the first Refs. [**13–16**]. Let us recall that all entries in the multiple Wronskian (5.3)–(5.5) are (derivatives of) EF's and adjoint EF's of the respective initial KP$_\alpha$ subsystems (2.11). In the case of "free" initial point on the DB orbit we have: $\overset{(\alpha)}{\partial}_s\overset{(\alpha)}{F}_b = (\overset{(\alpha)}{\partial}_1)^s\overset{(\alpha)}{F}_b$, $\overset{(\alpha)}{\partial}_s\overset{(\alpha\beta)}{\Phi}_i = (\overset{(\alpha)}{\partial}_1)^s\overset{(\alpha\beta)}{\Phi}_i$ and $\overset{(\alpha\beta)}{\Phi}_i = (-\overset{(\beta)}{\partial}_1)^{i-1}\overset{(\alpha\beta)}{\Phi}_1$.

As an example let us consider the general DB solution (5.3) in the case $N = 2$ and with a "free" initial $\tau^{(0;0)} = 1$:

$$\tau^{(-n;m)} = (-1)^{n-1}\det\begin{Vmatrix} \partial^{a-1}F_b & \partial^{a-1}F_{m-n+d} \\ \partial^{c-1}\overline{F}_b & \partial^{c-1}\overline{F}_{m-n+d} \end{Vmatrix}, \tag{5.6}$$
$$a,b = 1,\ldots,m-n;\; c,d = 1,\ldots,n,$$

where F_b,$\overline{F}_b$ are arbitrary free EF's of KP$_1$ and KP$_2$ one-component sub-hierarchies of two-component KP hierarchy, respectively: $\overset{(-)}{F}_b = \int d\lambda\, \overset{(-)}{\varphi}_b(\lambda) e^{\xi(\overset{(-)}{t},\lambda)}$. The tau-functions (5.6), taking into account (2.9), provide the following series of Wronskian solutions for DS system (1.13):

$$\begin{gathered} Q^{(-n;m)} = 4\partial_X^2 \ln\tau^{(-n;m)}, \quad \Phi_1^{(-n;m)} = \frac{\tau^{(-n+1;m)}}{\tau^{(-n;m)}}, \\ \Psi_1^{(-n;m)} = \frac{\tau^{(-n-1;m)}}{\tau^{(-n;m)}}. \end{gathered} \tag{5.7}$$

In the particular case $m = 2n$ and taking special forms of the spectral densities of the pertinent EF's in (5.6) $\overset{(-)}{\varphi}_b(\lambda) = \sum_{a=1}^{2n}\overset{(-)}{c}_{ba}\delta(\lambda-\lambda_a)$ with constant coefficients

$\overset{(-)}{c}_{ba}$, the series (5.7) with (5.6) contains the well-known n^2 (multi-)dromion solutions [**4, 12**] in the double-Wronskian form given in the first Refs. [**13–16**] (after making appropriate choice for the constant parameters in order to ensure reality properties).

More detailed analysis of the new series of multiple-Wronskian solutions (5.1)–(5.5) of N-component KP hierarchies, in particular, how other known soliton-type solutions are fitting there, will be given elsewhere[2].

6. Conclusions

In the present note we have shown that, given an ordinary one-component KP hierarchy KP_1, we can always construct a N-component matrix KP hierarchy, embedding the original one, in the following way. We choose $N-1$ infinite sets $\{\overset{(1\alpha)}{\Phi_k}, \overset{(1\alpha)}{\Psi_k}\}_{k=1}^{\infty}$, $\alpha = 2,\ldots,N$, of (adjoint) EF's of the initial KP_1 which we use to construct an infinite-dimensional Abelian algebra of additional ("ghost") symmetries (2.10)–(2.13), (3.3). The one-component KP hierarchy equipped with such additional symmetry structure turns out to be equivalent to the standard N-component matrix KP hierarchy. Namely, with the help of a subset of the "ghost" symmetry generating (adjoint) EF's of KP_1 we can define new tau-functions $\tau_{\alpha\beta} = \tau\varepsilon_{\alpha\beta}\overset{(1)}{\partial}{}^{-1}(\overset{(1\beta)(1\alpha)}{\Phi_1\,\Psi_1})$ (c.f. (2.9), (3.4)) with $\alpha \neq \beta, \alpha, \beta = 1,\ldots,N$, which together with τ (the tau-function of the initial KP_1 hierarchy) satisfy Hirota bilinear identities of N-component matrix KP hierarchy (3.1).

Furthermore, we have shown that there exists a special nonstandard Darboux–Bäcklund orbit of the initial KP_1 hierarchy (Section 4), which preserves the above mentioned additional ("ghost") symmetry structure and which thereby generates DB transformations of the whole N-component matrix KP hierarchy. This fact allows to use the well-known DB techniques within the context of the ordinary scalar KP subsystem KP_1 to generate soliton-like tau-function solutions of all higher N-component KP hierarchies which take the form of multiple Wronskians (Section 5). In particular, we find in this way new series of multiple-Wronskian solutions to well-known systems of integrable nonlinear soliton equations contained within the N-component matrix KP hierarchy: Davey–Stewartson system (for $N=2$, containing the previously obtained dromion solutions) and N'-wave system (for $N \geq 3$).

References

1. H. Aratyn, E. Nissimov, and S. Pacheva, *A new "dual" symmetry structure of the* KP *hierarchy*, Phys. Lett. A **244** (1998), 245–255, `solv-int/9712012`.
2. ______, *Method of squared eigenfunction potentials in integrable hierarchies of* KP *type*, Comm. Math. Phys. **193** (1998), 493–525, `solv-int/9701017`.
3. ______, *From one-component* KP *hierarchy to two-component* KP *hierarchy and back*, Topics in Theoret. Phys., vol. II (H. Aratyn, L. A. Ferreira, and J. F. Gomes, eds.), IFT-São Paulo, SP-1998, pp. 25–33, `solv-int/9808003`.
4. M. Boiti, J. Leon, L. Martina, and F. Pempinelli, *Scattering of localized solitons in the plane*, Phys. Lett. A **132** (1988), 432–439.

[2] Let us note that our multiple-Wronskian DB orbit differs from the DB orbit generated via matrix EF's [**19, 20**] within Sato matrix pseudo-differential operator formulation. The latter cannot be written compactly in a Wronskian form.

5. E. Date, M. Jimbo, M. Kashiwara, and T. Miwa, *Transformation groups for soliton equations. III. Operator approach to the Kadomtsev–Petviashvili equation*, J. Phys. Soc. Japan **50** (1981), 3806–3812.
6. A. Davey and K. Stewartson, *On three-dimensional packets of surface waves*, Proc. Roy. Soc. London Ser. A **338** (1974), 101–110.
7. L. Dickey, *Soliton equations and Hamiltonian systems*, Adv. Ser. Math. Phys., vol. 12, World Scientific Publishing Co., Inc., River Edge, NJ, 1991.
8. ———, *Lectures on classical W-algebras*, Acta Appl. Math. **47** (1997), 243–321.
9. A. Doliwa, M. Manas, L. Martinez Alonso, E. Medina, and P. Santini, *Charged free fermions, vertex operators and classical theory of conjugate nets*, J. Phys. A **32** (1999), 1197–1216, `solv-int/9803015`.
10. B. Enriquez, A. Orlov, and V. Rubtsov, *Dispersionful analogues of Benney's equations and N-wave systems*, Inverse Problems **12** (1996), 241–250, `solv-int/9510002`.
11. E. Ferapontov, *Laplace transformations of hydrodynamic type systems in Riemann invariants: Periodic sequences*, J. Phys. A **30** (1997), 6861–6878, `solv-int/9705017`; and references therein.
12. A. Fokas and P. Santini, *Coherent structures in multidimensions*, Phys. Lett. A **63** (1989), 1329–1333.
13. N. Freeman, *Soliton solutions of nonlinear evolution equations*, IMA J. Appl. Math. **32** (1984), 125–145.
14. C. Gilson and J. Nimmo, *A direct method for dromion solutions of the Davey–Stewartson equations and their asymptotic properties*, Proc. Roy. Soc. London Ser. A **435** (1991), 339–357.
15. F. Guil and M. Manas, *Darboux transformations for the Davey–Stewartson equations*, Phys. Lett. A **217** (1996), 1–6; and references therein.
16. J. Hietarinta and R. Hirota, *Multidromion solutions to the Davey–Stewartson equation*, Phys. Lett. A **145** (1990), 237–244.
17. M. Hisakado, *The Davey–Stewartson system and the Bäcklund transformations*, J. Phys. Soc. Japan **67** (1998), 3038–3043, `solv-int/9804009`.
18. V. Kac and J. van de Leur, *The n-component of* KP *hierarchy and representation theory*, Important Developments in Soliton Theory (A. Fokas and V.E. Zakharov, eds.), Springer, Berlin, 1993, pp. 302–334, `hep-th/9308137`.
19. W. Oevel, *Darboux theorems and Wronskians formulas for integrable systems. I. Constrained* KP *flows*, Phys. A **195** (1993), 533–576.
20. W. Oevel and W. Schief, *Squared eigenfunctions of the* (*modified*) KP *hierarchy and scattering problems of Loewner type*, Rev. Math. Phys. **6** (1994), 1301–1338.
21. A. Orlov, *Volterra operator algebra for zero curvature representation, universality of* KP, Nonlinear Processes in Physics, Proc. III Potsdam-V Kiev Workshop (A. S. Fokas *et al.*, eds.), Springer, Belin 1993, pp. 126–131.
22. A. Orlov and E. Schulman, *Additional symmetries for integrable equations and conformal algebra representation*, Lett. in Math. Phys. **12** (1986), 171–179.
23. K. Ueno and K. Takasaki, *Toda lattice hierarchy*, Group Representations and Systems of Differential Equations (Tokyo, 1982), Adv. Stud. Pure Math., vol. 4, North-Holland, Amsterdam-New York, 1984, pp. 1–95.
24. V. Zakharov, S. Manakov, S. Novikov, and L. Pitaevski, *Theory of Solitons. The inverse scattering method*, Nauka, Moscow, 1980; Engl. transl., Plenum, New York, 1984.

Department of Physics, University of Illinois at Chicago, 845 W. Taylor Street, Chicago, IL 60607-7059, USA.
E-mail address: `aratyn@uic.edu`

Institute of Nuclear Research and Nuclear Energy, Boul. Tsarigradsko Chausee 72, BG-1784 Sofia, Bulgaria and Department of Physics, Ben-Gurion University of the Negev, Box 653, IL-84105 Beer Sheva, Israel.
E-mail address, E. Nissimov: `nissimov@inrne.bas.bg; emil@bgumail.bgu.ac.il`
E-mail address, S. Pacheva: `svetlana@inrne.bas.bg; svetlana@bgumail.bgu.ac.il`

Centre de Recherches Mathématiques
CRM Proceedings and Lecture Notes
Volume **29**, 2001

The Darboux–Bäcklund Transformation and Clifford Algebras

Jan L. Cieśliński

ABSTRACT. We consider a class of spectral problems in Clifford algebras. The soliton surfaces approach is applied to produce interesting classes of immersions associated with integrable systems. We formulate the Darboux–Bäcklund transformation purely in terms of Clifford numbers without using any matrix representation. The construction is a straightforward generalization of the Darboux–Bianchi transformation for isothermic surfaces.

1. Soliton Surfaces Approach

We begin with presenting the soliton surfaces approach which associates a geometry with a given spectral problem of Zakharov–Shabat type. Then we discuss an extension of the approach on spectral problems in Clifford algebras which leads to an interesting class of immersions. Finally, we show how to construct the Darboux–Bäcklund transformation in terms of Clifford numbers rather than matrices.

The soliton surfaces approach [**31,33**] (see also [**3,11,12,20,23**]) is a systematic method to associate a geometry (a class of surfaces) to a given integrable system. By integrability we mean the existence of the spectral problem ("Lax pair", if systems for two independent variables are concerned): a system of linear partial differential equations with the spectral parameter [**37**]. The compatibility conditions for the linear equations are equivalent to the considered integrable system.

The main idea of the soliton surfaces approach consists in the following. Consider a Zakharov–Shabat type spectral problem

$$\Psi_{,x} = U\Psi, \quad \Psi_{,y} = V\Psi, \tag{1.1}$$

where U, V are matrices of a given degree m depending (in a prescribed way) on x, y through some fields, λ is the spectral parameter and Ψ is a fundamental solution

---

2000 *Mathematics Subject Classification.* 37K35, 37K25, 53A07, 15A66.

The research supported by the Polish State Committee for Scientific Research under the KBN grant 2 P03B 143 15.

I would like to thank Polish State Committee for Scientific Research and Organizers of the workshop for the financial support which enabled me to participate in the very interesting and stimulating meeting in Halifax.

This is the final form of the paper.

(i.e., a nondegenerate matrix whose columns are independent vector solutions). The compatibility conditions read

(1.2) $$U_{,y} - V_{,x} + [U, V] = 0.$$

Then the so called Sym–Tafel formula [**31**]

(1.3) $$F = \Psi^{-1}\Psi_{,\lambda}\,,$$

defines a λ-family of surfaces parameterized by coordinates x, y. The surfaces are immersed in a space of matrices which is obviously a linear space. The Gauss–Codazzi equations for a class of surfaces given by F are equivalent to the integrable system under consideration (1.2).

Given a spectral problem we can apply numerous techniques of the theory of solitons: inverse scattering transform, loop group approach, Riemann theta function solutions, conservation laws, Darboux–Bäcklund transformation etc. The Sym–Tafel formula enables one to use all these methods immediately in geometrical context provided that a geometrical interpretation of F is found. We point out that for $\mathbf{su}(2)$ spectral problems such an interpretation has been discovered in many interesting cases: pseudospherical surfaces, constant mean curvature surfaces, surfaces of positive Gaussian curvature, Bianchi surfaces, vortex filaments etc. [**3, 7, 16, 17, 26, 32**]. In this paper we are going to present more examples. They are associated with spectral problems of higher degree which can be conveniently represented in Clifford algebras [**8–10**].

From the technical point of view the Sym–Tafel formula seems to be so useful because of a very simple rule to compute derivatives of F. Indeed,

$$\begin{aligned} F_{,x} &= (\Psi^{-1})_{,x}\Psi_{,\lambda} + \Psi^{-1}\Psi_{,\lambda x} = -\Psi^{-1}\Psi_{,x}\Psi^{-1}\Psi_{,\lambda} + \Psi^{-1}(U\Psi)_{,\lambda} \\ &= -\Psi^{-1}U\Psi_{,\lambda} + \Psi^{-1}U_{,\lambda}\Psi + \Psi^{-1}U\Psi_{,\lambda} = \Psi^{-1}U_{,\lambda}\Psi, \end{aligned}$$

and, similarly,

$$F_{,y} = \Psi^{-1}V_{,\lambda}\Psi.$$

Second derivatives can be computed easily as well

$$\begin{aligned} F_{,xx} &= (\Psi^{-1}U_{,\lambda}\Psi)_{,\lambda} = \Psi^{-1}\Psi_{,x}\Psi^{-1}U_{,\lambda}\Psi + \Psi^{-1}U_{,\lambda x}\Psi + \Psi^{-1}U_{,\lambda}\Psi_{,x} \\ &= \Psi^{-1}(-UU_{,\lambda} + U_{,\lambda x} + U_{,\lambda}U)\Psi = \Psi^{-1}\big(U_{,\lambda x} + [U_{,\lambda}, U]\big)\Psi, \end{aligned}$$

and

$$F_{,xy} = \Psi^{-1}\big(U_{,\lambda y} + [U_{,\lambda}, V]\big)\Psi, \quad F_{,yy} = \Psi^{-1}\big(V_{,\lambda y} + [V_{,\lambda}, V]\big)\Psi.$$

Subtracting the tangent component from the second derivatives we obtain the normal part which obviously is a linear combination of some number of normal vectors $N^\mu = \Psi^{-1}n^\mu\Psi$, $\mu = 1, 2, \ldots$, where n^μ can be algebraically expressed by U, V and their derivatives. Fundamental forms $I = dF \cdot dF$ and $II^\mu = -dF \cdot dN^\mu$ do not depend explicitly on Ψ. The scalar product is standard, i.e. proportional to the trace. For example: $F_{,x} \cdot F_{,y} = c\,\mathrm{Tr}(U_{,\lambda}V_{,\lambda})$, where c is a constant.

Usually, U and V are matrices. In this paper we discuss the generalization of the soliton surfaces approach on U, V assuming values in Clifford algebras. The Clifford algebra $\mathcal{C}(p, q)$ is generated by elements $\mathbf{e}_1, \ldots, \mathbf{e}_p$, $\mathbf{e}_{p+1}, \ldots, \mathbf{e}_{p+q}$ ("Clifford numbers") which satisfy the relations

$$\mathbf{e}_j\mathbf{e}_k + \mathbf{e}_k\mathbf{e}_j = 2\eta_{jk},$$

where $\eta_{jk} = 0$ for $j \neq k$, $\eta_{kk} = 1$ for $1 \leq k \leq p$ and $\eta_{kk} = -1$ for $p+1 \leq k \leq p+q$. In the case $q = 0$ we will use the notation $\mathcal{C}(p,0) \equiv \mathcal{C}(p)$, moreover $\eta_{jk} \equiv \delta_{jk}$. The main automorphism α and the main anti-automorphism β of a Clifford algebra are defined as follows:

$$\begin{aligned} &\alpha(\mathbf{e}_j) = -\mathbf{e}_j, \quad \alpha(XY) = \alpha(X)\alpha(Y), \quad \alpha(c_1X + c_2Y) = \bar{c}_1\alpha(X) + \bar{c}_2\alpha(Y), \\ &\beta(\mathbf{e}_j) = \mathbf{e}_j, \qquad \beta(XY) = \beta(Y)\beta(X), \quad \beta(c_1X + c_2Y) = c_1\beta(X) + c_2\beta(Y). \end{aligned}$$

Notice that these properties define α and β uniquely. For more informations on Clifford algebras see, for instance, [**5, 30**].

The simplest example of a Clifford algebra is $\mathcal{C}(0,1)$ generated by $\mathbf{e}_1$ such that $\mathbf{e}_1^2 = -1$. Identifying $\mathbf{e}_1$ with i (imaginary unit) we see easily that $\mathcal{C}(0,1)$ is isomorphic to the field of complex numbers $\mathbb{C}$. More complicated but well known examples are obtained if $\mathbf{e}_k$ are represented by Pauli matrices or Dirac matrices. However, we will try to show in this paper that it is possible and even convenient to apply Clifford algebras without any specific representation.

In fact we will consider spectral problems of the form

$$\Psi_{,k} = U_k\Psi, \quad (k = 1, \ldots, n), \tag{1.4}$$

where $U_k = U_k(x_1, \ldots, x^n; \lambda)$ are elements of a given Clifford algebra. The soliton surfaces approach, presented above in the matrix case for $n = 2$, can be easily extended on the Clifford algebras case for any $n \geq 2$. In particular, the formulas to compute derivatives of F remain unchanged.

2. The Structure of Spectral Problems and Geometry

It is well known that the structure of the spectral problem defines (almost uniquely) the associated integrable system [**7**]. Let us recall some $\mathbf{su}(2)$ cases with the matrix U linear in λ. If V is quadratic in λ, then we obtain a slight generalization of the nonlinear Schrödinger equation. More interesting from geometrical point of view is V linear in λ^{-1} which is associated with the sine-Gordon equation and pseudospherical surfaces. What is even more interesting, if instead of $\lambda = \text{const}$ we consider $\lambda = \lambda(x, y; \kappa)$ ($\kappa = \text{const}$), then we have no freedom to choose the function λ. The compatibility conditions define it almost uniquely. The associated surfaces ("Bianchi surfaces") are a natural generalization of pseudospherical surfaces [**7, 26, 34**].

In the sequel we focus on spectral problems in Clifford algebras. Let us begin with the case which turned our attention into this direction [**8, 15**]. Let

$$\begin{aligned} \Psi_{,x} &= \mathbf{e}_1\big(B_1\mathbf{e}_2 + C_1\mathbf{e}_3 + \lambda(P_1\mathbf{e}_4 + Q_1\mathbf{e}_5)\big)\Psi, \\ \Psi_{,y} &= \mathbf{e}_2\big(A_2\mathbf{e}_1 + C_2\mathbf{e}_3 + \lambda(P_2\mathbf{e}_4 + Q_2\mathbf{e}_5)\big)\Psi, \end{aligned} \tag{2.1}$$

where A_k, B_k, C_k, P_k, Q_k $(k = 1, 2)$ are functions of x, y and $\mathbf{e}_1, \ldots, \mathbf{e}_5$ are generators of $\mathcal{C}(4,1)$, i.e.

$$\mathbf{e}_1^2 = \mathbf{e}_2^2 = \mathbf{e}_3^2 = \mathbf{e}_4^2 = -\mathbf{e}_5^2 = 1, \quad \mathbf{e}_j\mathbf{e}_k + \mathbf{e}_k\mathbf{e}_j = 0, \qquad (j \neq k).$$

If Ψ satisfies the Lax pair (2.1), then one can prove that F given by the Sym–Tafel formula, evaluated at $\lambda = 0$, yields a surface immersed into a 6-dimensional linear space. Namely:

$$\Psi^{-1}\Psi_{,\lambda}|_{\lambda=0} = \sum_{i=1}^{3}\sum_{\mu=4}^{5} f_{i\mu}\mathbf{e}_i\mathbf{e}_\mu,$$

where $f_{i\mu}$ are some functions of x, y. Moreover, it is not difficult to show [**8**] that the immersions

$$f(x,y) := 2\begin{pmatrix} f_{14}+f_{15} \\ f_{24}+f_{25} \\ f_{34}+f_{35} \end{pmatrix} \in \mathbf{R}^3, \quad f^*(x,y) := 2\begin{pmatrix} f_{14}-f_{15} \\ f_{24}-f_{25} \\ f_{34}-f_{35} \end{pmatrix} \in \mathbf{R}^3,$$

are both isothermic (f^* is dual to f). It means that they admit coordinates which are in the same time conformal and conjugate (i.e., the first fundamental form is proportional to $dx^2 + dy^2$ and the second fundamental form is diagonal). Coordinates x, y have exactly this property.

The case of isothermic surfaces prompted us to consider the following spectral problem [**10**]:

$$\Psi_{,j} = \mathbf{e}_j(\lambda \mathbf{a}_j + \mathbf{b}_j)\Psi, \quad (j = 1, \ldots, n), \tag{2.2}$$

where $\mathbf{a}_1, \ldots, \mathbf{a}_n \in W$ and $\mathbf{b}_1, \ldots, \mathbf{b}_n \in V$, and V, W are n-dimensional orthogonal vector spaces. Then, $\mathbf{e}_1, \ldots, \mathbf{e}_n$ and $\mathbf{e}_{n+1}, \ldots, \mathbf{e}_{2n}$ are orthonormal bases in V and W, respectively. We assume that $\mathbf{b}_j$ is orthogonal to $\mathbf{e}_j$ for $j = 1, \ldots, n$ (which is not particularly restrictive and means that the spectral problem is traceless). One can show that $F := \Psi^{-1}\Psi_{,\lambda}|_{\lambda=0}$ assumes values in n^2-dimensional space $V \wedge W$, i.e.

$$F = \sum_{i,j=1}^{n} f_{ij}\mathbf{e}_i\mathbf{e}_{n+j},$$

and, what is more, if we decompose F as follows

$$F = F^{(1)}\mathbf{e}_{n+1} + \ldots F^{(n)}\mathbf{e}_{2n},$$

then every function $F^{(k)}$ assumes values in V and explicitly describes orthogonal coordinates in $V \simeq \mathbf{R}^n$. These nets are related by the so called Combescure transformation (their corresponding tangent vectors are mutually parallel) [**18, 25**].

Another example is associated with isometric immersions of n-dimensional space forms (spaces of constant negative sectional curvature) in the Euclidean space of dimension $2n-1$ [**2, 35**]. The corresponding spectral problem is given by (compare [**1, 19**]):

$$\Psi_{,k} = \mathbf{e}_k\left(\lambda \mathbf{a}_k - \frac{1}{\lambda}\mathbf{e}_{n+1}\mathbf{a}_k\mathbf{e}_{n+1} + \mathbf{b}_k\right)\Psi, \quad (k = 1, \ldots, n), \tag{2.3}$$

where $\mathbf{a}_k$ and $\mathbf{b}_k$ assume values in mutually orthogonal spaces W and V respectively, $\mathbf{e}_1, \ldots, \mathbf{e}_n$ is an orthonormal basis in V while $\mathbf{e}_{n+1}, \ldots, \mathbf{e}_{2n}$ is an orthonormal basis in W. Moreover, $\mathbf{b}_k$ is orthogonal to $\mathbf{e}_k$ for $k = 1, \ldots, n$.

The Sym–Tafel formula, $F := \Psi^{-1}\Psi_{,\lambda}|_{\lambda=1}$, defines an immersion into the space $\mathbf{e}_{n+1} \wedge (V \oplus W)$ which can be identified with $\mathbf{R}^{2n-1}$ (see [**9, 12**]).

3. The Darboux–Bäcklund Transformation in the Matrix Case

In this section we consider matrix spectral problems of the form

$$\Psi_{,k} = U_k\Psi, \quad (k = 1, \ldots, n),$$

where $U_k = U_k(x^1, \ldots, x^n; \lambda)$ are given traceless matrices ($\mathrm{Tr}\, U_k = 0$) of degree m and Ψ is the fundamental solution. The structure of the matrices U_k is prescribed. In particular, their poles in λ-plane are given. Let us consider transformations

$$\widetilde{\Psi} = D\Psi, \tag{3.1}$$

(D is a matrix depending on $x^1, \ldots, x^n$ and λ) such that the structure of the spectral problem does not change, i.e., the matrices

$$\widetilde{U}_k = D_{,k} D^{-1} + D U_k D^{-1}, \tag{3.2}$$

have the same structure (the same form with respect to λ) as the matrices U_k. This requirement is very strong and the construction of any Darboux matrix is a nontrivial task. However, in the case of matrices with polynomial or rational dependence on λ there exist standard procedures to construct Darboux matrices [**7, 27–29, 37**]. The methods are closely related. Here we confine ourselves to the case of the one-soliton Darboux matrix [**7, 28**]

$$D = N\big(\lambda - \lambda_1 + (\lambda_1 - \mu_1)P\big),$$

where N is the normalization matrix (we assume $\det N \neq 0$), λ_1 and μ_1 are complex parameters, and P is a projector (i.e., $P^2 = P$) defined by

$$\ker P = \Psi(\lambda_1) V_{\text{ker}}, \quad \operatorname{im} P = \Psi(\mu_1) V_{\text{im}},$$

where $V_{\text{ker}}, V_{\text{im}}$ are fixed linear spaces such that $V_{\text{ker}} \oplus V_{\text{im}} = \mathbf{C}^m$. The matrix D can be written in a more standard form (with the factor $(\lambda - \lambda_1)^{-1}$) with no influence on (3.2). If matrices U_k are subject to constraints [**7, 8**]

$$U_k^T = -B U_k B^{-1}, \quad U_k^\dagger(\bar{\lambda}) = -H U_k(\lambda) H^{-1}, \quad U_k(-\lambda) = J U_k(\lambda) J^{-1}, \tag{3.3}$$

where B, H, J are some constant matrices, and the dagger means Hermitean conjugate, then we can put $\lambda_1 = -\mu_1 = i\kappa_1$ ($\kappa_1 \in \mathbb{R}$) and

$$D = N\lambda - \kappa_1 N A, \tag{3.4}$$

where $A := i(I - 2P)$. In the case of spectral problems linear in λ, i.e.,

$$U_k = u_{k0} + \lambda u_{k1},$$

the transformation (3.2) reads

$$\tilde{u}_{k1} = N u_{k1} N^{-1}, \quad \tilde{u}_{k0} = N\big(u_{k0} + \kappa_1 [u_{k1}, A]\big) N^{-1} + N_{,k} N^{-1}. \tag{3.5}$$

Moreover, A satisfies differential equations

$$A_{,k} = [u_{k0}, A] + \frac{1}{2}\kappa_1 \big[[u_{k1}, A], A\big], \tag{3.6}$$

which form a system of matrix Riccati equations (compare [**21, 36**]).

4. The Darboux–Bäcklund Transformation for Spectral Problems in Clifford Algebras

Let V and W be orthogonal unitary spaces ($\dim V = r, \dim W = q$). Consider spectral problems (1.4) of the form

$$\Psi_{,k} = \mathbf{e}_k(\lambda \mathbf{a}_k + \mathbf{b}_k)\Psi, \quad (k = 1, \ldots, n), \tag{4.1}$$

where $\mathbf{a}_k$ and $\mathbf{b}_k$ assume values in W and V respectively, $\mathbf{e}_1, \ldots, \mathbf{e}_r$ is an orthonormal basis in V and $\mathbf{e}_{r+1}, \ldots, \mathbf{e}_{r+q}$ is an orthogonal basis in W. The Sym–Tafel formula, $F = \Psi^{-1}\Psi_{,\lambda}|_{\lambda=0}$, yields an immersion into the space $V \wedge W$. Computing tangent vectors we have $F_{,k} = \Psi^{-1}\mathbf{e}_k \mathbf{a}_k \Psi$, which means, in particular, that $F_{,k}$ are pairwise orthogonal.

If $n = 2$, $r = 3$, $q = 2$, V is Euclidean space and W is 2-dimensional Minkowski space, then we obtain the spectral problem for isothermic surfaces in $\mathbf{R}^3$. The Euclidean case $r = q = n$ corresponds to some orthogonal nets in $\mathbf{R}^n$. In both cases one has to project F on appropriate subspaces. It seems that in general F describes a curved flat (compare [**6**]).

The linear equations $\Psi_{,k} = U_k\Psi$ of the form (4.1) satisfy the constraints

$$\beta(U_k) = -U_k, \quad \alpha\beta\big(U_k(\bar\lambda)\big) = -U_k(\lambda), \quad U_k(-\lambda) = \mathbf{e}U_k(\lambda)\mathbf{e}^{-1}, \tag{4.2}$$

where $\mathbf{e} := \mathbf{e}_{r+1}\ldots\mathbf{e}_{r+q}$. Notice that in the case of isothermic surfaces the constraints (3.3) and (4.2) coincide [**8, 13**].

Following the results of [**8**] and [**13**] let us propose the Darboux transformation in the form

$$D = \mathbf{e}_s(\lambda\mathbf{n} + \kappa_1\mathbf{p}), \tag{4.3}$$

where s is fixed$(1 \le s \le r)$, and $\mathbf{n} \in W$, $\mathbf{p} \in V$ are unit vectors ($\mathbf{p}^2 = \mathbf{n}^2 = 1$) depending on $x^1, \ldots, x^n$. Comparing (4.3) with (3.4) we see that

$$A = \mathbf{pn}, \quad N = \mathbf{e}_s\mathbf{n}.$$

One can prove that the above form of A, N is compatible with (4.2). Moreover, (3.6) implies that $\mathbf{n}$ and $\mathbf{p}$ satisfy the following differential equations

$$\mathbf{n}_{,k} = 2\kappa_1\langle\mathbf{p} \mid \mathbf{e}_k\rangle\big(\langle\mathbf{n} \mid \mathbf{a}_k\rangle\mathbf{n} - \mathbf{a}_k\big),$$

$$\mathbf{p}_{,k} = -2\langle\mathbf{p} \mid \mathbf{e}_k\rangle\mathbf{b}_k + 2\langle\mathbf{b}_k \mid \mathbf{p}\rangle\mathbf{e}_k + 2\kappa_1\langle\mathbf{n} \mid \mathbf{a}_k\rangle\big(\langle\mathbf{p} \mid \mathbf{e}_k\rangle\mathbf{p} - \mathbf{e}_k\big).$$

where the bracket $\langle\cdot \mid \cdot\rangle$ denotes the scalar product in W and V. The transformation (3.5) assumes the form

$$\tilde{\mathbf{a}}_k = \nu_{ks}\mathbf{n}\mathbf{a}_k\mathbf{n}^{-1},$$

$$\widetilde{\mathbf{b}_k} = \nu_{ks}\mathbf{e}_s\mathbf{b}_k\mathbf{e}_s^{-1} + 2\kappa_1\langle\mathbf{n} \mid \mathbf{a}_k\rangle\big(\langle\mathbf{p} \mid \mathbf{e}_k\rangle\mathbf{e}_k - \nu_{ks}\mathbf{e}_s\mathbf{p}\mathbf{e}_s^{-1}\big),$$

where $\nu_{ks} := 1$ for $k = s$ and $\nu_{ks} := -1$ for $k \neq s$. One can easily see that $\tilde{\mathbf{a}}_k \in W$ and $\tilde{\mathbf{b}}_k \in V$. Therefore D given by (4.3) generates the Darboux–Bäcklund transformation, indeed. Recently, we found convenient formulas which express $\mathbf{n}$ and $\mathbf{p}$ in terms of Φ evaluated at $i\kappa_1$ [**14**].

5. Conclusions

We have shown that in many interesting cases the structure of the spectral problem defines uniquely (through the Sym–Tafel formula) the associated geometry. On the other hand, the same structure is very important in the construction of the Darboux–Bäcklund transformation [**7**]. Recently, we have extended the range of applications of the Sym–Tafel formula on spectral problems in Clifford algebras. It turns out that in this case it is possible to construct the Darboux–Bäcklund transformation. We expressed the one-soliton Darboux matrix purely in terms of Clifford numbers. Comparing the matrix case [**8**] with the approach presented in this paper (see also [**13, 14**]) one can see practical advantage of the latter.

References

1. M. J. Ablowitz, R. Beals, and K. Tenenblat, *On the solution of the generalized wave and generalized sine-Gordon equations*, Stud. Appl. Math. **74** (1986), 177–203.
2. Yu. A. Aminov, *On immersions of regions of the n-dimensional Lobachevsky space into (2n − 1)-dimensional Euclidean space*, Dokl. Akad. Nauk. **236** (1977), 521–524. (Russian).
3. A. I. Bobenko, *Surfaces in terms of 2 by 2 matrices. Old and New Integrable Cases*, Harmonic Maps and Integrable Systems (A. P. Fordy and J. C. Wood, eds.), Aspects Math. , vol. E23, Vieweg, Brunswick, Braunschweig, 1994, pp. 81–127.

4. A. I. Bobenko and U. Pinkall, *Discrete isothermic surfaces*, J. Reine Angew. Math. **475** (1996), 187–208.
5. P. Budinich and A. Trautman, *The spinorial chessboard*, Trieste Notes in Physics, Springer-Verlag, Berlin-New York, 1988.
6. F. Burstall, U. Hertrich-Jeromin, F. Pedit, and U. Pinkall, *Curved flats and isothermic surfaces*, Math. Z. **225** (1997), 199–209.
7. J. Cieśliński, *An algebraic method to construct the Darboux matrix*, J. Math. Phys. **36** (1995), 5670–5706.
8. ———, *The Darboux–Bianchi transformation for isothermic surfaces. Classical results versus the soliton approach*, Differential Geom. Appl. **7** (1997), 1–28.
9. ———, *The spectral interpretation of n-spaces of constant negative curvature immersed in* R^{2n-1}, Phys. Lett. A **236** (1997), 425–430.
10. ———, *The cross ratio and Clifford algebras*, Adv. Appl. Clifford Algebras **7** (1997), 133–139.
11. ———, *A generalized formula for integrable classes of surfaces in Lie algebras*, J. Math. Phys. **38** (1997), 4255–4272.
12. ———, *The Darboux–Bianchi–Bäcklund transformations and soliton surfaces*, Nonlinearity and Geometry, (Warsaw, 1995) (D. Wójcik and J. Cieśliński, eds.), Polish Scientific Publ. PWN, Warsaw, 1998, pp. 81–107.
13. ———, *The Bäcklund transformation for discrete isothermic surfaces*, Symmetries and Integrability of Difference Equations, (Canterbury, 1996) (P. A. Clarkson and F. Nijhoff, eds.), London Math. Soc. Lecture Note Ser. , vol. 255, Cambridge Univ. Press, Cambridge, 1999, pp. 109–121.
14. ———, *The construction of the Darboux–Bäcklund transformation without using a matrix representation*, J. Phys. A **33** (2000), L363–L368.
15. J. Cieśliński, P. Goldstein, and A. Sym, *Isothermic surfaces in* E^3 *as soliton surfaces*, Phys. Lett. A **205** (1995), 37–43.
16. J. Cieśliński, P. K. H. Gragert, and A. Sym, *Exact solution to localized induction approximation equation modelling smoke ring motion*, Phys. Rev. Lett. **57** (1986), 1507–1510.
17. A. Doliwa and A. Sym, *Constant mean curvature surfaces in* E^3 *as an example of soliton surfaces*, Nonlinear Evolution Equations and Dynamical Systems (Baia Verde, 1991) (M. Boiti, et al. , eds.), World Sci. Publishing, River Edge, NJ, 1992, pp. 111–117,
18. L. P. Eisenhart, *A treatise on the differential geometry of curves and surfaces*, Dover Publications, Inc. , New York, 1960.
19. D. Ferus and F. Pedit, *Isometric immersions of space forms and soliton theory*, Math. Ann. **305** (1996), 329–342.
20. A. S. Fokas and I. M. Gel'fand, *Surfaces on Lie groups, on Lie algebras and their integrability*, Comm. Math. Phys. **177** (1996), 2030–220.
21. H. Gollek, *Bäcklund transformations, Matrix-Riccati systems and isometric immersions of space forms into space forms*, Colloq. Math. Soc. János Bolyai **56** (1989), 359–385.
22. C. H. Gu, *Bäcklund transformations and Darboux transformations*, Soliton Theory and Its Applications, Springer, Berlin, 1995, pp. 122–151, 393–395.
23. H. S. Hu, *Solitons and differential geometry*, Soliton Theory and Its Applications, Springer, Berlin, 1995, pp. 297–336.
24. A. R. Its, *Liouville's theorem and the method of the inverse problem* (Russian) Differential geometry, Lie groups and mechanics, VI. Zap. Nauchn. Sem. Leningrad. Otdel. Mat. Inst. Steklov. (LOMI) 133 (1984), 113–125.
25. B. G. Konopelchenko and W. K. Schief, *Lamé and Zakharov–Manakov systems: Combescure, Darboux and Bäcklund transformations*, Preprint AM93/9, Univ. New South Wales, Sydney, 1993.
26. D. Levi and A. Sym, *Integrable systems describing surfaces of nonconstant curvature*, Phys. Lett. A **149** (1990), 381–387.
27. V. B. Matveev and M. A. Salle, *Darboux transformations and solitons*, Springer Series in Nonlinear Dynamics, Springer-Verlag, Berlin, 1991.
28. A. V. Mikhailov, *The reduction problem and the inverse scattering method*, Physica D **3** (1981), 73–117.
29. G. Neugebauer and D. Kramer, *Einstein–Maxwell solitons*, J. Phys. A **16** (1983), 1927–1936.

30. I. R. Porteous, *Clifford algebras and the classical groups*, Cambridge Studies in Advanced Mathematics, vol. 50, Cambridge University Press, Cambridge, 1995.
31. A. Sym, *Soliton surfaces*, Lett. Nuovo Cimento **33** (1982), 394–400.
32. ———, *Soliton surfaces* II. *Geometric unification of solvable nonlinearities*, Lett. Nuovo Cimento **36** (1983), 307–312.
33. ———, *Soliton surfaces and their applications (soliton geometry from spectral problems)*, Geometric Aspects of the Einstein Equations and Integrable Systems (Scheveningen, 1984) (R. Martini, ed.), Lecture Notes in Phys. , vol. 239, Springer, Berlin–New York, 1985, pp. 154-231.
34. J. Tafel, *Surfaces in* $\mathbf{R}^3$ *with prescribed curvature*, J. Geom. Phys. **17** (1995), no. 4, 381–390.
35. K. Tenenblat and C. L. Terng, *Bäcklund theorem for n-dimensional submanifolds of* R^{2n-1}, Ann. of Math. (2) **111** (1980), no. 3, 477–490.
36. P. Winternitz, this volume.
37. V. E. Zakharov, S. V. Manakov, S. P. Novikov, and L. P. Pitaievsky, *Theory of solitons*, Nauka, Moscow, 1980. (Russian)

UNIW. W BIAŁYMSTOKU,INSTYTUT FIZYKI, UL. LIPOWA 41, 15-424 BIAŁYSTOK, POLAND.
E-mail address: janek@alpha.uwb.edu.pl

Centre de Recherches Mathématiques
CRM Proceedings and Lecture Notes
Volume **29**, 2001

On Discrete Painlevé Equations as Bäcklund Transformations

P. A. Clarkson, E. L. Mansfield, and H. N. Webster

ABSTRACT. In this paper we discuss a method for deriving difference equations, in particular the discrete Painlevé equations, from the Bäcklund transformations of the continuous Painlevé equations. Using this technique one is able to derive several of the known discrete Painlevé equations, in particular the first and second discrete Painlevé equations and some of their alternative versions. It is well-known that the Painlevé equations possess hierarchies of rational solutions and one-parameter families of solutions expressible in terms of the classical special functions, for special values of the parameters. Hence hierarchies of exact solutions for the associated discrete Painlevé equations can be generated. Thus solutions of the Painlevé equations simultaneously satisfy both a differential equation and a difference equation, in analogy with the classical special functions.

1. Introduction

The six Painlevé equations (PI–PVI) were discovered around the turn of the century by Painlevé and his colleagues whilst studying equations of the form $w'' = F(z, w, w')$, with $' \equiv \mathrm{d}/\mathrm{d}z$, where F is polynomial in w', rational in w and analytic in z (c.f. [**19**]). Although discovered from purely mathematical considerations, the Painlevé equations have attracted much interest since they arise in many physical situations and as reductions of the soliton equations which are solvable by inverse scattering (c.f. [**1**]).

The Painlevé equations can be thought of as nonlinear analogues of the classical special functions. Their general solutions are transcendental in the sense that they cannot be expressed in terms of known elementary functions. However, they possess many rational solutions and one-parameter families of solutions expressible in terms of special functions for certain values of the parameters, and they possess Bäcklund transformations which relate one solution to another solution either of the same equation, with different values of the parameters, or another equation (c.f. [**3, 4, 7, 15, 27–29, 35–38**] and the references therein).

2000 *Mathematics Subject Classification.* Primary: 39M55; Secondary: 39A10.
We thank Nalini Joshi for her helpful comments and illuminating discussions. The research of HNW was supported by an Institute of Mathematics & Statistics, University of Kent postgraduate studentship which is gratefully acknowledged.
This is the final form of the paper.

Recently there has been substantial interest in integrable mappings and discrete systems. The discrete Painlevé equations (dP1–dPVI), which have the form

$$x_{n+1} = \frac{f_1(x_n;n) + x_{n-1}f_2(x_n;n)}{f_3(x_n;n) + x_{n-1}f_4(x_n;n)}, \tag{1.1}$$

where the $f_j(x_n;n)$ are polynomials of degree at most four in x_n, have become popular. They occur in the theory of orthogonal polynomials (c.f. [**25**]) and in a variety of physical applications. For example dPI and dPII arise in two-dimensional quantum gravity (c.f. [**9**]) derived using the singularity confinement method [**39**], which relates the integrable character of discrete equations to their singularity structure. However Hietarinta and Viallet [**18**] have shown that singularity confinement is not sufficient for integrability. Jimbo and Sakai [**20**] derived a two-component form of dPVI, and Grammaticos and Ramani [**13**] a one-component form. In the continuous limit, the discrete Painlevé equations yield their corresponding continuous equation, though some of the discrete equations have limits to more than one Painlevé equation. The discrete Painlevé equations share a variety of other properties in common with the Painlevé equations including Lax pairs, bilinear representations, Bäcklund transformations and exact solutions for certain parameter values, expressible in terms of rational functions or discrete special functions (c.f. [**6, 11, 21, 30, 40, 48**] and the references therein).

The discrete Painlevé equations are in some respects much richer than the Painlevé equations. For example, they have several solutions which have no continuum limit and several other properties which are lost in the continuous limit. Perhaps the most fundamental difference is that the Painlevé equations have a **unique** canonical form whereas this is not true for the discrete Painlevé equations. For each discrete Painlevé equation there exists a number of different forms. Furthermore there are two distinct types of discrete Painlevé equations, namely those for which the discrete variable is additive, and the q-discrete Painlevé equations for which it is multiplicative.

Recently there have been several studies of the derivation of discrete Painlevé equations from Bäcklund transformations of the Painlevé equations. In particular Fokas *et al.* [**8**] (see also [**11, 12, 30**]) used an approach which is based on the Schlesinger transformations related the associated isomonodromy problem for the Painlevé equation, and Gromak and Tsegel'nik [**16, 17, 41, 42**], though the difference equations they derive are not identified as discrete Painlevé equations.

2. Deriving Discrete Painlevé Equations from Bäcklund Transformations for the Continuous Painlevé Equations

First we consider the special functions, using Bessel functions as an illustrative example.

EXAMPLE 2.1. The Bessel function $J_n(z)$ satisfies the differential equation

$$z^2 J_n''(z) + zJ_n'(z) + (z^2 - n^2)J_n(z) = 0, \tag{2.1}$$

and the recurrence relations

$$zJ_n'(z) = zJ_{n-1}(z) - nJ_n(z), \tag{2.2}$$

$$zJ_n'(z) = nJ_n(z) - zJ_{n+1}(z), \tag{2.3}$$

$$zJ_{n+1}(z) - 2nJ_n(z) + zJ_{n-1}(z) = 0, \tag{2.4}$$

(c.f. [**47**]). Eliminating $J_n'(z)$ between (2.2) and (2.3) yields the recurrence relation (2.4). Letting $n \to n+1$ in (2.2) and eliminating $J_{n+1}(z)$ using (2.3) gives the differential equation (2.1). Thus the Bessel function $J_n(z)$ satisfies both the differential equation (2.1) and the difference equation (2.4). As we now show, an analogous situation occurs for the Painlevé equations.

Suppose there are two Bäcklund transformations for a Painlevé equation in the form

$$w^{\pm}(z;\boldsymbol{\alpha}^{\pm}) = \mathbf{T}^{\pm}\big(w(z;\boldsymbol{\alpha})\big) = F^{\pm}\big(w(z;\boldsymbol{\alpha}), w'(z;\boldsymbol{\alpha}), z, \boldsymbol{\alpha}\big), \tag{2.5}$$

where $w(z;\boldsymbol{\alpha})$ and $w^{\pm}(z;\boldsymbol{\alpha}^{\pm})$ are solutions of the associated Painlevé equation corresponding to the parameters $\boldsymbol{\alpha} = (\alpha_1, \alpha_2, \ldots, \alpha_m)$ and $\boldsymbol{\alpha}^{\pm} = (\alpha_1^{\pm}, \alpha_2^{\pm}, \ldots, \alpha_m^{\pm})$, respectively. Eliminating $w'(z;\boldsymbol{\alpha})$ in (2.5) yields an algebraic relation, which is a recurrence relation, between three solutions $w^{+}(z;\boldsymbol{\alpha}^{+})$, $w(z;\boldsymbol{\alpha})$ and $w^{-}(z;\boldsymbol{\alpha}^{-})$ of the Painlevé equation. This algebraic relation can be thought of as a nonlinear superposition principle for solutions of the Painlevé equation, as an alternate form of the Bäcklund transformations or as a discrete equation of Painlevé type. The continuous variable z enters as a parameter and the independent variable of the discrete equation comes from the parameters $\boldsymbol{\alpha}$.

In order to do this consistently for a given Painlevé equation, we require two Bäcklund transformations $\mathbf{T}^{\pm}$ such that $\mathbf{T}^{+} = (\mathbf{T}^{-})^{-1}$, i.e. they are inverse transformations, and so the solutions and parameters are linked as follows

$$\big\{w^{-}(z;\boldsymbol{\alpha}^{-}), \boldsymbol{\alpha}^{-}\big\} \underset{\mathbf{T}^{-}}{\overset{\mathbf{T}^{+}}{\rightleftarrows}} \big\{w(z;\boldsymbol{\alpha}), \boldsymbol{\alpha}\big\} \underset{\mathbf{T}^{-}}{\overset{\mathbf{T}^{+}}{\rightleftarrows}} \big\{w^{+}(z;\boldsymbol{\alpha}^{+}), \boldsymbol{\alpha}^{+}\big\}.$$

Hence setting $w(z;\boldsymbol{\alpha}) = x_n$, $w^{\pm}(z;\boldsymbol{\alpha}^{\pm}) = x_{n\pm1}$, with $\boldsymbol{\alpha} = \boldsymbol{a}_n$ and $\boldsymbol{\alpha}^{\pm} = \boldsymbol{a}_{n\pm1}$, then we obtain the chain

$$\{x_{n-1}, \boldsymbol{a}_{n-1}\} \underset{\mathbf{T}^{-}}{\overset{\mathbf{T}^{+}}{\rightleftarrows}} \{x_n, \boldsymbol{a}_n\} \underset{\mathbf{T}^{-}}{\overset{\mathbf{T}^{+}}{\rightleftarrows}} \{x_{n+1}, \boldsymbol{a}_{n+1}\}.$$

We illustrate this procedure using PII as an example (see also [**8**]).

EXAMPLE 2.2. Suppose that $w(z;\alpha)$ is a solution of PII

$$w'' = 2w^3 + zw + \alpha, \tag{2.6}$$

where α is a constant. It is well-known (c.f. [**7, 24**]) that

$$\mathbf{T}^{\pm}: w(z;\alpha \pm 1) = -w(z;\alpha) - \frac{2\alpha \pm 1}{2w^2(z;\alpha) \pm 2w'(z;\alpha) + z}, \qquad \alpha \neq \mp\frac{1}{2}, \tag{2.7}$$

are also solutions of PII (2.6) for the parameters $\alpha \pm 1$. Eliminating $w'(z;\alpha)$ yields

$$\frac{2\alpha + 1}{w(z;\alpha+1) + w(z;\alpha)} + \frac{2\alpha - 1}{w(z;\alpha) + w(z;\alpha-1)} + 4w^2(z;\alpha) + 2z = 0. \tag{2.8}$$

Now if we set

$$a_{n+1} = \alpha + 1, \qquad a_n = \alpha, \qquad a_{n-1} = \alpha - 1, \tag{2.9}$$

then solving these difference equations yields $\alpha = a_n = n + \kappa - \frac{1}{2}$, where κ is an arbitrary constant (the factor $\frac{1}{2}$ is for convenience); since $\mathbf{T}^{\pm}$ are inverse transformations then (2.9) have a common solution. Next setting $x_m = w(z;a_m)$, in (2.8) gives the equation

$$\frac{n+\kappa}{x_{n+1}+x_n}+\frac{n-1+\kappa}{x_n+x_{n-1}}+2x_n^2+z=0, \tag{2.10}$$

where z is a parameter, which is an alternative form of dPI [**8**].

Thus solutions of PII (2.6) satisfy both a differential equation, namely PII itself, and a difference equation, namely a-dPI (2.10), which is completely analogous to the situation for Bessel equations discussed in Example 2.1.

EXAMPLE 2.3. Here we derive a system of discrete equations. Suppose $v(z;\alpha)$ and $w(z;\alpha)$ are solutions of PII and P34

$$v''=2v^3+zv+\alpha, \tag{2.11}$$

$$ww''+2w^3+\left(z-\frac{1}{2}\right)w^2+\frac{1}{2}\left(\alpha+\frac{1}{2}\right)^2=0, \tag{2.12}$$

respectively. The equation P34 (2.12) is so-called since it is equivalent to equation XXXIV of Chapter 14 in Ince [**19**]. It is well known (c.f. [**7**]) that solutions of PII (2.11) and P34 (2.12) are related through the following invertible transformation

$$v(z;\alpha)=\left[2w'(z;\alpha)+2\alpha+1\right]/\left[4w(z;\alpha)\right], \tag{2.13}$$

$$w(z;\alpha)=-v'(z;\alpha)-v^2(z;\alpha)-\frac{1}{2}z, \tag{2.14}$$

and furthermore the Bäcklund transformations for PII and P34 are

$$v(z;\alpha+1)=-v(z;\alpha)-(2\alpha+1)/\left[2v^2(z;\alpha)+2v'(z;\alpha)+z\right], \tag{2.15}$$

$$w(z;\alpha+1)=-w(z;\alpha)-\left[2w'(z;\alpha)-2\alpha-1\right]/w^2(z;\alpha), \tag{2.16}$$

respectively. Eliminating v' between (2.14) and (2.15), yields

$$v(z;\alpha+1)=-v(z;\alpha)+\left(\alpha+\frac{1}{2}\right)\Big/w(z;\alpha), \tag{2.17}$$

whilst eliminating w' between (2.13) and (2.16) yields

$$w(z;\alpha+1)=-w(z;\alpha)-z-2\left[v(z;\alpha)-\left(\alpha+\frac{1}{2}\right)\right]^2\Big/w^2(z;\alpha). \tag{2.18}$$

As in the previous example, solving the difference equations for the parameter relations yields $\alpha=a_n=n+\kappa-\frac{1}{2}$. Thus if we set $x_m=v(z;a_m)$ and $y_m=w(z;a_m)$ in then (2.17) and (2.18) we obtain the discrete system

$$x_{n+1}=-x_n+(n+\kappa)/y_n,\quad y_{n+1}=-y_n-z-2\left[x_n-(n+\kappa)/y_n\right]^2, \tag{2.19}$$

with z a parameter.

Now consider the system (2.19). Eliminating y_n yields a-dPI (2.10), whilst eliminating x_n yields the second-order second-degree discrete equation

$$\frac{(y_{n+1}-y_{n-1})^2y_n^2}{4(n+\kappa)^2}+y_{n+1}+2y_n+y_{n-1}+2z+\frac{(n+\kappa)^2}{y_n^2}=0. \tag{2.20}$$

It is straightforward to generate this discrete equation from the Bäcklund transformation (2.16) of P34 together with the associated inverse transformation

$$w(z;\alpha-1)=-w(z;\alpha)-2\left(\frac{w'(z;\alpha)+\alpha}{2w(z;\alpha)-z}\right)^2.$$

Thus we have generalized the procedure outlined above and obtained the discrete system (2.19). This system includes the discrete equations arising from the Bäcklund transformations of PII and P34, i.e. (2.10) and (2.20), respectively.

3. Discrete Equations Arising from Bäcklund Transformations of PIII

The third Painlevé equation (PIII) is given by

$$w'' = \frac{(w')^2}{w} - \frac{w'}{z} + \frac{\alpha w^2 + \beta}{z} + \gamma w^3 + \frac{\delta}{w}, \tag{3.1}$$

where α, β, γ and δ are arbitrary constants. In the generic case when $\gamma\delta \neq 0$ in (3.1), then without loss of generality we set $\gamma = 1$, and $\delta = -1$. All Bäcklund transformations of PIII with $\gamma = 1$ and $\delta = -1$, are known to be combinations of the following six transformations (see [**27**] for details)

$$\begin{aligned}
&\mathbf{T}_1\colon\ w_1(z;\alpha_1,\beta_1) = \frac{w(zw' + zw^2 + \alpha w + w + z)}{zw' + zw^2 - \beta w - w + z}, && \alpha_1 = -\beta - 2, && \beta_1 = -\alpha - 2,\\
&\mathbf{T}_2\colon\ w_2(z;\alpha_2,\beta_2) = \frac{w(zw' - zw^2 - \alpha w + w + z)}{zw' - zw^2 - \beta w - w + z}, && \alpha_2 = \beta + 2, && \beta_2 = \alpha - 2,\\
&\mathbf{T}_3\colon\ w_3(z;\alpha_3,\beta_3) = \frac{w(zw' + zw^2 + \alpha w + w - z)}{zw' + zw^2 + \beta w - w - z}, && \alpha_3 = \beta - 2, && \beta_3 = \alpha + 2,\\
&\mathbf{T}_4\colon\ w_4(z;\alpha_4,\beta_4) = \frac{w(zw' - zw^2 - \alpha w + w - z)}{zw' - zw^2 + \beta w - w - z}, && \alpha_4 = -\beta + 2, && \beta_4 = -\alpha + 2,\\
&\widehat{\mathbf{T}}\colon\ \widehat{w}(z;\widehat{\alpha},\hat{\beta}) = -w, && \widehat{\alpha} = -\alpha, && \hat{\beta} = -\beta,\\
&\widetilde{\mathbf{T}}\colon\ \widetilde{w}(z;\widetilde{\alpha},\tilde{\beta}) = 1/w, && \widetilde{\alpha} = -\beta, && \tilde{\beta} = -\alpha,
\end{aligned}$$

where $w \equiv w(z;\alpha,\beta)$ is a solution of (3.1) with $\gamma = 1$, and $\delta = -1$.

Within the Bäcklund transformations of PIII given by $\{\mathbf{T}_j, \widehat{\mathbf{T}}_j \equiv \widehat{\mathbf{T}} \circ \mathbf{T}_j, \widetilde{\mathbf{T}}_j \equiv \widetilde{\mathbf{T}} \circ \mathbf{T}_j, \mathbf{T}^*_j \equiv \widehat{\mathbf{T}} \circ \widetilde{\mathbf{T}} \circ \mathbf{T}_j\}_{j=1}^4$, there are four nontrivial pairs with compatible parameters and which are inverse transformations. In each case we may solve the recurrence relations for α and β, then eliminate w' and generate a discrete equation.

Example 3.1. $\{\widehat{\mathbf{T}}_1, \widehat{\mathbf{T}}_4\}$. Setting $a_{n+1} = \widehat{\alpha}_1$, $a_n = \alpha$, $a_{n-1} = \widehat{\alpha}_4$, $b_{n+1} = \hat{\beta}_1$, $b_n = \beta$, $b_{n-1} = \hat{\beta}_4$ in the parameter relations yield difference equations which have solution $a_n = \kappa + \mu(-1)^n + 2n$ and $b_n = \kappa - \mu(-1)^n + 2n$, with κ and μ arbitrary constants. Hence setting $x_n = w(z;a_n,b_n)$, $x_{n+1} = \widehat{w}_1(z;a_{n+1},b_{n+1})$ and $x_{n-1} = \widehat{w}_4(z;a_{n-1},b_{n-1})$, and eliminating w' gives

$$\frac{\kappa + 2n + 1}{x_{n+1} + x_n} + \frac{\kappa + 2n - 1}{x_n + x_{n-1}} + z - \frac{2n + \kappa - \mu(-1)^n}{x_n} + \frac{z}{x_n^2} = 0. \tag{3.2}$$

It can be shown that the continuum limit of this equation gives the special case of PIII (3.1) with $\alpha = \beta$ and $\gamma = \delta = 0$. Applying this procedure to the Bäcklund transformations $\{\widehat{\mathbf{T}}_2, \widehat{\mathbf{T}}_3\}$ yields a difference equation which is equivalent to (3.2).

Example 3.2. $\{\mathbf{T}^*_2, \mathbf{T}^*_3\}$. Setting $a_{n+1} = \alpha^*_2$, $a_n = \alpha$, $a_{n-1} = \alpha^*_3$, $b_{n+1} = \beta^*_2$, $b_n = \beta$, $b_{n-1} = \beta^*_3$ in the parameter relations yield difference equations which have solution $a_n = 2\kappa - 2n$ and $b_n = 2\mu + 2n$, with κ and μ arbitrary constants. Hence setting $x_n = w(z;a_n,b_n)$, $x_{n+1} = w^*_2(z;a_{n+1},b_{n+1})$ and $x_{n-1} = w^*_3(z;a_{n-1},b_{n-1})$, and eliminating w' gives

$$\frac{2n+1-\kappa+\mu}{x_n x_{n+1}+1}+\frac{2n-1-\kappa+\mu}{x_n x_{n-1}+1}+x_n z-2n+2\kappa-\frac{z}{x_n}=0, \tag{3.3}$$

which is an alternative dPII [**8, 30**]. Applying this procedure to the Bäcklund transformations $\{\widetilde{\mathbf{T}}_1, \widetilde{\mathbf{T}}_4\}$ yields a difference equation which is equivalent to the alternative dPII (3.3).

Hence the continuous Bäcklund transformations of PIII yield two discrete Painlevé equations given by (3.2) and (3.3). As with all discrete Painlevé equations, the naming of these equations will be based on the continuous equation obtained in the continuous limit, though, as we remarked above, this is not unique.

EXAMPLE 3.3. Consider the Bäcklund transformations $\widetilde{\mathbf{T}}_1$ and $\widehat{\mathbf{T}}_4$, given by

$$\widetilde{\mathbf{T}}_1\colon\ \widetilde{w}_1(z;\widetilde{\alpha}_1,\widetilde{\beta}_1)=\frac{zw'+zw^2-\beta w-w+z}{w(zw'+zw^2+\alpha w+w+z)},\qquad \widetilde{\alpha}_1=\alpha+2,\ \widetilde{\beta}_1=\beta+2.$$

$$\widehat{\mathbf{T}}_4\colon\ \widehat{w}_4(z;\widehat{\alpha}_4,\widehat{\beta}_4)=-\frac{w(zw'-zw^2-\alpha w+w-z)}{zw'-zw^2+\beta w-w-z},\qquad \widehat{\alpha}_4=\beta-2,\ \widehat{\beta}_4=\alpha-2.$$

Setting $a_{n+1}=\widetilde{\alpha}_1$, $a_n=\alpha$, $a_{n-1}=\widehat{\alpha}_4$, $b_{n+1}=\widetilde{\beta}_1$, $b_n=\beta$, $b_{n-1}=\widehat{\beta}_4$ in the parameter relations yield difference equations which have the solution $\alpha_n=\kappa+2n$ and $\beta_n=\kappa+2n$, where κ is an arbitrary constant. We remark that here there is only one arbitrary constant so the solution not as general as in the cases discussed above. Following the procedure given previously, eliminating w' in $\widetilde{\mathbf{T}}_1$ and $\widehat{\mathbf{T}}_4$, yields the discrete equation

$$zx_n+1+\frac{z}{x_n}+\frac{x_n(2n+\kappa-1)}{x_n+x_{n-1}}+\frac{2n+\kappa+1}{x_n x_{n+1}-1}=0. \tag{3.4}$$

Next we investigate the mapping (3.4) further by studying its effect on rational solutions of PIII (3.1) in which $\alpha=\beta$, $\gamma=1$ and $\delta=-1$. Applying the Bäcklund transformation $\widetilde{\mathbf{T}}_1$ to the simple rational solution $w(z;0,0)=1$, yields the rational solution $w(z;2,2)=(2z-1)/(2z+1)$. Then substituting $x_0=1$, $x_1=(2z-1)/(2z+1)$ and $\kappa=0$ into (3.4) gives

$$x_2=\frac{(2z+1)(32z^4-24z^3+4z^2+6z+1)}{(2z-1)(32z^4+24z^3+4z^2-6z+1)},$$

which is **not** a solution of (3.1), so (3.4) is not a mapping between solutions of PIII.

This poses the equation as to why the consistency of the parameters is insufficient? Consider the rational solution $w(z;2,2)=(2z-1)/(2z+1)$, which is in the "$w=1$" hierarchy (see [**27**]). Then applying the Bäcklund transformation $\widetilde{\mathbf{T}}_1$ followed by the transformation $\widehat{\mathbf{T}}_4$ to this solution gives the rational solution $w(z;2,2)=-(2z+1)/(2z-1)$, which is in the "$w=-1$" hierarchy. Hence for some sets of parameters there exists more than one exact solution of PIII (3.1), which explains the difficulty. Thus having consistent parameters is insufficient. It is necessary that the Bäcklund transformations are inverses.

It is straightforward to show that $\widehat{\mathbf{T}}_4(\widetilde{\mathbf{T}}_1(w))=-1/w$, so $\widehat{\mathbf{T}}_4\widetilde{\mathbf{T}}_1\neq\mathbf{I}$, the identity. We do not believe that (3.4) is a discrete Painlevé equation since it appears that it is **not** integrable. To demonstrate this, consider the singularity confinement of (3.4). Solving for x_{n+1} one finds that a divergence may occur if $x_n=0$ for some $n\in\mathbb{N}$. We set $x_n=\varepsilon$, instead of strictly zero, whilst keeping x_{n-1} arbitrary.

Then iterating (3.4) we find that $x_{n+1} = \varepsilon^{-1} + \mathrm{O}(1)$, $x_{n+3} = \varepsilon^{-1} + \mathrm{O}(1)$, $x_{n+2} = \varepsilon + \mathrm{O}(\varepsilon^2)$, $x_{n+4} = \varepsilon + \mathrm{O}(\varepsilon^2)$, etc. A further study shows that this pattern continues *ad infinitum*. Therefore the discrete equation (3.4) does not have the singularity confinement property and therefore is probably not integrable.

4. Discrete Equations Arising from Bäcklund Transformations of PIV

The fourth Painlevé equation (PIV) is given by

$$ww'' = \frac{1}{2}(w')^2 + \frac{3}{2}w^4 + 4zw^3 + 2(z^2 - \alpha)w^2 + \beta, \tag{4.1}$$

where α and β are arbitrary constants. The Bäcklund transformations for PIV are

$$\widetilde{\mathbf{W}}^{\pm} \ : \ \widetilde{w}^{\pm}(z;\widetilde{\alpha}^{\pm},\widetilde{\beta}^{\pm}) = \frac{w' - w^2 - 2zw \mp \sqrt{-2\beta}}{2w},$$

$$\widetilde{\alpha}^{\pm} = \frac{1}{4}(2 - 2\alpha \pm 3\sqrt{-2\beta}), \quad \widetilde{\beta}^{\pm} = -\frac{1}{2}\left(1 + \alpha \pm \frac{1}{2}\sqrt{-2\beta}\right)^2,$$

$$\widehat{\mathbf{W}}^{\pm} \ : \ \widehat{w}^{\pm}(z;\widehat{\alpha}^{\pm},\widehat{\beta}^{\pm}) = -\frac{w' + w^2 + 2zw \mp \sqrt{-2\beta}}{2w},$$

$$\widehat{\alpha}^{\pm} = -\frac{1}{4}(2 + 2\alpha \pm 3\sqrt{-2\beta}), \quad \widehat{\beta}^{\pm} = -\frac{1}{2}\left(1 - \alpha \pm \frac{1}{2}\sqrt{-2\beta}\right)^2,$$

$$\mathbf{W}^{\pm} \ : \ w^{\pm}(z;\alpha^{\pm},\beta^{\pm}) = w + \frac{2(1 - \alpha \mp 1/2\sqrt{-2\beta})w}{w' \pm \sqrt{-2\beta} + 2zw + w^2},$$

$$\alpha^{\pm} = \frac{3}{2} - \frac{1}{2}\alpha \mp \frac{3}{4}\sqrt{-2\beta}, \quad \beta^{\pm} = -\frac{1}{2}(1 - \alpha \pm 1/2\sqrt{-2\beta})^2,$$

$$\mathbf{W}^{\ddagger\pm}\!: \ w^{\ddagger\pm}(z;\alpha^{\ddagger\pm},\beta^{\ddagger\pm}) = w + \frac{2(1 + \alpha \pm 1/2\sqrt{-2\beta})w}{w' \mp \sqrt{-2\beta} - 2zw - w^2},$$

$$\alpha^{\ddagger\pm} = -\frac{3}{2} - \frac{1}{2}\alpha \mp \frac{3}{4}\sqrt{-2\beta}, \quad \beta^{\ddagger\pm} = -\frac{1}{2}\left(-1 - \alpha \pm \frac{1}{2}\sqrt{-2\beta}\right)^2,$$

$$\overline{\mathbf{W}} \ : \ \overline{w}(z;\overline{\alpha},\bar{\beta}) = -\mathrm{i}w(\mathrm{i}z;\alpha,\beta), \quad \overline{\alpha} = -\alpha, \bar{\beta} = \beta,$$

where $w \equiv w(z;\alpha,\beta)$ is a solution of PIV (4.1), valid for all solutions $w(z;\alpha,\beta)$ for which the denominators are nonzero. The Bäcklund transformations $\widetilde{\mathbf{W}}^{\pm}$ and $\widehat{\mathbf{W}}^{\pm}$ were derived by Lukashevich [**23**] and Gromak [**14**] and $\mathbf{W}^{\dagger\pm}$ and $\mathbf{W}^{\ddagger\pm}$ by Kitaev [**22**] (see also [**4, 28**]). Bassom *et al.* [**4**] showed that the transformations $\mathbf{W}^{\dagger\pm}$ and $\mathbf{W}^{\ddagger\pm}$ can be expressed in terms of $\widetilde{\mathbf{W}}^{\pm}$ and $\widehat{\mathbf{W}}^{\pm}$.

A difficulty with the Bäcklund transformations for PIV is the square root term $\sqrt{-2\beta}$ and so it is often simpler to let $\gamma = \sqrt{-2\beta}$, though one has to be careful about the choice of sign for γ. This has the effect of making the parameter relations linear, which is useful.

EXAMPLE 4.1. $\{\widetilde{\mathbf{W}}^{+},\widehat{\mathbf{W}}^{-}\}$. Here the parameter relations yield difference equations which have the solution $a_n = \frac{1}{2}\kappa - \frac{3}{2}\mu(-1)^n + \frac{1}{2}n$ and $c_n = \kappa + \mu(-1)^n + n$, with κ and μ arbitrary constants. Eliminating w' and setting $x_n = w(z;a_n,b_n)$, $x_{n+1} = \widetilde{w}^{+}(z;a_{n+1},b_{n+1})$ and $x_{n-1} = \widehat{w}^{-}(z;a_{n-1},b_{n-1})$, with $b_m = -\frac{1}{2}c_m^2$, gives

$$(x_{n+1} + x_n + x_{n-1})x_n = -2zx_n + -n + \kappa(-1)^n + \mu, \tag{4.2}$$

which is dPI. Applying the procedure to the Bäcklund transformations $\{\widetilde{\mathbf{W}}^{-},\widehat{\mathbf{W}}^{+}\}$ yields a difference equation which is equivalent to dPI (4.2).

The relationship between solutions of PIV (4.1) and solutions of dPI (4.2) was first noted by Fokas *et al.* [**9**] (see also [**8, 12, 16, 17**]). Bassom *et al.* [**4, 5**] derived hierarchies of simultaneous solutions of PIV (4.1) and dPI (4.2) in terms of parabolic cylinder functions. The relationship between solutions of PIV (4.1) and dPI (4.2) is reflected in the similarity of the results for dPI [**12**] to those for PIV [**4, 28**].

EXAMPLE 4.2. $\{\overline{\mathbf{W}}^{\dagger+} \equiv \overline{\mathbf{W}} \circ \mathbf{W}^{\dagger+}, \overline{\mathbf{W}}^{\ddagger+} \equiv \overline{\mathbf{W}} \circ \mathbf{W}^{\ddagger+}\}$. Here the parameter relations yield difference equations which have the solution $a_n = -\frac{3}{2}\kappa + \frac{1}{2}\mu(-1)^n - \frac{3}{2}n$ and $c_n = -\kappa - \mu(-1)^n - n$, with κ and μ arbitrary constants. Eliminating w' and setting $x_n = w(z; a_n, c_n)$, $x_{n+1} = \overline{w}^{\dagger+}(z; a_{n+1}, c_{n+1})$ and $x_{n-1} = \overline{w}^{\ddagger+}(z; a_{n-1}, c_{n-1})$ gives

$$\frac{2n+1+2\kappa}{x_{n+1}-x_n} + \frac{2n-1+2\kappa}{x_{n-1}-x_n} + x_n + 2z + \frac{n+\kappa+\mu(-1)^n}{x_n} = 0, \tag{4.3}$$

which is an alternate form of dPI. Applying the procedure to the Bäcklund transformations $\{\overline{\mathbf{W}}^{\dagger-} \equiv \overline{\mathbf{W}} \circ \mathbf{W}^{\dagger-}, \overline{\mathbf{W}}^{\ddagger-} \equiv \overline{\mathbf{W}} \circ \mathbf{W}^{\ddagger-}\}$ yields a difference equation which is equivalent to (4.3).

EXAMPLE 4.3. Murata [**28**] derived the Bäcklund transformations $\mathbf{T}_\pm$ for PIV (4.1) given by

$$\mathbf{T}_\pm\colon\ w_\pm(z;\alpha_\pm,\beta_\pm) = \frac{\left(w' \mp w^2 \mp 2zw\right)^2 + 2\beta}{2w\left[w' \pm w^2 - 2zw \pm 2(\alpha \pm 1)\right]}, \quad \alpha_\pm = \alpha \pm 2,\ \beta_\pm = \beta,$$

which are the transformations $\mathbf{R}_5$ and $\mathbf{R}_7$ given by Fokas *et al.* [**10**]. Bassom *et al.* [**4**] showed that $\mathbf{T}_\pm$ can be expressed in terms of the transformations $\widetilde{\mathbf{W}}^\pm$ and $\widehat{\mathbf{W}}^\pm$. Setting $\alpha_\pm = a_n^\pm$, $\alpha = a_n$, and solving the difference equations gives $\alpha_n = \kappa + 2n$, with κ an arbitrary constant. Letting $x_{n\pm1} = w_\pm(z; a_n^\pm, \beta)$ and $x_n = w(z; a_n, \beta)$, and eliminating w' yields the second-order second-degree difference equation

$$\begin{aligned}
x_{n+1}^2\{2x_n^4 &+ 8x_n^3 z + 4(2z^2-\alpha-2n-1)x_n^2 - 8z(\alpha+2n+1)x_n \cdot \\
&+ 2(\alpha+2n+1)^2 + \beta - \left[2x_n^2 - 4zx_n - 4(\alpha+2n)\right]x_n x_{n-1}\} \\
&+ 2x_n^6 + 4x_n^5(4z + x_{n+1} - x_{n-1}) + 24zx_n^4(x_{n+1}-x_{n-1}) - 6x_n^4 x_{n+1}x_{n-1} \\
&+ 48z^2x_n^3(x_{n+1}-x_{n-1}) - 4x_n^3(x_{n-1}+x_{n+1}) - 4(\alpha+2n)x_n^3(x_{n+1}-x_{n-1}) \\
&- 24zx_n^3x_{n+1}x_{n-1} + 8(\alpha+2n-3z^2)x_n^2x_{n+1}x_{n-1} \\
&+ 16z(\alpha+2n)x_nx_{n+1}x_{n-1} + \left[4(\alpha+2n)^2 - 4 - 2\beta\right]x_{n+1}x_{n-1} \\
&- 16z(\alpha+2n+1-2z^2)x_n^2x_{n+1} + 16z(\alpha+2n-1-2z^2)x_n^2x_{n-1} \\
&- 16z^2(\alpha+2n+1)x_nx_{n+1} + 16z^2(\alpha+2n-1)x_nx_{n-1} \\
&+ 48z^2x_n^4 + 64z^3x_n^3 + 32z^4 + 4\beta(x_n+2z)^2 + 4\beta(x_n+2z)(x_{n+1}-x_{n-1}) \\
&+ x_{n-1}^2\{2x_n^4 + 8x_n^3z + 4(2z^2-\alpha-2n+1)x_n^2 - 8z(\alpha+2n-1)x_n \\
&\qquad + 2(\alpha+2n-1)^2 + \beta + \left[2x_n^2 + 4zx_n - 4(\alpha+2n)\right]x_nx_{n+1}\} = 0.
\end{aligned}$$

5. Discussion

We have illustrated that there is a close relationship between the Bäcklund transformations and the solution hierarchies for the Painlevé and discrete Painlevé equations. Hierarchies of solutions of Painlevé equations satisfy both a differential equation and a difference equation. Using Bäcklund transformations of Painlevé equations one can derive various difference equations, including discrete Painlevé equations, second-degree difference equations and systems of difference equations.

Hierarchies of rational and one-parameter families of exact solutions of PII, PIII and PIV are well-known, as remarked above. Since there is an explicit relationship between PII, PIII, PIV and some discrete Painlevé equations, then these solution hierarchies also satisfy difference equations as well as ordinary differential equations. This is entirely analogous to the situation for the classical special functions. This is further evidence that the Painlevé equations may be thought of as nonlinear special functions and that there is a fundamental relationship between special functions, Painlevé equations and discrete Painlevé equations (see also [**40**]).

The procedure described here can be generalised to derive partial difference equations. Consider two Bäcklund transformations $\mathbf{T}$ and $\mathbf{S}$ of a Painlevé equation such that $\mathbf{TS} = \mathbf{ST}$, with respective inverses $\mathbf{T}^{-1}$ and $\mathbf{S}^{-1}$, so that the solutions $w(z;\boldsymbol{\alpha}_{n,m})$ of the Painlevé equation are related as follows

$$
\begin{array}{ccc}
w(z;\boldsymbol{\alpha}_{n,m}) & \underset{\mathbf{T}^{-1}}{\overset{\mathbf{T}}{\rightleftarrows}} & w(z;\boldsymbol{\alpha}_{n,m+1}) \\
\mathbf{S}\downarrow\uparrow\mathbf{S}^{-1} & & \mathbf{S}\downarrow\uparrow\mathbf{S}^{-1} \\
w(z;\boldsymbol{\alpha}_{n+1,m}) & \underset{\mathbf{T}^{-1}}{\overset{\mathbf{T}}{\rightleftarrows}} & w(z;\boldsymbol{\alpha}_{n+1,m+1})
\end{array}
\tag{5.1}
$$

Eliminating w' between $\mathbf{T}\big(w(z;\boldsymbol{\alpha}_{n,m})\big)$ and $\mathbf{S}\big(w(z;\boldsymbol{\alpha}_{n,m})\big)$ yields a partial difference equation relating $x_{n,m} \equiv w(z;\boldsymbol{\alpha}_{n,m})$, $x_{n+1,m} \equiv w(z;\boldsymbol{\alpha}_{n+1,m})$ and $x_{n,m+1} \equiv w(z;\boldsymbol{\alpha}_{n,m+1})$. We illustrate this in the following example (see also [**46**]).

EXAMPLE 5.1. Consider the Bäcklund transformations $\mathbf{B}_1 \equiv \widetilde{\mathbf{T}} \circ \mathbf{T}_1$ and $\mathbf{B}_2 \equiv \widetilde{\mathbf{T}} \circ \widehat{\mathbf{T}} \circ \mathbf{T}_2$ of PIII with respective inverses $\mathbf{B}_4 \equiv \widetilde{\mathbf{T}} \circ \mathbf{T}_4$ and $\mathbf{B}_3 \equiv \widetilde{\mathbf{T}} \circ \widehat{\mathbf{T}} \circ \mathbf{T}_3$. The parameter relations yield partial difference equations which have the solution $a_{n,m} = 2n - 2m + \kappa$, $b_{n,m} = 2n + 2m + \mu$ with κ and μ arbitrary constants. Eliminating w' between $\mathbf{B}_1$ and $\mathbf{B}_2$, substituting $\alpha = a_{n,m}$ and $\beta = b_{n,m}$ and setting $w_1 = x_{n-1,m}$, $w = x_{n,m}$, $w_2 = x_{n,m+1}$ gives the partial difference equation

$$
\frac{2n+\mu+\kappa+1}{x_{n+1,m}x_{n,m}-1} + \frac{2m+\mu-\kappa+1}{x_{n,m+1}x_{n,m}+1} + z x_{n,m} + 2n - 2m + 2\kappa = 0.
$$

To conclude we state some open questions:

1. Are all the difference equations derived from Bäcklund transformations of Painlevé equations "integrable"? If so, can the Lax pair for the discrete equation be obtained algorithmically from the Lax pair (monodromy problem) and Bäcklund transformation of the associated Painlevé equation?
2. Can all the discrete Painlevé equations be obtained from Bäcklund transformations of the Painlevé equations? In particular, can this method be used to derive the q-discrete Painlevé equations? To date, only discrete Painlevé equations of difference type have been derived using this procedure.
3. Okamoto [**35–38**] has shown that PII–PVI admit symmetries under affine Weyl groups of type $\tilde{A}_1$, $\widetilde{C_2}$, $\tilde{A}_2$, $\tilde{A}_3$ and $\widetilde{D_4}$, respectively, which are related to the associated Bäcklund transformations. The relationship between the

affine Weyl groups, rational solutions of PII–PVI and discrete equations is studied in [**31–34, 43–45**] (see also [**2**]). This relationship warrants further investigation.

References

1. M. J. Ablowitz and P. A. Clarkson, *Solitons, nonlinear evolution equations and inverse scattering*, London Math. Soc. Lecture Note Ser. **149**, Cambridge University Press, Cambridge, 1991.
2. V. E. Adler, *Nonlinear chains and Painlevé equations*, Phys. D **73** (1994), 335–351.
3. H. Airault, *Rational solutions of Painlevé equations*, Stud. Appl. Math. **61** (1979), 31–53.
4. A. P. Bassom, P. A. Clarkson, and A. C. Hicks, *Bäcklund transformations and solution hierarchies for the fourth Painlevé equation*, Stud. Appl. Math. **95** (1995), 1–71.
5. ———, *On the application of solutions of the fourth Painlevé equation to various physically motivated nonlinear partial differential equations*, Adv. Differential Equations **1** (1995), 175–198.
6. P. A. Clarkson and F.W. Nijhoff (eds.), *Symmetries and integrability of difference equations*, London Math. Soc. Lecture Note Ser. **255**, Cambridge University Press, Cambridge, 1999.
7. A. S. Fokas and M. J. Ablowitz, *On a unified approach to transformations and elementary solutions of Painlevé equations*, J. Math. Phys. **23** (1982), 2033–2042.
8. A. S. Fokas, B. Grammaticos, and A. Ramani, *From continuous to discrete Painlevé equations*, J. Math. Anal. Appl. **180** (1993), 342–360.
9. A. S. Fokas, A. R. Its, and A. V. Kitaev, *Discrete Painlevé equations and their appearance in quantum-gravity*, Commun. Math. Phys. **142** (1991), 313–344.
10. A. S. Fokas, U. Mugan, and M. J. Ablowitz, *A method of linearisation for Painlevé equations: Painlevé* IV, V, Phys. D **30** (1988), 247–283.
11. B. Grammaticos, F. W. Nijhoff, and A. Ramani, *Discrete Painlevé equations*, The Painlevé Property, One Century Later (R. Conte, ed.), CRM Series in Math. Phys., Springer-Verlag, Berlin, 1999 (to appear).
12. B. Grammaticos and A. Ramani, *From continuous Painlevé* IV *to the asymmetric discrete Painlevé* I, J. Phys. A **31** (1998), 5787–5798.
13. ———, *On a novel q-discrete analog of the Painlevé* VI *equation*, Phys. Lett. A **257** (1999), 288–292.
14. V. I. Gromak, *One-parameter systems of solutions of Painlevé's equations*, Differential Equations **14** (1978), 1510–1513.
15. ———, *Bäcklund transformations of Painlevé equations and their applications*, The Painlevé Property, One Century Later (R. Conte, ed.), CRM Series in Math. Phys., Springer-Verlag, Berlin, 1999 (to appear).
16. V. I. Gromak and V. V. Tsegel'nik, *Functional relations between solutions to P-type equations*, Differential Equations **30** (1994), 1037–1043.
17. ———, *Some properties of the Bäcklund transformations of the Painlevé equations*, Differential Equations **32** (1996), 1024–1029.
18. J. Hietarinta and C. Viallet, *Singularity confinement and chaos in discrete systems*, Phys. Rev. Lett. **81** (1998), 325–328.
19. E. L. Ince, *Ordinary differential equations*, Dover, New York, 1956.
20. M. Jimbo and H. Sakai, *A q-analog of the sixth Painlevé equation*, Lett. Math. Phys. **38** (1996), 145–154.
21. N. Joshi, A. Ramani, and B. Grammaticos, *A bilinear approach to discrete Miura transformations*, Phys. Lett. A **249** (1998), 59–62.
22. A. V. Kitaev, private communication, 1991.
23. N. A. Lukashevich, *Theory of the fourth Painlevé equation*, Differential Equations **3** (1967), 395–399.
24. ———, *The second Painlevé equation*, Differential Equations **7** (1971), 853–854.
25. A. P. Magnus, *Painlevé type differential equations for the recurrence coefficients of semi-classical orthogonal polynomials*, J. Comput. Appl. Anal. **57** (1995), 215–237.
26. E. L. Mansfield and H. N. Webster, *On one-parameter families of Painlevé* III, Stud. Appl. Math. **101** (1998), 321–341.

27. A. E. Milne, P. A. Clarkson, and A. P. Bassom, *Bäcklund transformations and solution hierarchies for the third Painlevé equation*, Stud. Appl. Math. **98** (1997), 139–194.
28. Y. Murata, *Rational solutions of the second and the fourth Painlevé equations*, Funkcial. Ekvac. **28** (1985), 1–32.
29. ———, *Classical solutions of the third Painlevé equations*, Nagoya Math. J. **139** (1995), 37–65.
30. F. W. Nijhoff, J. Satsuma, K. Kajiwara, B. Grammaticos, and A. Ramani, *A study of the alternative discrete Painlevé* II *equation*, Inverse Problems **12** (1996), 697–716.
31. M. Noumi, S. Okada, K. Okamota, and H. Umemura, *Special polynomials associated with the Painlevé equations* II, Integrable Systems and Algebraic Geometry (M.-H. Saito, Y. Shimizu, and R. Ueno, eds), World Scientific, Singapore, 1998, pp. 349–372.
32. M. Noumi and Y. Yamada, *Affine Weyl groups, discrete dynamical systems and Painlevé equations*, Comm. Math. Phys. **199** (1998), 281–295.
33. ———, *Unemura polynomials for Painlevé* V *equation*, Phys. Lett. A **247** (1998), 65–69.
34. ———, *Symmetries in the fourth Painlevé equation and Okamoto polynomials*, Nagoya Math. J. **153** (1999), 53–86.
35. K. Okamoto, *Studies on the Painlevé equations* I. *Sixth Painlevé equation* $P_{\rm VI}$, Ann. Mat. Pura Appl. (4) **146** (1987), 337–381.
36. ———, *Studies on the Painlevé equations* II. *Fifth Painlevé equation* $P_{\rm V}$, Japan. J. Math. (N.S.) **13** (1987), 47–76.
37. ———, *Studies on the Painlev equations.* III. *Second and fourth Painlev equations,* $P_{\rm II}$ *and* $P_{\rm IV}$, Math. Ann. **275** (1986), no. 2, 221–255.
38. ———, *Studies on the Painlevé equations* IV. *Third Painlevé equation,* $P_{\rm III}$, Funkcial. Ekvac. **30** (1987), no. 2-3, 305–332.
39. A. Ramani, B. Grammaticos, and J. Hietarinta, *Discrete versions of the Painlevé equations*, Phys. Rev. Lett. **67** (1991), 1829–1832.
40. K. M. Tamizhmani, A. Ramani, B. Grammaticos, and K. Kajiwara, *Coalescence cascades and special-function solutions for the continuous and discrete Painlevé equations*, J. Phys. A **31** (1998), 5799–5810.
41. V. V. Tsegel'nik, *On a relationship between solutions of the* 3rd *Painlevé equation*, Theoret. and Math. Phys. **102** (1995), 265–266.
42. ———, *Analytic properties of solutions of a nonlinear P-type equation*, Differential Equations **32** (1996), 1433–1435.
43. H. Umemura, *Special polynomials associated with the Painlevé equations.* I, Proceedings of the Workshop on Painlevé Transcendents, Montréal, Canada, 1996 (to appear).
44. H. Umemura and H. Watanabe, *Solutions of the second and fourth Painlevé equations.* I, Nagoya Math. J. **148** (1997), 151–198.
45. ———, *Solutions of the third Painlevé equation.* I, Nagoya Math. J. **151** (1998), 1–24.
46. H. N. Webster, *On the continuous and discrete third Painlevé equation*, Ph.D. Thesis, Institute of Mathematics and Statistics, University of Kent at Canterbury, UK, 1999.
47. E. E. Whittaker and G. M. Watson, *Modern analysis*, 4th ed., Cambridge University Press, Cambridge, 1927.
48. P. Winternitz and D. Levi, and L. Vinet (eds.), *Symmetries and integrability of difference equations*, (Estérel, 1994), CRM Proc. & Lecture Notes, vol. 9, Amer. Math. Soc. Providence, RI, 1996.

INSTITUTE OF MATHEMATICS AND STATISTICS, UNIVERSITY OF KENT, CANTERBURY, CT2 7NF, UK.
E-mail address, P. A. Clarkson: `P.A.Clarkson@ukc.ac.uk`

E-mail address, E. L. Mansfield: `E.L.Mansfield@ukc.ac.uk`

METEOROLOGICAL OFFICE, LONDON ROAD, BRACKNELL, RG12 2SZ, UK.
E-mail address, H. N. Webster: `hnwebster@meto.gov.uk`

Centre de Recherches Mathématiques
CRM Proceedings and Lecture Notes
Volume **29**, 2001

A Bäcklund Transformation for Timelike Surfaces of Constant Mean Curvature in $\mathbb{R}^{1,2}$

J. N. Clelland

ABSTRACT. We show that there is a Bäcklund transformation between timelike surfaces of constant mean curvature in $\mathbb{R}^{1,2}$. In the minimal case we describe the transformation explicitly and give examples.

1. Introduction

In this paper we use the method of moving frames to prove a theorem similar to Bäcklund's original theorem for surfaces of constant negative Gauss curvature in $\mathbb{R}^3$. Bäcklund's theorem may be stated as follows:

THEOREM 1.1 (Bäcklund). *Let $U \subset \mathbb{R}^2$ be an open set, and let $X, \overline{X}\colon U \to \mathbb{R}^3$ be oriented immersed surfaces. Let $N(p), \overline{N}(p)$ denote the unit normals to $X, \overline{X}$ at $X(p)$ and $\overline{X}(p)$, respectively. Suppose that X and $\overline{X}$ satisfy the following conditions:*

1. $\langle X(p) - \overline{X}(p), N(p)\rangle = \langle X(p) - \overline{X}(p), \overline{N}(p)\rangle = 0$, *i.e., the line between $X(p)$ and $\overline{X}(p)$ is tangent to both surfaces.*
2. $r = |X(p) - \overline{X}(p)|$ *is a nonzero constant that is independent of p.*
3. *The angle α between $N(p)$ and $\overline{N}(p)$ is a nonzero constant that is independent of p.*

Then X and $\overline{X}$ both have constant negative Gauss curvature $K = -\sin^2\alpha/r^2$. Moreover, if $X\colon U \to \mathbb{R}^3$ is any such surface, then for any $p \in X$ and any unit vector $v \in T_pX$ which is not in a principal direction, there exists another such surface $\overline{X}$ such that $\overline{X}(p) = X(p) + rv$ and $X, \overline{X}$ satisfy the conditions above. Furthermore, the surface $\overline{X}$ can be constructed from X by solving ordinary differential equations.

A proof of this theorem using the method of moving frames may be found in [**1**].

We will consider timelike surfaces in $\mathbb{R}^{1,2}$. The main result of this paper is

THEOREM 1.2 (c.f. Theorem 3.1). *Let $U \subset \mathbb{R}^2$ be an open set, and let X, $\overline{X}\colon U \to \mathbb{R}^{1,2}$ be oriented immersed timelike surfaces. Let $N(p)$, $\overline{N}(p)$ denote the*

2000 *Mathematics Subject Classification.* Primary: 58G37, 53C50, 53A10; Secondary: 58A15.

This research was supported in part by NSF Grant DMS-9627403.

This is the final form of the paper.

unit normals to $X, \overline{X}$ *at* $X(p)$ *and* $\overline{X}(p)$*, respectively. Suppose that* X *and* $\overline{X}$ *satisfy the following conditions:*

1. $r = \langle X(p) - \overline{X}(p), N(p)\rangle = \langle X(p) - \overline{X}(p), \overline{N}(p)\rangle$ *is a nonzero constant that is independent of* p.
2. $\langle X(p) - \overline{X}(p), X(p) - \overline{X}(p)\rangle = -r^2$.
3. $\alpha = \langle N(p), \overline{N}(p)\rangle \neq -1$ *is a constant that is independent of* p.

Then X *has constant mean curvature* $H = (\alpha - 1)/(2r)$ *and* $\overline{X}$ *has constant mean curvature* $\overline{H} = (1-\alpha)/(2r)$*. Moreover, if* $X\colon U \to \mathbb{R}^{1,2}$ *is any such surface, then for any* $p \in X$ *and any null vector* $v \in T_pX$*, there exists another such surface* $\overline{X}$ *such that* $\overline{X}(p) = X(p) + rN(p) - rv$ *and* $X, \overline{X}$ *satisfy the conditions above. Furthermore, the surface* $\overline{X}$ *can be constructed from* X *by solving ordinary differential equations.*

In particular, when $\alpha = 1$, X and $\overline{X}$ are both timelike minimal surfaces. In this case we can express the transformation explicitly in terms of the solution of a scalar ODE, and we will describe some examples.

2. Local Geometry of Timelike Surfaces in $\mathbb{R}^{1,2}$

$\mathbb{R}^{1,2}$ is defined to be the vector space $\mathbb{R}^3$ with the Lorentzian inner product

$$\langle x, y\rangle = x^1y^1 - x^2y^2 - x^3y^3,$$

where $x = (x^1, x^2, x^3)$, $y = (y^1, y^2, y^3)$. A *timelike surface* $X\colon U \to \mathbb{R}^{1,2}$ is an immersed surface for which the induced metric has signature $(1, 1)$.

Given a timelike surface $X \subset \mathbb{R}^{1,2}$, an *orthonormal frame field* along X is a choice of an orthonormal (with respect to the Lorentzian metric) basis $\{e_1, e_2, e_3\}$ of the tangent space $T_p\mathbb{R}^{1,2}$ for each $p \in X$ that varies smoothly with p. We say that the frame field is *adapted* to X if for each $p \in X$, the vectors e_1, e_2 form a basis for the tangent space T_pX and

$$\langle e_1, e_1\rangle = 1, \quad \langle e_2, e_2\rangle = -1.$$

For any such frame field, e_3 is the unit normal vector to X and $\langle e_3, e_3\rangle = -1$. If $\{e_1, e_2, e_3\}$ is an adapted orthonormal frame field, then any other adapted orthonormal frame field $\{\tilde{e}_1, \tilde{e}_2, \tilde{e}_3\}$ has the form

$$\begin{aligned}
\tilde{e}_1 &= \pm(\cosh\tau e_1 + \sinh\tau e_2),\\
\tilde{e}_2 &= \pm(\sinh\tau e_1 + \cosh\tau e_2),\\
\tilde{e}_3 &= \pm e_3,
\end{aligned}$$

for some function τ on X.

Given an adapted orthonormal frame field along X, there exist unique 1-forms ω^i, ω^i_j, $1 \le i, j \le 3$ on X such that

$$dX = \sum_{i=1}^{3} e_i\omega^i, \quad de_i = \sum_{j=1}^{3} e_j\omega^j_i.$$

The ω^i are called the *dual forms* of the frame field, and the ω^i_j are called the *connection forms.* Differentiating these equations shows that

$$d\omega^i = -\sum_{j=1}^{3} \omega^i_j \wedge \omega^j, \quad d\omega^i_j = -\sum_{k=1}^{3} \omega^i_k \wedge \omega^k_j.$$

These are called the *structure equations*. Differentiating the equations

$$\langle e_i, e_j \rangle = \begin{cases} 0 & i \neq j, \\ 1 & i = j = 1, \\ -1 & i = j = 2, 3, \end{cases}$$

shows that $\omega_i^i = 0$ for $i = 1, 2, 3$, and

$$\omega_1^2 = \omega_2^1, \quad \omega_3^1 = \omega_1^3, \quad \omega_3^2 = -\omega_2^3.$$

Since e_1, e_2 form a basis for T_pX at each point $p \in X$, the 1-form dX must be a linear combination of e_1 and e_2; therefore, $\omega^3 = 0$. Moreover, the 1-forms ω^1, ω^2 are linearly independent and so form a basis for the 1-forms on X. Differentiating the equation $\omega^3 = 0$ yields

$$0 = d\omega^3 = -\omega_1^3 \wedge \omega^1 - \omega_2^3 \wedge \omega^2.$$

By Cartan's Lemma, there exist functions h_{11}, h_{12}, h_{22} on X such that

$$\omega_1^3 = h_{11}\omega^1 + h_{12}\omega^2, \quad \omega_2^3 = h_{12}\omega^1 + h_{22}\omega^2.$$

The structure equations for the dual forms can now be written in the form

$$d\omega^1 = -\omega_2^1 \wedge \omega^2, \quad d\omega^2 = -\omega_2^1 \wedge \omega^1, \tag{2.1}$$

where ω_2^1 is the Levi–Civita connection form for the metric on X. The Gauss equation is

$$d\omega_2^1 = -\omega_1^3 \wedge \omega_2^3 = -(h_{11}h_{22} - h_{12}^2)\omega^1 \wedge \omega^2,$$

and the Codazzi equations are

$$d\omega_1^3 = -\omega_2^3 \wedge \omega_2^1, \quad d\omega_2^3 = -\omega_1^3 \wedge \omega_2^1.$$

The first and second fundamental forms of X are

$$I = \langle dX, dX \rangle = (\omega^1)^2 - (\omega^2)^2,$$
$$II = \langle dX, de_3 \rangle = h_{11}(\omega^1)^2 + 2h_{12}\omega^1\omega^2 + h_{22}(\omega^2)^2.$$

The mean curvature H of X is defined to be one-half of the trace of II with respect to I, i.e.,

$$H = \frac{1}{2}(h_{11} - h_{22}).$$

This quantity is independent (up to sign) of the choice of adapted orthonormal frame field on X. Note that

$$\omega_1^3 \wedge \omega^2 - \omega^1 \wedge \omega_2^3 = (h_{11} - h_{22})\omega^1 \wedge \omega^2 = 2H\omega^1 \wedge \omega^2.$$

3. A Bäcklund Transformation Between Surfaces of Constant Mean Curvature

In this section we will prove the following theorem.

THEOREM 3.1. *Let $U \subset \mathbb{R}^2$ be an open set, and let $X, \overline{X} \colon U \to \mathbb{R}^{1,2}$ be oriented immersed timelike surfaces. Let $N(p)$, $\overline{N}(p)$ denote the unit normals to X, $\overline{X}$ at $X(p)$ and $\overline{X}(p)$, respectively. Suppose that X and $\overline{X}$ satisfy the following conditions:*

1. $r = \big\langle X(p) - \overline{X}(p), N(p) \big\rangle = \big\langle X(p) - \overline{X}(p), \overline{N}(p) \big\rangle$ *is a nonzero constant that is independent of p.*
2. $\big\langle X(p) - \overline{X}(p), X(p) - \overline{X}(p) \big\rangle = -r^2$.
3. $\alpha = \big\langle N(p), \overline{N}(p) \big\rangle \neq -1$ *is a constant that is independent of p.*

Then X has constant mean curvature $H = (\alpha - 1)/(2r)$ and $\overline{X}$ has constant mean curvature $\overline{H} = (1-\alpha)/(2r)$. Moreover, if $X\colon U \to \mathbb{R}^{1,2}$ is any such surface, then for any $p \in X$ and any null vector $v \in T_pX$, there exists another such surface $\overline{X}$ such that $\overline{X}(p) = X(p)+rN(p)-rv$ and $X, \overline{X}$ satisfy the conditions above. Furthermore, the surface $\overline{X}$ can be constructed from X by solving ordinary differential equations.

PROOF. Let $\{e_1, e_2, e_3\}$ be an adapted orthonormal frame field along X and $\{\bar{e}_1, \bar{e}_2, \bar{e}_3\}$ an adapted orthonormal frame field along $\overline{X}$ such that e_3, $\bar{e}_3$ agree with the oriented unit normals $N, \overline{N}$, respectively. By conditions 1 and 2 (and by choosing different adapted orthonormal frame fields if necessary), we can assume that

$$X - \overline{X} = r(e_1 + e_2 - e_3) = r(\bar{e}_1 + \bar{e}_2 - \bar{e}_3). \tag{3.1}$$

This condition determines the frame vectors e_i, $\bar{e}_i$ uniquely. Now at each point $p \in U$ there exists a matrix $A = [a_{ij}] \in O(1,2)$ such that

$$\begin{bmatrix} \bar{e}_1 & \bar{e}_2 & \bar{e}_3 \end{bmatrix} = \begin{bmatrix} e_1 & e_2 & e_3 \end{bmatrix} A. \tag{3.2}$$

Moreover, by condition 3 we have $a_{33} = -\alpha$. Substituting (3.2) into equation (3.1) and to the equations

$$\langle \bar{e}_i, \bar{e}_j \rangle = \begin{cases} 0 & i \neq j, \\ 1 & i = j = 1, \\ -1 & i = j = 2, 3, \end{cases}$$

yields a system of equations for the a_{ij} whose solution is

$$A = \begin{bmatrix} \frac{\alpha^2+\alpha+1}{\alpha+1} & \frac{\alpha}{\alpha+1} & \alpha \\ -\frac{\alpha}{\alpha+1} & \frac{\alpha}{\alpha+1} & -1 \\ -\alpha & -1 & -\alpha \end{bmatrix}.$$

Now let ω^i, ω^i_j be the dual and connection forms for the frame $\{e_1, e_2, e_3\}$ on X and $\overline{\omega}^i$, $\overline{\omega}^i_j$ the dual and connection forms for the frame $\{\bar{e}_1, \bar{e}_2, \bar{e}_3\}$ on $\overline{X}$. According to the structure equations, we have

$$\begin{aligned} d\overline{X} &= \bar{e}_1\overline{\omega}^1 + \bar{e}_2\overline{\omega}^2 \\ &= \left(\frac{\alpha^2+\alpha+1}{\alpha+1}e_1 - \frac{\alpha}{\alpha+1}e_2 - \alpha e_3\right)\overline{\omega}^1 + \left(\frac{\alpha}{\alpha+1}e_1 + \frac{\alpha}{\alpha+1}e_2 - e_3\right)\overline{\omega}^2 \\ &= \left(\frac{\alpha^2+\alpha+1}{\alpha+1}\overline{\omega}^1 + \frac{\alpha}{\alpha+1}\overline{\omega}^2\right)e_1 + \left(-\frac{\alpha}{\alpha+1}\overline{\omega}^1 + \frac{\alpha}{\alpha+1}\overline{\omega}^2\right)e_2 \\ &\qquad + (-\alpha\overline{\omega}^1 - \overline{\omega}^2)e_3. \end{aligned}$$

But we also have

$$\begin{aligned} d\overline{X} &= dX - r(de_1 + de_2 - de_3) \\ &= (\omega^1 - r\omega^1_2 + r\omega^3_1)e_1 + (\omega^2 - r\omega^1_2 - r\omega^3_2)e_2 + (-r\omega^3_1 - r\omega^3_2)e_3. \end{aligned}$$

Comparing coefficients yields

$$\overline{\omega}^1 = \frac{1}{\alpha+1}(\omega^1 - \omega^2 + r\omega^3_1 + r\omega^3_2), \quad \overline{\omega}^2 = \frac{1}{\alpha+1}(-\alpha\omega^1 + \alpha\omega^2 + r\omega^3_1 + r\omega^3_2),$$

together with the equation

$$\omega_2^1 = \frac{\alpha}{r(\alpha+1)}\omega^1 + \frac{1}{r(\alpha+1)}\omega^2 - \omega_2^3. \tag{3.3}$$

Similarly, differentiating equation (3.2) shows that

$$\begin{aligned}
\overline{\omega}_2^1 &= \frac{\alpha^2}{r(\alpha+1)}\omega^1 + \frac{\alpha}{r(\alpha+1)}\omega^2 - \omega_1^3, \\
\overline{\omega}_1^3 &= -\frac{\alpha}{r(\alpha+1)}\omega^1 - \frac{1}{r(\alpha+1)}\omega^2 - \frac{\alpha}{\alpha+1}\omega_1^3 + \frac{1}{\alpha+1}\omega_2^3, \\
\overline{\omega}_2^3 &= -\frac{\alpha^2}{r(\alpha+1)}\omega^1 - \frac{\alpha}{r(\alpha+1)}\omega^2 + \frac{\alpha}{\alpha+1}\omega_1^3 - \frac{1}{\alpha+1}\omega_2^3.
\end{aligned}$$

Now since $\overline{\omega}^3 = 0$ we have

$$\begin{aligned}
0 &= -d\overline{\omega}^3 \\
&= \overline{\omega}_1^3 \wedge \overline{\omega}^1 + \overline{\omega}_2^3 \wedge \overline{\omega}^2 \\
&= \omega_1^3 \wedge \omega^2 - \omega^1 \wedge \omega_2^3 - \frac{\alpha-1}{r}\omega^1 \wedge \omega^2.
\end{aligned}$$

Therefore the mean curvature of X is $H = (\alpha-1)/(2r)$, as claimed. An analogous argument shows that the mean curvature of $\overline{X}$ is $\overline{H} = (1-\alpha)/(2r)$.

To prove the remainder of the theorem, it suffices to show that given a timelike surface X with $H = (\alpha-1)/(2r)$, a point $p \in X$, and a null vector $v \in T_pX$, we can find an adapted orthonormal coframing $\{e_1, e_2, e_3\}$ on X such that $v = e_1 + e_2$ and the surface $\overline{X} = X - r(e_1 + e_2 - e_3)$ has mean curvature $\overline{H} = (1-\alpha)/(2r)$. Clearly there exists a unique such coframing at the point p, and equation (3.3) defines a system of differential equations for e_1 and e_2 (e_3 is uniquely determined by the requirement that it be the unit normal vector at each point) via the structure equations (2.1). Moreover, equation (3.3) is completely integrable, since

$$\begin{aligned}
&d\left(\omega_2^1 - \frac{\alpha}{r(\alpha+1)}\omega^1 - \frac{1}{r(\alpha+1)}\omega^2 + \omega_2^3\right) \\
&= -\omega_1^3 \wedge \omega_2^3 + \frac{\alpha}{r(\alpha+1)}\omega_2^1 \wedge \omega^2 + \frac{1}{r(\alpha+1)}\omega_2^1 \wedge \omega^1 - \omega_1^3 \wedge \omega_2^1 \\
&= \omega_2^1 \wedge \left(\frac{1}{r(\alpha+1)}\omega^1 + \frac{\alpha}{r(\alpha+1)}\omega^2 + \omega_1^3\right) - \omega_1^3 \wedge \omega_2^3 \\
&= \left(\frac{\alpha}{r(\alpha+1)}\omega^1 + \frac{1}{r(\alpha+1)}\omega^2 - \omega_2^3\right) \wedge \left(\frac{1}{r(\alpha+1)}\omega^1 + \frac{\alpha}{r(\alpha+1)}\omega^2 + \omega_1^3\right) \\
&\qquad - \omega_1^3 \wedge \omega_2^3 \text{(by equation (3.3))} \\
&= -\frac{\alpha}{r(\alpha+1)}(\omega_1^3 \wedge \omega^1 + \omega_2^3 \wedge \omega^2) \\
&\qquad - \frac{1}{r(\alpha+1)}\left(\omega_1^3 \wedge \omega^2 - \omega^1 \wedge \omega_2^3 - \frac{\alpha-1}{r}\omega^1 \wedge \omega^2\right) \\
&= 0,
\end{aligned}$$

where the first term vanishes because it is a multiple of $d\omega^3$ and the second term vanishes because X has mean curvature $H = (\alpha-1)/(2r)$. The result now follows directly from the Frobenius Theorem. □

4. The Minimal Case

When $\alpha = 1$, Theorem 3.1 says that X and $\overline{X}$ both have mean curvature zero and so are timelike minimal surfaces in $\mathbb{R}^{1,2}$. A Weierstrass-type representation for such surfaces is described in [**3**], and we will use this representation to describe the Bäcklund transformation explicitly. Given a timelike minimal surface $X \subset \mathbb{R}^{1,2}$ there exist null coordinates u, v on X such that X has a parametrization of the form

$$\begin{aligned} X(u,v) &= U(u) + V(v) \\ &= \int_{u_0}^{u} \frac{1}{\sqrt{2}}\big(1, \cos A(t), \sin A(t)\big)\, dt + \int_{v_0}^{v} \frac{1}{\sqrt{2}}\big(1, \cos B(t), \sin B(t)\big)\, dt, \end{aligned}$$

where A, B are functions from some open intervals in $\mathbb{R}$ to the unit circle S^1. (This parametrization is chosen, somewhat arbitrarily, so that the vectors $U'(u)$, $V'(v)$ have *Euclidean* length 1.) X is immersed if and only if the images of A and B are disjoint in S^1. Any adapted orthonormal frame on X has the form

$$\begin{aligned} e_1 &= \pm(\cosh\tau\,\underline{e}_1 + \sinh\tau\,\underline{e}_2), \\ e_2 &= \pm(\sinh\tau\,\underline{e}_1 + \cosh\tau\,\underline{e}_2), \\ e_3 &= \pm\frac{1}{1-\cos(A(u)-B(v))}\big(\sin\big(B(v)-A(u)\big), \\ &\qquad\qquad \sin B(v) - \sin A(u), \cos A(u) - \cos B(v)\big), \end{aligned}$$

where $\tau(u,v)$ is an arbitrary function on X and

$$\begin{aligned} \underline{e}_1 &= \frac{1}{\sqrt{1-\cos(A(u)-B(v))}}\big(2, \cos A(u) + \cos B(v), \sin A(u) + \sin B(v)\big), \\ \underline{e}_2 &= \frac{1}{\sqrt{1-\cos(A(u)-B(v))}}\big(0, \cos A(u) - \cos B(v), \sin A(u) - \sin B(v)\big). \end{aligned}$$

Without loss of generality (switching u and v if necessary), we can take $+$ signs in each choice above.

The dual forms of this frame are

$$\begin{aligned} \omega^1 &= \frac{1}{2}\sqrt{1-\cos(A(u)-B(v))}(e^{-\tau}\,du + e^{\tau}\,dv), \\ \omega^2 &= \frac{1}{2}\sqrt{1-\cos(A(u)-B(v))}(e^{-\tau}\,du - e^{\tau}\,dv). \end{aligned}$$

From the structure equations it is straightforward to compute that

$$\begin{aligned} \omega^1_2 &= d\tau + \frac{\sin(A(u)-B(v))}{2-2\cos(A(u)-B(v))}\big(A'(u)\,du + B'(v)\,dv\big), \\ \omega^3_1 &= \frac{1}{\sqrt{2-2\cos(A(u)-B(v))}}\big(e^{\tau}A'(u)\,du - e^{-\tau}B'(v)\,dv\big), \\ \omega^3_2 &= \frac{1}{\sqrt{2-2\cos(A(u)-B(v))}}\big(e^{\tau}A'(u)\,du + e^{-\tau}B'(v)\,dv\big). \end{aligned}$$

Since $\alpha = 1$, equation (3.3) takes the form

$$\omega^1_2 = \frac{1}{2r}\omega^1 + \frac{1}{2r}\omega^2 - \omega^3_2,$$

which is equivalent to the two equations

$$\begin{aligned}\tau_u &= \frac{e^{-\tau}\sqrt{1-\cos(A(u)-B(v))}}{2r} \\ &\qquad - \left(\frac{e^{\tau}}{\sqrt{2-2\cos(A(u)-B(v))}} + \frac{\sin(A(u)-B(v))}{2-2\cos(A(u)-B(v))}\right)A'(u), \\ \tau_v &= -\left(\frac{e^{-\tau}}{\sqrt{2-2\cos(A(u)-B(v))}} + \frac{\sin(A(u)-B(v))}{2-2\cos(A(u)-B(v))}\right)B'(v).\end{aligned}$$

This is a compatible, overdetermined system of equations for the function τ, and so it can be solved by ODE methods. Explicitly, if we set $T = e^{\tau}$, this system takes the form

$$\begin{aligned}T_u &= \frac{\sqrt{1-\cos(A(u)-B(v))}}{2r} \\ &\qquad - \left(\frac{T^2}{\sqrt{2-2\cos(A(u)-B(v))}} + \frac{T\sin(A(u)-B(v))}{2-2\cos(A(u)-B(v))}\right)A'(u), \\ T_v &= -\left(\frac{1}{\sqrt{2-2\cos(A(u)-B(v))}} + \frac{T\sin(A(u)-B(v))}{2-2\cos(A(u)-B(v))}\right)B'(v).\end{aligned} \tag{4.1}$$

The general solution of the second equation is

$$T(u,v) = -\frac{\sin(A(u)-B(v))}{\sqrt{2-2\cos(A(u)-B(v))}} + \frac{\sqrt{1-\cos(A(u)-B(v))}}{\sqrt{2}}f(u),$$

where f is an arbitrary function of one variable. Substituting this expression into the first equation of (4.1) yields the ordinary differential equation

$$f'(u) = -\frac{1}{2}\big(f(u)^2+1\big)A'(u) + \frac{1}{r\sqrt{2}}, \tag{4.2}$$

for f. It is not difficult to show that this equation has no general solution of the form

$$f(u) = F\big(u, A(u)\big),$$

but it may be possible to solve the equation explicitly for some particular functions $A(u)$.

The transformed surface $\overline{X}$ is given by

$$\begin{aligned}\overline{X} &= X - r(e_1+e_2-e_3) \\ &= X - r(T\underline{e}_1 + T\underline{e}_2 - e_3) \\ &= \int_{u_0}^{u}\frac{1}{\sqrt{2}}\big(1,\cos A(t),\sin A(t)\big)\,dt + \frac{1}{\sqrt{2}}\int_{v_0}^{v}\big(1,\cos B(t),\sin B(t)\big)\,dt \\ &\qquad - r\left[\frac{1}{2}\left(-\frac{\sin(A(u)-B(v))}{1-\cos(A(u)-B(v))} + f(u)\right)(2, 2\cos A(u), 2\sin A(u)) \right. \\ &\qquad\qquad - \frac{1}{1-\cos(A(u)-B(v))}\big(\sin\big(B(v)-A(u)\big), \sin B(v) - \sin A(u), \\ &\qquad\qquad\qquad \left. \cos A(u) - \cos B(v)\big)\right] \\ &= \int_{u_0}^{u}\frac{1}{\sqrt{2}}\big(1,\cos A(t),\sin A(t)\big)\,dt + \frac{1}{\sqrt{2}}\int_{v_0}^{v}\big(1,\cos B(t),\sin B(t)\big)\,dt \\ &\qquad - r\big(f(u), f(u)\cos A(u) + \sin A(u), f(u)\sin A(u) - \cos A(u)\big).\end{aligned}$$

Therefore, up to a reparametrization of the form $\bar{u} = \bar{u}(u)$, the Weierstrass-type representation of $\overline{X}$ is

$$\overline{X} = \overline{U}(\overline{u}) + V(v),$$

where

$$\overline{U}(\bar{u}) = U(u) - r\big(f(u), f(u)\cos A(u) + \sin A(u), f(u)\sin A(u) - \cos A(u)\big).$$

Note that the transformation fixes one of the null curves that generate X and transforms the other, so it may be thought of as a transformation of null curves in $\mathbb{R}^{1,2}$.

It is straightforward to compute that

$$\begin{aligned}\frac{d\overline{U}}{du} = \frac{1}{2}rA'(u)\big(f(u)^2+1,\ \big(f(u)^2-1\big)\cos A(u) + 2f(u)\sin A(u),\\ \big(f(u)^2-1\big)\sin A(u) - 2f(u)\cos A(u)\big).\end{aligned}$$

Therefore $\overline{U}'(u) = 0$ whenever $A'(u) = 0$, and when $A'(u) \neq 0$,

$$\begin{aligned}\frac{d\overline{U}}{d\overline{u}} &= \frac{1}{|\overline{U}'(u)|_{\text{Euc}}}\frac{d\overline{U}}{du}\\ &= \frac{1}{\sqrt{2}\big(f(u)^2+1\big)}\big(f(u)^2+1,\ \big(f(u)^2-1\big)\cos A(u) + 2f(u)\sin A(u),\\ &\qquad\qquad \big(f(u)^2-1\big)\sin A(u) - 2f(u)\cos A(u)\big).\end{aligned}$$

So assuming $A'(u) \neq 0$, the standard Weierstrass-type parametrization of $\overline{X}$ is given by

$$\begin{aligned}X(u,v) &= \overline{U}(\bar{u}) + V(v)\\ &= \int_{\bar{u}_0}^{\bar{u}} \frac{1}{\sqrt{2}}\big(1, \cos\bar{A}(t), \sin\bar{A}(t)\big)\,dt + \int_{v_0}^{v}\frac{1}{\sqrt{2}}\big(1, \cos B(t), \sin B(t)\big)\,dt,\end{aligned}$$

where

$$\begin{aligned}\bar{A}(t) &= \sin^{-1}\left(\frac{\big(f(t)^2-1\big)\sin A(t) - 2f(t)\cos A(t)}{f(t)^2+1}\right)\\ &= \cos^{-1}\left(\frac{\big(f(t)^2-1\big)\cos A(t) + 2f(t)\sin A(t)}{f(t)^2+1}\right).\end{aligned}$$

This procedure can be iterated; a computation shows that

$$\bar{A}'(u) = \frac{\sqrt{2}}{r\big(f(u)^2+1\big)},$$

so the Bäcklund transform of $\overline{X}$ is defined by a formula similar to that for the Bäcklund transform of X, with A replaced by $\bar{A}$ and f replaced by $\bar{f}$, where $\bar{f}$ is the solution of the differential equation

$$\begin{aligned}\bar{f}'(u) &= -\frac{1}{2}\big(\bar{f}(u)^2+1\big)\frac{\sqrt{2}}{r(f(u)^2+1)} + \frac{1}{r\sqrt{2}}\\ &= \frac{1}{r\sqrt{2}}\frac{f(u)^2 - \bar{f}(u)^2}{f(u)^2+1}.\end{aligned}$$

5. Examples

Since the Bäcklund transformation only acts nontrivially on one of the two null curves that generate a surface X, we will give examples of its action on null curves rather than on surfaces.

EXAMPLE 1. Suppose that $A(u) \equiv A_0$ is constant, so that the null curve

$$U(u) = \int_{u_0}^{u} \frac{1}{\sqrt{2}}(1, \cos A_0, \sin A_0)\, dt,$$

is a straight line. Since $A'(u) \equiv 0$, the transformed curve $\overline{U}$ has $\overline{U}'(u) = 0$ and so degenerates to a point.

EXAMPLE 2. Let $r = 1$ and suppose that

$$A(u) = \frac{1}{\sqrt{2}} u.$$

The corresponding null curve is

$$\begin{aligned} U(u) &= \int_{0}^{u} \frac{1}{\sqrt{2}}\big(1, \cos A(u), \sin A(u)\big)\, dt \\ &= \frac{1}{\sqrt{2}}\left(u, \sqrt{2}\sin\left(\frac{1}{\sqrt{2}}u\right), -\sqrt{2}\cos\left(\frac{1}{\sqrt{2}}u\right) + \sqrt{2}\right). \end{aligned}$$

This curve is a helix, shown in Fig. 1.

To find the transformed curve, we need to solve the ODE

$$f'(u) = -\frac{1}{2\sqrt{2}}\big(f(u)^2 - 1\big).$$

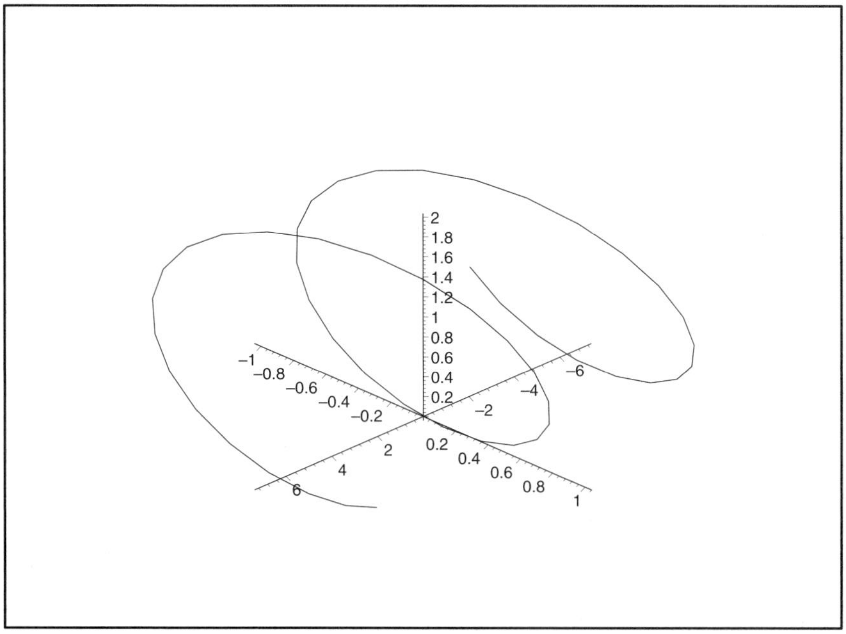

FIGURE 1

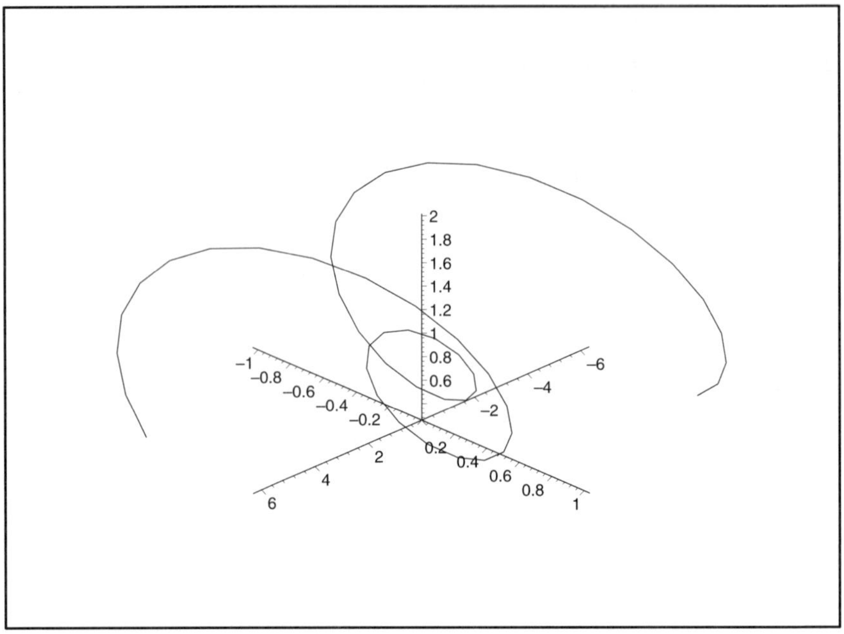

FIGURE 2

If we impose the initial condition $f(0) = 0$, then the solution is

$$f(u) = \tanh\left(\frac{1}{2\sqrt{2}}t\right).$$

The transformed curve (up to a reparametrization) is

$$\begin{aligned}\overline{U}(u) &= U(u) - r\big(f(u), f(u)\cos A(u) + \sin A(u), f(u)\sin A(u) - \cos A(u)\big)\\ &= \Bigg(\frac{1}{\sqrt{2}}u - \tanh\left(\frac{1}{2\sqrt{2}}u\right), -\cos\left(\frac{1}{\sqrt{2}}u\right)\tanh\left(\frac{1}{2\sqrt{2}}u\right),\\ &\qquad\qquad 1 - \sin\left(\frac{1}{\sqrt{2}}u\right)\tanh\left(\frac{1}{2\sqrt{2}}u\right)\Bigg).\end{aligned}$$

This is a helix-like curve which is shown in Fig. 2.

This process could be iterated, but first we would have to reparametrize the curve $\overline{U}$ so that its derivative has Euclidean length 1. This is probably best done numerically and is left as an exercise for the reader.

References

1. S.-S. Chern and C.-L. Terng, *An analogue of Bäcklund's theorem in affine geometry*, Rocky Mountain J. Math. **10** (1980), 105–124.
2. P. A. Griffiths, *On Cartan's method of Lie groups and moving frames as applied to existence and uniqueness questions in differential geometry*, Duke Math J. **41** (1974), 775–814.
3. T. K. Milnor, *Entire timelike minimal surfaces in $E^{3,1}$*, Michigan Math. J. **37** (1990), 163–177.

DEPARTMENT OF MATHEMATICS, 395 UCB, UNIVERSITY OF COLORADO, BOULDER, CO 80309-0395, USA.

E-mail address: `Jeanne.Clelland@colorado.edu`

Centre de Recherches Mathématiques
CRM Proceedings and Lecture Notes
Volume **29**, 2001

On Ribaucour Transformations

A. V. Corro, W. Ferreira, and *K. Tenenblat*

Abstract. Minimal surfaces in R^3 are associated to solutions $\phi(u_1, u_2)$ of the nonlinear differential equation $\Delta\phi = e^{-2\phi}$. We consider Ribaucour transformations for minimal surfaces in R^3. The analytic interpretation of this transformation provides solutions for the differential equation from a given solution. By applying the theory to the catenoid and to Enneper's surface, we generate 3-parameter families of minimal surfaces and their corresponding solutions ϕ.

1. Introduction

Ribaucour transformations for hypersurfaces, parametrized by lines of curvature, were classically studied by Bianchi [**1**]. Imposing certain conditions, they can be applied to obtain surfaces of constant Gaussian curvature from a given such surface. Analytically, it corresponds to composing appropriate Bäcklund transformations for the classical sine–Gordon equation and the elliptic sinh–Gordon equation.

Similarly, by using Ribaucour transfomations one may obtain minimal surfaces from a given such surface. In this case the analytic interpretation provides solutions for the differential equation $\Delta\phi = e^{-2\phi}$. Although the transfomation for minimal surfaces is well known, it has not been sufficiently explored.

In this paper, we exhibit 3-parameter families of minimal surfaces associated to Enneper's surface and to the catenoid, by applying Ribaucour transformations. Moreover, we provide the analytic interpretation of our geometric results in terms of explicit solutions for the equation $\Delta\phi = e^{-2\phi}$. Detailed proofs of our results will appear elsewhere.

2. Ribaucour Transformation for Minimal Surfaces

In this section, we first recall the theory of Ribaucour transformation for surfaces. For the proofs and more details see [**2**] and [**1**]. We then restrict ourselves to Ribaucour transformations between minimal surfaces.

A *sphere congruence* in R^3 is a 2-parameter family of spheres, with a differentiable radius function, whose centers lie on a surface M_0 contained in R^3. An

2000 *Mathematics Subject Classification.* Primary: 53A10, 53C42; Secondary: 37K35.
Partially supported by CNPQ and FAPDF.
This is the final form of the paper.

involute of a sphere congruence is a surface M of R^3 such that M, at each point, is tangent to a sphere of the sphere congruence. Two surfaces M and $\widetilde{M}$ are said to be *associated by a sphere congruence* if there is a difeomorphism $\psi : M \to \widetilde{M}$ such that at corresponding points p and $\psi(p)$ the sufaces are tangent to the same sphere of the sphere congruence. It follows from this definition that the normal lines at corresponding points intersect at an equidistant point on the center surface M_0. An important special case occurs when ψ preserves lines of curvature.

Let M and $\widetilde{M}$ be orientable surfaces of R^3. We denote by N and $\widetilde{N}$ their Gauss map. We say that M and $\widetilde{M}$ are *associated by a Ribaucour transformation*, if and only if, there exists a differentiable function h defined on M and a diffeomorphism $\psi : M \to \widetilde{M}$ such that

a) $p + h(p)N(p) = \psi(p) + h(p)\widetilde{N}(\psi(p))$, for all $p \in M$.
b) The subset $p + h(p)N(p)$, $p \in M$ is an 2-dimensional surface.
c) ψ preserves lines of curvature.

One can also consider surfaces locally associated by Ribaucour transformations or parametrized surfaces associated by such transformations.

The following theorem gives a characterization of Ribaucour transformations. For the proof and more details for hypersurfaces M^n of R^{n+1} see [**2**].

THEOREM 2.1. *Let M be an orientable surface of R^3. Let e_i, $1 \leq i \leq 2$ be orthonormal principal directions, λ^i the corresponding principal curvatures. A surface $\widetilde{M}$ is associated to M by a Ribaucour transformation, if and only if, M and $\widetilde{M}$ are associated by a sphere congruence whose radius function $h : M \to R$ satisfies*

$$dZ^j(e_i) + Z^i \omega_{ij}(e_i) - Z^i Z^j \lambda^i = 0, \quad 1 \leq i \neq j \leq 2, \tag{2.1}$$

where

$$Z^i = \frac{dh(e_i)}{1 + h\lambda^i}, \tag{2.2}$$

and ω_{ij} are the connection forms of the frame e_i.

REMARK 2.2. Locally, whenever $X(u_1, u_2)$ is a parametrization of the surface M by lines of curvature, the function $h(u_1, u_2)$ is a differentiable function which satisfies the following equation, equivalent to (2.1),

$$h_{,12} - \frac{1 + h\lambda^1}{1 + h\lambda^2}\Gamma^2_{12} h_{,2} - \frac{1 + h\lambda^2}{1 + h\lambda^1}\Gamma^1_{12} h_{,1} - \left(\frac{\lambda^2}{1 + h\lambda^2} + \frac{\lambda^1}{1 + h\lambda^1}\right) h_{,1} h_{,2} = 0,$$

where Γ^i_{12} are the Christoffel symbols and $h_{,1}$ and $h_{,12}$ denote the partial derivatives of h with respect to u_1 and to u_1 u_2 respectively.

One can linearize the problem of obtaining the function h, as we will see in the following proposition. We will consider e_1, e_2 to be orthonormal principal directions and ω_1, ω_2 the dual frame.

PROPOSITION 2.3. *Suppose that h is a nonvanishing function which satisfies equation* (2.1) *then*

$$\psi = \frac{1}{h} \sum_{i=1}^{2} Z^i \omega_i,$$

is a closed 1*-form and there exists a nonvanishing function* Ω *defined on a simply connected domain such that*

$$d\Omega(e_i) = \frac{\Omega}{h} Z^i.$$

For each nonvanishing function h, which is a solution of (2.1), we consider Ω as above and we define

$$\Omega_i = d\Omega(e_i), \quad W = \frac{\Omega}{h}.$$

With this notation, equation (2.1) is equivalent to a linear system given in the following result.

PROPOSITION 2.4. *A nonvanishing function* h *is a solution of* (2.1) *defined on a simply connected domain, if and only if,* $h = \Omega/W$ *where* Ω *and* W *are nonvanishing functions which satisfy*

$$d\Omega_i(e_j) = \Omega_j \omega_{ij}(e_j), \quad \textit{for } i \neq j, \tag{2.3}$$

$$d\Omega = \sum_{i=1}^{n} \Omega_i \omega_i, \tag{2.4}$$

$$dW = -\sum_{i=1}^{n} \Omega_i \lambda^i \omega_i. \tag{2.5}$$

For each solution Ω_i, $1 \leq i \leq 2$, of (2.3), there exists a 2-parameter family of solutions of the system (2.4), (2.5). In fact, equation (2.3) is the integrability condition of the system of equations (2.4), (2.5) for Ω and W.

The Ribaucour transformation of a surface is given in terms of the solutions of the above system (see [**2**] for details).

THEOREM 2.5. *Let* M *be an orientable surface of* R^3 *parametrized by* $X : U \subset R^2 \to M$. *Assume* e_i, $1 \leq i \leq 2$ *are orthonormal principal directions,* λ^i *the corresponding principal curvatures and* N *is a unit vector field normal to* M. *A surface* $\widetilde{M}$ *is locally associated to* M, *by a Ribaucour transformation, if and only if, there exist differentiable functions* W, Ω, $\Omega_i : V \subset U \to R$, *which satisfy* (2.3)–(2.5) *and* $\widetilde{X} : V \subset R^2 \to \widetilde{M}$, *is a parametrization of* $\widetilde{M}$ *given by*

$$\widetilde{X} = X - \frac{2\Omega}{\sum_i (\Omega_i)^2 + W^2} \left(\sum_i \Omega_i e_i - WN \right). \tag{2.6}$$

We now recall a classical result which gives a sufficient condition for a Ribaucour transformation to transform a minimal surface into another minimal surface.

THEOREM 2.6. *Let* M *and* $\widetilde{M}$ *be surfaces of* R^3, *with no parabolic points, which are associated by a Ribaucour transformation, such that the normal lines intersect at a distance function* h. *If* $h = \Omega/W$ *is not constant along the lines of curvature and the functions* Ω_i, Ω *and* W *satisfy* (2.3)–(2.5) *and the additional relation*

$$\Omega_1^2 + \Omega_2^2 + W^2 = 2c\Omega W,$$

where $c \neq 0$ *is a real constant, then* $\widetilde{M}$ *is a minimal surface, if and only if* M *is minimal.*

One can also prove a permutability theorem for Ribaucour transformations between minimal surfaces. We will give its analytic interpretation as a superposition formula for equation (2.8).

We conclude this section, by recalling the well-known correspondence between minimal surfaces and solutions of the differential equation $\Delta\phi = e^{-2\phi}$ (see **[1]**,**[4]**,**[5]**).

PROPOSITION 2.7. *Minimal surfaces in R^3 with no parabolic points are in correspondence with solutions of the equation*

$$\Delta\phi = e^{-2\phi}. \tag{2.7}$$

Simple examples are given by the catenoid, which is associated to the function

$$\phi = \log(\cosh u_1),$$

and the Enneper's surface that corresponds to the function

$$\phi = \log\left[1 + \frac{u_1^2}{4} + \frac{u_2^2}{4}\right].$$

The analytic interpretation of Ribaucour transformation for minimal surfaces is given by the following result.

THEOREM 2.8. *Let $\phi(u_1, u_2)$ be a solution of the equation*

$$\Delta\phi = e^{-2\phi}. \tag{2.8}$$

Then for each constant $c \neq 0$, the following system of equations for Ω_i, Ω and W

$$\begin{aligned} \Omega_{i,u_j} &= \Omega_j \phi_{,u_i} \quad 1 \leq i \neq j \leq 2, \\ \Omega_{,u_i} &= \Omega_i e^{\phi}, \\ W_{,u_i} &= (-1)^{i+1}\Omega_i e^{-\phi}, \\ \Omega_1^2 + \Omega_2^2 + W^2 &= 2c\Omega W, \end{aligned} \tag{2.9}$$

is integrable and a new solution of (2.8) is given by the function $\tilde{\phi}$ defined by

$$e^{\tilde{\phi}} = e^{-\phi}\frac{\Omega}{W}.$$

The solution $\tilde{\phi}$ is said to be *associated to ϕ by the Ribaucour transformation* and the constant c.

THEOREM 2.9 (Superposition formula). *Let ϕ be a solution of (2.8). Let ϕ' and ϕ'' be solutions of (2.8) associated to ϕ by the Ribaucour transformation (2.9) for two distinct nonzero constants c' and c'' respectively. Let Ω', W' and Ω'', W'' be the corresponding solutions of (2.9). Then ϕ^* defined algebraically by*

$$e^{\phi^*} = e^{\phi}\left(\frac{-B + c'e^{\phi''} + c''e^{\phi'}}{-B + c'e^{\phi'} + c''e^{\phi''}}\right),$$

where

$$B = e^{-\phi} + e^{\phi}\langle\nabla(\log W'), \nabla(\log W'')\rangle,$$

is a new solution of (2.8).

3. Families of Minimal Surfaces Associated to Enneper's Surface and to the Catenoid

In this section, by using Ribaucour transformations, we will provide 3-parameter families of minimal surfaces from Enneper's surface and from the catenoid. Moreover, we will give families of solutions of equation (2.8) associated to these families of minimal surfaces.

PROPOSITION 3.1. *Consider Enneper's surface parametrized by*

$$X = \left(u_1 - \frac{u_1^3}{3} + u_1 u_2^2, u_2 - \frac{u_2^3}{3} + u_2 u_1^2,\ u_1^2 - u_2^2\right), \tag{3.1}$$

$\widetilde{X}$ *is a minimal surface associated to* X *as in Theorem* 2.5, *if and only if, up to a rigid motion of* R^3,

$$\widetilde{X} = X + \frac{1}{c}(-u_1, u_2, 1) - \frac{1}{2c(f+g)}(f' X_{u_1} - g' X_{u_2}), \tag{3.2}$$

where

$$f(u_1) = a_1 e^{2\sqrt{c} u_1} + a_2 e^{-2\sqrt{c} u_1},$$
$$g(u_2) = b_1 \cos(2\sqrt{c} u_2) + b_2 \sin(2\sqrt{c} u_2),$$

the constants are real numbers satisfying $c > 0$, $a_1^2 + a_2^2 \neq 0$ *and* $b_1^2 + b_2^2 - 4a_1 a_2 = 0$.

REMARK 3.2. The family of minimal surfaces given by (3.2) includes Enneper's surface. In fact, if we choose the constants $a_1 = b_1 = b_2 = 0$ and $a_2 \neq 0$, by considering the variable $\overline{u} = u + 1/\sqrt{c}$, we obtain the Enneper surface translated by $(3/c^{3/2}, 0, 0)$. A similar result holds for the case $a_2 = b_1 = b_2 = 0$ and $a_1 \neq 0$.

Similar results are obtained by considering the catenoid.

PROPOSITION 3.3. *Consider the catenoid parametrized by*

$$X = (\cos u_2 \cosh u_1, \sin u_2 \cosh u_1, u_1),$$

$\widetilde{X}$ *is a minimal surface associated to* X *by a Ribaucour transformation as in Theorem* 2.8, *if and only if,*

$$\widetilde{X} = X - \frac{\cosh u_1}{c}(\cos u_2, \sin u_2, 0) + \frac{1}{c(f+g)}(f' X_{u_1} - g' X_{u_2}), \tag{3.3}$$

where $c \neq 0$, $f(u_1)$ *and* $g(u_2)$ *are solutions of the equations*

$$f'' + (2c-1)f = g'' - (2c-1)g = c_1,$$

c_1 *is a real constant,* $f + g$ *does not vanish and the initial conditions for* f *and* g *must satisfy*

$$\left((f')^2 + (g')^2 + (2c-1)(f^2 - g^2) - 2c_1(f+g)\right)(u_1^0, u_2^0) = 0.$$

REMARK 3.4. The catenoid is included in the family of surfaces given by (3.3). In fact, we choose $c = 1/2$, $c_1 = 0$, $f = a_1$, $g = a_2$, $a_1 + a_2 \neq 0$.

The analytic interpretation of the results given above are the following:

a) We consider the solution ϕ of equation (2.8) associated to the catenoid,

$$\phi = \log(\cosh u_1).$$

For any $c \neq 0$, the solutions of (2.9) are given by

$$\begin{aligned}\Omega_1 &= -\tanh u_1(f+g) + f',\\ \Omega_2 &= -g',\\ W &= (f+g)/\cosh u_1,\\ \Omega &= (\cosh^2 u_1 \Delta W)/2c + W/c,\end{aligned}$$

where $f(u_1)$ and $g(u_2)$ satisfy the differential equation

$$f'' + (2c-1)f = g'' - (2c-1)g.$$

Hence, the function $\widetilde{\phi}$

$$e^{\widetilde{\phi}} = \frac{\Omega}{f+g}$$

is a new solution for equation (2.8). More solutions are explicitly obtained by using the superposition formula of Theorem 2.9.

b) We consider the function

$$\phi = \log\left[1 + \frac{u_1^2}{4} + \frac{u_2^2}{4}\right],$$

which is a solution of (2.8), associated to Enneper's surface. For any $c \neq 0$, the solutions of (2.9) are given by

$$\begin{aligned}\Omega_1 &= -u_1(f+g)/(2\varphi) + f',\\ \Omega_2 &= u_2(f+g)/(2\varphi) - g',\\ W &= (f+g)/\varphi\\ \Omega &= \varphi^2 \Delta W/2c + W/c\end{aligned}$$

where $\varphi = 1 + u_1^2/4 + u_2^1/4$, $f(u_1)$ and $g(u_2)$ satisfy the differential equation

$$f'' - 2cf = g'' + 2cg.$$

Hence, the function

$$e^{\widetilde{\phi}} = \frac{\Omega}{f+g},$$

is a new solution of (2.8). More explicit solutions are obtained by using the superposition formula.

We refer to [**3**] for the proofs of Theorems 3.1, 3.2 and for geometric properties of the families of minimal surfaces given in these theorems. The reader is also referred to [**6**] for a computer animated video which provides illustrative examples of these minimal surfaces.

References

1. L. Bianchi, Lezioni di Geometria Differenciale (Bologna Nicola Zanichelli, ed.), 1927.
2. A. V. Corro, W. Ferreira, and K. Tenenblat, *On Ribaucour transformations for hypersurfaces*, Mat. Contemp. **17** (1999), 137–160.
3. ———, *Minimal surfaces obtained by Ribaucour transformations*, (to appear).
4. L. P. Eisenhart, *Transformations of Surfaces*, Chelsea Publishing Co., New York, 1962.
5. K. Tenenblat, *Transformations of manifolds and applications to differential equations*, Addison Wesley Longman, Pitman Monographs and Surveys in Pure and Applied Mathematics **93**, 1998, 224 pp.
6. ———, *Minimal surfaces associated to the catenoid and to Enneper's surface*, Video, Universidade de Brasília, 1999.

Instituto de Matemática e Estatística, Universidade Federal de Goiás, Goiânia, GO, Brazil.

E-mail address, A. V. Corro: `corro@mat.ufg.br`

E-mail address, W. Ferreira: `walter@mat.ufg.br`

Departamento de Matemática, Universidade de Brasília, 70910-900, Brasília, DF, Brazil.

E-mail address, K. Tenenblat: `keti@mat.unb.br`

Centre de Recherches Mathématiques
CRM Proceedings and Lecture Notes
Volume **29**, 2001

The Ribaucour Congruences of Spheres Within Lie Sphere Geometry

Adam Doliwa

ABSTRACT. The Ribaucour congruences of spheres provide transformations of curvature nets on surfaces. Such transformation of surfaces, called in this context the Ribaucour transformation, is a classical example of the Darboux-type transformation of a geometric origin. The classical result states that within the approach of Lie's geometry of spheres the Ribaucour congruences are described by conjugate nets in the quadric of Lie $\mathcal{Q}^4_L$. In this paper we generalize the above-mentioned approach to the Ribaucour transformation of circular lattices, which are the integrable discrete analogs of curvature nets. In particular, we show that the discrete Ribaucour congruences of spheres are described by quadrilateral lattices (the integrable discrete analogs of conjugate nets) in the quadric of Lie $\mathcal{Q}^4_L$.

1. Introduction

The *curvature net* on a surface in Euclidean space $\mathbb{E}^3$ can be characterized as follows **[20]**: when the normal $\boldsymbol{\nu}$ to the surface is driven along a line of curvature then it forms a *developable surface* (see Fig. 1), i.e., such one-parameter family of normal lines is made out of tangent lines to a curve, called the *edge of regression* of the developable surface.

The normals $\boldsymbol{\nu}$ to the curvature net provide transformation between two surfaces made out of the corresponding edges of regression. Two families of spheres with centers on the edges of regression and tangent to the curvature net in corresponding points are called the *principal spheres.*

The coordinate system on a surface is called a *conjugate net* if the tangents to one parametric lines driven along the other parameter form a developable surface (see Fig. 2). This definition makes sense also if the ambient space is the projective space $\mathbb{P}^N$, $N \geq 2$.

It turns out **[15]** that the curvature nets can be equivalently characterized as conjugate and orthogonal nets in $\mathbb{E}^3$.

In his study of cyclic systems Ribaucour considered two-parameter families of spheres in $\mathbb{E}^3$ upon the sheets of whose envelope the lines of curvature correspond **[22]**. The relation between the two sheets has been called a *transformation*

2000 *Mathematics Subject Classification.* Primary: 37K25; Secondary: 51B25, 37K60.

I would like to thank the organizers of the AARMS–CRM Workshop on Bäcklund & Darboux Transformations for invitation and support.

This is the final form of the paper.

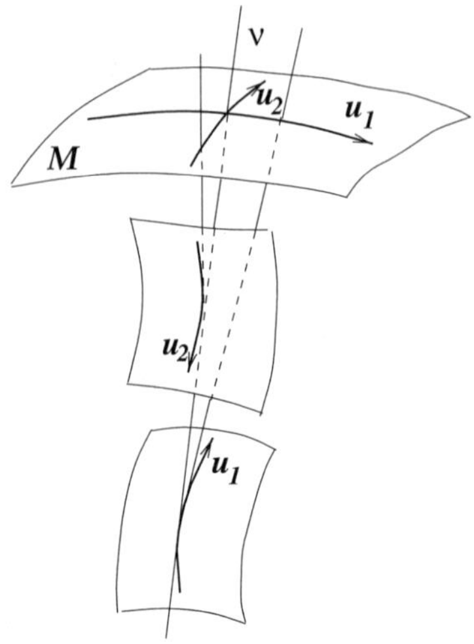

FIGURE 1. The normal congruence and its focal nets.

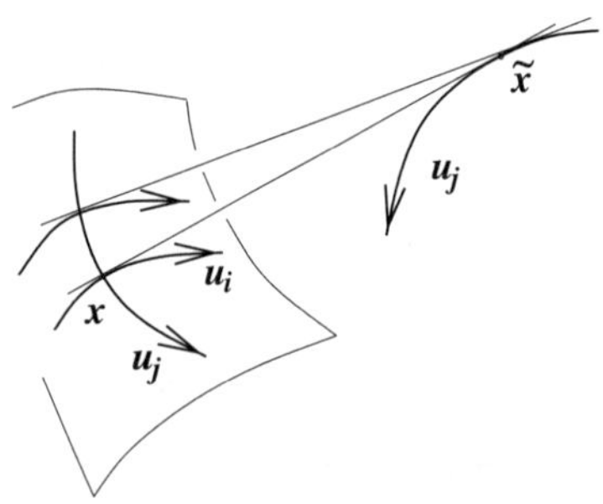

FIGURE 2. Conjugate net.

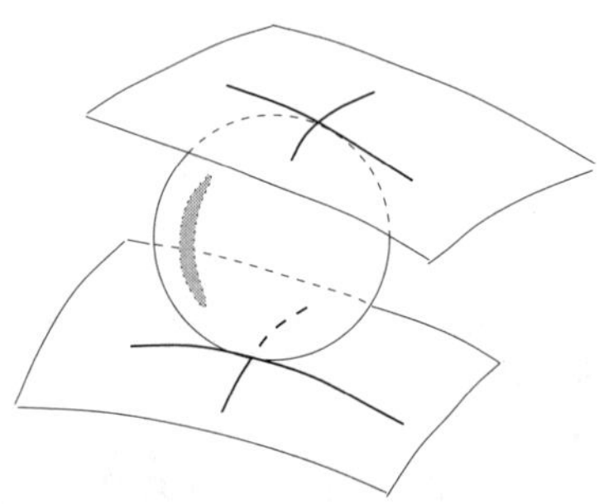

FIGURE 3. Ribaucour congruence of spheres.

of Ribaucour and such family of spheres has been called a *Ribaucour congruence of spheres* (see Fig. 3).

Oriented spheres in $\mathbb{E}^3$ can be represented by points of four-dimensional quadric $\mathcal{Q}_L^4 \subset \mathbb{P}^5$, called the quadric of Lie [**3**]. The Ribaucour congruence of spheres defines therefore a surface in $\mathcal{Q}^4_L$. When the curvature lines on the envelops are chosen as parametric lines on such a surface then we obtain the following characterization of the Ribaucour congruences.

THEOREM 1.1 ([**16**]). *In Lie's representation the Ribaucour congruences of spheres are given by conjugate nets in the quadric* $\mathcal{Q}^4_L$.

The goal of this paper is to show how the above classical results can be generalized to the level of a *discrete geometry.* In the discrete geometry the role of manifolds is played by graphs, in particular one can study discrete surfaces made of two dimensional integer lattices $\mathbb{Z}^2$ in $\mathbb{E}^3$. The *discrete integrable geometry* is concerned with lattices, which can be described by solutions of *integrable discrete equations.* A distinguished feature of such *integrable lattices* is existence of analogs of classical Darboux-type transformations between such lattices.

The discrete analogs of curvature nets on surfaces have been introduced in [**19, 21**] as maps $\mathbb{Z}^2 \to \mathbb{E}^3$ with all elementary quadrilaterals inscribed in circles, and are called therefore *circular lattices* (see also [**1,2,4**]). The Ribaucour transformations of such lattices have been defined in [**17**]. The circular lattice is a reduction of a more fundamental lattice consisting of planar quadrilaterals [**8, 13, 23**]; such lattice is the integrable discrete analog of a conjugate net. The theory of quadratic reductions of quadrilateral lattices, which generalizes the corresponding theory of quadratic reductions of conjugate nets [**16**] to the discrete level, has been constructed in [**9**]. In particular, the theory of circular lattices and their transformations provides the simplest example of such reduction [**9**]. In this paper we present the main results of the theory of discrete analogs of the Ribaucour congruences of spheres studied in [**6**], where we also sent the interested reader for more details. In particular, we consider such families of spheres as primary objects and we show that they are represented by quadrilateral lattices in the quadric of Lie $\mathcal{Q}_L$. In [**6**] one can find also generalization of the results presented here to multidimensional case.

The paper is constructed as follows. In Section 2 we collect basic results of the Lie geometry of spheres, which provides the appropriate setting for subsequent discussion of geometric properties of the Ribaucour congruences. In Section 3 we recall necessary material from the theory of quadrilateral lattices and rectilinear congruences. Section 4 is devoted to geometric theory of two-dimensional circular lattices. Finally, in Section 5 we present the theory of Ribaucour congruences of spheres.

To make the presentation self-contained let us recall some definitions and useful properties of objects appearing in the classical *projective differential geometry* (the details can be found in [**16, 18**]).

The *rectilinear congruence* is a two-parameter family of straight lines such that through each line pass two developable surfaces of the family. For example, the normals $\boldsymbol{\nu}$ to a surface in $\mathbb{E}^3$ form a congruence (the so called *normal congruence*) whose developables cut the surface along the curvature lines. Another example of a congruence is provided by the tangents to one family of parametric lines of a conjugate net. The edges of regression of developables of a rectilinear congruence form two surfaces, called the *focal surfaces* of the congruence. The developables of the congruence cut the focal surfaces along conjugate nets.

We finally remark that the one can define multidimensional conjugate nets and congruences [**5, 11**], which provide geometric interpretation [**11**] to basic objects

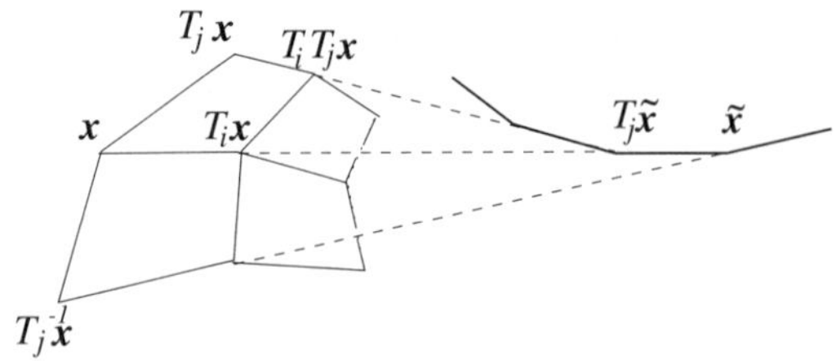

FIGURE 4. Discrete conjugate net (quadrilateral lattice).

appearing in the multicomponent Kadomtsev–Petviashvilii hierarchy of integrable equations.

2. The Geometry of Spheres

In this section we present basic results concerning geometry of oriented spheres in $\mathbb{E}^3$. We adopt the convention that a positive signed radius corresponds to the orientation of the sphere given by the inward field of unit normals. In our presentation we follow the monograph [**3**] where we also send the reader for proofs and more details.

Consider the quadric hyper-surface $\mathcal{Q}_L^4$ in five-dimensional projective space $\mathbb{P}^5$ given in homogeneous coordinates by the equation

$$\langle \boldsymbol{y}, \boldsymbol{y} \rangle = -y_1^2 + y_2^2 + y_3^2 + y_4^2 + y_5^2 - y_6^2 = 0, \tag{2.1}$$

this quadric $\mathcal{Q}_L^4$ is called the *Lie quadric*. Points of $\mathcal{Q}_L^4$ are in the bijective correspondence between the *Lie spheres* in $\mathbb{E}^3$ (oriented spheres, oriented planes and points of $\mathbb{E}^3$ together with the infinity point):

- *point* $\boldsymbol{u} \in \mathbb{E}^3$

$$\left[\left(\frac{1+\boldsymbol{u}\cdot\boldsymbol{u}}{2}, \frac{1-\boldsymbol{u}\cdot\boldsymbol{u}}{2}, \boldsymbol{u}, 0\right)\right], \tag{2.2}$$

- *the improper point* ∞

$$\left[(1, -1, \boldsymbol{0}, 0)\right], \tag{2.3}$$

- *the plane* $\boldsymbol{u}\cdot\boldsymbol{N} = h$ *with unit normal* $\boldsymbol{N}$

$$\left[(h, -h, \boldsymbol{N}, 1)\right], \tag{2.4}$$

- *the sphere with center* $\boldsymbol{p}$ *and signed radius* r

$$\left[\left(\frac{1+\boldsymbol{p}\cdot\boldsymbol{p}-r^2}{2}, \frac{1-\boldsymbol{p}\cdot\boldsymbol{p}+r^2}{2}, \boldsymbol{p}, r\right)\right]. \tag{2.5}$$

3. Quadrilateral Lattices and Rectilinear Congruences

The discrete analogs of conjugate nets on surfaces have been introduced in [**23**] as maps $\mathbb{Z}^2 \to \mathbb{E}^3$ with all elementary quadrilaterals planar. The neighbouring tangents of one parameter intersect, which is the discrete analog of the basic property of tangents on a surface with conjugate-line parametrization (compare Fig. 4 with Fig. 2).

The definition of discrete conjugate net makes sense if the ambient space is the projective space $\mathbb{P}^N$, $N > 2$ [**8**]. Algebraically, the planarity condition can be expressed in terms of the Laplace equation satisfied by the affine coordinates $\boldsymbol{x}\colon \mathbb{Z}^2 \to \mathbb{R}^N$ of the lattice

$$\Delta_1\Delta_2\boldsymbol{x} = (T_1A_{12})\Delta_1\boldsymbol{x} + (T_2A_{21})\Delta_2\boldsymbol{x}, \tag{3.1}$$

where T_i is the translation operator in the i direction, $i = 1, 2$, $\Delta_i = T_i - 1$ is the corresponding partial difference operator, and A_{12}, A_{21} are $\mathbb{R}$-valued functions on $\mathbb{Z}^2$.

Due to the fundamental role played by planarity of elementary quadrilaterals in the integrability scheme in multidimensions, the discrete conjugate nets are called lattices of planar quadrilaterals, or *quadrilateral lattices* [**13**].

The *discrete rectilinear congruence* is a $\mathbb{Z}^2$-parametrized family of straight lines such that any two neighbouring lines intersect. Such intersection points form the *focal lattices* of the congruence, which are quadrilateral lattices as well. Any family of tangents of a quadrilateral lattice are examples of discrete rectilinear congruences. The detailed theory of discrete congruences and transformations of quadrilateral lattices is presented in [**14**].

4. The Circular Lattices

The discrete analogs of curvature nets on surfaces have been introduced in [**19**, **21**] as maps $\mathbb{Z}^2 \to \mathbb{E}^3$ with all elementary quadrilaterals inscribed in circles. Notice that the circular lattice is a reduction of the quadrilateral lattice.

Algebraically, circular lattices can be characterized as follows:

PROPOSITION 4.1 ([**17**]). *The vector function $\boldsymbol{x}\colon \mathbb{Z}^2 \to \mathbb{E}^3$ of the circular lattice and the scalar function $\boldsymbol{x}\cdot\boldsymbol{x}\colon \mathbb{Z}^2 \to \mathbb{R}$ satisfy the same Laplace equation* (3.1).

COROLLARY 4.2. *This characterization is a simple consequence of the natural observation that the circular lattices are mapped by stereographic projection into quadrilateral lattices in the Möbius sphere* [**9**].

Denote by $\boldsymbol{C}$ the circle passing through $\boldsymbol{x}$, $T_1\boldsymbol{x}$, $T_2\boldsymbol{x}$, and by $\boldsymbol{\nu}$ denote the line normal to the plane of the circle and passing through its center. It is not difficult to check that the family of normals $\boldsymbol{\nu}$ is a rectilinear congruence. This property provides the discrete analog of the basic relation between normals on a surface in $\mathbb{E}^3$ and the curvature lines (compare Fig. 5 with Fig. 1).

Consider one-parameter family of spheres containing the circle $\boldsymbol{C}$ (the pencil of spheres with the base circle $\boldsymbol{C}$) and similar family containing the circle $T_i^{-1}\boldsymbol{C}$. These two pencils have in common the sphere $\boldsymbol{S}_i$, called the i-th *principal sphere* of the circular lattice. The center of $\boldsymbol{S}_i$ is the intersection point of the two corresponding normals $\boldsymbol{\nu}$ and $T_i^{-1}\boldsymbol{\nu}$ to the circular lattice $\boldsymbol{x}$. The centers of the principal spheres $\boldsymbol{S}_i$ form focal lattices of the normal congruence.

5. The Ribaucour Transformation of Circular Lattices and the Ribaucour Congruences of Spheres

Given circular lattice $\boldsymbol{x}\colon \mathbb{Z}^2 \to \mathbb{E}^3$, its *Ribaucour transform* $\mathcal{R}(\boldsymbol{x})$ is another circular lattice such that the quadrilaterals with vertices $\boldsymbol{x}$, $\mathcal{R}(\boldsymbol{x})$, $T_i\boldsymbol{x}$ and $\mathcal{R}(T_i\boldsymbol{x})$, $i = 1, 2$, are also circular (see Fig, 6).

The algebraic characterization of the Ribaucour transformations of circular lattices is presented in the following Theorem:

FIGURE 5. Discrete normal congruence.

THEOREM 5.1 ([**9,17**]). *Given circular lattice* $\boldsymbol{x}\colon \mathbb{Z}^2 \to \mathbb{E}^3$, *its Ribaucour transformation is given by*

$$\mathcal{R}(\boldsymbol{x}) = \boldsymbol{x} - \frac{\phi}{\phi'}\boldsymbol{x}', \tag{5.1}$$

where

(i) $\boldsymbol{x}'$ *is a solution of the system*

$$\Delta_i \boldsymbol{x}' = (T_i\sigma_i)\Delta_i \boldsymbol{x}, \tag{5.2}$$

where, due to the compatibility of the system (5.2) *the functions* σ_i *satisfy*

$$\Delta_j \sigma_i = A_{ij}(T_j\sigma_j - T_j\sigma_i), \quad i \neq j, \tag{5.3}$$

(ii) $\phi' = \boldsymbol{x}' \cdot \boldsymbol{x}'$, *and*

(iii) ϕ *is a solution of the equation*

$$\Delta_i \phi' = (T_i\sigma_i)\Delta_i \phi. \tag{5.4}$$

It is not difficult to see that both circles $\boldsymbol{C}$ and $\mathcal{R}(\boldsymbol{C})$ are contained in a sphere $\boldsymbol{S}$, which can be called the sphere of the Ribaucour transformation. The $\mathbb{Z}^2$-parameter family of such spheres is called the (discrete analog of the) *Ribaucour*

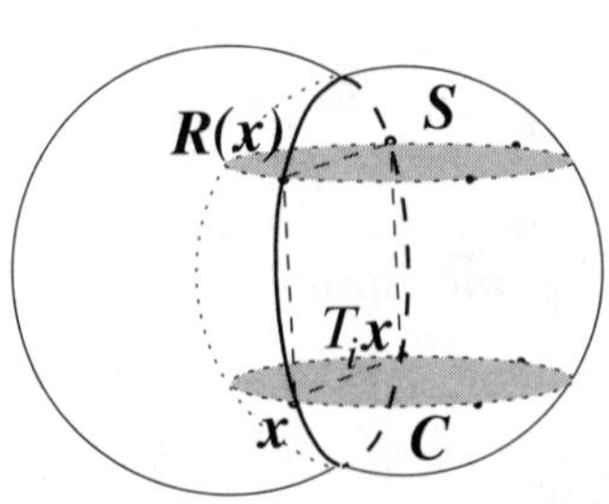

FIGURE 6. Circles of the Ribaucour congruence.

congruence of spheres [**6**]. Notice that the center of the sphere $\boldsymbol{S}$ is the intersection of the normal lines $\boldsymbol{\nu}$ and $\mathcal{R}(\boldsymbol{\nu})$ of the circular lattices.

Our aim is to characterize the $\mathbb{Z}^2$-parameter families of spheres in $\mathbb{E}^3$, which are the Ribaucour congruences of spheres. We consider the Ribaucour congruences of spheres as primary objects. Characterization of such families of spheres is provided by the following theorem, which is a direct analog of the classical result stated in Theorem 1.1.

THEOREM 5.2 ([**6**]). *The Ribaucour congruence of spheres is represented in the Lie sphere geometry by a quadrilateral lattice in* $\mathcal{Q}^4_L$.

Taking into account the formulas (2.5) describing the Lie representation of the sphere with center $\boldsymbol{p}$ and signed radius r we obtain the following algebraic properties of the Ribaucour congruences of spheres.

COROLLARY 5.3. *If* $\mathbb{Z}^2$*-parameter family of spheres* $\boldsymbol{S}(\boldsymbol{p}, r)$ *in* $\mathbb{E}^3$ *is the Ribaucour congruences of spheres then following conditions hold*:

(i) *the centers* $\mathbf{p}$ *of the spheres form a quadrilateral lattice,*
(ii) *the scalar functions* $\mathbf{p}\cdot\mathbf{p} - r^2 \colon \mathbb{Z}^2 \to \mathbb{R}$ *and* $r\colon \mathbb{Z}^2 \to \mathbb{R}$ *satisfy the same Laplace equation as the vector function* $\boldsymbol{p}\colon \mathbb{Z}^2 \to \mathbb{E}^3$.

6. Conclusion

We presented the basic results [**6**] of the theory of discrete Ribaucour congruences of spheres in $\mathbb{E}^3$. We have shown that the proper geometric setting for this theory is the geometry of oriented spheres developed by Sophus Lie. Within this approach the Ribaucour congruences of spheres turn out to be quadrilateral lattices in Lie's quadric $\mathcal{Q}^4_L$. In this way the theory of the Ribaucour congruences forms a part of the theory of *quadratic reductions* of the quadrilateral lattices [**9**].

It should be mentioned that similar results concerning the discrete analogs of the Weingarten congruences, asymptotic lattices, the Plücker geometry of straight lines, and the Plücker–Klein quadric have been obtained recently in [**7**]. These two new examples show once again the crucial role played by quadrilateral lattices and their reductions (see [**8–10, 12–14**]) in the integrable discrete geometry.

References

1. A. Bobenko, *Discrete conformal maps and surfaces*, Symmetries and Integrability of Difference Equations (Canterbury, 1996) (P. Clarkson and F. Nijhoff, eds.), London Math. Soc. Lecture Note Ser., vol. 255, Cambridge University Press, Cambridge, 1999, pp. 97–108.
2. A. Bobenko and U. Pinkall, *Discrete isothermic surfaces*, J. Reine Angew. Math. **475** (1996), 187–208.
3. Y. E. Cecil, *Lie sphere geometry*, Springer-Verlag, New York–Berlin, 1992.
4. J. Cieśliński, A. Doliwa, and P. M. Santini, *The integrable discrete analogues of orthogonal coordinate systems are multidimensional circular lattices*, Phys. Lett. A **235** (1997), 480–488.
5. G. Darboux, *Leçons sur les systèmes orthogonaux et les coordonnées curvilignes*, Gauthier–Villars, Paris, 1910.
6. A. Doliwa, *Circular lattices and the Ribaucour congruences of spheres within the sphere geometry of Lie*, to be published.
7. ______, *Discrete asymptotic nets and W-congruences within the Plücker line geometry*, J. Geom. Phys. **39** (2001), 9–29.
8. ______, *Geometric discretisation of the Toda system*, Phys. Lett. A **234** (1997), 187–192.
9. ______, *Quadratic reductions of quadrilateral lattices*, J. Geom. Phys. **30** (1999), 169–186.
10. A. Doliwa, S. V. Manakov, and P. M. Santini, $\bar{\partial}$*-reductions of the multidimensional quadrilateral lattice*: *The multidimensional circular lattice*, Comm. Math. Phys. **196** (1998), 1–18.

11. A. Doliwa, M. Mañas, L. Martínez Alonso, E. Medina, and P. M. Santini, *Multicomponent* KP *hierarchy and classical transformations of conjugate nets*, J. Phys. A **32** (1999), 1197–1216.
12. A. Doliwa and P. M. Santini, *The symmetric, D-invariant and Egorov reductions of the quadrilateral lattice*, J. Geom. Phys. **36** (2000), 60–102.
13. ———, *Multidimensional quadrilateral lattices are integrable*, Phys. Lett. A **233** (1997), 365–372.
14. A. Doliwa, P. M. Santini, and M. Mañas, *Transformations of quadrilateral lattices*, J. Math. Phys. **41** (2000), no. 2, 944–990, `solv-int/9712017`.
15. L. P. Eisenhart, *A treatise on the differential geometry of curves and surfaces*, Ginn and Company, Boston, 1909.
16. ———, *Transformations of surfaces*, Princeton University Press, Princeton, 1923.
17. B. G. Konopelchenko and W. K. Schief, *Three-dimensional integrable lattices in Euclidean spaces: Conjugacy and orthogonality*, Proc. Roy. Soc. London Ser. A **454** (1998), 3075–3104.
18. E. P. Lane, *Projective differential geometry of curves and surfaces*, Univ. Chicago Press, Chicago, 1932.
19. R. R. Martin, J. de Pont, and T. J. Sharrock, *Cyclic surfaces in computer aided design*, The Mathematics of Surfaces (J. A. Gregory, ed.), Clarendon Press, Oxford, 1986, pp. 253–268.
20. G. Monge, *Application de l'analyse à la géometrie*, Bachalier, Paris, 1850.
21. A. W. Nutbourne, *The solution of a frame matching equation*, The Mathematics of Surfaces (J. A. Gregory, ed.), Clarendon Press, Oxford, 1986, pp. 233–252.
22. A. Ribaucour, *Mémoire sur la théorie générale des surfaces courbes*, J. Math. Pures Appl. (9) **7** (1891), 5–108, 219–270.
23. R. Sauer, *Differenzengeometrie*, Springer, Berlin, 1970.

Inst. Fizyki Teoretycznej, Uniwersytet Warszawski, ul. Hoża 69, 00-681 Warszawa, Poland.

E-mail address: `doliwa@fuw.edu.pl`

Centre de Recherches Mathématiques
CRM Proceedings and Lecture Notes
Volume **29**, 2001

Bäcklund Transformations for the Reduced Maxwell–Bloch Equations

N. M. Ercolani

1. Introduction

The scales at which we can perceive the dynamics of nature are ultimately limited by the temporal resolution of the instruments we use. Currently, modern laser technology is defining this limit. Ultrafast lasers can resolve processes at the femtosecond scale (1 femtosecond $= 10^{-15}$ sec.) which is enough to allow direct observation of vibrational molecular dynamics [**8**]. At these scales, one of the principal analytical tools for modeling nonlinear optics, the slowly varying envelope approximation (SVEA) [**7**], is expected to break down. Indeed these technological developments have reawakened interest in electromagnetic solitary waves for which the polarizations of the electric and magnetic fields do not change direction within the pulse—i.e. ultra-short pulses of less than single cycle duration [**6**]. The term electromagnetic bubble (EMB) has been coined to emphasize the nonoscillating nature of such a pulse and the inapplicability of envelope models to describe these states. The earliest example of a reduced semiclassical model whose derivation did not employ the SVEA appeared in [**3**] in the context of self-induced transparency.

In this note we will present an overview of some recent work done on a two-level reduction of the semi-classical model of the interaction of light with matter. These reduced Maxwell–Bloch (RMB) equations are derived without recourse to the SVEA and are a generalization of the reduced equations derived in [**3**]. The full details concerning the results announced here appear in article [**2**]. We show, for the first time, that such reduced equations are completely integrable by deriving a Lax pair for them. Furthermore, and this will be the main emphasis of this note, we use the integrable structure to construct Darboux–Bäcklund transformations from which we can deduce explicit formulae for physically relevant one and two soliton solutions. Finally we numerically evaluate the latter to display the effects of permanent dipole moments of various strengths. We also illustrate parameter ranges over which the electric field displays internal oscillations at one extreme and what appear to be reasonable candidates for EMB's at the other.

2000 *Mathematics Subject Classification.* 70H06, 78A60, 81V80.

Support from the National Science Foundation (DMS-9626306) and the Air Force Office of Scientific Research (F49620-94-1-0144) is gratefully acknowledged.

This is the final form of the paper.

2. Derivation of the RMB Equations

The starting point for our study is a semi-classical model of the interaction of light with matter

$$\varepsilon_\infty \vec{E}_{tt} + c^2 \vec{\nabla} \times \vec{\nabla} \times \vec{E} = -\mathcal{N}/\varepsilon_0 \operatorname{Tr}\{\vec{\mu}\rho\}, \tag{2.1}$$

$$\frac{\partial \rho}{\partial t} = \frac{-i}{\hbar}\big[(H_0 - \vec{\mu}\cdot\vec{E}), \rho\big]. \tag{2.2}$$

The dynamical variables here are the classical electric field $\vec{E}$ and the electronic density matrix ρ. These equations involve time-independent constitutive variables: the dipole interaction matrix $\vec{\mu}$ whose entries are vectors so that its dot product with $\vec{E}$ is a matrix with scalar entries and the bare atomic Hamiltonian matrix H_0. They also depend on parameters: the density of optically active atoms $\mathcal{N}$ and the vacuum (resp. high frequency) permitivity ε_0 (resp. ε_∞).

The first equation is the classical Maxwell wave equation (Eq. (2.1)), in the absence of a magnetic response, within a medium. Classically, the right-hand side of this equation would be proportional to the *polarization field* of the medium. Here it is expressed in terms of the the quantum mechanical density matrix ρ which encodes the quantum-optical properties of the medium. The second equation is the accompanying Liouville equation which in turn prescribes ρ's $\vec{E}$-dependent evolution. For a system which can be modelled as having N bare electronic states (a so-called N-level atom), the matrix ρ (as well as μ) would be an $N \times N$ matrix.

In this semiclassical picture, in which one is modelling a medium that consists of many atoms, an implicit mesoscopic averaging has been made. Thus the above equations contain no information regarding the location of specific atoms. The material Hamiltonian $H_{\text{material}} = H_0 - \vec{\mu}\cdot\vec{E}$ could still reflect macroscopic spatial inhomogeneities as well as other complications. However, we will assume that this is not the case so that H_{material} can be viewed as the Hamiltonian describing a single representative atom. The eigenvectors of H_{material} are then interpreted as the states of the atom which possess observable energies. The latter are the eigenvalues of the matrix. In the absence of a field $\vec{E}$, these eigenvectors are the bare electronic states mentioned before. A general state of the system can always be described as a superposition of these bare states.

Similarly the contraction of the density matrix ρ on a general state is interpreted as the probability that the atom can be found in that state (although it would probably be more accurate in this semiclassical setting to regard ρ as encoding the actual relative number of atoms in a mesoscopic volume which appear in various purely quantum mechanical states). It follows from this that ρ is Hermitean with nonnegative eigenvalues that sum to 1.

It is the electric dipole matrix $\vec{\mu}$ that determines the strength and nature of the interaction of an atomic system with an applied electric field. We are working here within a dipole approximation in which the spatial variation of the electric field $\vec{E}$ is assumed to be negligible over atomic dimensions; this accounts for the simplicity of the interaction term within the material Hamiltonian. Electric dipole transitions are generally responsible for the absorption and dispersion properties of materials in ultraviolet to infrared portions of the optical spectrum.

A significant distinction here from other derivations of reduced Maxwell–Bloch equations is that we will not assume that $\vec{\mu}$ has vanishing diagonal entries. Such a vanishing would be a consequence of the eigenstates of H_0 possessing a definite parity. For example, a system with a center of symmetry or a many-body system in which the potential depends only on the distance between particles, would have an H_0 with parity invariance. However, if this invariance does not hold, then all matrix elements of the dipole operator $\vec{\mu}$ may be nonzero and a permanent electric dipole moment, corresponding to a nonvanishing diagonal element, may exist.

We introduce some further assumptions which, in appropriate regimes, are physically realizable and through which we arrive at a system which is analytically tractable.

ASSUMPTION 1. The material is isotropic so that one may take the electric field $\vec{E}$ and the polarization field $\mathcal{N}\,\mathrm{Tr}\{\vec{\mu}\rho\}$ to be parallel (say to the x-axis). In this case we can take the vector field E to be a scalar and, correspondingly, the electric dipole matrix $\vec{\mu}$ becomes a Hermitean matrix μ with scalar entries.

ASSUMPTION 2. We restrict consideration to solutions which are propagating along the z-direction, independent of x and y. Then the system becomes $1-D$ as well as scalar.

ASSUMPTION 3. The electric field is predominantly traveling in one direction (say to the right). This is justifiable in the low density regime; i.e., for $\mathcal{N}$ sufficiently small one may neglect the effects of backscatter. For example, at typical values of the polarization, 10^{-18} cgs units, and for atomic densities, $\mathcal{N}$, on the order of 10^{12} atoms per cubic cm, one is neglecting terms on the order of one part in 10^8 [**3**].

ASSUMPTION 4. We consider a material which is well approximated as a two-level atom. Then H_0 has two eigenstates with real eigenvalues $\hbar\omega_1 > \hbar\omega_2$. The coefficients ρ_{ij}, μ_{ij} of the matrices ρ and μ, are taken with respect this basis of eigenstates.

Under these assumptions and truncating at first order in the low density approximation, one formally derives the Reduced Maxwell–Bloch (RMB) equations for a two-level atom:

$$\frac{\partial E}{\partial t} = -\frac{c}{\sqrt{\varepsilon_\infty}}\frac{\partial E}{\partial z} + i\frac{\mathcal{N}\omega}{2\varepsilon_0\varepsilon_\infty}(\mu_{12}^*\rho_{12} - \mu_{12}\rho_{12}^*), \tag{2.3}$$

$$\frac{\partial \rho_{12}}{\partial t} = -i\left[\left(\omega - \frac{\Delta\mu}{\hbar}E\right)\rho_{12} + \frac{\mu_{12}}{\hbar}E\Delta\rho\right], \tag{2.4}$$

$$\frac{\partial \Delta\rho}{\partial t} = \frac{-2i}{\hbar}E(\mu_{12}^*\rho_{12} - \mu_{12}\rho_{12}^*), \tag{2.5}$$

where $\omega = \omega_1 - \omega_2$ (the atomic transition frequency), $\Delta\mu = \mu_{11} - \mu_{22}$, $\Delta\rho = \rho_{11} - \rho_{22}$, and the conservation of probability

$$\frac{\partial(\rho_{11}+\rho_{22})}{\partial t} = \frac{\partial\,\mathrm{Tr}\{\rho\}}{\partial t} = \mathrm{Tr}\left\{\frac{\partial\rho}{\partial t}\right\} = -\frac{i}{\hbar}\mathrm{Tr}\{[H,\rho]\} = 0,$$

so that $\rho_{11} + \rho_{22} = 1$. Since ρ_{11} and ρ_{22} are both probabilities, they are both nonnegative and the population difference must satisfy $-1 \le \Delta\rho \le 1$.

At this point we pause briefly to discuss some of the effects that are retained in this model but which are not kept in other reductions of the Maxwell–Bloch equations.

The approximation of a slowly varying envelope around a definite frequency was not used. Thus the effects of arbitrary detuning are retained in these equations.

We also retain the effect of a nonvanishing permanent dipole moment. Within the regime of validity of the dipole approximation, the usual assumption (mentioned earlier) that this effect is absent corresponds to an assumption on the spatial structure of the two bare atomic eigenstates. This would require that both electronic levels have definite parity with respect to inversion through the axis defining the dipole coupling. This simplifying assumption is not made here.

We will see that the effect of the atom-field detuning and that of the permanent dipole enter the equations at nearly the same level and are, therefore, somewhat similar in effect. The effect of each is to destroy perfect energy exchange between medium and field. On the other hand, the major (somewhat paradoxical) distinction between the two effects is that while that of atomic detuning is, in a sense, permanent, that of the permanent dipole is locally field dependent and, therefore, transitory for anything other than continuous operation. More specifically, the effect of the permanent dipole is to "dress" the bare electronic energy levels in a spatially dependent way by establishing a so-called a.c. Stark shift proportional to the local field amplitude. This dressing of the levels decreases or increases the energy separation between the electronic levels depending, respectively, on whether the field points with or against the permanent dipole.

We introduce the dimensionless variables d, n, e, ε, τ and ζ through the following substitutions:

$$\begin{aligned} \mu &= (e^- \lambda_0) d, & \varepsilon_0 &= \left(\frac{{e^-}^2}{2\pi\hbar c}\right)\varepsilon = \left(\frac{{e^-}^2}{\hbar\omega\lambda_0}\right)\varepsilon, \\ \mathcal{N} &= \left(\frac{\varepsilon\varepsilon_\infty}{\lambda_0^3}\right) n, & t &= \left(\frac{1}{\omega}\right)\tau, \\ E &= \left(\frac{\hbar\omega}{\lambda_0 e^-}\right) n|d_{12}|e, & z &= \left(\frac{1}{k}\right)\zeta = \left(\frac{c}{\sqrt{\varepsilon_\infty}\omega}\right)\zeta, \end{aligned}$$

where e^- is the electron charge, $\lambda_0 = 2\pi c/\omega$ is the deBroglie length associated with the transition from level 2 to level 1, and k is the wave number in the host medium for this transition. There is an arbitrary phase in these equations which can be pinned down by writing $d_{12} = e^{i\theta} d$, $d \geq 0$ and, with an abuse of notation, replacing ρ_{12} by $e^{i\theta}\rho_{12}$. Then the RMB equations become, in dimensionless form,

$$\frac{\partial e}{\partial \tau} = -\frac{\partial e}{\partial \zeta} + \frac{i}{2}(\rho_{12} - \rho_{12}^*), \tag{2.6}$$

$$\frac{\partial \rho_{12}}{\partial \tau} = -i\big[\big(1 - 2\Omega\sin(\theta)e\big)\rho_{12} + \Omega\cos(\theta)e\Delta\rho\big], \tag{2.7}$$

$$\frac{\partial \Delta\rho}{\partial \tau} = -2i\Omega\cos(\theta)e(\rho_{12} - \rho_{12}^*). \tag{2.8}$$

Here $\Omega\cos(\theta) = nd^2$, $\Omega\sin(\theta) = nd\Delta d/2$, $\Delta d = d_{11} - d_{22}$, and we can choose $\Omega = nd\sqrt{d^2 + (\Delta d/2)^2}$. $H = H_0 - \hbar\omega nde$.

The physical significance of these rescaled parameters can be summarized by observing that Ω is a composite parameter containing information about the number of optically active atoms per unit volume and the magnitude of the dipole moment of an individual atom. It measures the effectiveness of the coupling between the radiation and the material. The parameter θ measures the relative strengths of the permanent and induced dipole moments (whereas Ω is proportional to the total contribution of both types of dipoles). When there is no permanent dipole moment, $\theta = 0$.

3. The Loop Algebra for the RMB Equations

In [2], the RMB equations are shown to be completely integrable through the derivation of a Lax pair for the PDE system (2.6)–(2.8). The Lax pair is a pair A, L of linear differential operators having the form

$$\begin{aligned} A &= -\partial_\tau + \lambda(h_0\mathcal{H} + f_0\mathcal{F}) + e_0\mathcal{E}, \\ L &= \partial_\zeta + \lambda(h_0\mathcal{H} + f_0\mathcal{F}) + e_0\mathcal{E} + (1-\lambda^2)^{-1}\big(\lambda(h_1\mathcal{H} + f_1\mathcal{F}) + e_1\mathcal{E}\big), \end{aligned} \tag{3.1}$$

where

$$\mathcal{H} = \begin{pmatrix} i & 0 \\ 0 & -i \end{pmatrix}, \quad \mathcal{F} = \begin{pmatrix} 0 & 1 \\ -1 & 0 \end{pmatrix}, \quad \mathcal{E} = \begin{pmatrix} 0 & i \\ i & 0 \end{pmatrix},$$

is the usual Pauli basis of su(2) and

$$\begin{aligned} h_0 &= \frac{\cos(\theta)}{2}, & h_1 &= \frac{-\Omega}{2}(\Delta\rho - 2\tan(\theta)\operatorname{Re}\rho_{12}), \\ f_0 &= 0, & f_1 &= \Omega\sec(\theta)\operatorname{Im}\rho_{12}, \\ e_0 &= -\frac{\sin(\theta)}{2} + \Omega e, & e_1 &= \Omega\left(\frac{\tan(\theta)}{2}\Delta\rho + \operatorname{Re}\rho_{12}\right). \end{aligned} \tag{3.2}$$

Note that when $\theta = 0$; i.e., the dipole is "off", the above coefficients simplify to

$$\begin{aligned} h_0 &= 1/2, & h_1 &= \frac{-\Omega}{2}\Delta\rho, \\ f_0 &= 0, & f_1 &= \Omega\operatorname{Im}\rho_{12}, \\ e_0 &= \Omega e, & e_1 &= \Omega\operatorname{Re}\rho_{12}. \end{aligned}$$

The existence of such a Lax Pair which is rational in the spectral parameter suggests that one should be able to interpret the associated dynamics as a flow on a subspace of a generalized Lie algebra commonly referred to as a *loop algebra* [4, 7]. The elements of a loop algebra are formal partial fractions expansions whose coefficients are elements of a given Lie algebra or, more concretely, matrices corresponding to some representation of the Lie algebra. In this setting one expects the RMB flow to be one among an infinite hierarchy of commuting flows. The above structure suggests that the general form of the loop subspace on which the RMB flow evolves should be

$$Q = \sum_{k=0}^{\infty}(1-\lambda^2)^{-k}\big(\lambda(h_k\mathcal{H} + f_k\mathcal{F}) + e_k\mathcal{E}\big). \tag{3.3}$$

The Lax pair for the nth flow is given by A and

$$L_n = \partial_{\zeta_n} + Q^{(n)},$$

where

$$Q^{(n)} = \sum_{k=0}^{n}(1-\lambda^2)^{-k}\big(\lambda(h_k\mathcal{H} + f_k\mathcal{F}) + e_k\mathcal{E}\big). \tag{3.4}$$

We'll refer to a loop of the form $Q^{(n)}$ as a *height n* loop. The requirement that $[A, L_n] = 0$ determines the dependence of h_k, f_k, e_k on τ and ζ_n. Specifically, this equation may be written

$$Q^{(0)}_{\zeta_n} + Q^{(n)}_{\tau} - [Q^{(0)}, Q^{(n)}] = 0.$$

The fact that these flows pairwise commute with one another follows from a Hamiltonian interpretation of these flows associated to a natural Lie–Poisson structure on the loop algebra [**1**] which we will not pursue here.

One may interpret these ζ_n-flows as flows on a space of functions of τ by "freezing" the ζ dependence; i.e., we look for solutions that are stationary in ζ. These then satisfy the stationary equations $Q^{(n)}_{\tau} = [Q^{(0)}, Q^{(n)}]$. These in turn can all be viewed as polynomial truncations of a single master equation [**4**]:

$$Q_{\tau} = [Q^{(0)}, Q], \tag{3.5}$$

where Q is the infinite series defined earlier. Expanding on the Pauli basis one has $Q = h\mathcal{H}+f\mathcal{F}+e\mathcal{E}$ where h, f, e are rational functions of λ and $Q^{(0)} = \lambda h_0\mathcal{H}+e_0\mathcal{E}$. It is not hard to arrive at a general solution of equation (3.5). Let Ψ be a fundamental solution of $\partial_{\tau}\Psi = Q^{(0)}\Psi$. Then it is straightforward to check that

$$Q = \Psi\mathcal{H}\Psi^{-1}, \tag{3.6}$$

solves equation (3.5). If we consider fundamental solutions which only have poles at $\lambda = \pm 1$ then (3.6) presents the loop algebra subspace, on which the RMB flows evolve, as an orbit under conjugation by the "loop group" elements Ψ. Taking a partial fractions expansion of Q with respect to these two poles reproduces the original presentation of this orbit as given in (3.3).

In this presentation the coefficients, h_k, e_k, f_k become dependent functions of τ and this subspace can be naturally interpreted as a phase space for PDE flows (in ζ_k). Because it appears as an orbit under conjugation it has formal similarities to the *co-adjoint orbits* of finite dimensional Lie algebras [**5**] and so we will conveniently refer to it as the "co-adjoint orbit".

4. Darboux–Bäcklund Transformations

The commutation of the operators A and L implies that they have a simultaneous fundamental solution Ψ; i.e., $\Psi = \Psi(\tau, \zeta)$ simultaneously solves $A\Psi = L\Psi = 0$. In fact since the operators L_n pairwise commute, Ψ can be extended to a function of infinitely many variables τ and ζ_n which simultaneously solves $L_n\Psi = 0$.

The notion of a general dressing transformation is based on the picture, introduced in the last section, that the phase space, or co-adjoint orbit as we called it, for the RMB flows is a subspace of a loop algebra on which certain invertible loop elements act by conjugation. A *dressing* transformation is induced by a gauge transformation acting simultaneously on the operators L_k. As loop elements they can depend on ζ, τ and λ as well as the point Q at which they act. Let us denote such a loop element by $G(\tau, \zeta; \lambda)$. Then this action is given by

$$L_k \to L_k^{\mathrm{new}} = GL_kG^{-1}.$$

It is too much to ask that G be globally invertible; we only require that it be invertible for generic values of the parameters and re-express the above as

$$GL_k = L_k^{\mathrm{new}}G. \tag{4.1}$$

A dressing transformation differs from a general gauge transformation in that we require that it should preserve the RMB phase space; in other words, Q_k^{new}

should again be a loop of the form presented in (3.3). To say that G preserves the form of Q is equivalent to demanding

(i) the singularity structure of the loop is preserved; i.e., no new poles are introduced, and
(ii) the real structure of the loop is preserved. Precisely, the real structure of the loop is that the coefficients h_k, f_k, e_k should all be real and that under the involution induced by $\lambda \to -\lambda$, $(h_k, f_k, e_k) \to (-h_k, -f_k, e_k)$.

We will reserve the term Darboux–Bäcklund transformation for a dressing transformation that is linear in λ; i.e., it is of the form $G = \lambda - M$, where M is of the form

$$M = N \begin{pmatrix} \nu & 0 \\ 0 & \bar{\nu} \end{pmatrix} N^{-1} = \mathrm{Re}(\nu) + \mathrm{Im}(\nu) N \mathcal{H} N^{-1},$$

with $\mathrm{Im}(\nu) \neq 0$ and N an invertible matrix. Thus we may set

$$G = N \begin{pmatrix} \lambda - \nu & 0 \\ 0 & \lambda - \bar{\nu} \end{pmatrix} N^{-1}, \tag{4.2}$$

and try to determine N as a function of τ, ζ and Q so that (i) and (ii) are satisfied. Note first that $G(\lambda)$ has a nontrivial kernel only if $\lambda = \nu$ or $\bar{\nu}$. The kernel of $G(\nu)$ is one-dimensional and, as is clear from definition (4.2), it is spanned by Ne_1 where e_1 is the standard basis vector $(1,0)^T$. By the commutation relation (4.1) it then follows that $L_k(\nu)Ne_1$ is in the kernel of $G(\nu)$ and so there is a constant γ_1 such that

$$L_k(\nu)Ne_1 = \gamma_1 Ne_1.$$

Similarly, one shows that there is a constant γ_2 such that

$$L_k(\bar{\nu})Ne_2 = \gamma_2 Ne_2.$$

From this it is straightforward to show that the first and second columns of

$$N \begin{pmatrix} e^{-\gamma_1 \zeta_k} & 0 \\ 0 & e^{-\gamma_2 \zeta_k} \end{pmatrix},$$

respectively solve $L_k(\nu)\vec{\phi} = 0$ and $L_k(\bar{\nu})\vec{\phi} = 0$, since

$$\begin{pmatrix} e^{-\gamma_1 \zeta_k} & 0 \\ 0 & e^{-\gamma_2 \zeta_k} \end{pmatrix},$$

commutes with

$$\begin{pmatrix} \lambda - \nu & 0 \\ 0 & \lambda - \bar{\nu} \end{pmatrix}.$$

In this way we arrive at a necessary condition on N in order for G to be a Darboux–Bäcklund transformation: one must be able to choose N in the definition of G so that $L_k(\nu)Ne_1 = 0$ and $L_k(\bar{\nu})Ne_2 = 0$. We next give a recipe for constructing a dressing transformation that automatically satisfies this necessary condition.

- Let $\Psi = (\vec{\psi}_+, \vec{\psi}_-)$ be a fundamental solution of $L_k(\lambda)$ associated to a particular loop $Q^{(k)}(\lambda)$.
- Associate to a complex vector $\vec{c} = (c_+, c_-)$ a particular solution

$$\vec{\phi} = \Psi(\lambda)\vec{c} = c_+\vec{\psi}_+ + c_-\vec{\psi}_-.$$

• Define N associated to $\vec{c}$ and a particular eigenvalue $\lambda = \nu$ by

$$N := \begin{pmatrix} \phi_1(\nu) & -\overline{\phi_2(\nu)} \\ \phi_2(\nu) & \overline{\phi_1(\nu)} \end{pmatrix} = (\vec{\phi}(\nu), -\mathcal{F}\overline{\vec{\phi}(\nu)}), \tag{4.3}$$

where $(\bar{\cdot})$ denotes complex conjugation and F is the Pauli matrix defined in (3.1).

• By symmetries of the operator L_k, it is straightforward to check directly that if $L_k(\nu)\vec{\phi}(\nu) = 0$ then $L_k(\bar{\nu})\big(-\mathcal{F}\overline{\vec{\phi}(\nu)}\big) = 0$ and so the necessary condition deduced earlier is satisfied by this choice of N. As defined, this N is almost unitary:

$$N^{-1} = \frac{1}{\det N} N^*,$$

which will help to insure that the real structure of the loop is preserved.

From now on, N will always denote the matrix constructed by the above recipe. In many cases we can show that the associated dressing transformation is a Darboux–Bäcklund transformation. The new fundamental solution associated to the dressed operator, $L_k^{\text{new}} = GL_kG^{-1}$, is

$$\Psi_{\text{new}}(\tau, \zeta, \lambda) = G(\lambda, \nu, \vec{c})\Psi(\tau, \zeta, \lambda). \tag{4.4}$$

The candidate for a new loop element is then determined from this by

$$Q^{(k)}_{\text{new}} = \dot{\Psi}_{\text{new}}\Psi^{-1}_{\text{new}}, \tag{4.5}$$

where $\dot{}$ denotes $-\partial/\partial\zeta_k$.

We can now state our results.

THEOREM 4.1. *$Q^{(k)}_{\text{new}}$ has the same singularities (in λ) as the original $Q^{(k)}$.*

This may be seen directly from the fundamental formula (derived in [**2**])

$$\begin{aligned} N^{-1}&Q^{(k)}_{\text{new}}(\lambda)N \\ &= \left[N^{-1}Q^{(k)}(\nu)Ne_1, N^{-1}Q^{(k)}(\bar{\nu})Ne_2\right] + ((\lambda - \operatorname{Re}(\nu)) - \operatorname{Im}(\nu)\mathcal{H}) \\ &\quad\times \left[N^{-1}\left(\frac{Q^{(k)}(\lambda) - Q^{(k)}(\nu)}{\lambda - \nu}\right)Ne_1, N^{-1}\left(\frac{Q^{(k)}(\lambda) - Q^{(k)}(\bar{\nu})}{\lambda - \bar{\nu}}\right)Ne_2\right], \end{aligned}$$

where $\{e_1, e_2\}$ are the standard basis vectors in $\mathbb{R}^2$. If $Q^{(k)}(\lambda)$ has poles only at $\lambda = \pm 1$, then this formula shows that the only possible locations for new poles would be at $\lambda = \nu$ or $\bar{\nu}$. However, as $\lambda \to \nu$ (resp. $\lambda \to \bar{\nu}$) the difference quotients go to finite limits since $Q^{(k)}(\lambda)$ is differentiable.

For height zero loops, which correspond to the A operator in the Lax pair for the RMB equations, we can state precise conditions for our dressing transformation to be Darboux–Bäcklund. Given Theorem 4.1, this is just a question of finding conditions under which $Q^{(0)}_{\text{new}}(\lambda)$ continues to preserve the real structure of the loops.

THEOREM 4.2. *$Q^{(0)}_{\text{new}}(\lambda)$ remains in the loop orbit iff* $\operatorname{Re}(\phi_1\overline{\phi_2}) = 0$.

This result is an immediate consequence of the following explicit formula.

$$Q^{(0)}_{\text{new}}(\lambda) - Q^{(0)}(\lambda) = -4\operatorname{Im}(\nu)h_0\left\{\frac{\operatorname{Re}(\phi_1\overline{\phi_2})\mathcal{F} + \operatorname{Im}(\phi_1\overline{\phi_2})\mathcal{E}}{|\phi_1|^2 + |\phi_2|^2}\right\}. \tag{4.6}$$

We next consider the special but important case of constant coefficient loops of height 1 with

$$\begin{aligned} h_0 &= \frac{\cos\theta}{2}, & h_1 &= \frac{-\Omega}{2}\sec^2(\theta)\Delta\rho, \\ f_0 &= 0, & f_1 &= 0, \\ e_0 &= 0, & e_1 &= 0, \end{aligned} \tag{4.7}$$

where $-1 \le \Delta\rho \le 1$ is a real constant. We refer to such a loop as a *vacuum loop*. We have shown, in [**2**], that

THEOREM 4.3. *If $Q^{(1)}(\lambda)$ is a vacuum loop, then $Q^{(1)}_{\text{new}}(\lambda)$ remains in the co-adjoint orbit iff $N\mathcal{H}N^{-1} \in \operatorname{span}\{\mathcal{H}, \mathcal{F}\}$ and ν is pure imaginary.*

We will call a loop which arises in this way, from a Darboux–Bäcklund transformation of a vacuum loop, a *one-soliton* loop and denote it $Q^{(1)}_{\text{sol}}(\lambda)$.

Using the formulae for the Darboux–Bäcklund transformation and equation (3.2) one can solve for $e^{\text{sol}}, \Delta\rho^{\text{sol}}, \operatorname{Re}\rho_{12}^{\text{sol}}$ and $\operatorname{Im}\rho_{12}^{\text{sol}}$ to obtain a complete set of solutions of the pde system (2.6)–(2.8):

$$\begin{aligned} e^{\text{sol}} &= \frac{2mh_0}{\Omega}\operatorname{sech}\big(x(\tau,\zeta)\big) + \frac{\sin(\theta)}{2\Omega}, \\ \Delta\rho^{\text{sol}} &= \left[1 - \left(\frac{2m^2}{m^2+1}\right)\operatorname{sech}^2\big(x(\tau,\zeta)\big) - \left(\frac{2m}{m^2+1}\right)\tan(\theta)\operatorname{sech}\big(x(\tau,\zeta)\big)\right]\Delta\rho^{\text{span}}, \\ \operatorname{Re}\rho_{12}^{\text{sol}} &= \left[\tan(\theta)\left(\frac{-1}{2} + \left(\frac{m^2}{1+m^2}\right)\operatorname{sech}^2\big(x(\tau,\zeta)\big)\right)\right. \\ &\qquad \left. - \left(\frac{m}{m^2+1}\right)\cos(\theta)\operatorname{sech}\big(x(\tau,\zeta)\big)\right]\Delta\rho^{\text{span}}, \\ \operatorname{Im}\rho_{12}^{\text{sol}} &= \frac{m^2}{m^2+1}\sec(\theta)\operatorname{sech}\big(x(\tau,\zeta)\big)\tanh\big(x(\tau,\zeta)\big)\Delta\rho^{\text{span}}, \end{aligned}$$

where,

$$x(\tau,\zeta) = 2mh_0\tau - 2\alpha\zeta + \log(c_2/c_1), \quad \alpha = mh_0 + \left(\frac{m}{m^2+1}\right)h_1.$$

These results suggest that Bäcklund transformations for pure imaginary ν and with $\operatorname{Re}(\phi_1\overline{\phi_2}) = 0$ should preserve the real structure of a loop in the co-adjoint orbit. Of course it is possible that a composite (i.e., iterated) Bäcklund transformation could preserve the real structure of a loop even though the individual factors of this composite do not. We expect that this will be the case and that there are dressing transformations associated to eigenvalues which are pure imaginary but which still preserve the the real structure of loops. We also expect that a general dressing transformation can always be factored into an iteration of Bäcklund transformations although the factors may not individually preserve reality. Indeed, we have constructed explicit examples of this in the case when $\theta = 0$.

A double iteration of the Bäcklund transformation at $\nu_1 = im_1$ and $\nu_2 = im_2$, yields the expression for the electric field of a 2-soliton

$$\begin{aligned} & e_0^{2-\text{soliton}} \\ &= \left(\frac{m_1^2 - m_2^2}{m_1^2 + m_2^2}\right)\frac{2m_1\,h_0\operatorname{sech}(x_1) - 2m_2\,h_0\operatorname{sech}(x_2)}{1 + ((2m_1m_2)/(m_1^2+m_2^2))(\tanh(x_1)\tanh(x_2) - \operatorname{sech}(x_1)\operatorname{sech}(x_2))}, \end{aligned} \tag{4.8}$$

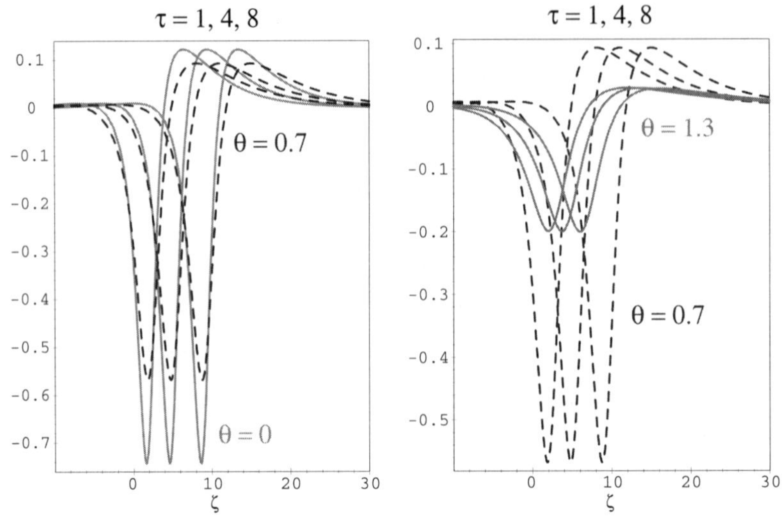

(a) $\theta = 0$ (solid), $\theta = 0.7$ (dashed)

(b) $\theta = 0.7$ (dashed), $\theta = 1.3$ (solid)

FIGURE 1. Propagation of two-soliton solution ($\nu_1 = -0.2i$, $\nu_2 = -0.6i$, $\theta_1 = -1$, $\theta_2 = 0$ and $\Delta\rho^{\text{span}} = 0.5$) for varying dipole strengths. The amplitude of the pulse decreases as θ increases.

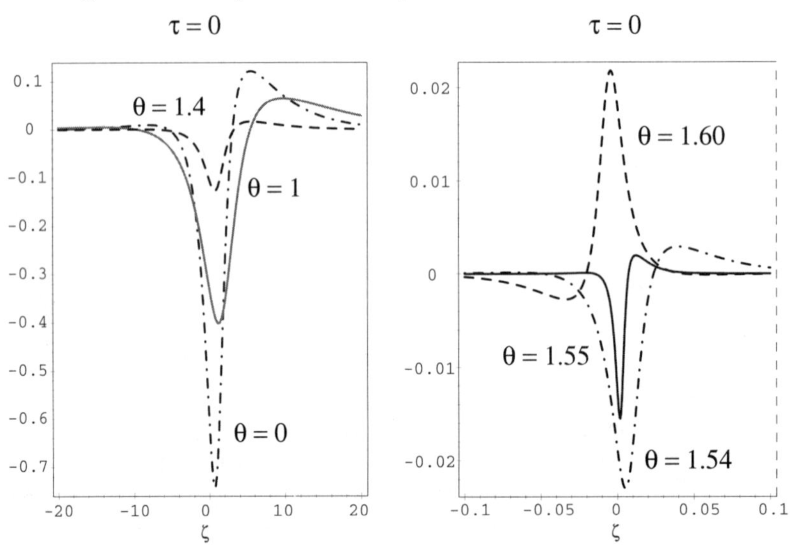

(a) $\theta = 0$ (dash-dot), 1 (solid), 1.4 (dashed)

(b) $\theta = 1.54$ (dash-dot), 1.55 (solid), 1.60 (dashed)

FIGURE 2. Simultaneous snapshots of two-soliton solutions ($\nu_1 = -0.2i$, $\nu_2 = -0.6i$, $\theta_1 = -1$, $\theta_2 = 0$ and $\Delta\rho^{\text{span}} = 0.5$) for varying dipole strengths. Observe that for a sufficiently large permanent dipole moment ($\theta = 1.6$) the pulse is inverted.

where,

$$x_1 = 2m_1h_0\tau - 2\alpha_1\zeta + \theta_1, \quad x_2 = 2m_2h_0\tau - 2\alpha_2\zeta + \theta_2,$$

$\alpha_i = m_ih_0 + \big(m_i/(m_i^2+1)\big)h_1$ and $\theta_1 = \log(c_-^1/c_+^1)$ and $\theta_2 = \log(c_-^2/c_+^2)$.

Figures 1 and 2 show the effect of the permanent dipole moment on the speed and the amplitude of the electric field pulse. In particular we note how this dipole moment tunes the amplitude of the field even causing it to invert for sufficiently strong moments.

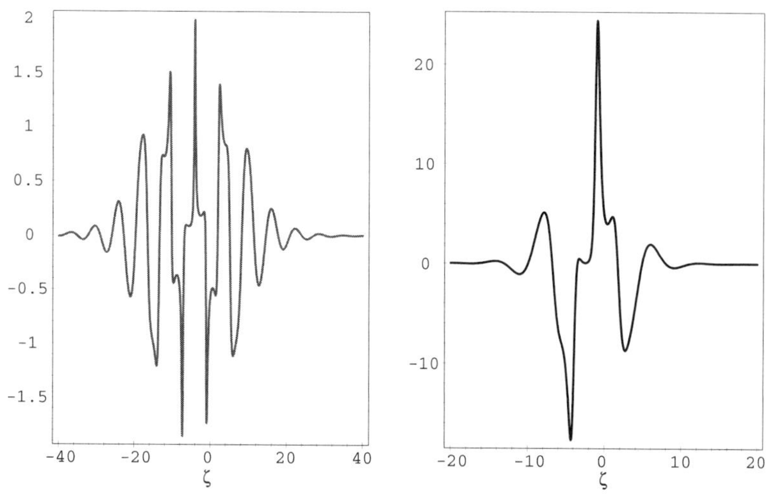

(a) $\nu_1 = -1 - 0.2\mathrm{i}$, $\nu_2 = 1 - 0.2i$ (b) $\nu_1 = -1 - 0.5i$, $\nu_2 = 1 - 0.5i$

FIGURE 3. Localized pulses with internal oscillations ($\theta_1 = \theta_2 = 0$, $\Delta\rho^{\text{span}} = 0.5$).

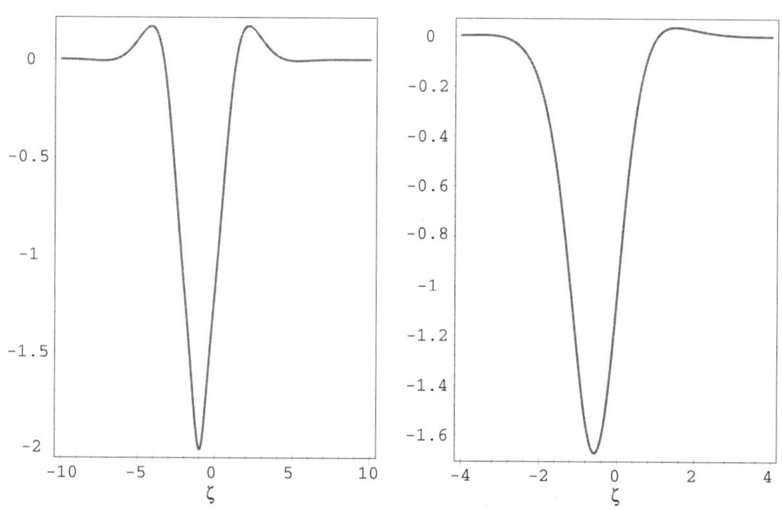

(a) $\nu_1 = -1 - i$, $\nu_2 = 1 - i$ (b) $\nu_1 = -1 - 2i$, $\nu_2 = 1 - 2i$

FIGURE 4. Electromagnetic bubbles ($\theta_1 = \theta_2 = 0$, $\Delta\rho^{\text{span}} = 0.5$).

At $\theta = 0$ if, in the above formula, one replaces m_1 and m_2 respectively by $-i\nu_1$ and $-i\nu_2$, for complex ν_i, such that $\nu_1 = -\nu_2^*$ and $\theta_1 = -\theta_2^*$, the corresponding 2-soliton solution will be real. The other elements of the solution can also be shown to satisfy the reality constraints. (In fact these solutions recapture two-soliton solutions found by [**3**].) Thus for iterated Bäcklund transformations the constraint of pure imaginary spectrum can be relaxed.

By varying these spectral parameters one may produce localized pulses with internal oscillations at one extreme, Fig 3, and electromagnetic bubbles at the other, Fig 4.

References

1. M. Agrotis, *Integrable structure of reductions of the Maxwell–Bloch equations*, Ph.D dissertation, University of Arizona, in preparation.
2. M. Agrotis, N. M. Ercolani, S. A. Glasgow, and J. V. Moloney, *Complete integrability of the reduced Maxwell-Bloch equations with permanent dipole*, Phys. D **138** (2000), no. 1-2, 134–162.
3. J. C. Eilbeck, J. D. Gibbon, P. J. Caudrey, and R. K. Bullough, *Solitons in nonlinear optics,* I. *A more accurate description of the* 2π *piles in self-induced transparency*, J. Phys. A **6** (1973), 1337–1347.
4. H. Flaschka, A. C. Newell, and T. Ratiu, *Kac–Moody algebras and soliton equations,* II. *Lax equations associated with* $A_1^{(1)}$, Phys. D **9** (1983), 300–323.
5. V. Guillemin and S. Sternberg, *Symplectic techniques in physics*, Cambridge University Press, Cambridge, 1984.
6. A. E. Kaplan and P. L. Shkolnikov, *Subfemtosecond high-intensity unipolar electromagnetic solitons and shockwaves*, J. Nonlinear Opt. Phys. and Materials **4** (1995), 831–841.
7. A. C. Newell, *Solitons in mathematics and physics*, CBMS–NSF Regional Conf. Ser. **48**, SIAM Press, 1985.
8. G. Steinmeyer, D. H. Sutter, L. Gallmann, N. Matuschek, and U. Geller, *Frontiers in ultrashort pulse generation: Pushing the limits in linear and nonlinear optics*, Science **286** (1999), 1507–1512.

Mathematics Department, University of Arizona, Tucson, Arizona 85721, USA.
E-mail address: `ercolani@math.arizona.edu`

Centre de Recherches Mathématiques
CRM Proceedings and Lecture Notes
Volume **29**, 2001

Transformations of Quasilinear Systems Originating from the Projective Theory of Congruences

E. V. Ferapontov

ABSTRACT. We continue the investigation of the correspondence between systems of conservation laws

$$u^i_t = f^i(u)_x, \quad i = 1, \dots, n,$$

and n-parameter families of lines (congruences of lines) in A^{n+1} defined by the equations

$$y^i = u^i y^0 - f^i(u).$$

Relationship between "additional" conservation laws

$$h(u)_t = g(u)_x,$$

and hypersurfaces conjugate to a congruence is established. This construction allows us to introduce, in a purely geometric way, the Lévy transformations of semi-Hamiltonian systems. Correspondence between commuting flows

$$u^i_\tau = q^i(u)_x, \quad i = 1, \dots, n,$$

and certain n-parameter families of planes containing the lines of the congruence is pointed out. In the particular case $n = 2$ this construction provides an explicit parametrization of surfaces, harmonic to a given congruence. Adjoint Lévy transformations of semi-Hamiltonian systems are discussed. Explicit formulae for the Lévy and adjoint Lévy transformations of the characteristic velocities are set down.

A closely related construction of the Ribaucour congruences of spheres is discussed in the Appendix.

1. Systems of Conservation Laws

We consider systems of conservation laws

$$u^i_t = f^i(u)_x = v^i_j(u) u^j_x, \quad v^i_j = \frac{\partial f^i}{\partial u^j}. \tag{1.1}$$

Eigenvalues λ^i of the matrix v^i_j are called the characteristic velocities of system (1.1). In what follows system (1.1) is assumed to be strictly hyperbolic, so that λ^i

2000 *Mathematics Subject Classification.* 53A25, 53B50, 35L65.
I would like to thank M. V. Pavlov for useful discussions. This research was partially supported by INTAS 96-0770 and the RFFI Grant 99-01-00010.
This is the final form of the paper.

are real and pairwise distinct. Let $\big(\vec{\xi_i} = (\xi_i^1(u), \xi_i^n(u)\big)^t$ be the corresponding right eigenvectors:

$$v\vec{\xi_i} = \lambda^i \vec{\xi_i}, \text{ or, in the components, } v_k^s \xi_i^k = \lambda^i \xi_i^s.$$

We denote by $L_i = \xi_i^k \partial/\partial u^k$ the Lie derivative along the vector field $\vec{\xi_i}$ and introduce the commutator expansions

$$[L_i, L_j] = L_i L_j - L_j L_i = c_{ij}^k L_k,$$

where c_{ij}^k are certain functions of u. Let $h(u)_t = g(u)_x$, be any conservation law of system (1.1). Its density h and flux g satisfy the equations

$$\frac{\partial g}{\partial u^k} = \frac{\partial h}{\partial u^s} v_k^s.$$

Contraction with $\vec{\xi_i} = (\xi_i^1, \dots, \xi_i^n)^t$ results in

$$\frac{\partial g}{\partial u^k} \xi_i^k = \frac{\partial h}{\partial u^s} v_k^s \xi_i^k,$$

or, equivalently,

$$L_i g = \lambda^i L_i h, \quad i = 1, \dots, n. \tag{1.2}$$

Equations (1.2) are the defining equations for the conserved densities h and the corresponding fluxes g. The compatibility conditions of (1.2) imply the following linear second-order system for the conserved densities h:

$$L_i L_j h = \frac{L_j \lambda^i}{\lambda^j - \lambda^i} L_i h + \frac{L_i \lambda^j}{\lambda^i - \lambda^j} L_j h + c_{ij}^k \frac{\lambda^i - \lambda^k}{\lambda^i - \lambda^j} L_k h, \quad i \neq j. \tag{1.3}$$

In particular, $h = u^1, \dots, u^n$ satisfy (1.3). In what follows, we will make use of the equations satisfied by the ratio $\varphi = g/h$, which can be obtained by rewriting (1.2) in the form $L_i(\varphi h) = \lambda^i L_i h$, or, equivalently,

$$L_i \ln h = \frac{L_i \varphi}{\lambda^i - \varphi}. \tag{1.4}$$

The compatibility conditions of (1.4) imply the following nonlinear second-order system for φ:

$$\begin{aligned} L_i L_j \varphi = \left(\frac{1}{\varphi - \lambda^i} + \frac{1}{\varphi - \lambda^j} \right) L_i \varphi \, L_j \varphi + \frac{L_j \lambda^i}{\lambda^j - \lambda^i} \frac{\varphi - \lambda^j}{\varphi - \lambda^i} L_i \varphi \\ + \frac{L_i \lambda^j}{\lambda^i - \lambda^j} \frac{\varphi - \lambda^i}{\varphi - \lambda^j} L_j \varphi + c_{ij}^k \frac{\lambda^i - \lambda^k}{\lambda^i - \lambda^j} \frac{\varphi - \lambda^j}{\varphi - \lambda^k} L_k \varphi. \end{aligned} \tag{1.5}$$

Formula (1.4) establishes an equivalence between the linear system (1.3) and the nonlinear system (1.5). The ratio $\varphi = g/h$ naturally arises in projective differential geometry (describing surfaces conjugate to a congruence—see Section 2), and in the Lie sphere geometry (parametrizing Ribaucour congruences of spheres—see the Appendix).

2. Diagonalizable Systems of Conservation Laws

Let us assume that all coefficients c_{ij}^k (with distinct i, j, k) are zero. In this case one can normalize eigenvectors $\vec{\xi}_i$ in such a way that the Lie derivatives L_i will pairwise commute: $[L_i, L_j] = 0$, so that the remaining coefficients c_{ij}^j will also be zero. The commutativity of L_i implies the existence of the coordinates $R^1(u), \dots, R^n(u)$, such that L_i become partial derivatives: $L_i = \partial_i = \partial/\partial R^i$. In the coordinates R^i equations (1.1) assume diagonal form

$$R_t^i = \lambda^i(R) R_x^i, \quad i = 1, \dots, n. \tag{2.1}$$

Variables R^i are called the Riemann invariants of system (1.1). Systems (1.1), possessing Riemann invariants, are called diagonalizable. Let $u_t = f_x$ be a conservation law of system (2.1). In the diagonalizable case equations (1.2) assume the form

$$\partial_i f = \lambda^i \partial_i u, \quad i = 1, \dots, n, \tag{2.2}$$

while system (1.3) for the conserved densities u simplifies to

$$\partial_i \partial_j u = a_{ij} \partial_i u + a_{ji} \partial_j u, \quad i \neq j, \tag{2.3}$$

where $a_{ij} = (\partial_j \lambda^i)/(\lambda^j - \lambda^i)$. The compatibility conditions of system (2.3) are of the form

$$\partial_k a_{ij} = a_{ik} a_{kj} + a_{ij} a_{jk} - a_{ij} a_{ik}, \quad i \neq j \neq k, \tag{2.4}$$

they must be identically satisfied if we want system (2.3) to possess n functionally independent solutions $u = u^1, \dots, u^n$. In fact, conditions (2.4) imply the existence of infinitely many conservation laws parametrized by n arbitrary functions of one variable. Systems (2.1) satisfying (2.4) are called semi-Hamiltonian. We refer to [**5, 13, 14**] for further information concerning integrability, differential geometry and applications of semi-Hamiltonian systems of conservation laws. Semi-Hamiltonian systems possess infinitely many commuting flows

$$R_\tau^i = \mu^i R_x^i,$$

with the characteristic velocities μ^i governed by the equations

$$\frac{\partial_j \mu^i}{\mu^j - \mu^i} = \frac{\partial_j \lambda^i}{\lambda^j - \lambda^i} = a_{ij}, \quad i \neq j.$$

We point out that any semi-Hamiltonian system possesses infinitely many different conservative representations.

3. Systems of Conservation Laws and Line Congruences. Hypersurfaces Conjugate to a Congruence

With any system of conservation laws

$$u_t^i = f^i(u)_x,$$

we associate an n-parameter family of lines

$$\begin{aligned} y^1 &= u^1\, y^0 - f^1(u), \\ &\cdots\cdots\cdot \\ y^n &= u^n\, y^0 - f^n(u), \end{aligned} \tag{3.1}$$

in the $(n+1)$-dimensional space A^{n+1} with the coordinates $y^0, y^1, \dots, y^n$. We refer to [**1, 2**] for motivation and the most important properties of this correspondence. In the case $n = 2$ we obtain two-parameter family of lines, or a congruence of lines in A^3. From the beginning of the 19th century theory of congruences was one of the most popular chapters of classical differential geometry—see e.g. [**9**]. We keep the name "congruence" for n-parameter family of lines (3.1). Any congruence possesses n focal hypersurfaces $\vec{\mathbf{r}}_i$, $i = 1, \ldots, n$, with the parametric equations

$$\vec{\mathbf{r}}_i = (y^0, y^1, \ldots, y^n) = (\lambda^i, u^1\lambda^i - f^1, \ldots, u^n\lambda^i - f^n),$$

where $\lambda^1, \ldots, \lambda^n$ are the characteristic velocities of system (1.1)—see [**1, 2**]. A line (3.1) is tangent to $\vec{\mathbf{r}}_i$ in the point with $y^0 = \lambda^i$. Let us consider a hypersurface M^n with the radius-vector $\vec{\mathbf{r}}$ given by

$$\vec{\mathbf{r}} = (y^0, y^1, \ldots, y^n) = (\varphi, u^1\varphi - f^1, \ldots, u^n\varphi - f^n), \tag{3.2}$$

where $\varphi(u)$ is an arbitrary function which is assumed to be different from λ^i, so that M^n is not focal. Line (3.1) meets M^n in the point with $y^0 = \varphi$. Obviously, any hypersurface $M^n \in A^{n+1}$ can be parametrized in the form (3.2) for an appropriate function φ. We say that hypersurface M^n is conjugate to congruence (3.1) if and only if

$$L_iL_j\vec{\mathbf{r}} \in TM^n \quad \text{for any} \quad i \neq j.$$

Geometrically, this means that the developable surfaces of congruence (3.1) meet M^n in the curves of a conjugate net. In 3-space the notion of conjugacy between a surface and a congruence was introduced by Guichard (see [**6**, chap. 1], [**9**]).

THEOREM 3.1. *Hypersurface* (3.2) *is conjugate to a congruence if and only if* φ *is representable in the form* $\varphi = g/h$, *where* $h_t = g_x$ *is a conservation law of system* (1.1).

PROOF. The tangent space of M^n is spanned by the vectors

$$L_j\vec{\mathbf{r}} = (L_j\varphi)\vec{\mathbf{U}} + (\varphi - \lambda^j)L_j\vec{\mathbf{U}}, \tag{3.3}$$

where $\vec{\mathbf{U}}$ denotes the $(n+1)$-vector $(1, u^1, \ldots, u^n)$. Hence,

$$L_j\vec{\mathbf{U}} = \frac{L_j\varphi}{\lambda^j - \varphi}\vec{\mathbf{U}} \mod TM^n. \tag{3.4}$$

Let us compute $L_iL_j\vec{\mathbf{r}}$:

$$L_iL_j\vec{\mathbf{r}} = (L_iL_j\varphi)\vec{\mathbf{U}} + (L_j\varphi)L_i\vec{\mathbf{U}} + L_i(\varphi - \lambda^j)L_j\vec{\mathbf{U}} + (\varphi - \lambda^j)L_iL_j\vec{\mathbf{U}}.$$

Inserting here $L_iL_j\vec{\mathbf{U}}$ from (1.3) and keeping in mind (3.4), we arrive at

$$L_iL_j\vec{\mathbf{r}} = \Bigg(L_iL_j\varphi + L_j\varphi\frac{L_i\varphi}{\lambda^i - \varphi} + L_i(\varphi - \lambda^j)\frac{L_j\varphi}{\lambda^j - \varphi} + (\varphi - \lambda^j) \\ \times \left(\frac{L_j\lambda^i}{\lambda^j - \lambda^i}\frac{L_i\varphi}{\lambda^i - \varphi} + \frac{L_i\lambda^j}{\lambda^i - \lambda^j}\frac{L_j\varphi}{\lambda^j - \varphi} + c^k_{ij}\frac{\lambda^i - \lambda^k}{\lambda^i - \lambda^j}\frac{L_k\varphi}{\lambda^k - \varphi}\right)\Bigg)\vec{\mathbf{U}} \mod TM^n.$$

Hence, $L_iL_j\vec{\mathbf{r}} \in TM^n$ if and only if the coefficient at $\vec{\mathbf{U}}$ vanishes. The resulting system for φ identically coincides with (1.5).

Thus, hypersurfaces conjugate to a congruence (3.1) are parametrized by conservation laws of system (1.1). □

4. Surfaces Harmonic to a Congruence

In this section we consider 2-component systems of conservation laws

$$u^1_t = f^1_x, \quad u^2_t = f^2_x, \tag{4.1}$$

and the associated congruences of lines in A^3:

$$y^1 = u^1\, y^0 - f^1, \quad y^2 = u^2\, y^0 - f^2. \tag{4.2}$$

Let

$$u^1_\tau = q^1_x, \quad u^2_\tau = q^2_x, \tag{4.3}$$

be a commuting flow of system (4.1). In Riemann invariants R^1, R^2 (we point out that any two-component system is diagonalizable) equations (4.1) and (4.3) assume the forms

$$R^1_t = \lambda^1\, R^1_x, \quad R^2_t = \lambda^2\, R^2_x,$$

and

$$R^1_\tau = \mu^1\, R^1_x, \quad R^2_\tau = \mu^2\, R^2_x,$$

respectively. The densities $u = (u^1, u^2)$ and the fluxes $f = (f^1, f^2)$, $q = (q^1, q^2)$ satisfy the equations

$$\partial_i f = \lambda^i \partial_i u, \quad \partial_i q = \mu^i \partial_i u, \qquad i = 1, 2.$$

With the commuting flow (4.3) we associate 2-parameter family of planes in A^3 defined by the equations

$$\frac{y^1 - u^1\, y^0 + f^1}{q^1} = \frac{y^2 - u^2\, y^0 + f^2}{q^2}. \tag{4.4}$$

The family of planes (4.4) has the following remarkable properties:

1. Each plane π from family (4.4) contains a line l of congruence (4.2).
2. The congruence of lines $l_1 = \pi \cap \partial_1 \pi$ is conjugate to the focal surface $\vec{\mathbf{r}}_1$ of congruence (4.2). Similarly, the congruence of lines $l_2 = \pi \cap \partial_2 \pi$ is conjugate to $\vec{\mathbf{r}}_2$. Lines l_1 and l_2 are called the characteristics of plane π. Characteristic l_1 (resp., l_2), meets the line l in the point of tangency of l with the focal surface $\vec{\mathbf{r}}_1$ (resp., $\vec{\mathbf{r}}_2$).

The proof follows from the explicit parametrization of congruences l_1, l_2:

- Congruence l_1:

$$y^1 = \left(u^1 - \frac{q^1}{\mu^1}\right) y^0 - \left(f^1 - \frac{\lambda^1 q^1}{\mu^1}\right),$$
$$y^2 = \left(u^2 - \frac{q^2}{\mu^1}\right) y^0 - \left(f^2 - \frac{\lambda^1 q^2}{\mu^1}\right).$$

- Congruence l_2:

$$y^1 = \left(u^1 - \frac{q^1}{\mu^2}\right) y^0 - \left(f^1 - \frac{\lambda^2 q^1}{\mu^2}\right),$$
$$y^2 = \left(u^2 - \frac{q^2}{\mu^2}\right) y^0 - \left(f^2 - \frac{\lambda^2 q^2}{\mu^2}\right).$$

Obviously, the line l_1 passes through the point

$$(y^0, y^1, y^2) = (\lambda^1, u^1\lambda^1 - f^1, u^2\lambda^1 - f^2),$$

of the focal surface $\vec{\mathbf{r}}_1$. Similarly, the line l_2 passes through the point

$$(y^0, y^1, y^2) = (\lambda^2, u^1\lambda^2 - f^1, u^2\lambda^2 - f^2),$$

of the focal surface $\vec{\mathbf{r}}_2$. The point of intersection $l_1 \cap l_2 \in \pi$ has the coordinates

$$\begin{aligned} y^0 &= \frac{\lambda^2\mu^1 - \lambda^1\mu^2}{\mu^1 - \mu^2}, \\ y^1 &= \frac{\lambda^2\mu^1 - \lambda^1\mu^2}{\mu^1 - \mu^2} u^1 + \frac{\lambda^1 - \lambda^2}{\mu^1 - \mu^2} q^1 - f^1, \\ y^2 &= \frac{\lambda^2\mu^1 - \lambda^1\mu^2}{\mu^1 - \mu^2} u^2 + \frac{\lambda^1 - \lambda^2}{\mu^1 - \mu^2} q^2 - f^2, \end{aligned} \tag{4.5}$$

and sweeps a surface in A^3. By a construction, surface (4.5) is the envelope of the family of planes (4.4). It has the following remarkable properties:

1. Each tangent plane π of surface (4.5) contains a line l of congruence (4.2). By a construction, π and $l \in \pi$ correspond to the same values of parameters R^1, R^2. Thus, one can speak of the correspondence between lines (4.2) and points of surface (4.5).
2. The net R^1, R^2 on the surface (4.5) is conjugate. In other words, developable surfaces of congruence (4.2) correspond to a conjugate net on the surface (4.5).

Surfaces, satisfying the properties 1, 2, are called harmonic to congruence (4.2)—see [**9**, p. 251]. Formulae (4.5) provide an explicit parametrization of surfaces, harmonic to congruence (4.2), by commuting flows of system (4.1). Conversely, any surface harmonic to congruence (4.2) is representable in the form (4.5).

5. Lévy Transformations of Semi-Hamiltonian Systems

Let us consider semi-Hamiltonian system (2.1) in Riemann invariants:

$$R^i_t = \lambda^i(R)R^i_x, \quad i = 1, \dots, n,$$

whose conservation laws $u_t = f_x$, satisfy equations (2.2), (2.3). Let us choose particular conservation law $h_t = g_x$, of system (2.1) and introduce the new variable U by the formula

$$U = u - \frac{h}{\partial_\alpha h}\partial_\alpha u, \tag{5.1}$$

where α is fixed. Transformations of this type originate from projective differential geometry of conjugate nets and are known as the transformations of Lévy [**12**], [**10**, p. 94], [**6**, Chapter 1]. In paper [**4**] transformations of Lévy have been identified with the vertex operators of the multicomponent KP hierarchy. Their geometric interpretation will be clarified in the second half of this section. We will refer to (5.1) as to the transformation of Lévy $\mathcal{L}_\alpha$. Direct calculation shows that $U = \mathcal{L}_\alpha(u)$ satisfies the equations of the same form as u:

$$\partial_i\partial_j U = A_{ij}\partial_i U + A_{ji}\partial_j U, \tag{5.2}$$

where the new coefficients $A = \mathcal{L}_\alpha(a)$ are given by the formulae

$$A_{\alpha i} = \left(1 - \frac{a_{i\alpha}h}{\partial_\alpha h}\right)\frac{\partial_i h}{h}, \qquad i \neq \alpha, \tag{5.3}$$

$$A_{ij} = a_{ij} + \partial_j \ln\left(1 - \frac{a_{i\alpha}h}{\partial_\alpha h}\right), \quad i \neq \alpha, \quad j \text{ is arbitrary.} \tag{5.4}$$

Transformations $\mathcal{L}_\alpha$ can be pulled back to the transformations of the corresponding hydrodynamic type systems: let us introduce the system

$$R^i_T = \Lambda^i(R)\, R^i_X, \quad i = 1, \dots, n, \tag{5.5}$$

with the characteristic velocities

$$\Lambda^\alpha = \frac{g}{h}, \quad \Lambda^i = \frac{\lambda^i \partial_\alpha h - a_{i\alpha} g}{\partial_\alpha h - a_{i\alpha} h}, \qquad i \neq \alpha. \tag{5.6}$$

THEOREM 5.1. *Conservation laws* $U_T = F_X$, *of system* (5.5), (5.6) *are the* $\mathcal{L}_\alpha$*-transforms of conservation laws* $u_t = f_x$ *of system* (2.1):

$$U = \mathcal{L}_\alpha(u) = u - \frac{h}{\partial_\alpha h}\partial_\alpha u, \quad F = \mathcal{L}_\alpha(f) = f - \frac{g}{\partial_\alpha g}\partial_\alpha f.$$

Formally, the proof of this theorem follows from the identities

$$\partial_i F = \Lambda^i \partial_i U, \quad A_{ij} = \frac{\partial_j \Lambda^i}{\Lambda^j - \Lambda^i},$$

which can be verified by a direct calculation. Geometric constructions underlying these formulae will be discussed below. System (5.5), (5.6) will be called the $\mathcal{L}_\alpha$-transform of system (2.1). Obviously, transformations $\mathcal{L}_\alpha$ preserve the semi-Hamiltonian property.

Lévy transformations of hydrodynamic type systems in Riemann invariants are closely related to the transformations of Laplace discussed recently in [**7, 11**]. We recall that Laplace transformation $S_{\alpha\beta}$ of system (2.3) is defined by the formula

$$U = S_{\alpha\beta}(u) = u - \frac{\partial_\alpha u}{a_{\beta\alpha}},$$

where both indices $\alpha \neq \beta$ are fixed. Laplace transformations also induce transformations of the characteristic velocities λ^i, the explicit form of which has been set down in [**7**]. One can check directly that the Lévy transformation $\mathcal{L}_\alpha$ of system (2.1) is related to its Lévy transformation $\mathcal{L}_\beta$ via the Laplace transformation $S_{\alpha\beta}$:

$$\mathcal{L}_\alpha = S_{\alpha\beta} \circ \mathcal{L}_\beta.$$

To clarify geometric picture underlying transformations $\mathcal{L}_\alpha$ we choose an arbitrary conservative representation $u^i_t = f^i_x$, of system (2.1) and introduce the associated congruence

$$y^i = u^i y^0 - f^i.$$

Let M^n be a hypersurface conjugate to this congruence. Following Section 3, we represent the radius-vector $\vec{\mathbf{r}}$ of M^n in the form

$$\vec{\mathbf{r}} = (\varphi, u^1\varphi - f^1, \dots, u^n\varphi - f^n), \quad \varphi = \frac{g}{h},$$

where $h_t = g_x$ is a conservation law of system (2.1). Coordinate system $R^1, \dots, R^n$ on M^n is conjugate, so that $\partial_i\partial_j\vec{\mathbf{r}} \in TM^n$ for any $i \neq j$. Let us introduce a new congruence, formed by the tangents to the R^α-curves on hypersurface M^n. Parametrically, its lines can be represented in the form $\vec{\mathbf{r}} + t\partial_\alpha \vec{\mathbf{r}}$, or, in the components,

$$y^0 = \varphi + t\partial_\alpha\varphi,$$
$$y^i = u^i\varphi - f^i + t(u^i\partial_\alpha\varphi + (\varphi - \lambda^\alpha)\partial_\alpha u^i), \quad i = 1, \dots, n.$$

Inserting $t = y^0 - \varphi/\partial_\alpha\varphi$ in the last n equations, we arrive at the new congruence

$$y^i = U^i y^0 - F^i, \tag{5.7}$$

where

$$U^i = u^i + \frac{\varphi - \lambda^\alpha}{\partial_\alpha\varphi}\partial_\alpha u^i, \quad F^i = f^i + \varphi\frac{\varphi - \lambda^\alpha}{\partial_\alpha\varphi}\partial_\alpha u^i.$$

Since $\varphi - \lambda^\alpha/\partial_\alpha\varphi = -h/\partial_\alpha h$, these formulae can be rewritten in the form

$$U = u - \frac{h}{\partial_\alpha h}\partial_\alpha u, \quad F = f - \frac{g}{\partial_\alpha g}\partial_\alpha f.$$

Congruence (5.7) will be called the $\mathcal{L}_\alpha$-transform of the initial congruence. The corresponding system of conservation laws $U^i_T = F^i_X$, has the same Riemann invariants $R^1, \dots, R^n$: $R^i_T = \Lambda^i R^i_X$, where Λ^i can be computed by the formula $\Lambda^i = \partial_i F/\partial_i U$. A direct calculation results in formulae (5.6). Note that the final expressions for Λ^i do not depend on the particular conservative representation $u^i_t = f^i_x$ of system (2.1). If, for M^n, we choose any of the focal hypersurfaces (which are all conjugate to a congruence if the system possesses Riemann invariants), the above construction gives transformations of Laplace.

6. The Adjoint Transformations of Lévy

We again consider semi-Hamiltonian systems (2.1)

$$R^i_t = \lambda^i(R)R^i_x,$$

with conservation laws $u_t = f_x$, satisfying the equations (2.2), (2.3). Let

$$R^i_\tau = \mu^i(R)R^i_x, \tag{6.1}$$

be a commuting flow of system (2.1):

$$\frac{\partial_j\mu^i}{\mu^j - \mu^i} = \frac{\partial_j\lambda^i}{\lambda^j - \lambda^i} = a_{ij}.$$

Let q be the flux of density u, corresponding to this commuting flow: $u_\tau = q_x$. The flux q and the density u satisfy the equations $\partial_i q = \mu^i\partial_i u$. Let us introduce the new variable U by the formula

$$U = u - \frac{q}{\mu^\alpha}, \tag{6.2}$$

where α is fixed. We will refer to (6.2) as to the adjoint transformation of Lévy $\mathcal{L}^*_\alpha$. A direct calculation shows that $U = \mathcal{L}^*_\alpha(u)$ satisfies the equations of the same form as u:

$$\partial_i\partial_j U = A_{ij}\partial_i U + A_{ji}\partial_j U,$$

where the new coefficients $A = \mathcal{L}_\alpha^*(a)$ are given by the formulae

$$A_{\alpha i} = a_{\alpha i} + \partial_i \ln \frac{\partial_\alpha \mu^\alpha}{\mu^\alpha}, \qquad i \neq \alpha,$$

$$A_{ij} = a_{ij} + \partial_j \ln\left(1 - \frac{\mu^i}{\mu^\alpha}\right), \quad i \neq \alpha, \quad j \text{ is arbitrary.}$$

Transformations $\mathcal{L}_\alpha^*$ can be pulled back to the transformations of the corresponding hydrodynamic type systems: let us introduce the system

$$R_T^i = \Lambda^i(R) R_X^i, \quad i = 1, \ldots, n, \tag{6.3}$$

with the characteristic velocities

$$\Lambda^\alpha = \frac{\lambda^\alpha \partial_\alpha \mu^\alpha - \mu^\alpha \partial_\alpha \lambda^\alpha}{\partial_\alpha \mu^\alpha}, \quad \Lambda^i = \frac{\lambda^i \mu^\alpha - \lambda^\alpha \mu^i}{\mu^\alpha - \mu^i}, \qquad i \neq \alpha. \tag{6.4}$$

THEOREM 6.1. *Conservation laws* $U_T = F_X$, *of system* (6.3), (6.4) *are the* $\mathcal{L}_\alpha^*$*-transforms of conservation laws* $u_t = f_x$ *of system* (2.1):

$$U = \mathcal{L}_\alpha^*(u) = u - \frac{q}{\mu^\alpha}, \quad F = \mathcal{L}_\alpha^*(f) = f - \frac{\lambda^\alpha q}{\mu^\alpha}.$$

Formally, the proof of this theorem follows from the identities

$$\partial_i F = \Lambda^i \partial_i U, \quad A_{ij} = \frac{\partial_j \Lambda^i}{\Lambda^j - \Lambda^i},$$

which can be verified by a direct calculation. Geometric constructions underlying these formulae will be discussed below. System (6.3), (6.4) will be called the $\mathcal{L}_\alpha^*$-transform of system (2.1). Obviously, transformations $\mathcal{L}_\alpha^*$ preserve the semi-Hamiltonian property.

Transformations $\mathcal{L}_\alpha^*$ and the Laplace transformations $S_{\alpha\beta}$ satisfy the identities

$$\mathcal{L}_\alpha^* = \mathcal{L}_\beta^* \circ S_{\beta\alpha}.$$

To clarify geometric picture underlying transformations $\mathcal{L}_\alpha^*$ we choose an arbitrary conservative representation $u_t^i = f_x^i$ of system (2.1) and introduce the associated congruence

$$y^i = u^i y^0 - f^i.$$

Let $u_\tau^i = q_x^i$ be a commuting flow of system (2.1) with the characteristic velocities μ, so that $\partial_i q = \mu^i \partial_i u$, (the last identity holding for any $q = q^k$, $u = u^k$). Let us introduce n-parameter family of 2-planes in A^{n+1} defined by the equations

$$\frac{y^1 - u^1 y^0 + f^1}{q^1} = \cdots = \frac{y^n - u^n y^0 + f^n}{q^n}. \tag{6.5}$$

The family of planes (6.5) possesses the following three important properties:

1. Each plane π from family (6.5) contains a line l of the initial congruence.
2. Each plane π intersects the plane $\partial_i \pi$ along a line l_i:

$$l_i = \pi \cap \partial_i \pi,$$

(we point out that two planes in A^{n+1} do not necessarily intersect along a line unless $n = 2$). Geometrically, this property implies that each 1-parameter subfamily of (6.5), specified by fixing the values of R^k, $k \neq i$, envelopes a developable surface in A^{n+1}. Lines l_i, $i = 1, \ldots, n$, are called the characteristics of plane π.

3. Congruence l_i is conjugate to the ith focal hypersurface

$$\vec{\mathbf{r}_i} = (\lambda^i, u^1\lambda^i - f^1, \dots, u^n\lambda^i - f^n),$$

of the initial congruence l.

Conversely, one can show that any n-parameter family of 2-planes satisfying the properties 1–3 is necessarily of the form (6.5) for an appropriate commuting flow $u^i_\tau = q^i_x$.

Congruence l_α will be called the $\mathcal{L}^*_\alpha$-transform of the initial congruence l. A direct calculation shows that l_α is representable in the form

$$y^i = U^i y^0 - F^i,$$

where

$$U^i = u^i - \frac{q^i}{\mu^\alpha}, \quad F^i = f^i - \frac{\lambda^\alpha q^i}{\mu^\alpha},$$

(compare with Theorem 7.1). Line l_α meets the focal hypersurface $\vec{\mathbf{r}_\alpha}$ in the point

$$(\lambda^\alpha, u^1\lambda^\alpha - f^1, \dots, u^n\lambda^\alpha - f^n).$$

The corresponding system of conservation laws $U^i_T = F^i_X$ has the same Riemann invariants $R^1, \dots, R^n$: $R^i_T = \Lambda^i R^i_X$, (in fact, this is the analytic manifestation of the above Property 3), where the transformed characteristic velocities $\Lambda^i = \partial_i F/\partial_i U$ coincide with (6.4). Note that the final expressions for Λ^i do not depend on the particular conservative representation $u^i_t = f^i_x$ of system (2.1).

Obviously, the inverse transformation $l_\alpha \to l$ is the transformation $\mathcal{L}_\alpha$ of Lévy. Indeed, l_α is conjugate to hypersurface $\vec{\mathbf{r}_\alpha}$, while the initial congruence l consists of the R^α-tangents to hypersurface $\vec{\mathbf{r}_\alpha}$. Thus, transformations of Lévy $\mathcal{L}_\alpha$ are the inverses of $\mathcal{L}^*_\alpha$.

7. Appendix: Ribaucour Congruences of Spheres

Let M^n be a hypersurface in the Euclidean space E^{n+1} parametrized by coordinates $u^1, \dots, u^n$. Let $\vec{\mathbf{r}}$ and $\vec{\mathbf{n}}$ be the radius-vector and the unit normal of M^n, respectively. The Weingarten formulae

$$\frac{\partial \vec{\mathbf{n}}}{\partial u^j} = w^i_j(u)\frac{\partial \vec{\mathbf{r}}}{\partial u^i},$$

define the so-called Weingarten (shape) operator of hypersurface M^n. Its eigenvalues and eigenvectors are called the principal curvatures and the principal directions of M^n, respectively. Let us consider a hypersphere S of radius R and the centre $\vec{\mathbf{r}} - R\vec{\mathbf{n}}$, which is tangent to M^n at the point $\vec{\mathbf{r}}$. Specifying R as a function of u, we obtain n-parameter family of hyperspheres (or a congruence of hyperspheres) enveloped by hypersurface M^n. Let $\widetilde{M}^n$ be the second sheet of the envelope. Clearly, there exists a point correspondence between both sheets M^n and $\widetilde{M}^n$: a point $p \in M^n$ corresponds to $\tilde{p} \in \widetilde{M}^n$ if p and $\tilde{p}$ are the two points of tangency of one and the same hypersphere from the family $S(u)$.

DEFINITION. Family of hyperspheres $S(u)$ is called the family of Ribaucour if the principal distributions of M^n correspond to the principal distributions of $\widetilde{M}^n$.

Let us introduce the system of hydrodynamic type

$$u_t^i = w_j^i(u)u_x^j, \tag{7.1}$$

where w_j^i is the Weingarten operator of M^n. We refer to [**8**] for the general discussion of the correspondence between hypersurfaces and systems of hydrodynamic type. Let $h(u)_t = g(u)_x$ be a conservation law of system (7.1).

THEOREM 7.1. *Congruence $S(u)$ is the congruence of Ribaucour if and only if $R(u)$ is representable in the form*

$$R(u) = \frac{h(u)}{g(u)},$$

for some conservation law of system (7.1).

In the case $n = 2$ this result (stated in a somewhat different form) can be found in [**6**]. It should be emphasized that this theorem equally applies to hypersurfaces which do not possess a curvature-line parametrization (for $n = 2$ such parametrization is always possible). We hope to present the details elsewhere.

References

1. S. I. Agafonov and E. V. Ferapontov, *Systems of conservation laws from the point of view of the projective theory of congruences*, Izv. Ross. Akad. Nauk Ser. Mat. **60** (1996), no. 6, 3–30.
2. ———, *Theory of congruences and systems of conservation laws*, Geometry 4, J. Math. Sci. (New York) **94** (1999), no. 5, 1748–1794.
3. M. A. Akivis and V. V. Goldberg, *Projective differential geometry of submanifolds*, North-Holland Mathematical Library, vol. 49, North-Holland Publishing Co., Amsterdam, 1993.
4. A. Doliwa, M. Mañas, L. M. Alonso, E. Medina, and P. M. Santini, *Charged free fermions, vertex operators and classical theory of conjugate nets*, J. Phys. A **32** (1999), no. 7, 1197–1216.
5. B. A. Dubrovin and S. P. Novikov, *Hydrodynamics of weakly deformed soliton lattices. Differential geometry and Hamiltonian theory*, Uspekhi Mat. Nauk **44** (1989), no. 6 (270), 29–98 (Russian); English transl., Russian Math. Surveys **44** (1989), no. 6, 35–124.
6. L. P. Eisenhart, *Transformations of surfaces*, 2nd ed., Chelsea Publishing Co., New York, 1962.
7. E. V. Ferapontov, *Laplace transformations of hydrodynamic type systems in Riemann invariants: Periodic sequences*, J. Phys. A **30** (1997), no. 19, 6861–6878.
8. ———, *Dupin hypersurfaces and integrable Hamiltonian systems of hydrodynamic type which do not possess Riemann invariants*, Differential Geom. Appl. **5** (1995), 121–152.
9. S. P. Finikov, *Theory of congruences*, Gosudarstv. Izdat. Tehn.-Teor. Lit., Moscow-Leningrad, 1950 (Russian).
10. A. R. Forsyth, *Theory of differential equations*, vol. 6, Cambridge University Press, London, 1906.
11. N. Kamran and K. Tenenblat, *Laplace transformations in higher dimensions*, Duke Math. J. **84** (1996), no. 1, 237–266.
12. M. L. Lévy, *Quelques équations linéaires aux dérivés partielles du second ordre*, J. École Polytechnique **56** (1886), 63–67.
13. B. Sévennec, *Géométrie des systèmes hyperboliques de lois de conservation*, Mém. Soc. Math. France (N.S.), no. 56 (1994) (French); Suppl. au Bulletin de la Société Mathématique de France **122** (1994), 1–125.
14. S. P. Tsarev, *The geometry of Hamiltonian systems of hydrodynamic type. The generalized hodograph method*, Izv. Akad. Nauk SSSR Ser. Mat. **54** (1990), no. 5, 1048–1068 (Russian); English transl., Math. USSR-Izv. **37** (1991), no. 2, 397–419.

Department of Mathematical Sciences, Loughborough University, Loughborough, Leicestershire LE11 3TU, United Kingdom, and Centre for Nonlinear Studies, Landau Institute of Theoretical Physics, Academy of Science of Russia, Kosygina 2, 117940 Moscow, GSP-1, Russia.

E-mail address: E.V.Ferapontov@lboro.ac.uk

Centre de Recherches Mathématiques
CRM Proceedings and Lecture Notes
Volume **29**, 2001

Bäcklund Links Between Different Analytic Descriptions of Constant Mean Curvature Surfaces

E. V. Ferapontov and *A. M. Grundland*

ABSTRACT. Bäcklund transformations between different analytic descriptions of constant mean curvature surfaces are established. In particular, it is demonstrated that the system

$$\partial\psi_1 = (|\psi_1|^2 + |\psi_2|^2)\psi_2, \quad \bar{\partial}\psi_2 = -(|\psi_1|^2 + |\psi_2|^2)\psi_1,$$

descriptive of CMC surfaces within the framework of the generalized Weierstrass representation, decouples into a direct sum of the elliptic Sh-Gordon and Laplace equations. Connections of this system with the sigma model equations are established. It is pointed out, that the instanton solutions correspond to different Weierstrass parametrizations of the standard sphere $S^2 \subset E^3$.

1. Introduction

We investigate the system

$$\partial\psi_1 = \big(|\psi_1|^2 + |\psi_2|^2\big)\psi_2, \quad \bar{\partial}\psi_2 = -\big(|\psi_1|^2 + |\psi_2|^2\big)\psi_1, \tag{1.1}$$

which has been derived in [**5**] and governs constant mean curvature (CMC) surfaces in the conformal parametrization z, $\bar{z}$ ($\partial = \partial_z, \bar{\partial} = \partial_{\bar{z}}$). This system was subsequently discussed in [**2, 3, 7**]. In this paper, we demonstrate that system (1.1) can be decoupled into a direct sum of the elliptic Sh-Gordon and Laplace equations. Firstly, we change from ψ_1, ψ_2 to the new dependent variables Q, R

$$Q = 2(\psi_2\partial\overline{\psi}_1 - \overline{\psi}_1\partial\psi_2), \quad R = 2\big(|\psi_1|^2 + |\psi_2|^2\big)^2/|Q|.$$

Secondly, we introduce new independent variables η, $\bar{\eta}$ according to

$$d\eta = \sqrt{Q}\,dz, \quad \bar{\eta} = \sqrt{\overline{Q}}\,d\bar{z},$$

2000 *Mathematics Subject Classification.* 93C20.

The first author would like to thank Centre de Recherches Mathématiques, Université de Montréal for generous hospitality and financial support during the period when this investigation was performed. This work was supported by the research grant from NSERC of Canada and the Fonds FCAR du Gouvernement du Québec. We would like to thank Professor B. G. Konopelchenko for drawing our attention to this subject and for helpful discussions.

This is the final form of the paper.

(these formulae are correct since Q is holomorphic). In the new variables $Q, R, \eta, \bar{\eta}$, system (1.1) assumes the decoupled form

$$(\ln R)_{\eta\bar{\eta}} = \frac{1}{R} - R, \quad Q_{\bar{\eta}} = \overline{Q}_{\eta} = 0.$$

This transformation is an immediate corollary of the known properties of CMC surfaces. Connection of system (1.1) with the sigma-model equations

$$\partial\bar{\partial}\rho - \frac{2\bar{\rho}}{1+|\rho|^2}\partial\rho\bar{\partial}\rho = 0,$$

($\rho = i\overline{\psi}_1/\psi_2$) is also discussed. In terms of ρ, our transformation adopts the form:

$$Q = \frac{2\partial\rho\partial\bar{\rho}}{(1+|\rho|^2)^2}, \quad R = \left|\frac{\partial\bar{\rho}}{\partial\rho}\right|.$$

2. Generalized Weierstrass Representation of Surfaces in $\mathbb{R}^3$

Following the results of [**5**], with any solution ψ_1, ψ_2 of the Dirac equations

$$\partial\psi_1 = p\psi_2, \quad \bar{\partial}\psi_2 = -p\psi_1, \tag{2.1}$$

($p(z,\bar{z})$ is a real potential), we associate a surface $M^2 \subset E^3$ with the radius-vector $\mathbf{r}(z,\bar{z})$ defined by the formulae

$$\begin{aligned}\partial\mathbf{r} &= \left(i(\psi_2^2 + \overline{\psi}_1^2), \overline{\psi}_1^2 - \psi_2^2, -2\psi_2\overline{\psi}_1\right),\\ \bar{\partial}\mathbf{r} &= \left(-i(\overline{\psi}_2^2 + \psi_1^2), \psi_1^2 - \overline{\psi}_2^2, -2\psi_1\overline{\psi}_2\right).\end{aligned}$$

The latter are compatible by virtue of (2.1). The unit normal $\mathbf{n}$ of the surface M^2 can be calculated according to the formula $\mathbf{n} = (-i)\partial\mathbf{r}\times\bar{\partial}\mathbf{r}/|\partial\mathbf{r}\times\bar{\partial}\mathbf{r}|$ and is represented as follows:

$$\mathbf{n} = \frac{1}{q}\left(i(\overline{\psi}_1\overline{\psi}_2 - \psi_1\psi_2), \overline{\psi}_1\overline{\psi}_2 + \psi_1\psi_2, \psi_1\overline{\psi}_1 - \psi_2\overline{\psi}_2\right), \quad q = |\psi_1|^2 + |\psi_2|^2.$$

One can verify directly that the scalar products in E^3 are $(\mathbf{n},\mathbf{n}) = 1$, $(\mathbf{n},\partial\mathbf{r}) = (\mathbf{n},\bar{\partial}\mathbf{r}) = 0$. Equations of motion of the complex frame $\partial\mathbf{r}$, $\bar{\partial}\mathbf{r}$, $\mathbf{n}$ are of the form

$$\begin{aligned}\partial\begin{pmatrix}\partial\mathbf{r}\\ \bar{\partial}\mathbf{r}\\ \mathbf{n}\end{pmatrix} &= \begin{pmatrix}2\frac{\partial q}{q} & 0 & Q\\ 0 & 0 & 2Hq^2\\ -H & \frac{Q}{2q^2} & 0\end{pmatrix}\begin{pmatrix}\partial\mathbf{r}\\ \bar{\partial}\mathbf{r}\\ \mathbf{n}\end{pmatrix},\\ \bar{\partial}\begin{pmatrix}\partial\mathbf{r}\\ \bar{\partial}\mathbf{r}\\ \mathbf{n}\end{pmatrix} &= \begin{pmatrix}0 & 0 & 2Hq^2\\ 0 & 2\frac{\bar{\partial}q}{q} & \overline{Q}\\ -\frac{\overline{Q}}{2q^2} & -H & 0\end{pmatrix}\begin{pmatrix}\partial\mathbf{r}\\ \bar{\partial}\mathbf{r}\\ \mathbf{n}\end{pmatrix},\end{aligned} \tag{2.2}$$

where $Q = 2(\psi_2\partial\overline{\psi}_1 - \overline{\psi}_1\partial\psi_2)$ and $H = p/q$ is the mean curvature. Formulae (2.2) are compatible with the scalar products

$$(\mathbf{n},\mathbf{n}) = 1, \quad (\partial\mathbf{r},\bar{\partial}\mathbf{r}) = 2q^2,$$

(all other scalar products being equal to zero). Using (2.2), one can derive the following useful equation for the unit normal $\mathbf{n}$:

$$\partial\bar{\partial}\mathbf{n} + (\partial\mathbf{n},\bar{\partial}\mathbf{n})\mathbf{n} + \bar{\partial}H\partial\mathbf{r} + \partial H\bar{\partial}\mathbf{r} = 0. \tag{2.3}$$

The first fundamental form $\mathrm{I} = (d\mathbf{r}, d\mathbf{r})$ and the second fundamental form $\mathrm{II} = (d^2\mathbf{r}, \mathbf{n})$ of the surface M^2 are given by

$$\mathrm{I} = 4q^2\,dz\,d\bar{z}, \quad \mathrm{II} = Q\,dz^2 + 4Hq^2\,dz\,d\bar{z} + \overline{Q}\,d\bar{z}^2.$$

The quantity $Q\,dz^2$ is called the Hopf differential. The corresponding Gauss–Codazzi equations which are the compatibility conditions for (2.2), are of the form

$$\partial\bar{\partial}(\ln q^2) = \frac{1}{2}\frac{Q\overline{Q}}{q^2} - 2H^2q^2, \quad \bar{\partial}Q = 2q^2\partial H, \quad \partial\overline{Q} = 2q^2\bar{\partial}H. \tag{2.4}$$

In fact, Eqs. (2.4) are a direct differential consequence of (2.1), as can be checked by a straightforward calculation. Gauss–Codazzi equations of surfaces in conformal parametrization z, $\bar{z}$ have been discussed in [**1**]. We recall also that the Gaussian curvature K of the surface M^2 can be calculated as follows:

$$K = -\frac{1}{q^2}\partial\bar{\partial}(\ln q). \tag{2.5}$$

The unit normal $\mathbf{n} = (n_1, n_2, n_3)$, maps a surface M^2 onto the unit sphere S^2. Combining this map with the stereographic projection, we obtain a map ρ of the surface M^2 onto the complex plane, called the complex Gauss map. In our notation, ρ assumes the form

$$\rho = \frac{n_1 + in_2}{1 - n_3} = i\frac{\overline{\psi}_1}{\psi_2}. \tag{2.6}$$

According to the results of [**4**], Gauss map ρ satisfies the nonlinear equation

$$\left(\partial\bar{\partial}\rho - \frac{2\bar{\rho}}{1+|\rho|^2}\partial\rho\bar{\partial}\rho\right)H = \partial H\bar{\partial}\rho, \tag{2.7}$$

which formally can be viewed as a differential consequence of (2.1). In terms of ρ and H, the initial data ψ_1, ψ_2 and p assume the form

$$\psi_1 = \frac{\bar{\rho}}{\sqrt{H}}\frac{\sqrt{i\partial\rho}}{1+|\rho|^2}, \quad \psi_2 = \frac{1}{\sqrt{H}}\frac{\sqrt{i\partial\bar{\rho}}}{1+|\rho|^2}, \quad p = \frac{|\partial\bar{\rho}|}{1+|\rho|^2},$$

while the expressions for q and Q take the form

$$q = \frac{1}{H}\frac{\bar{\partial}\rho\partial\bar{\rho}}{1+|\rho|^2}, \quad Q = \frac{2}{H}\frac{\partial\rho\partial\bar{\rho}}{(1+|\rho|^2)^2}.$$

REMARK 2.1. Linear problem (2.2) can be rewritten in terms of ψ_1, ψ_2 as follows. First of all, we point out that

$$\partial q = \psi_1\partial\overline{\psi}_1 + \overline{\psi}_2\partial\psi_2.$$

Combining this equation with the definition of Q:

$$\frac{1}{2}Q = \psi_2\partial\overline{\psi}_1 - \overline{\psi}_1\partial\psi_2,$$

and solving these two equations for $\partial\overline{\psi}_1$, $\partial\psi_2$, we can "close" system (2.1) as follows:

$$\begin{aligned}\partial\begin{pmatrix}\psi_1\\ \psi_2\end{pmatrix} &= \begin{pmatrix}0 & qH\\ -\frac{Q}{2q} & \frac{\partial q}{q}\end{pmatrix}\begin{pmatrix}\psi_1\\ \psi_2\end{pmatrix},\\ \bar{\partial}\begin{pmatrix}\psi_1\\ \psi_2\end{pmatrix} &= \begin{pmatrix}\frac{\bar{\partial}q}{q} & \frac{\overline{Q}}{2q}\\ -qH & 0\end{pmatrix}\begin{pmatrix}\psi_1\\ \psi_2\end{pmatrix}.\end{aligned} \tag{2.8}$$

The compatibility conditions for system (2.8) coincide with (2.4). We point out that the 2×2 matrix approach to surfaces in E^3 has been extensively developed in [**1**]. From the point of view of the theory of integrable systems, linear system (2.2) can be regarded as the squared eigenfunction equations corresponding to (2.8) (indeed, $\partial \mathbf{r}$, $\bar\partial \mathbf{r}$ and $\mathbf{n}$ are quadratic expressions in ψ_1 and ψ_2).

3. CMC-1 Surfaces

The class of CMC-1 surfaces is characterized by the constraint $H = 1$ or, equivalently, $p = q$. Introducing this ansatz in (2.1), we arrive at the nonlinear system (1.1)

$$\partial\psi_1 = \left(|\psi_1|^2 + |\psi_2|^2\right)\psi_2, \quad \bar\partial\psi_2 = -\left(|\psi_1|^2 + |\psi_2|^2\right)\psi_1,$$

which is the main subject of our study. According to the previous section, system (1.1) is equivalent to

$$\partial\bar\partial \ln q^2 = \frac{1}{2}\frac{Q\overline{Q}}{q^2} - 2q^2, \quad \bar\partial Q = \partial\overline{Q} = 0, \tag{3.1}$$

where $q = |\psi_1|^2 + |\psi_2|^2$, $Q = 2(\psi_2\partial\overline{\psi}_1 - \overline{\psi}_1\partial\psi_2)$. Thus, for CMC surfaces the Hopf differential $Q\,dz^2$ is holomorphic. Applying to system (3.1) the reciprocal transformation

$$d\eta = \sqrt{Q}\,dz, \quad d\bar\eta = \sqrt{\overline{Q}}\,d\bar z,$$

(that is, changing from z, $\bar z$ to the new independent variables η, $\bar\eta$ which are correctly defined in view of the holomorphicity of Q) and introducing

$$R = \frac{2q^2}{|Q|},$$

we transform system (3.1) into the decoupled form

$$(\ln R)_{\eta\bar\eta} = \frac{1}{R} - R, \quad Q_{\bar\eta} = \overline{Q}_\eta = 0. \tag{3.2}$$

This result provides the rationale for the change of variables which we introduce in Section 1.

REMARK 3.1. System (1.1) is invariant under the SU(2)-symmetry

$$\phi_1 = \alpha\psi_1 + \beta\overline{\psi}_2, \quad \phi_2 = -\beta\overline{\psi}_1 + \alpha\psi_2, \tag{3.3}$$

where α, β are complex constants subject to the constraint $\alpha\bar\alpha + \beta\bar\beta = 1$. One can check directly, that the quantities Q and R are invariant under transformations (3.3), so that the surfaces, corresponding to (ψ_1, ψ_2) and (ϕ_1, ϕ_2), have coincident fundamental forms. Thus, they are identical up to a rigid motion in E^3. The passage from ψ_1, ψ_2 to Q, R can thus be viewed as a passage to the differential invariants of the point symmetry group (3.3).

REMARK 3.2. For CMC-1 surfaces, system (2.8) allows the introduction of a spectral parameter

$$\partial\begin{pmatrix}\psi_1\\ \psi_2\end{pmatrix} = \begin{pmatrix}0 & q\\ -\lambda\frac{Q}{2q} & \frac{\partial q}{q}\end{pmatrix}\begin{pmatrix}\psi_1\\ \psi_2\end{pmatrix}, \quad \bar\partial\begin{pmatrix}\psi_1\\ \psi_2\end{pmatrix} = \begin{pmatrix}\frac{\bar\partial q}{q} & \frac{1}{\lambda}\frac{\overline{Q}}{2q}\\ -q & 0\end{pmatrix}\begin{pmatrix}\psi_1\\ \psi_2\end{pmatrix}, \tag{3.4}$$

where λ is a unitary constant, $|\lambda| = 1$. The gauge transformation

$$\widetilde{\psi}_1 = \psi_1, \quad \widetilde{\psi}_2 = q^{-1}\psi_2,$$

reduces linear spectral problem (3.4) to the SL(2) form

$$\partial\begin{pmatrix}\widetilde{\psi}_1\\ \widetilde{\psi}_2\end{pmatrix}=\begin{pmatrix}0 & q^2\\ -\lambda\frac{Q}{2q^2} & 0\end{pmatrix}\begin{pmatrix}\widetilde{\psi}_1\\ \widetilde{\psi}_2\end{pmatrix}\bar{\partial}\begin{pmatrix}\widetilde{\psi}_1\\ \widetilde{\psi}_2\end{pmatrix}=\begin{pmatrix}\frac{\bar{\partial}q}{q} & \frac{\overline{Q}}{2\lambda}\\ -1 & -\frac{\bar{\partial}q}{q}\end{pmatrix}\begin{pmatrix}\widetilde{\psi}_1\\ \widetilde{\psi}_2\end{pmatrix}. \tag{3.5}$$

The compatibility conditions for both systems (3.4) and (3.5) coincide with (3.1).

Linear system (3.5) can be readily rewritten in terms of ψ_1, ψ_2. Indeed, observing that (3.5) implies

$$\partial\ln q=\partial\ln\left(|\psi_1|^2+|\psi_2|^2\right),\quad \bar{\partial}\ln q=\bar{\partial}\ln\left(|\psi_1|^2+|\psi_2|^2\right),$$

we can take $q=c\left(|\psi_1|^2+|\psi_2|^2\right)$, $c\in\mathbb{C}$, which, upon substitution in (3.5), produces system (1.1). Thus transformation from (3.1) to (1.1) consists of rewriting (3.5) in terms of the ψ. Representations in terms of ψ are called eigenfunction equations, and are fundamental in soliton theory—see e.g. [**6**]. The Lax pair for system (1.1) is of the form [**2**]

$$\partial\begin{pmatrix}\phi_1\\ \phi_2\end{pmatrix}=\frac{2}{\lambda+1}\begin{pmatrix}-\overline{\psi}_1\psi_2+\frac{Q}{2q^2}\psi_1\overline{\psi}_2 & -\overline{\psi}_1^2-\frac{Q}{2q^2}\overline{\psi}_2^2\\ \psi_2^2+\frac{Q}{2q^2}\psi_1^2 & \overline{\psi}_1\psi_2-\frac{Q}{2q^2}\psi_1\overline{\psi}_2\end{pmatrix}\begin{pmatrix}\phi_1\\ \phi_2\end{pmatrix},$$

$$\bar{\partial}\begin{pmatrix}\phi_1\\ \phi_2\end{pmatrix}=\frac{2}{\lambda-1}\begin{pmatrix}-\psi_1\overline{\psi}_2+\frac{\overline{Q}}{2q^2}\overline{\psi}_1\psi_2 & \overline{\psi}_2^2+\frac{\overline{Q}}{2q^2}\overline{\psi}_1^2\\ -\psi_1^2-\frac{\overline{Q}}{2q^2}\psi_2^2 & \psi_1\overline{\psi}_2-\frac{\overline{Q}}{2q^2}\overline{\psi}_1\psi_2\end{pmatrix}\begin{pmatrix}\phi_1\\ \phi_2\end{pmatrix}.$$

It is interesting to note that the compatibility conditions for this linear system, which is of the first order in the derivatives of ψ, give us exactly system (1.1). The latter is also of the first order in ψ.

4. CMC Surfaces and Sigma Model Equations

For CMC surfaces, Eqs. (2.7) imply the nonlinear sigma model

$$\partial\bar{\partial}\rho-\frac{2\bar{\rho}}{1+|\rho|^2}\partial\rho\bar{\partial}\rho=0, \tag{4.1}$$

descriptions of the stationary two-dimensional SU(2) magnet. The transformation of the system (4.1) into the decoupled form (3.2) now assumes the form

$$Q=2\frac{\partial\rho\partial\bar{\rho}}{(1+|\rho|^2)^2},\qquad R=\left|\frac{\partial\bar{\rho}}{\partial\rho}\right|,$$
$$d\eta=\sqrt{Q}\,dz,d\bar{\eta}=\sqrt{\overline{Q}}\,d\bar{z}. \tag{4.2}$$

In terms of the unit normal vector $\mathbf{n}$, Eq. (2.3) adopts the form of the SO(3) sigma model

$$\bar{\partial}\partial\mathbf{n}+(\partial\mathbf{n},\bar{\partial}\mathbf{n})\mathbf{n}=0,\quad (\mathbf{n},\mathbf{n})=1. \tag{4.3}$$

Formula (2.6) establishes a link between sigma models (4.1) and (4.3). The topological charge

$$\frac{1}{4\pi}\iint\left(\mathbf{n},[\partial\mathbf{n}\times\bar{\partial}\mathbf{n}]\right)dz\wedge d\bar{z},$$

can be written as

$$\frac{1}{2\pi i}\iint\partial\bar{\partial}\ln q\,dz\wedge d\bar{z},$$

which, in view of (2.5), is the topologically invariant integral curvature of the surface M^2.

Instanton solutions of system (4.3) are specified by the ansatz

$$\partial\mathbf{n} = \pm i\mathbf{n} \times \partial\mathbf{n}, \quad \bar{\partial}\mathbf{n} = \mp i\mathbf{n} \times \bar{\partial}\mathbf{n},$$

which, after a simple calculation, implies $Q = 0$. Solutions of system (1.1) specified by a constraint $Q = 0$, can be represented in the form

$$\psi_1 = \frac{\rho\sqrt{\bar{\partial}\bar{\rho}}}{1+|\rho|^2}, \quad \psi_2 = \frac{\sqrt{\partial\rho}}{1+|\rho|^2}, \quad p = \frac{|\partial\rho|}{1+|\rho|^2}, \tag{4.4}$$

where $\rho(z)$ is an arbitrary holomorphic function. In the case when the energy

$$E = \iint \frac{\partial\rho\bar{\partial}\rho}{1+|\rho|^2}\, dz \wedge d\bar{z},$$

is finite the function $\rho(z)$ is rational in z [**8**].

Geometrically, instanton solutions (4.4) parametrize the standardly embedded sphere $S^2 \subset E^3$. This is the consequence of the fact that in case $Q = 0$, we have the proportionality of fundamental forms I and II. This example shows that different Weierstrass data (ψ_1, ψ_2, p) can correspond to different parametrizations of one and the same surface $M^2 \subset E^3$.

Introducing the two-component complex vector

$$\mathbf{N} = \left(\frac{\psi_1}{\sqrt{q}}, \frac{\overline{\psi}_2}{\sqrt{q}}\right), \quad q = |\psi_1|^2 + |\psi_2|^2,$$

one can check that $\mathbf{N}$ satisfies the equations of the $\mathbb{C}P^1$ sigma model

$$(\mathbf{N}, \overline{\mathbf{N}}) = 1, \quad \partial\bar{\partial}\mathbf{N} = (\overline{\mathbf{N}}, \bar{\partial}\mathbf{N})\partial\mathbf{N} + (\overline{\mathbf{N}}, \partial\mathbf{N})\,\bar{\partial}\mathbf{N} - \lambda\mathbf{N}, \tag{4.5}$$

where

$$\lambda = -2(\overline{\mathbf{N}}, \partial\mathbf{N})(\mathbf{N}, \bar{\partial}\overline{\mathbf{N}}) + \frac{1}{2}(\partial\mathbf{N}, \bar{\partial}\overline{\mathbf{N}}) + \frac{1}{2}(\bar{\partial}\mathbf{N}, \partial\overline{\mathbf{N}}).$$

Equations (4.5) are associated with the Lagrangian

$$L = \iint \{(\bar{\partial}\mathbf{N}, \partial\overline{\mathbf{N}}) + (\partial\mathbf{N}, \bar{\partial}\overline{\mathbf{N}}) + 2(\overline{\mathbf{N}}, \partial\mathbf{N})(\overline{\mathbf{N}}, \bar{\partial}\mathbf{N}) - 2\lambda\big[(\mathbf{N}, \overline{\mathbf{N}}) - 1\big]\big\}\, dz\, d\bar{z},$$

where λ is the Lagrange multiplier.

References

1. A. I. Bobenko, *Surfaces in terms of* 2×2 *matrices. Old and new integrable cases*, Harmonic Maps and Integrable Systems (A. Fordy and J. Wood, eds.), Vieweg, Braunschweig, 1994, pp. 193–202.
2. P. Bracken, A. M. Grundland, and L. Martina, *The Weierstrass–Enneper system for constant mean curvature surfaces and the completely integrable sigma model*, J. Math. Phys. **7** (1999), 3379–3403.
3. R. Carrol and B. G. Konopelchenko, *Generalized Weierstrass-Enneper inducing, conformal immersions, and gravity*, Internat. J. Modern Phys. A **11** (1996), no. 7, 1183–1216.
4. K. Kenmotsu, *Weierstrass formula for surfaces of prescribed mean curvature*, Math. Ann. **245** (1979), 89–99.
5. B. G. Konopelchenko, *Induced surfaces and their integrable dynamics*, Stud. Appl. Math. **96**, (1996), 9–51.
6. ———, *Soliton eigenfunction equations: The* IST *integrability and some properties*, Rev. Math. Phys. **2** (1990), 399–426.
7. B. G. Konopelchenko and I. A. Taimanov, *Constant mean curvature surfaces via an integrable dynamical system*, J. Phys. A **29** (1996), 1261–1265.

8. W. Zakrzewski, *Low dimensional sigma-models*, Hilfger, London, 1989.

Centre for Nonlinear Studies, Landau Institute for Theoretical Physics, Academy of Sciences of Russia, Kosygina 2, 117940 Moscow, GSP-1, Russia

Current address: Department of Mathematical Sciences, Loughborough University, Loughborough, Leicestershire LE11 3TU, United Kingdom.

E-mail address: E.V.Ferapontov@lboro.ac.uk

Centre de Recherches Mathematiques, Université de Montréal, C.P. 6128, Succ. Centre-ville, Montréal, Québec H3C 3J7, Canada.

E-mail address: grundlan@crm.umontreal.ca

Centre de Recherches Mathématiques
CRM Proceedings and Lecture Notes
Volume **29**, 2001

On the Integrability of Weingarten Surfaces

F. Finkel

Abstract. The integrability of the Gauss–Mainardi–Codazzi (GMC) equations for Weingarten surfaces immersed in $\mathbb{R}^3$ is discussed. Our results suggest that the GMC equations are integrable only in the case of linear Weingarten surfaces.

1. Introduction

A surface whose Gaussian and mean curvatures satisfy a functional relation of the form $r(K, H) = 0$ is known as a *Weingarten* surface. The geometric interest of this type of surfaces was already recognized by some of the most eminent geometers of the nineteenth century, starting with Weingarten and followed among others by Lie, Ribacour, Beltrami and Darboux; classical references are [**3**] and [**6**]. Surfaces of constant Gaussian or mean curvature, and surfaces of revolution are well-known particular examples of Weingarten surfaces.

A surface immersed into the 3-dimensional Euclidean space is determined up to rigid motions by its first and second fundamental forms. The coefficients of these forms satisfy a system of three nonlinear partial differential equations (PDEs) called the Gauss–Mainardi–Codazzi (GMC) equations. For a Weingarten surface, the GMC equations reduce to a single nonlinear PDE known as the Gauss equation. In a few remarkable cases, the Gauss equation belongs to a special class of equations generically known as integrable equations. This class of equations possess several characteristic features enabling the computation of explicit solutions. An essential property of integrable equations is the existence of an infinite hierarchy of generalized symmetries which can be obtained recursively [**7**]. In every known case, the existence of a single generalized symmetry implies the existence of an infinite number. We shall thus consider a PDE (or more generally, a system of PDEs) to be *integrable* if it possesses a generalized symmetry.

2000 *Mathematics Subject Classification.* Primary: 35Q53, 53A05; Secondary: 35Q55, 35Q58.

It is a pleasure to thank Prof. A. S. Fokas for helpful discussions and his kind support. The author was supported in part by DGES Grant PB98-0821.

This is the final form of the paper.

The problem we address in this paper is to determine which Weingarten surfaces are associated to an integrable Gauss equation[1]. Particular examples thereof are surfaces of constant mean curvature or constant positive Gaussian curvature, whose corresponding Gauss equation is the sinh-Gordon equation, or surfaces with constant negative Gauss curvature, associated to the sine-Gordon equation. The parallel deformations of these surfaces are characterized by a linear relation between the Gaussian and mean curvatures, and are associated to the same integrable Gauss equation as the corresponding undeformed surface [**2,4,5**]. Our results suggest that the converse is also true, i.e., the linearity in K and H of the relation $r(K,H)=0$ is a necessary condition for the Gauss equation to be integrable. If the Gauss equation for a Weingarten surface is not integrable, it may still possess integrable reductions. An interesting class of reductions correspond to surfaces of revolution, whose reduced Gauss equation can be solved in closed form.

The outline of the rest of paper is as follows. In Section 2 we briefly recall the Lie-algebraic representation of surfaces immersed into the 3-dimensional Euclidean space introduced in [**4**]. This formalism is especially suited to study families of surfaces with integrable GMC equations. Indeed, the Gauss–Weingarten equations correspond in this case to a Lax pair for the GMC equations [**1**]. In Section 3 we derive a new explicit representation of the Gauss equation for a Weingarten surface. In particular, we shall see that the Gauss equation for a Weingarten surface of the form $K=k(H)\neq \frac{1}{4}(H^2-c^2)$, $c\in\mathbb{R}$, parameterized by its lines of curvature, is given by

$$\left(\left(\Psi''-\frac{1}{G}\right)e^{\Psi'}G_v\right)_v-\left(\left(\Psi''+\frac{1}{G}\right)e^{-\Psi'}G_u\right)_u+\frac{1}{2G}\left((\Psi'G-\Psi)^2-G^2\right)=0,$$

where $G(u,v)$ is defined in terms of the mean curvature H by $G=\sqrt{H^2-4k(H)}$, and $\Psi(G)$ is related to the inverse of this relation. We conjecture that any Weingarten surface associated to an integrable Gauss equation is necessarily a linear Weingarten surface. We finally examine a certain family of quadratic Weingarten surfaces of revolution, presenting some pictures thereof.

2. Surfaces Immersed in $\mathbb{R}^3$ Parameterized by the Lines of Curvature

A surface immersed into the 3-dimensional Euclidean space can be identified with a surface on the Lie algebra su(2), according to the following result [**4**].

THEOREM 2.1. *Consider the Lie algebra* su(2), *with scalar product defined by*

$$\langle g_1,g_2\rangle=-\tfrac{1}{2}\,\mathrm{trace}(g_1g_2),\qquad g_1,g_2\in\mathrm{su}(2). \tag{2.1}$$

An orthonormal basis is spanned by $e_j=-i\sigma_j$, $j=1,2,3$, *where* σ_j *are the usual Pauli matrices. Let* $U(u,v)$, $V(u,v)$, $A(u,v)$, *and* $B(u,v)$ *be* su(2)*-valued smooth functions of* (u,v) *in a domain* $\Omega\subset\mathbb{R}^2$. *Assume that* U, V, A, *and* B *satisfy the nonlinear PDEs*

$$U_v-V_u+[U,V]=0, \tag{2.2}$$

$$A_v-B_u+[A,V]+[U,B]=0, \tag{2.3}$$

---

[1] The terms *integrable surface* and *integrable geometry* are commonly used in the literature to refer to surfaces with integrable GMC equations.

where $[\,,\,]$ *is the usual commutator. Then there exists an* su(2)*-valued surface with immersion function* $F(u,v)$ *defined by*

$$F_u = \Phi^{-1} A \Phi, \quad F_v = \Phi^{-1} B \Phi, \tag{2.4}$$

where the SU(2)*-valued function* $\Phi(u,v)$ *satisfies*

$$\Phi_u = U\Phi, \quad \Phi_v = V\Phi. \tag{2.5}$$

The surface $F(u,v) = -i\sum_1^3 F_j \sigma_j$ *is identified with the surface in* $\mathbb{R}^3$ *defined by*

$$x_j = F_j(u,v), \quad j = 1,2,3. \tag{2.6}$$

A frame on this surface is

$$\Phi^{-1} A \Phi, \quad \Phi^{-1} B \Phi, \quad \Phi^{-1} C \Phi, \tag{2.7}$$

where

$$C = \frac{[A,B]}{|[A,B]|}, \quad |A| = \sqrt{\langle A, A\rangle}.$$

The first and second fundamental forms are respectively given by

$$\langle A, A\rangle (du)^2 + 2\langle A, B\rangle\, du\, dv + \langle B, B\rangle (du)^2, \tag{2.8}$$

$$\begin{aligned}\Big\langle \frac{\partial A}{\partial u} + [A,U], C\Big\rangle (du)^2 + 2\Big\langle \frac{\partial A}{\partial v} + [A,V], C\Big\rangle\, du\, dv \\ + \Big\langle \frac{\partial B}{\partial v} + [B,V], C\Big\rangle (dv)^2.\end{aligned} \tag{2.9}$$

REMARK 2.2. Equations (2.2) are the GMC equations of the surface (2.6). They are the compatibility condition of the Gauss–Weingarten equations (2.5).

The usual parametrizations can be obtained as particular cases of the general setting described in Theorem 2.1 [**4**]. The study of Weingarten surfaces is best performed using parametric lines of curvature.

EXAMPLE 2.3. Let

$$U = \begin{pmatrix} -\frac{ia_v}{2b} & -\frac{af}{2} \\ \frac{af}{2} & \frac{ia_v}{2b} \end{pmatrix}, \quad V = \begin{pmatrix} \frac{ib_u}{2a} & \frac{ibh}{2} \\ \frac{ibh}{2} & -\frac{ib_u}{2a} \end{pmatrix}, \tag{2.10}$$

where a, b, e and f are functions of u and v. The GMC equations (2.2) are

$$\left(\frac{a_v}{b}\right)_v + \left(\frac{b_u}{a}\right)_u + abfh = 0, \tag{2.11}$$

$$(bh)_u - fb_u = 0, \quad (af)_v - ha_v = 0. \tag{2.12}$$

Consider the surface constructed from U, V, and

$$A = \begin{pmatrix} 0 & -ia \\ -ia & 0 \end{pmatrix}, \quad B = \begin{pmatrix} 0 & -b \\ b & 0 \end{pmatrix}. \tag{2.13}$$

The first and second fundamental forms are respectively given by

$$a^2(du)^2 + b^2(dv)^2, \quad a^2 f(du)^2 + b^2 h(dv)^2. \tag{2.14}$$

The Gaussian and mean curvatures are

$$K = fh, \quad H = f + h.$$

3. Weingarten Surfaces

In the rest of the paper we shall restrict ourselves to Weingarten surfaces immersed in $\mathbb{R}^3$. We shall assume that the functional relation defining the surface can be written as $K = k(H)$, where k is a smooth function[2]. Let us choose the principal curvatures f and h so that

$$2f = H - G, \quad 2h = H + G,$$

where

$$G = \sqrt{H^2 - 4k(H)}. \tag{3.1}$$

We first assume that this relation can be (locally) inverted. Let $H = \phi(G)$ be the corresponding inverse function. The Codazzi relations (2.12) then reduce to

$$2Gb_u + (1 + \phi')bG_u = 0, \quad 2Ga_v + (1 - \phi')aG_v = 0,$$

with general solution given by

$$a = \frac{a_0(u)}{\sqrt{G}}e^{\psi/2}, \quad b = \frac{b_0(v)}{\sqrt{G}}e^{-\psi/2}, \quad \psi(G) = \int^G \frac{\phi'(G)}{G}\,dG. \tag{3.2}$$

Taking without loss of generality $a_0(u)$ and $b_0(v)$ as unity, and introducing the new function

$$\Psi(G) = \int^G \psi(G)\,dG,$$

the Gauss equation (2.11) reduces to the nonlinear PDE

$$\left(\left(\Psi'' - \frac{1}{G}\right)e^{\Psi'}G_v\right)_v - \left(\left(\Psi'' + \frac{1}{G}\right)e^{-\Psi'}G_u\right)_u + \frac{1}{2G}\left((\Psi'G - \Psi)^2 - G^2\right) = 0. \tag{3.3}$$

This PDE characterizes (almost) all Weingarten surfaces satisfying a given functional relation $K = k(H)$. Our goal is to identify all Weingarten surfaces whose corresponding Gauss equation (3.3) is integrable, in the sense that it possesses a generalized symmetry. In order to study this problem, let us assume that $G\Psi'' - 1 \neq 0$ (otherwise note that $G\Psi'' + 1 \neq 0$ and proceed in the same way), and rewrite equation (2.11) as

$$G_{vv} + \mathcal{A}G_{uu} + \mathcal{B}(G_v)^2 + \mathcal{C}(G_u)^2 + \mathcal{D} = 0, \tag{3.4}$$

where the coefficients $\mathcal{A}$, $\mathcal{B}$, $\mathcal{C}$, $\mathcal{D}$ are functions of G, given explicitly by

$$\mathcal{A} = \frac{1 + G\Psi''}{1 - G\Psi''}e^{-2\Psi'}, \qquad \mathcal{B} = \Psi'' - \frac{G\Psi''' + 1/G}{1 - G\Psi''},$$

$$\mathcal{C} = \frac{G\Psi''' - 1/G - \Psi''(1 + G\Psi'')}{1 - G\Psi''}e^{-2\Psi'}, \ \mathcal{D} = \frac{(1 - (\Psi')^2)G^2 + 2\Psi\Psi'G - \Psi^2}{2(1 - G\Psi'')}e^{-\Psi'}.$$

The following result is proved by straightforward computation.

---

[2]This excludes surfaces of constant mean curvature unless K is also a constant. Surfaces of constant mean curvature are associated to the hyperbolic sine-Gordon equation (if H is a nonzero constant) or the Liouville equation (if H vanishes identically), which are both integrable PDEs [**1**, **4**].

THEOREM 3.1. *Let F be a Weingarten surface of the form $K = k(H)$, parameterized by its lines of curvature. The function k is linear, i.e.,*

$$k(H) = \alpha + \beta H, \quad \alpha, \beta \in \mathbb{R}, \tag{3.5}$$

if and only if $\mathcal{A}' = 0$. In this case the Gauss equation (3.4) is integrable.

REMARK 3.2. The following three cases arise in connection with Theorem 3.1:

(i) $\mathcal{A}$ is a positive constant if and only if $\alpha + \beta^2 > 0$. In this case equation (3.4) is equivalent to the sinh-Gordon equation.
(ii) $\mathcal{A}$ is a negative constant if and only if $\alpha + \beta^2 < 0$. The equation (3.4) is then equivalent to the sine-Gordon equation.
(iii) $\mathcal{A} = 0$ if and only if $\alpha + \beta^2 = 0$. The equation (3.4) then reduces to an ordinary differential equation for G.[3]

REMARK 3.3. It can be verified directly that equation (3.4) does not possess generalized symmetries up to order 3, unless $\mathcal{A}' = 0$. In fact, we believe that this equation admits no generalized symmetries if $\mathcal{A}' \neq 0$. Indeed, as the order of a possible symmetry grows, the function $\mathcal{A}(G)$ is required to satisfy an increasingly more complicated system of ODEs, which is most likely to be incompatible if $\mathcal{A}' \neq 0$.

If the function G in equation (3.1) is a constant $c \neq 0$,[4] i.e., if the principal curvatures of the Weingarten surface differ by a constant, the Codazzi relations (2.12) can be integrated in terms of the mean curvature H. The Gauss equation (2.11) is then

$$\left(e^{H/c}H_v\right)_v - \left(e^{-H/c}H_u\right)_u + \frac{c}{2}(H^2 - c^2) = 0. \tag{3.6}$$

Just as in the previous case, we conjecture that equation (3.6) does not possess generalized symmetries of any order. Let us summarize the above discussion with the following

CONJECTURE 3.4. *Let F a Weingarten surface. The* GMC *equations are integrable if and only if F is a linear Weingarten surface, i.e.,*

$$\alpha + \beta H + \gamma K = 0, \quad \alpha, \beta, \gamma \in \mathbb{R}.$$

Note that the "if" part of this conjecture is certainly true.

Even if the Gauss equation for a Weingarten surface is not integrable, it can admit integrable reductions. A simple reduction of equation (3.3) is obtained if we assume that G does not depend on one of the variables[5], say u. The resulting ODE can always be integrated explicitly. Indeed, if $G = G(v)$, the Gauss equation (3.3) becomes

$$\left(\mathcal{R}(G)G_v\right)_v + \mathcal{S}(G) = 0, \tag{3.7}$$

where

$$\mathcal{R}(G) = \left(\Psi'' - \frac{1}{G}\right)e^{\Psi'}, \quad \mathcal{S}(G) = \frac{1}{2G}\left((\Psi' G - \Psi)^2 - G^2\right). \tag{3.8}$$

---

[3]Note that in this case one of the principal curvatures is constant. For this reason these surfaces are also called "channel" surfaces [**3**].

[4]Note that the case $c = 0$ corresponds to a sphere.

[5]It is easy to show in this respect that equation (3.6) admits no other type of reduction.

A first integral of equation (3.7) is

$$G_v^2 = -\frac{2}{(\mathcal{R}(G))^2}\int^G \mathcal{R}(G)\mathcal{S}(G)\,dG, \tag{3.9}$$

which can be immediately reduced to quadratures. The corresponding surfaces are surfaces of revolution.

EXAMPLE 3.5. We consider Weingarten surfaces of revolution satisfying the quadratic relation $K = k(H) = \lambda H^2$, with $0 \neq \lambda < \frac{1}{4}$. The function G defined by equation (3.1) is thus proportional to the mean curvature,

$$H = \phi(G) = \mu G, \quad \mu = \pm(1-4\lambda)^{-1/2}. \tag{3.10}$$

We choose the constant of integration in Ψ so that

$$\mathcal{R}(G) = G^{\mu-1}, \quad \mathcal{S}(G) = \frac{\mu+1}{2}G.$$

The first integral (3.9) then leads to the relations

$$\pm v+\hat{c} = \frac{2}{(\mu-1)G}\left(c\,\mathrm{F}\left(\frac{-1}{\mu+1};\frac{1}{2};\frac{\mu}{\mu+1};\frac{G^{\mu+1}}{c^2}\right)-\sqrt{c^2-G^{\mu+1}}\right),\ 0<G<c^{2/\mu+1},$$

where c, $\hat{c}$ are constants and F is the hypergeometric function. Let $g = g_\pm(c,\hat{c},\mu;v)$ be the inverse function defined by either of these relations. The first and second fundamental forms can be expressed directly in terms of g. Indeed, from Eqs. (2.14),

FIGURE 1. Weingarten surfaces of revolution $K = \frac{3}{16}H^2$, with parameters $c = 1$, $\hat{c} = 0$, and $\mu = 2$ (left), $\mu = -2$ (right).

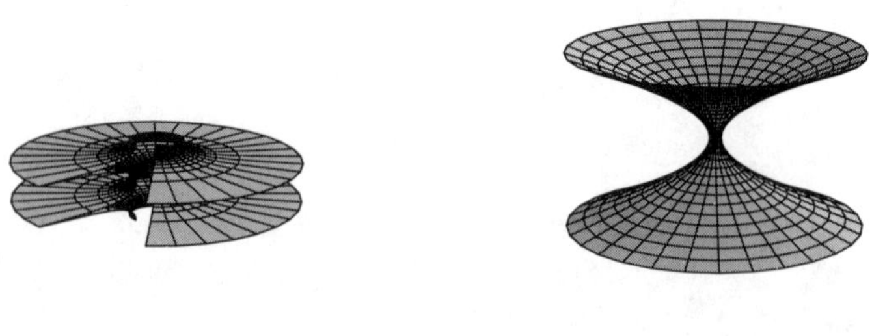

FIGURE 2. Weingarten surfaces of revolution $K = -\frac{3}{4}H^2$, with parameters $c = 1$, $\hat{c} = 0$, and $\mu = \frac{1}{2}$ (left), $\mu = -\frac{1}{2}$ (right). A portion of the surface for $\mu = 2$ has been removed to show its inner side.

(3.2), and (3.10), we obtain, respectively,

$$g^{\mu-1}(du)^2 + g^{-\mu-1}(dv)^2, \quad \frac{\mu-1}{2}g^{\mu}(du)^2 + \frac{\mu+1}{2}g^{-\mu}(dv)^2.$$

The surface described by the immersion function $F(u,v)$ is determined up to its position in space by Eqs. (2.4), (2.5), (2.6), (2.10), (2.13), and (3.2). A solution is given by

$$x = \frac{2}{c(\mu-1)}\sin\left(\frac{c(\mu-1)}{2}u\right)g^{(\mu-1)/2},$$
$$y = \frac{2}{c(\mu-1)}\cos\left(\frac{c(\mu-1)}{2}u\right)g^{(\mu-1)/2},$$
$$z = \frac{v}{c}.$$

We present here the corresponding plots for $c = 1$, $\hat{c} = 0$, and $\mu = \pm 2$ (Figure 1), $\mu = \pm\frac{1}{2}$ (Figure 2).

References

1. A. I. Bobenko, *Surfaces in terms of 2 by 2 matrices. Old and new integrable cases*, Harmonic Maps and Integrable Systems (A. P. Fordy and J. C. Wood, eds.), Vieweg, 1994, 83–127.
2. J. Cieśliński, *A generalized formula for integrable classes of surfaces in Lie algebras*, J. Math. Phys. **38** (1997), 4255–4272.
3. G. Darboux, *Leçons sur la théorie générale des surfaces et les applications géométriques du calcul infinitésimal*, vol. III, Gauthier-Villars, France, 1894.
4. A. S. Fokas and I. M. Gel'fand, *Surfaces on Lie groups, on Lie algebras and their integrability*, Comm. Math. Phys. **177** (1996), 203–220.
5. A. S. Fokas, I. M. Gel'fand, F. Finkel, and Q. M. Liu, *A formula for constructing infinitely many surfaces on Lie algebras and integrable equations*, Preprint, Selecta Math. (N.S.) (in press).
6. A. R. Forsyth, *Lectures on the differential geometry of curves and surfaces*, Cambridge Univ. Press, Cambridge, 1912.
7. P. J. Olver, *Applications of Lie groups to differential equations*, Springer, Berlin, 1993.

Department of Mathematics, Imperial College, London SW7 2BZ, UK.
E-mail address: f.finkel@ic.ac.uk

Centre de Recherches Mathématiques
CRM Proceedings and Lecture Notes
Volume **29**, 2001

A New Immersion Formula for Surfaces on Lie Algebras and Integrable Equations

F. Finkel and *A. S. Fokas*

ABSTRACT. A new immersion formula for constructing large families of surfaces in Lie algebras with integrable Gauss–Mainardi–Codazzi equations is presented. The formula is applied to construct new integrable surfaces immersed into the 3-dimensional Euclidean space, including the list of surfaces recently presented by Bobenko and some special deformations thereof.

1. Introduction

The immersion of a 2-dimensional surface into the n-dimensional Euclidean space can be formulated in terms of an n-dimensional semisimple Lie algebra and its corresponding Lie group [**6**]. This immersion is characterized by $n-1$ fundamental forms, whose coefficients satisfy a set of nonlinear partial differential equations (PDEs) known as the Gauss–Mainardi–Codazzi (GMC) equations. For very distinguished families of surfaces, this set of PDEs belongs to a special type of equations known generically as integrable equations. This observation provides a link between classical differential geometry and the rapidly developing field of integrable systems, and has been the subject of a considerable amount of research over the last decade; see [**1–4, 6, 7, 9, 10, 12–14**] and references therein.

Given the immersion function of a surface it is straightforward to compute its associated fundamental forms. The usual approach in the references above is precisely the opposite. Consider for definiteness the case $n = 3$, corresponding to immersions into the 3-dimensional Euclidean space. The starting point to construct surfaces characterized by a particular geometrical property is a suitable parametrization of the first and second fundamental forms. For example, hyperbolic surfaces are best analyzed using asymptotic coordinates [**9**]. In the second step, one finds a subset of the class of surfaces under study such that the corresponding GMC equations are integrable. Their explicit solution by standard analytical methods typically yields an explicit formula for the immersion functions

2000 *Mathematics Subject Classification.* Primary: 35Q53, 53A05; Secondary: 35Q55, 35Q58.

The first author was supported in part by DGES Grant PB98-0821.
The second author was supported in part by EPSRC Grant GG/J71885.
This is the final form of the paper.

of the surfaces. For example, the so-called Bianchi surfaces are a particular type of hyperbolic surfaces with integrable GMC equations admitting solitonic solutions.

Given a class of surfaces with integrable GMC equations, a natural problem is to construct deformations associated with the same integrable GMC equations as the initial undeformed surfaces. A few implicit results for parallel deformations[1] first appeared in [**1**]. Some more general deformations using a Lie-algebraic formulation were presented by Fokas and Gel'fand in [**6**], and then subsequently generalized by Cieśliński in [**3**]. In this paper, we report a new immersion formula for integrable surfaces which provides a further generalization of the previous deformations [**7**]. It should be emphasized that although this formula enables one to construct large families of surfaces associated to a given set of integrable GMC equations, it is nontrivial to select those admitting an invariant geometric characterization. A list of such surfaces was published by Bobenko in [**1**]. As our second goal we examine the parallel deformations of these surfaces, referring to [**7**] for additional details.

The rest of the paper is organized as follows. In Section 2 we define the concepts of surfaces on Lie algebras, integrable PDEs, and symmetries which are needed for our study. The main result describing the new immersion formula for integrable surfaces on Lie algebras is presented in Section 3, Theorem 3.1. Using an appropriate Lie algebra, these surfaces can be identified with surfaces immersed into the n-dimensional Euclidean space. We use the immersion formula to construct a class of surfaces in $\mathbb{R}^3$ associated with symmetries of the modified Korteweg–de Vries (mKdV) equation. Finally, in Section 4 we review Bobenko's list of integrable surfaces and their corresponding parallel deformations. The family of surfaces with harmonic inverse mean curvature is discussed in detail.

2. Surfaces on Lie Algebras, Integrable PDEs, and Symmetries

In this section we introduce the basic notions and definitions that shall be used in the sequel.

Let G be a Lie group and $\mathcal{G}$ its corresponding Lie algebra, with $\dim \mathcal{G} = n < \infty$. Assume that $\mathcal{G}$ is endowed with a nondegenerate invariant scalar product. Let $\{e_j\}_1^n$ be an orthogonal basis in $\mathcal{G}$.

Let $\{\theta_j(u,v)\}_1^N$ be a set of scalar differentiable functions of (u,v) in a domain $\Omega \subset \mathbb{R}^2$. Let $\mathcal{R}^{(N)}$ be the ring of smooth functions of u, v, θ_j and its u- and v-partial derivatives. Elements of the ring $\mathcal{R}^{(N)}$ will be denoted by $f\{\theta_1,\dots,\theta_N\}$. If these functions also depend smoothly on a complex parameter λ, they will be denoted by $f\{\theta_1,\dots,\theta_N;\lambda\}$, and the corresponding ring by $\mathcal{R}_\lambda^{(N)}$.

PROPOSITION 2.1. *Let $U(u,v)$ and $V(u,v)$ be $\mathcal{G}$-valued differentiable functions of (u,v) in $\Omega \subset \mathbb{R}^2$. Assume that U and V satisfy the nonlinear PDE*

$$U_v - V_u + [U,V] = 0, \tag{2.1}$$

where $[\ ,\ ]$ is the Lie algebra commutator. Then there is a G-valued differentiable function $\Phi(u,v)$ such that

$$\Phi_u = U\Phi, \quad \Phi_v = V\Phi. \tag{2.2}$$

[1]Recall that two surfaces are parallel if they can be obtained from each other by a constant translation along the normal direction.

Let $A(u,v)$ and $B(u,v)$ be $\mathcal{G}$-valued differentiable functions of $(u,v) \in \Omega$. Assume that A and B satisfy the equation

$$A_v - B_u + [A,V] + [U,B] = 0, \tag{2.3}$$

with U and V satisfying (2.1). *Then there exists a $\mathcal{G}$-valued immersion function $F(u,v)$ defined by the equations*

$$F_u = \Phi^{-1} A \Phi, \quad F_v = \Phi^{-1} B \Phi. \tag{2.4}$$

PROOF. See [**6**]. □

DEFINITION 2.2. Let λ be a complex parameter. Let $U(u,v,\lambda)$ and $V(u,v,\lambda)$ be $\mathcal{G}$-valued functions satisfying Eq. (2.1). Assume that U and V can be parametrized in terms of elements of the ring $\mathcal{R}^{(N)}$ and the parameter λ, namely

$$U = \sum_1^n U_j\{\theta_1, \dots, \theta_N; \lambda\}\, e_j, \quad V = \sum_1^n V_j\{\theta_1, \dots, \theta_N; \lambda\}\, e_j. \tag{2.5}$$

In addition, suppose that the $\mathcal{G}$-valued Eq. (2.1) is equivalent to a system of N nonlinear PDEs which is independent of λ, i.e.

$$M_j\{\theta_1, \dots, \theta_N\} = 0, \quad j = 1, \dots, N. \tag{2.6}$$

Then we say that Eq.(2.1) is an *integrable* nonlinear PDE, and Eqs. (2.2) are the corresponding *Lax pair.*

REMARK 2.3. It has become customary to use the term *integrable surfaces* to refer to surfaces associated with integrable GMC equations [**1, 6**].

DEFINITION 2.4. Let $\vartheta\{\theta_1, \dots, \theta_N\}$ and $\vartheta_j\{\theta_1, \dots, \theta_N\}$, $j = 1, \dots, N$, be elements of the ring $\mathcal{R}^{(N)}$. The *Fréchet derivative* of ϑ in the direction ϑ_j is

$$\left(\frac{D\vartheta}{D\theta_j}\right)\vartheta_j = \frac{\partial}{\partial \varepsilon}\vartheta\{\theta_1, \dots, \theta_{j-1}, \theta_j + \varepsilon\vartheta_j, \theta_{j+1}, \dots, \theta_N\}\bigg|_{\varepsilon=0}. \tag{2.7}$$

DEFINITION 2.5. Suppose that the set of differentiable functions $\{\theta_j(u,v)\}_1^N$ satisfy the system (2.6) of PDEs. The set of scalar functions $\{\varphi_j\}_1^N$ is called a *symmetry* of the system of Eqs. (2.6) if $\{\varphi_j\}_1^N$ belong to the ring $\mathcal{R}^{(N)}$ and they satisfy the system of linear equations

$$\sum_1^N \left(\frac{DM_k}{D\theta_j}\right)\varphi_j = 0, \quad k = 1, \dots, N. \tag{2.8}$$

EXAMPLE 2.6. Let $\mathcal{G} = \mathrm{su}(2)$, $e_j = -i\sigma_j$, $j = 1, 2, 3$, where σ_j are the usual Pauli matrices. Let

$$U = \frac{i}{2}(-\theta_u\sigma_1 + \lambda\sigma_3), \quad V = \frac{i}{2\lambda}(\sin\theta\sigma_2 - \cos\theta\sigma_3), \tag{2.9}$$

i.e.

$$U_1 = \frac{\theta_u}{2}, \quad U_2 = 0, \quad U_3 = -\frac{\lambda}{2}, \quad V_1 = 0, \quad V_2 = -\frac{1}{2\lambda}\sin\theta, \quad V_3 = \frac{1}{2\lambda}\cos\theta.$$

Then Eq.(2.1) is equivalent to the sine-Gordon equation

$$\theta_{uv} - \sin\theta = 0. \tag{2.10}$$

A symmetry of Eq.(2.10) is an element φ in $\mathcal{R}^{(1)}$ satisfying the linear PDE

$$\varphi_{uv} - (\cos\theta)\varphi = 0. \tag{2.11}$$

For example, $\varphi = \theta_u$ is a symmetry of the sine-Gordon equation. Indeed, φ is an element of $\mathcal{R}^{(1)}$ and φ solves Eq. (2.11) (this follows by differentiating Eq.(2.10) with respect to u).

REMARK 2.7. An essential property of integrable equations is the existence of infinitely many symmetries, which can be obtained recursively; see [**5, 11**]. For example,

$$\varphi_k = \Delta^k \theta_u, \quad \Delta = \partial_u^2 + \theta_u^2 - \theta_u \partial_u^{-1} \theta_{uu}, \tag{2.12}$$

is a symmetry of the sine-Gordon Eq.(2.10) for any nonnegative integer k.

3. The Generalized Immersion Formula

In this section we present the main result, whose proof can be found in [**7**].

THEOREM 3.1. *Let $U(u,v,\lambda)$ and $V(u,v,\lambda)$ be $\mathcal{G}$-valued functions satisfying Eq.(2.1). Assume that this equation is integrable, that is, U and V can be parametrized as in* (2.5) *and Eq.* (2.1) *is equivalent to a system of nonlinear PDEs of the form* (2.6). *Let the G-valued function $\Phi(u,v,\lambda)$ satisfy the Lax pair Eqs.* (2.2). *Let the $\mathcal{G}$-valued functions $A(u,v,\lambda)$ and $B(u,v,\lambda)$ be defined by*

$$A = \alpha U_\lambda + S_u + [S,U] + \sum_1^N \left(\frac{DU}{D\theta_j}\right)\varphi_j, \tag{3.1}$$

$$B = \alpha V_\lambda + S_v + [S,V] + \sum_1^N \left(\frac{DV}{D\theta_j}\right)\varphi_j, \tag{3.2}$$

where $\alpha(\lambda)$ is an arbitrary scalar function of λ, $S(u,v,\lambda)$ is an arbitrary $\mathcal{G}$-valued function of u, v, λ, and the set of scalar functions $\{\varphi_j\}_1^N$ is a symmetry of the system (2.6). *Then, there exists a $\mathcal{G}$-valued surface with immersion function $F(u,v,\lambda)$ defined in terms of A, B and Φ by Eqs.* (2.4). *Furthermore, F is given up to an additive $\mathcal{G}$-valued constant $C(\lambda)$ by the formula*

$$F = \Phi^{-1}\left(\alpha\Phi_\lambda + S\Phi + \sum_1^N \left(\frac{D\Phi}{D\theta_j}\right)\varphi_j\right). \tag{3.3}$$

REMARK 3.2. The particular case of Eq. (3.3) with $S = \{\varphi_j\}_1^N = 0$, i.e.

$$F = \alpha \Phi^{-1}\Phi_\lambda, \tag{3.4}$$

is called the Sym (or the Sym–Tafel) formula[2] [**12, 13**]. This formula has been extensively employed in the solution of geometric problems by analytical methods. A generalization of the Sym formula, namely Eq. (3.3) with a particular form of S and $\{\varphi_j\}_1^N = 0$, was introduced in [**6**]. A further generalization with arbitrary S and $\{\varphi_j\}_1^N = 0$ was given in [**10**] and [**3**].

One can use Theorem 3.1 and Proposition 2.1 to construct integrable surfaces immersed into the 3-dimensional Euclidean space, according to the following result[3].

[2]Apparently the Sym formula was not known to the classical geometers of the last century. Its main advantage lies in the fact that it allows to determine the immersion function F without the need of integrating Eqs. (2.4) explicitly.

[3]The n-dimensional generalization of Theorem 3.3 can be found in [**6**].

THEOREM 3.3. *Let $\mathcal{G} = \mathrm{su}(2)$, with scalar product defined by*

$$\langle g_1, g_2\rangle = -\frac{1}{2}\,\mathrm{trace}(g_1 g_2), \quad g_1, g_2 \in \mathrm{su}(2). \tag{3.5}$$

Consider the orthonormal basis spanned by $e_j = -i\sigma_j$, $j = 1, 2, 3$, where σ_j are the usual Pauli matrices. Let $U(u,v,\lambda)$, $V(u,v,\lambda)$, $A(u,v,\lambda)$, and $B(u,v,\lambda)$ be su(2)*-valued functions satisfying Eqs.* (2.1) *and* (2.3). *Then there exists a surface in $\mathbb{R}^3$,*

$$x_j = F_j(u,v), \quad j = 1,2,3, \tag{3.6}$$

with immersion function $F(u,v) = -i\sum_1^3 F_j\sigma_j$ defined by Eqs. (2.2) *and* (2.4). *A frame on this surface is*

$$\Phi^{-1}A\Phi, \quad \Phi^{-1}B\Phi, \quad \Phi^{-1}C\Phi, \tag{3.7}$$

where

$$C = \frac{[A,B]}{|[A,B]|}, \quad |A| = \sqrt{\langle A, A\rangle}.$$

The first and second fundamental forms are respectively given by

$$\langle A, A\rangle(du)^2 + 2\langle A, B\rangle\, du\, dv + \langle B, B\rangle(du)^2, \tag{3.8}$$

$$\left\langle \frac{\partial A}{\partial u} + [A,U], C\right\rangle(du)^2 + 2\left\langle \frac{\partial A}{\partial v} + [A,V], C\right\rangle du\, dv + \left\langle \frac{\partial B}{\partial v} + [B,V], C\right\rangle(dv)^2. \tag{3.9}$$

If the first and second fundamental forms are denoted by

$$g_{11}(du)^2 + 2g_{12}\, du\, dv + g_{22}(dv)^2, \quad d_{11}(du)^2 + 2d_{12}\, du\, dv + d_{22}(dv)^2,$$

the Gaussian and mean curvatures are

$$K = \frac{d_{11}d_{22} - d_{12}^2}{g_{11}g_{22} - g_{12}^2}, \quad H = \frac{d_{11}g_{22} - 2d_{12}g_{12} + d_{22}g_{11}}{g_{11}g_{22} - g_{12}^2}. \tag{3.10}$$

PROOF. See [**6**]. □

REMARK 3.4. The generalized immersion formula (3.3) was applied in [**2**] to construct a family of surfaces immersed in $\mathbb{R}^3$ associated to symmetries of the sine-Gordon Eq. (2.10). It was also shown that if a surface in this family is regular, oriented, and compact connected, then it is homeomorphic to a sphere.

Let us illustrate Theorems 3.1 and 3.3 with a family of integrable surfaces associated to the mKdV equation.

EXAMPLE 3.5. Let

$$U = \frac{i}{2}\begin{pmatrix} \lambda & -\theta \\ -\theta & \lambda \end{pmatrix}, \quad V = -\frac{i}{2}\begin{pmatrix} -\frac{\lambda}{2}\theta^2 + \lambda^3 & v_1 - iv_2 \\ v_1 + iv_2 & \frac{\lambda}{2}\theta^2 - \lambda^3 \end{pmatrix}, \tag{3.11}$$

where $\theta(u,v)$ is a scalar real-valued function and

$$v_1 = \theta_{uu} + \frac{\theta^3}{2} - \lambda^2\theta, \quad v_2 = -\lambda\theta_u. \tag{3.12}$$

In this case, Eq. (2.1) is equivalent to the mKdV equation

$$\theta_v = \theta_{uuu} + \frac{3}{2}\theta^2\theta_u. \tag{3.13}$$

Let φ be a symmetry of the mKdV equation, i.e. an element of the ring $\mathcal{R}^{(1)}$ satisfying the linear PDE

$$\varphi_v = \varphi_{uuu} + 3\theta\theta_u\varphi + \frac{3}{2}\theta^2\varphi_u. \tag{3.14}$$

Consider the surface constructed from U, V, and A, B determined by Eqs. (3.1) and (3.2) with $\alpha = S = 0$ and symmetry 2φ,

$$A = 2\left(\frac{DU}{D\theta}\right)\varphi = -i\begin{pmatrix} 0 & \varphi \\ \varphi & 0 \end{pmatrix}, \tag{3.15}$$

$$\begin{aligned} B &= 2\left(\frac{DV}{D\theta}\right)\varphi \\ &= -i\begin{pmatrix} -\lambda\theta\varphi & \varphi_{uu} + i\lambda\varphi_u + \left(\frac{3}{2}\theta^2 - \lambda^2\right)\varphi \\ \varphi_{uu} - i\lambda\varphi_u + \left(\frac{3}{2}\theta^2 - \lambda^2\right)\varphi & \lambda\theta\varphi \end{pmatrix}. \end{aligned} \tag{3.16}$$

The first fundamental form is given by

$$\left[\varphi\, du + \left(\varphi_{uu} + \left(\frac{3}{2}\theta^2 - \lambda^2\right)\varphi\right) dv\right]^2 + \lambda^2(\theta^2\varphi^2 + \varphi_u^2)(dv)^2,$$

and the second one is given by

$$\begin{aligned} &\frac{\lambda}{\sqrt{\theta^2\varphi^2 + \varphi_u^2}}\Bigg[\theta\varphi^2\,(du)^2 \\ &\quad + \left[(\theta^2 - 2\lambda^2)\theta\varphi + 2\theta_u\varphi_u\right]\varphi\, du\, dv \\ &\quad + \Bigg[\left(\frac{1}{4}(\theta^2 - 2\lambda^2)^2 - 3\theta_u^2 - 4\theta\theta_{uu}\right)\theta\varphi^2 + (\theta_{uuu} - \lambda^2\theta_u)\varphi\varphi_u + \left(\theta(\theta^2 + \lambda^2) - \theta_{uu}\right)\varphi_u^2 \\ &\qquad + \left(\theta_u\varphi_u - (\theta^2 + \lambda^2)\theta\varphi\right)\varphi_{uu} + \theta\varphi_u\varphi_{uuu} - \theta\varphi\varphi_{uuuu}\Bigg](dv)^2\Bigg]. \end{aligned}$$

We have used Eqs. (3.13) and (3.14) to simplify the expression of the second fundamental form. The Gaussian and mean curvatures are given by

$$\begin{aligned} K = \frac{1}{(\theta^2\varphi^2 + \varphi_u^2)^2}\Big[&-(3\theta_u^2 + 4\theta\theta_{uu})\theta^2\varphi^2 + \left((\lambda^2 - \theta^2)\theta_u + \theta_{uuu}\right)\theta\varphi\varphi_u \\ &+ \left((\theta^2 + \lambda^2)\theta^2 - \theta_u^2 - \theta\theta_{uu}\right)\varphi_u^2 + \left(\theta_u\varphi_u - (\theta^2 + \lambda^2)\theta\varphi\right)\theta\varphi_{uu} \\ &+ \theta^2(\varphi_u\varphi_{uuu} - \varphi\varphi_{uuuu})\Big], \end{aligned}$$

$$\begin{aligned} H = \frac{1}{\lambda(\theta^2\varphi^2 + \varphi_u^2)^{3/2}}\Big[&\left((\theta^2 + \lambda^2)\theta^2 - 3\theta_u^2 - 4\theta\theta_{uu}\right)\theta\varphi^2 + \left((\lambda^2 - 3\theta^2)\theta_u + \theta_{uuu}\right)\varphi\varphi_u \\ &+ \left((\theta^2 + 2\lambda^2)\theta - \theta_{uu}\right)\varphi_u^2 + \left((\theta^2 - \lambda^2)\theta\varphi - \theta_u\varphi_u\right)\varphi_{uu} \\ &+ \theta(\varphi_{uu}^2 + \varphi_u\varphi_{uuu} - \varphi\varphi_{uuuu})\Big]. \end{aligned}$$

A moving frame on the surface is given by (3.7), with A, B given by (3.15) and (3.16), C given by

$$C = \frac{1}{\sqrt{\theta^2\varphi^2 + \varphi_u^2}}\begin{pmatrix} i\varphi_u & -\theta\varphi \\ \theta\varphi & -i\varphi_u \end{pmatrix},$$

and Φ a solution of Eqs. (2.2). Finally, the immersion function is given by Eq. (3.3) with $\alpha = S = 0$, $N = 1$, and $\varphi_1 = 2\varphi$.

4. Integrable Surfaces with an Invariant Geometric Characterization

The generalized immersion formula (3.3) can be used to generate large families of integrable surfaces. This is also its main limitation, for it is difficult to tell a priori whether a surface thus obtained possesses an invariant geometric characterization. The following list of integrable surfaces characterized by some intrinsic geometric property was presented by Bobenko in 1994 [**1**]:

(0) Minimal surfaces: $H = 0$.
(1) Surfaces with constant mean curvature: $H = \text{const}$.
(2) Surfaces with constant positive Gaussian curvature: $K = \text{const} > 0$.
(3) Surfaces with constant negative Gaussian curvature: $K = \text{const} < 0$.
(4) Surfaces with harmonic inverse mean curvature: $(1/H)_{z\bar{z}} = 0$, where z is a conformal coordinate of the first fundamental form and $\bar{z}$ is the complex conjugate of z.
(5) Bianchi surfaces: $(1/\sqrt{-K})_{uv} = 0$, where u and v are asymptotic coordinates.
(6) Bianchi surfaces of positive curvature: $(1/\sqrt{K})_{z\bar{z}} = 0$, where z is a conformal coordinate of the second fundamental form.
(7) Bonnet surfaces: members of a nontrivial family of isometric surfaces with the same principal curvatures.

REMARK 4.1. Minimal surfaces are associated with the Liouville equation [**6**],

$$\theta_{uu} + \theta_{vv} - e^{-2\theta} = 0,$$

which is linearizable by an explicit change of coordinates. Both minimal surfaces and the Liouville equation have been extensively studied in the literature and thus will not be considered here. Surfaces (1) and (2) of the list are associated with the sinh-Gordon equation, whereas surfaces (3) are associated with the sine-Gordon equation. Surfaces (4)–(6) are associated with certain integrable systems of PDEs. Bonnet surfaces (7) are associated with a certain nonlinear ODE of the second order.

The parallel deformations of the surfaces (1)–(7) were studied in [**7**]. These deformed surfaces are associated to the same integrable GMC equations as the corresponding undeformed surface, and admit an invariant geometric characterization. Indeed, if $\widetilde{K}$ and $\widetilde{H}$ denote the Gaussian and mean curvatures of a deformed surface, the deformations of surfaces (1)–(7) have the following properties:

$(\tilde{1})$, $(\tilde{2})$ The deformation of surfaces with constant mean curvature or constant positive Gaussian curvature are linear Weingarten surfaces[4]

$$c_1\widetilde{H} + c_2\widetilde{K} + 1 = 0, \quad c_1, c_2 \in \mathbb{R}, \tag{4.1}$$

such that $c_2 < c_1^2$.

$(\tilde{3})$ The deformation of surfaces with constant negative Gaussian curvature are linear Weingarten surfaces of the form (4.1) with $c_2 > c_1^2$. This case is discussed in [**6**] and [**3**].

[4]Recall that a Weingarten surface is defined by relation of the form $F(K, H) = 0$.

($\tilde{4}$) The deformation of surfaces with harmonic inverse mean curvature satisfy the invariant condition

$$\frac{\partial^2}{\partial z\partial\bar{z}}\left(\frac{1+\mu\widetilde{H}+\mu^2\widetilde{K}}{\widetilde{H}+2\mu\widetilde{K}}\right)=0,\quad \mu\in\mathbb{R}.$$

($\tilde{5}$) The deformation of Bianchi surfaces are characterized by the invariant condition

$$\frac{\partial 2}{\partial u\partial v}\sqrt{-\left(\frac{1}{\widetilde{\widetilde{K}}}+\mu\frac{\widetilde{H}}{\widetilde{\widetilde{K}}}+\mu^2\right)}=0,\quad \mu\in\mathbb{R}.$$

($\tilde{6}$) The deformation of Bianchi surfaces of positive curvature satisfy the invariant condition

$$\frac{\partial 2}{\partial z\partial\bar{z}}\sqrt{\frac{1}{\widetilde{\widetilde{K}}}+\mu\frac{\widetilde{H}}{\widetilde{\widetilde{K}}}+\mu^2}=0,\quad \mu\in\mathbb{R}.$$

($\tilde{7}$) The principal curvatures of the deformed surfaces in a Bonnet family coincide, but the surfaces are *not* isometric.

REMARK 4.2. It may be shown that the linear Weingarten surfaces of the form (4.1) with $c_2=c_1^2$ are associated with a single linearizable ODE. This implies that *all* linear Weingarten surfaces of the form (4.1) are integrable. There is some evidence suggesting that integrable Weingarten surfaces are necessarily linear. Note however that one can still construct nonlinear Weingarten surfaces associated to an integrable *reduction* of the GMC equations [**2**].

For some of the families of surfaces in the above list, it is convenient to employ a complex parametrization with variables $(z,\bar{z})$, where $z=u+iv$. This parametrization is completely equivalent to the real parametrization (u,v), according to the following elementary result [**7**].

LEMMA 4.3. *Let $U(u,v)$, $V(u,v)$, $A(u,v)$, $B(u,v)$ and $F(u,v)$ be $\mathcal{G}$-valued differentiable functions of $(u,v)\in\Omega\subset\mathbb{R}^2$. Let $\Phi(u,v)$ be a G-valued differentiable function of $(u,v)\in\Omega$. Let $z=u+iv$. The functions U, V, A, B, Φ and F satisfy Eqs.* (2.1)–(2.4) *if and only if the functions*

$$\mathcal{U}=\frac{1}{2}(U-iV),\quad \mathcal{V}=\frac{1}{2}(U+iV),\quad \mathcal{A}=\frac{1}{2}(A-iB),\quad \mathcal{B}=\frac{1}{2}(A+iB),$$

and the functions Φ and F, satisfy the equations

$$\mathcal{U}_{\bar{z}}-\mathcal{V}_z+[\mathcal{U},\mathcal{V}]=0,\tag{4.2}$$

$$\Phi_z=\mathcal{U}\Phi,\quad \Phi_{\bar{z}}=\mathcal{V}\Phi,\tag{4.3}$$

$$\mathcal{A}_{\bar{z}}-\mathcal{B}_z+[\mathcal{A},\mathcal{V}]+[\mathcal{U},\mathcal{B}]=0,\tag{4.4}$$

$$F_z=\Phi^{-1}\mathcal{A}\Phi,\quad F_{\bar{z}}=\Phi^{-1}\mathcal{B}\Phi\,.\tag{4.5}$$

Moreover, Eqs. (3.1) *and* (3.2) *hold with U, V, A, B, u, v replaced by $\mathcal{U}$, $\mathcal{V}$, $\mathcal{A}$, $\mathcal{B}$, z, $\bar{z}$.*

EXAMPLE 4.4. Let us examine the family of surfaces with harmonic inverse mean curvature and their corresponding parallel deformations. Let

$$\mathcal{U}=\begin{pmatrix}\frac{1}{2}\theta_{1z} & -e^{-\theta_1}\theta_2\\ \frac{(1-2i\lambda\bar{h})e^{\theta_1}}{2(1+2i\lambda h)(h+\bar{h})} & -\frac{1}{2}\theta_{1z}\end{pmatrix},\quad \mathcal{V}=\begin{pmatrix}-\frac{1}{2}\theta_{1\bar{z}} & -\frac{(1+2i\lambda h)\,e^{\theta_1}}{2(1-2i\lambda\bar{h})(h+\bar{h})}\\ e^{-\theta_1}\bar{\theta}_2 & \frac{1}{2}\theta_{1\bar{z}}\end{pmatrix},\tag{4.6}$$

where h is an arbitrary holomorphic function of z, while the real-valued scalar function $\theta_1(z,\bar z)$ and the complex-valued scalar function $\theta_2(z,\bar z)$ satisfy the system of integrable nonlinear PDEs

$$\theta_{1z\bar z} - e^{-2\theta_1}|\theta_2|^2 + \frac{e^{2\theta_1}}{4(h+\bar h)^2} = 0, \quad \theta_{2\bar z} = -\frac{e^{2\theta_1}\,h_z}{2(h+\bar h)^2}.$$

(a) The immersion function constructed from $\mathcal{U}$, $\mathcal{V}$, $\mathcal{A}=\mathcal{U}_\lambda$, $\mathcal{B}=\mathcal{V}_\lambda$ describes a conformally parametrized surface with harmonic inverse mean curvature. Indeed, the mean curvature is given by

$$H = \frac{2}{h^\lambda + \overline{h^\lambda}}, \quad h^\lambda = \frac{h}{1+2i\lambda h}, \tag{4.7}$$

and the first and second fundamental forms are

$$e^{2\theta_1^\lambda}\,dz\,d\bar z, \quad \theta_2^\lambda (dz)^2 + \frac{H}{2}e^{2\theta_1^\lambda}\,dz\,d\bar z + \overline{\theta_2^\lambda}(d\bar z)^2,$$

where

$$e^{2\theta_1^\lambda} = \frac{e^{2\theta_1}}{(1-2i\lambda\bar h)^2(1+2i\lambda h)^2}, \quad \theta_2^\lambda = \frac{\theta_2}{(1+2i\lambda h)^2}. \tag{4.8}$$

The Gaussian curvature is given by

$$K = \left((h+\bar h)^{-2} - 4|\theta_2|^2 e^{-4\theta_1}\right)(1-2i\lambda\bar h)^2(1+2i\lambda h)^2. \tag{4.9}$$

The immersion function F is given by the Sym formula (3.4) with $\alpha=1$ and Φ a solution of Eqs. (4.3) for $\mathcal{U}$ and $\mathcal{V}$ given by (4.6).

(b) The immersion function constructed from $\mathcal{U}$, $\mathcal{V}$, and

$$\tilde{\mathcal{A}} = \frac{\partial\mathcal{U}}{\partial\lambda} - i\mu[\sigma_3,\mathcal{U}], \quad \tilde{\mathcal{B}} = \frac{\partial\mathcal{V}}{\partial\lambda} - i\mu[\sigma_3,\mathcal{V}], \quad \mu\in\mathbb{R},$$

describes a surface characterized by the invariant condition

$$\frac{\partial 2}{\partial z\partial\bar z}\left(\frac{1+\mu\widetilde{H}+\mu^2\widetilde{K}}{\widetilde{H}+2\mu\widetilde{K}}\right) = 0.$$

The first and second fundamental forms are

$$\mu\theta_2^\lambda(\mu H-2)(dz)^2 + \left(e^{2\theta_1^\lambda}\left(1-\frac{\mu H}{2}\right)^2 + 4\mu^2|\theta_2^\lambda|^2 e^{-2\theta_1^\lambda}\right)dz\,d\bar z + \mu\overline{\theta_2^\lambda}(\mu H-2)(d\bar z)^2,$$

$$\theta_2^\lambda(1-\mu H)(dz)^2 + \left(\frac{H}{2}e^{2\theta_1^\lambda}\left(1-\frac{\mu H}{2}\right) - 4\mu|\theta_2^\lambda|^2 e^{-2\theta_1^\lambda}\right)dz\,d\bar z + \overline{\theta_2^\lambda}(1-\mu H)(d\bar z)^2,$$

with H given by (4.7), and $e^{2\theta_1^\lambda}$, θ_2^λ given by (4.8). The Gaussian and mean curvatures are obtained from Eqs. (4.7) and (4.9) by means of the classical formulae [**8**]

$$\widetilde{K} = \frac{K}{1-\mu H+\mu^2 K}, \quad \widetilde{H} = \frac{H-2\mu K}{1-\mu H+\mu^2 K}. \tag{4.10}$$

The immersion function $\widetilde{F}$ is given by Eq. (3.3) for $N=2$, with α and Φ as in Case (a), $S=-i\mu\sigma_3$, and $\varphi_1=\varphi_2=0$.

Note that the parametrization is not conformal unless $\mu=0$.

References

[1] A. I. Bobenko, *Surfaces in terms of 2 by 2 matrices. Old and new integrable cases*, Harmonic Maps and Integrable Systems (A. P. Fordy and J. C. Wood, eds.), Aspects Math., vol. E23, Vieweg, Braunschweig, 1994, pp. 83–127.

[2] Ö. Ceyhan, A. S. Fokas, and M. Gürses, *Deformations of surfaces associated with integrable Gauss–Mainardi–Codazzi equations*, J. Math. Phys. **41** (2000), no. 4, 2251–2270.

[3] J. Cieśliński, *A generalized formula for integrable classes of surfaces in Lie algebras*, J. Math. Phys. **38** (1997), 4255–4272.

[4] R. K. Dodd, *Soliton immersions*, Comm. Math. Phys. **197** (1998), 641–665.

[5] A. S. Fokas, *Symmetries and integrability*, Stud. Appl. Math. **77** (1987), 253–299.

[6] A. S. Fokas and I. M. Gel'fand, *Surfaces on Lie groups, on Lie algebras and their integrability*, Comm. Math. Phys. **177** (1996), 203–220.

[7] A. S. Fokas, I. M. Gel'fand, F. Finkel, and Q. M. Liu, *A formula for constructing infinitely many surfaces on Lie algebras and integrable equations*, Preprint, Selecta Math. (N.S.) (in press).

[8] A. R. Forsyth, *Lectures on the differential geometry of curves and surfaces*, Cambridge, 1912.

[9] D. Levi and A. Sym, *Integrable systems describing surfaces of nonconstant curvature*, Phys. Lett. A **149** (1990), 381–387.

[10] Q. M. Liu, *Generalized conditional symmetries, asymptotic integrability and integrable surfaces*, Ph.D. Thesis, Clarkson University, 1995.

[11] P. J. Olver, *Applications of Lie groups to differential equations*, Springer-Verlag, Berlin-New York, 1993.

[12] A. Sym, *Soliton surfaces*, Lett. Nuovo Cimen. **33** (1982), 394–400.

[13] ———, *Soliton surfaces and their applications* (*soliton geometry from spectral problems*), Geometric Aspects of the Einstein Equations and Integrable Systems (Scheveningen, 1984), (R. Martini, ed.), Lecture Notes in Phys., vol. 239, Springer-Verlag, Berlin-New York, 1985, pp. 154–231.

[14] J. Tafel, *Surfaces in $\mathbb{R}^3$ with prescribed curvature*, J. Geom. Phys. **17** (1995), 381–390.

Department of Mathematics, Imperial College, London SW7 2BZ, UK.
E-mail address: f.finkel@ic.ac.uk

Department of Mathematics, Imperial College, London SW7 2BZ, United Kingdom. Also, Department of Mathematics and Computer Science, Clarkson University, Potsdam, NY 13699, USA.
E-mail address: a.fokas@ic.ac.uk

Centre de Recherches Mathématiques
CRM Proceedings and Lecture Notes
Volume **29**, 2001

Difficulties With the SDiff(2) Toda Equation

J. D. Finley, III

Abstract. The SDiff(2) Toda equation appears in several fields of physics. Its algebra of symmetries is the area-preserving diffeomorphisms of a 2-surface, isomorphic to the algebra of Poisson brackets generated by Hamiltonian vector fields in 2 variables. These facts have allowed various researchers to create "zero-curvature representations" and even to create a formal presentation of a general solution; however, none of this seems particularly useful in the search for new, explicit solutions.

A new attempt to find a more direct method to "buy" new solutions from old ones begins with the Bäcklund transformation for the (2-dimensional) Toda lattice built over the Cartan matrix for the Lie algebra $\mathbf{A}_n$ or $\mathbf{A}_n^{(1)}$, and considers the limit as $n \to \infty$. While this limit gives the desired (second-order) pde's, the associated limit of the (first-order) pde's that define the Bäcklund transformation gives only trivial results. Reasons for this are analyzed, but a satisfactory solution has yet to be found. As well the known "Lax pairs" are analyzed in this language, suggesting that they are not useful for this purpose since the representations of SDiff(2) involved are over the defining manifold rather than over (infinite-dimensional) tangent manifolds in additional (pseudopotential) variables.

1. The Equation Itself, and the Goals

An equation of interest in several fields of theoretical physics is the SDiff(2) Toda equation, for one unknown function of three variables, with differentiation shown by a subscript:

$$u_{,z\tilde{z}} = e^{u_{,tt}} \iff v_{,z\tilde{z}} = (e^v)_{,tt}, \quad v \equiv u_{,tt}. \tag{1.1}$$

One derivation was given by myself and Charles Boyer [**BF**] in 1982. It begins with the Plebański equation for general self-dual vacuum solutions of the Einstein field equations. All such solutions that admit at least one rotational Killing vector may be determined by solutions of this equation. The angular variable associated with that Killing vector combines with the other three above to constitute a local

2000 *Mathematics Subject Classification.* Primary: 58G37, 35Q75; Secondary: 35Q58.

This work has been greatly facilitated by many discussions with John K. McIver. In addition some conversations with Wolfgang Schief were helpful.

This is the final form of the paper.

coordinate chart for the 4-dimensional space in question. The name I have chosen has also been used by Mikhail Saveliev [**KS**] and by Kanehisa Takasaki and T. Takebe [**TT, TT2**]. It emphasizes the resemblance to the usual Toda lattice equations, which involve a family of functions, u^a, of two variables:

$$u^a_{,xy} = e^{K^a_b u^b} \iff v^a_{,xy} = K^a_b e^{v^b}, \quad v^a \equiv K^a_b u^b, \tag{1.2}$$

where K^a_b is the Cartan matrix for some appropriate Lie algebra, which is then a symmetry algebra for the equations.

The second purpose of this name is then to remind us that, in some appropriate limit, the symmetry algebra for our equation is SDiff(2), the algebra for the group of area-preserving diffeomorphisms of a 2-surface. In order that such a limit exist, reasonable choices for the original symmetry algebra would be $\mathbf{A}_n$, or $\mathbf{A}_\infty$, or perhaps $\mathbf{A}_n^{(1)}$; here we will simply refer to the algebra as $\mathcal{G}$, deferring an actual choice. When one begins to write the system as a *first-order system*, to effect the prolongation, there is still considerable gauge freedom remaining. We therefore use that freedom, in the manner given by Leznov and Saveliev [**LS**], to arbitrarily divide the (original) unknown functions into two parts:

$$\begin{aligned} u^a &\equiv m^a + n^a, \quad Z^a = m^a_{,x}, \quad W^a = -n^a_{,y}, \\ a^a &\equiv K^a_b m^b, \quad b^a \equiv K^a_b n^b \implies v^b = a^a + b^a. \end{aligned} \tag{1.3}$$

Since many people have already struggled with this equation, toward various goals, it seems important to say what I want from it, and/or why I am not sufficiently happy with the results that already exist? I want a "solution generator", which will "buy" new solutions from old ones. I would say that I want a Bäcklund transform, [**PR**] although the more general description above would satisfy my desires if that were all that was available.

2. Prolongation Structures for the (2-Dimensional) Toda Lattice

Our (geometric) approach to finding new solutions, via nonlocal symmetries, begins [**FM**] by thinking of a kth order partial differential equation as a variety, Y, of a finite jet bundle, $J^{(k)}(M, N)$, with M the independent- and N the dependent-variables, which is then prolonged to a fiber space (over $J^{(\infty)}$) with fibers W, where vertical flows map solution spaces of one pde into another pde, which has as its dependent variables the quantities w^A, that coordinatize the fibers. The compatibility conditions for such flows to exist are *"zero-curvature conditions"* for the fiber space, required to vanish when restricted to the infinite prolongation, $Y^\infty \times W$, of the fiber space over the original pde. We may express these conditions by first lifting $\{\partial_x, \partial_y\}$, a basis for the tangent vectors over M, to the total derivative operators, $\{D_x, D_y\}$ over the infinite jet, which still commute, and then prolonging them further into the fibers W.

This act requires the addition of some vertical vector fields, $\mathbf{F} = \sum F^A(\partial/\partial w^A)$ and $\mathbf{G} = \sum G^A(\partial/\partial w^A)$. As the total derivatives are of the form $D_x = \partial_x + u_x \partial_u + u_{xx}\partial_{u_x} + \ldots$, the additional terms denoted by, for instance, $\mathbf{F}$, should have the same interpretation with respect to the fiber variables, i.e., we eventually want to think of the w^A also as functions of the independent variables that will satisfy some pde's. This means that the satisfaction of the prolongation requirements should allow us to interpret F^A as w^A_x and G^A as w^A_y, where, for instance F^A originally are functions of all the jet variables and the fiber variables. Such an interpretation of course requires that the prolonged total derivatives should commute, at least on the restricted manifold:

$$0 = [D_x + \mathbf{F}, D_y + \mathbf{G}]\Big|_{Y^\infty \times W} = \{\overline{D}_x(G^C) - \overline{D}_y(F^C)\}\frac{\partial}{\partial w^C} + [\mathbf{F}, \mathbf{G}], \tag{2.1}$$

where the overbar indicates the restriction in question. As the construction gives **F** and **G** the "form" of the components of a connection, these equations are referred to as "zero-curvature" requirements for non local symmetries for the original pde. This is a slight generalization of the usual approach since they are still only elements of an **abstract** Lie algebra of vector fields, with neither coordinates, nor even the number of those coordinates yet determined. Nonetheless, at least in two variables, the general solution for **F** and **G** describes all possible Bäcklund transformations for this pde.

Solutions to the prolongation equations, for **F** and **G**, have two requisites, namely the determination of the (smallest) Lie algebra which satisfies the requirements for the commutators of these connections, and an acceptable decision for the number of pseudopotentials to be involved in a realization of this algebra. The constraints in question are induced by the requirement that the restricted zero-curvature equation vanishes. For now, we follow the conventional choice for a Lie algebra that satisfies the constraints, namely the same Lie algebra that defines the system of pde's itself. If one then also follows the thoughts of Leznov and Saveliev, who insist that **F** and **G** should be a linear combination of elements chosen from the Cartan subalgebra, $\mathcal{G}_0$, and its simple roots, $\mathcal{G}_{\pm 1}$, spanned by $\{\mathbf{h}_a\}_1^n$, $\{\mathbf{e}_b\}_1^n$ and $\{\mathbf{f}_c\}_1^n$, respectively, the unique solution is the following:

$$\begin{aligned} \mathbf{F} &= Z^a \mathbf{h}_a + e^{n^b(\mathrm{ad}\mathbf{h}_b)} \sum_c \mathbf{e}_c = Z^a \mathbf{h}_a + e^{b^c} \mathbf{e}_c, \\ \mathbf{G} &= W^a \mathbf{h}_a + e^{-m^b(\mathrm{ad}\mathbf{h}_b)} \sum_c \mathbf{f}_c = W^a \mathbf{h}_a + e^{a^c} \mathbf{f}_c. \end{aligned} \tag{2.2}$$

One then requires a choice for the realization. A very reasonable approach to the number of pseudopotentials takes it as the same number as the original dependent variables v^a, thereby making it reasonably simple to hope to find an auto-Bäcklund transformation. A choice of presentation of this standard solution may then be written:

$$\begin{aligned} \mathbf{F} &= \{K_b^a Z^b - e^{w^a + b^a} + e^{w^{a-1} + b^{a-1}}\}\partial_{w^a}, \\ \mathbf{G} &= \{K_b^a W^b + e^{-w^a + a^a} - e^{-w^{a+1} + a^{a+1}}\}\partial_{w^a}. \end{aligned} \tag{2.3}$$

The first-order pde's for the explicit Bäcklund transformation are then simply

$$\{w^a - a^a\}_{,x} = -e^{w^a + b^a} + e^{w^{a-1} + b^{a-1}}, \quad \{w^a + b^a\}_{,y} = e^{-w^a + a^a} - e^{-w^{a+1} + a^{a+1}}. \tag{2.4}$$

The zero-curvature conditions, where we subtract the extended derivatives of **F** and **G**, are then exactly the original Toda equations, Eqs. (1.2), in the variables v^a, as desired. Moreover, if instead one adds the two derivatives, and inserts the form for $b^a_{,xy}$ from the Toda equations, the result is a system of pde's that must be satisfied by the pseudopotential variables:

$$w^a_{,xy} = a^a_{,xy} - e^{v^a} + e^{v^{a-1}} + e^{\ell^a} - e^{\ell^{a-1}}, \tag{2.5}$$

$$\text{where } \ell^a_{,xy} \equiv w^a_{,xy} - w^{a+1}_{,xy} + a^{a+1}_{,xy} + b^a_{,xy}.$$

The Bäcklund transformation so created does indeed determine a mapping between the pde's satisfied by the v^a's and the pde's satisfied by the w^a's. However, it is surely **not** in an optimal form since the new pde's contain not only the new variables but also the old ones. This difficulty can be resolved by using, instead, the new variables $\{\ell^a\}_1^{n-1}$ that occur here in a very natural way, although there are in fact only $n-1$ of them. (This is not a problem in the limit as n goes off to infinity.) We find the very simple result that these new variables must also satisfy the Toda equations, but for $\mathbf{sl}(n)$ instead of $\mathbf{sl}(n+1)$:

$$\ell^a_{,xy} = 2e^{\ell^a} - e^{\ell^{a-1}} - e^{\ell^{a+1}} = (\underset{n-1}{K})^a_b e^{\ell^b}, \quad a, b = 1, \ldots, n-1. \tag{2.6}$$

3. Continuous Limit of the Toda Lattice Equations

We now consider limits on the Bäcklund transformation equations, as we change from discrete indices to functions of a continuous variable. We use the system of functions v^a to define a new function that also depends on a continuous variable, t, which varies from, say, 0 to β. We superpose on this domain a lattice of n points, a distance δ apart, and then fill in the space between the lattice points by taking the limit as $n \to \infty$, with β fixed, which is the same as taking the limit as $\delta \to 0$:

$$(3.1)\quad V(z,\tilde{z},t)\big|_{t=a\delta} \equiv v^a(z/\delta, \tilde{z}/\delta), \quad a = 1, \ldots, n, \quad \delta = \beta/(n-1),$$
$$\Longrightarrow V(z,\tilde{z},t) \equiv \lim_{\delta\to 0} v^{[t/\delta]}(z/\delta, \tilde{z}\delta),$$

where the square brackets indicate the integer part of the quotient within them. Following Park's example, [**P**] we have re-scaled the original two independent variables, to have appropriate differences of the exponentials of the v^a's to create second derivatives, with respect to t. However, the u^b's need a scaling of their own, to create their second derivatives:

$$U(z,\tilde{z},t)\big|_{t=a\delta} \equiv \delta^2\{u^a(z/\delta, \tilde{z}/\delta)\}. \tag{3.2}$$

We suppose the new functions are sufficiently continuous to expand in a Taylor series about the point $t = a\delta$. Suppressing the dependence on $z \equiv x\delta$ and $\tilde{z} \equiv y\delta$, we may write

$$v^{a\pm 1} \to V(t\pm\delta) = V(t) \pm \delta\{V_{,t}(t)\} + \frac{1}{2}\delta^2\{V_{,tt}(t)\} + O(\delta^3), \tag{3.3}$$

which give us the desired differences, including $K^a_b v^b \equiv -v^{a+1} + 2v^a - v^{a-1}$:

$$(3.4)\quad \begin{aligned} v^{a+1} - v^a &\to V(t+\delta) - V(t) = \delta\{V_{,t}(t)\} + \frac{1}{2}\delta^2\{V_{,tt}(t)\} + O(\delta^3),\\ v^a - v^{a-1} &\to V(t) - V(t-\delta) = \delta\{V_{,t}(t)\} - \frac{1}{2}\delta^2\{V_{,tt}(t)\} + O(\delta^3),\\ -K^a_b v^b &\to V(t+\delta) - 2V(t) + V(t-\delta) = \delta^2\{V_{,tt}(t)\} + O(\delta^3). \end{aligned}$$

We may now take limits of the Toda equations, which give the desired, and expected, results:

$$U_{,z\tilde{z}} = e^{-U_{,tt}}, \quad V_{,z\tilde{z}} = -\partial_t^2 e^V. \tag{3.5}$$

However, we also want to take the same limits on the prolongation equations themselves. Rewriting the prolongation equations in terms of the w^a's in Eqs. (2.4), we have the equations and their integrability conditions:

Looking at these equations, we *agree* to treat the gauged parts, a^a and b^a, the same as their sum v^a, and also the pseudopotentials w^a, which gives us the following limiting forms:

$$\begin{gathered} W(z,\tilde{z},t)\big|_{t=a\delta} \equiv w^a(z/\delta,\tilde{z}/\delta), \\ A(z,\tilde{z},t)\big|_{t=a\delta} \equiv a^a(z/\delta,\tilde{z}/\delta), \quad B(z,\tilde{z},t)\big|_{t=a\delta} \equiv b^a(z/\delta,\tilde{z}/\delta), \\ \Longrightarrow (W-A)_{,z} = -\partial_t e^{W+B}, \quad (W+B)_{,\tilde{z}} = -\partial_t e^{-(W-A)}, \end{gathered} \tag{3.6}$$

where this last line is then the limit of the original Bäcklund transformation equations. Their integrability conditions are then

$$\begin{aligned} (W-A)_{,z\tilde{z}} &= -\partial_t\{e^{W+B}(-\partial_t e^{-(W-A)})\} = -\partial_t\{e^V\partial_t(W-A)\}, \\ (W+B)_{,\tilde{z}z} &= -\partial_t\{e^{-(W-A)}\partial_t e^{(W+B)}\} \\ &= -\partial_t\{e^{-(W-A)}e^{(W+B)}\partial_t(W+B)\} \\ &= -\partial_t\{e^V\partial_t(W+B)\}, \end{aligned} \tag{3.7}$$

where the sum of A and B that appears creates the V shown. Adding and subtracting these two equations we finally acquire

$$V_{,z\tilde{z}} = -\partial_t e^V\partial_t V = -\partial_t^2 e^V, \quad \text{and} \quad (2W+B-A)_{,z\tilde{z}} = -\partial_t\{e^V\partial_t(2W+B-A)\}. \tag{3.8}$$

The first of these equations is of course what we expect; however, the second is surely not what we expect/want. In particular, since $B-A$ is independent of $B+A=V$, then the second equation is simply *linear* in the unknown function, $2W+B-A$. Of course in the symmetric gauge choice, where $a^a = \frac{1}{2}v^a = b^a$, this would just be $2W$. [It is amusing that this pair is just the system LeBrun finds to determine his "weak heavens", but then we would have to treat this as a system, and again try to determine its Bäcklund transformation.]

In the discrete case, instead, a re-formulation of this equation turned out to be the Toda lattice equations again, for the new dependent variables, ℓ^a. In the current case, however, the limit of that particular combination of pseudopotentials and original dependent variables that satisfies the same pde as the original variables, namely ℓ^a, is the same as that of v^a:

$$L(t) = W(t) + B(t) - \{W(t) - A(t)\} = V(t). \tag{3.9}$$

The two satisfy the same equations, **but**, unfortunately they are not different since the limit $W(t)$ of the pseudopotential variables simply cancels out of the combination. A way of looking at "why" this happens is to note that for the finite lattice case the difference was $w^a - w^{a+1}$. What can one do about this, if anything?

4. Some other Work Toward Finding a Solution Generator

Various other groups have considered this equation, and even provided interesting results. Foremost in my mind is that provided by Mikhail Saveliev and A. M. Vershik [**SV, SV2**]. They used this equation as a beginning place to guide them toward a general theory of continuum Lie algebras. In fact they published several papers explaining the form of the "general solution" to the equation as determined by choices of initial conditions. Unfortunately, from my viewpoint, this form is very complicated, and not at all workable for obtaining manageable and interesting, specific, new solutions.

In a different direction, in Britain, R. S. Ward [**W, W2, W3**], and separately, in Japan, K. Takasaki and various coworkers [**TT, TT2**] have created objects they refer to as Lax pairs for this equation, although in the process they have changed the usual matrix/vector field commutator in the Lax equation into a Poisson bracket for functions. As well, their approach has generated an infinite hierarchy of associated equations—in the spirit of the KP hierarchy of equations; nonetheless their methods also do not appear to me to allow for the determination of interesting, specific, new solutions. I have some specific thoughts on their approach that I will describe below.

There are also some groups that have, instead, been looking directly at the properties of the full Plebański equation, before the reduction—via one rotational Killing vector—to our equation was made. Obviously they could also be useful for the goals I have in mind. Early work was by Boyer and Plebański [**BP, BP2**], looking for nonlinear superposition principles underlying the solution space for the Plebański equation. Their viewpoint on the equation amounts to lifting curves of solutions into an infinite jet bundle. It may be that they are at least very close to what I want. They give an algorithm that could perhaps be followed to generate new solutions from old. For a case that begins with simple *pp*-waves, they actually do it. Can it be generalized? I don't know, but we are working on that.

Then Boyer and Pavel Winternitz [**BW**] found the complete set of 1-, 2-, and 3-dimensional subgroups of the group of symmetries of the Plebański equation. The reduction to the SDiff(2) Toda equation is clear and other lower-dimensional reductions exist; however, this approach has not yet been used to acquire previously-unknown solutions.

More recently various twistorial researchers have considered these problems from both sides. Firstly, they have created quite a clever approach to describing a Poisson-bracket based Lax pair for Grant's form [**G**] of the Plebański equation, which allows Ian Strachan [**S**] to describe an infinite hierarchy of symmetries of this equation in a very straightforward way. On the other end of the complications they have also been rather involved in finding classes of solutions to the equation when it has been reduced to a system of ode's by insisting on a $\mathbf{sl}(2)$ group of symmetries [**T**].

5. (ZS) Zero-Curvature Relations, (EW) Prolongation Structures, and the (T/W) Poisson Bracket

It is worth giving some explanation of why the currently-known "Lax pairs" are interesting, but are not useful toward the goals of finding specific new solutions. Takasaki's version of the "Lax pair" for this equation is written in terms of a pair of functions of 4 variables, with the usual commutator being replaced by a Poisson bracket in two of those variables:

$$g_{,\tilde{z}} - f_{,z} + \{f, g\} = 0. \tag{5.1}$$

The Poisson bracket uses the (phase-space) variables t and $p \equiv \ln \lambda$. Three of these variables, $\{z, \tilde{z}, t\}$, come from the original pde, and thus should be expected, although obviously they have not been all treated in the same manner. However, the fourth is an extra variable that has been added, although it is not a pseudopotential, or fiber, variable; Takasaki refers to λ as a "spectral parameter". In fact, it ends up playing two different roles. Since it is a "parameter", instead of an independent variable, they feel justified in specifying in advance the dependence of f and g on λ. This creates a polynomial in λ, so that the vanishing of the commutator engenders three different equations, which must then be combined to re-create the original pde. However, quite a different role is its use in "modifying" the usual vector-field (or matrix) commutator, in the zero-curvature form, to a Poisson bracket in t and p. While this is perhaps unexpected, from a physical point of view, the purpose of our pde is to determine a potential for a spacetime with a metric that always admits one Killing vector. Therefore the metric itself will be independent of the fourth coordinate, and it is reasonable that the potential will have rather simple dependence on that variable.

I find it informative to put their approach back into the EW approach already outlined above, by the use of Hamiltonian vector fields generated by their functions. Recall that an arbitrary Hamiltonian vector field lives on the tangent space to some appropriate phase space, is generated by some (non constant) function of the coordinates on that space, say $f = f(q,p) \equiv f(q^i)$, and has the following standard properties:

$$\begin{gathered} \mathcal{H}_f \equiv f_{,q}\partial_p - f_{,p}\partial_q = J^{ij} f_{,i}\partial_{,j}, \\ \mathcal{H}_{af+g} = a\mathcal{H}_f + \mathcal{H}_g, \\ \mathcal{H}_f(g) = \{f,g\}, \quad [\mathcal{H}_f,\mathcal{H}_g] = \mathcal{H}_{\{f,g\}}. \end{gathered} \tag{5.2}$$

Here J^{ij} are the components of the symplectic 2-form, $\mathbf{J}$, for that phase space; in our two dimensions it may be taken as the (2-dimensional) Levi–Civita symbol, ε^{ij}. We use $\mathcal{H}_f$ for our earlier $\mathbf{F}$ and $\mathcal{H}_g$ for $\mathbf{G}$, and we choose t and p for our phase-space variables. We may then re-formulate their language in our own:

$$\begin{aligned} [D_x + \mathcal{H}_f, D_y + \mathcal{H}_g] &= \mathcal{H}_{g_{,x}} - \mathcal{H}_{f_{,y}} + [\mathcal{H}_f,\mathcal{H}_g] \\ &= \mathcal{H}_{g_{,x}} - \mathcal{H}_{f_{,y}} + \mathcal{H}_{\{f,g\}} = \mathcal{H}_{g_{,x} - f_{,y} + \{f,g\}}. \end{aligned} \tag{5.3}$$

As our algebra has been divided out by the set of all constant functions, we may claim that a Hamiltonian vector field vanishes if and only if its generating function vanishes. Therefore, the requirement of "zero-curvature" is the same as the vanishing of the generating function for the (single) Hamiltonian vector field on the far right-hand side of Eq. (5.3):

$$[D_x + \mathcal{H}_f, D_y + \mathcal{H}_g] = 0 \implies g_{,x} - f_{,y} + \{f,g\} = 0. \tag{5.4}$$

Takasaki's explicit λ-dependence, for his f and g, is the following, where the other quantities, $\{a, \bar{a}, b, \bar{b}\}$, depend on the remaining three variables:

$$f = -\lambda a - b, \quad g = -\lambda\bar{a} - \bar{b}. \tag{5.5}$$

The three equations that result do re-create our original SDiff(2) Toda equation.

The very important difference between this and the desired prolongation equations is the lack of fiber variables, to be used to generate another pde. We recall that the zero-curvature equations restrict the prolongations $\mathbf{F}$ and $\mathbf{G}$ to lie in some sub-algebra of all the vector fields over a fiber W. If we now take $\{\mathbf{A}^\mu\}$ as a set of vector fields that constitute a basis for that algebra, we could write out

$$F^A \frac{\partial}{\partial w^A} = \mathbf{F} \equiv \mathcal{F}_\mu \mathbf{A}^\mu, \tag{5.6}$$

where the $\mathcal{F}_\mu$ depend on the original jet variables, only. The dependence on the pseudopotentials is in the basis vector-fields $\mathbf{A}^\mu$. (Recall Eqs. (2.2).) In Takasaki's analogues of a "Lax pair", the $\mathbf{A}^\mu$ are vector fields directly over the independent variable t and the spectral parameter. We need, instead to create a related approach, but with a realization of SDiff(2) "lifted" to some additional (pseudopotential) variables. We had hoped this limiting procedure might accomplish this; unfortunately our efforts so far, as already presented, have failed.

References

[BF] C. P. Boyer and J. D. Finley, III, *Killing vectors in self-dual, Euclidean Einstein spaces*, J. Math. Phys. **23** (1982), 1126–1130.

[BP] C. P. Boyer and J. F. Plebański, *An infinite hierarchy of conservation laws and nonlinear superposition principles for self-dual Einstein spaces*, J. Math. Phys. **26** (1985), 229–234.

[BP2] ———, *Heavens and their integral manifolds*, J. Math. Phys. **18** (1977), 1022–1031.

[BW] C. P. Boyer and P. Winternitz, *Symmetries of the self-dual Einstein equations.* I. *The infinite-dimensional symmetry group and its low-dimensional subgroups*, J. Math. Phys. **30** (1989), 1081–1094.

[FM] J. D. Finley, III and John K. McIver, *Prolongations to higher jets of Estabrook–Wahlquist coverings for* PDE*'s*, Acta Appl. Math. **32** (1993), 197–225.

[G] James D. E. Grant, *On self-dual gravity*, Phys. Rev. D **48** (1993), 2606–2612.

[KS] R. M. Kashaev, M. V. Saveliev, S. A. Savelieva, and A. M. Vershik, *On nonlinear equations associated with Lie algebras of diffeomorphism groups of two-dimensional manifolds*, Ideas and Methods in Mathematical Analysis, Stochastics, and Applications (S. Albeverio, J. E. Fenstad, H. Holden and T. Lindstrom, eds.), vol. 1, Cambridge University Press, Cambridge, UK, 1992, pp. 295–307.

[LS] A. N. Leznov and M. V. Saveliev, *Group-theoretical methods for integration of nonlinear dynamical systems*, Birkhäuser Verlag, Basel, 1992.

[P] Q-H. Park, *Extended conformal symmetries in real heavens*, Phys. Lett. B **236** (1990), 429–32.

[PR] F. Pirani, D. Robinson, and W. Shadwick, *Local jet bundle formulation of Bäcklund transformations*, Math. Phys. Stud., vol. 1, Reidel, Dordrecht, 1979.

[SV] M. V. Saveliev and A. M. Vershik, *Continuum analogues of contragredient Lie algebras*, Comm. Math. Phys. **126** (1989), 367–378.

[SV2] ———, *New examples of continuum graded Lie algebras*, Phys. Lett. A **143** (1990), 121–128.

[S] I. A. B. Strachan, *The symmetry structure of the anti-self-dual Einstein hierarchy*, J. Math. Phys. **36** (1995), 3566–3573.

[TT] K. Takasaki and T. Takebe, SDiff(2) Toda equation—hierarchy, tau function and symmetries, Lett. Math. Phys. **23** (1991), 205–214.

[T] K. P. Tod, *Scalar-flat Kähler and hyper-Kähler metrics from Painlevé* III, Classical Quantum Gravity **12** (1995), 1535–1547, is an example with other references therein.

[TT2] K. Ueno and K. Takasaki, *Toda lattice hierarchy*, Adv. Stud. Pure Math. **4** (1984), 1–95.

[W] R. S. Ward, *Multi-dimensional integrable systems*, Springer Lecture Notes in Physics **280** (1987).

[W2] ———, *Linearization of the* SU(∞) Nahm equations, Phys. Lett. B **234** (1990), 81–4.

[W3] ———, *Infinite-dimensional gauge groups and special nonlinear gravitons*, J. Geom. Phys. **8** (1992), 317–25.

Department of Physics and Astronomy, Univ. of New Mexico, Albuquerque, NM 87131, USA.

E-mail address: `finley@tagore.phys.unm.edu`

Centre de Recherches Mathématiques
CRM Proceedings and Lecture Notes
Volume **29**, 2001

Superposition Formulas Based on Nonprimitive Group Action

M. Havlíček, S. Pošta and P. Winternitz

ABSTRACT. The general solution of a certain system of nonlinear differential equations is given in terms of a finite number of particular solutions. The system of equations and its superposition formula are based on a nonlinear and nonprimitive action of the Lie group $\mathrm{SL}(N,\mathbb{C})$ on a homogenous space. Explicit examples are given for $N = 2$ and $N = 3$.

1. Introduction

Let us consider an r-dimensional Lie group G, some manifold $M \subset \mathbb{R}^n$, $(0,\dots,0) \in M$ and some left group action $a\colon G \times M \to M$, where $a \in C^\infty(\mathbb{R}^{r+n})$. For each $y \in M$, g, $g' \in G$ it satisfies

$$a(e,y) = y, \quad a\big(g', a(g,y)\big) = a(g'g, y),$$

where e is the identity in the group G. Let $F \subset C^\infty(\mathbb{R}^n)$ be linear space consisting of functions $f\colon M \to \mathbb{C}$. Let T be linear representation of G on F given by the formula

$$(T_g f)(y) := f\big(a(g^{-1}, y)\big). \tag{1.1}$$

Choose r one-parametric subgroups $g_1,\dots,g_r$ in G ($g_j\colon \mathbb{R} \to G$) and consider the corresponding induced Lie algebra L

$$[X^j, X^k] = c_l^{jk} X^l, \quad j = 1,\dots,r,$$

where

$$(X^j f)(y) = \frac{d}{dt}\big(T_{g_j(t)} f\big)(y)\,\Big|_{t=0} = \sum_{\mu=1}^{n} A_\mu^j(y)\frac{\partial f(y)}{\partial y_\mu}. \tag{1.2}$$

The theory developed by S. Lie (1893) [**7**], who recognized the connection with representation theory of Lie groups and algebras, says that for any path $X\colon \mathbb{R} \to L$ in the Lie algebra L,

$$X(t) := \sum_{j=1}^{r} z_j(t) X_j,$$

2000 *Mathematics Subject Classification.* Primary: 34A26; Secondary: 22E60.
This is the final form of the paper.

and any initial condition $x(0) \in M$ there exists a path $g\colon \mathbb{R} \to G$ such that $g(0) = e$ and a function $x\colon \mathbb{R} \to M$ defined using the left action

$$x(t) := a\big(g(t), x(0)\big), \tag{1.3}$$

which solves the differential equation (dot denotes differentiation with respect to t)

$$\sum_{j=1}^{r} z_j(t) A^j\big(x(t)\big) = \dot{x}(t). \tag{1.4}$$

A sufficient number m of equations of type (1.3), i.e.

$$x^{(k)}(t) := a\big(g(t), x^{(k)}(0)\big), \quad k = 1, \ldots, m,$$

allows us to express all elements of matrix g in (1.3) in terms of m particular solutions $x^{(k)}$. That means we construct the superposition formula

$$x(t) = F\big(x^{(1)}(t), \ldots, x^{(m)}(t), x(0)\big), \tag{1.5}$$

where $x^{(1)}, \ldots, x^{(m)}\colon \mathbb{R} \to M$ are particular solutions, $x\colon \mathbb{R} \to M$ is the general solution and $x(0) \in M$ its initial condition. The particular solutions are subject to certain independence conditions that must be specified in each case.

S. Lie proved that the number of solutions m needed in the superposition formula (1.5) satisfies the relation

$$mn \geq r, \tag{1.6}$$

where n is the number of equations (dimension of $\mathbb{R}^n$) and r is the dimension of the Lie algebra L. It means that we need at least $\lceil r/n \rceil$ particular solutions for the construction of a superposition formula.

2. Example 1: SL(2, $\mathbb{R}$) and One Variable

Let us take $G = \mathrm{SL}(2, \mathbb{R})$ and representation T on linear space F defined as (see (1.1))

$$(T_g f)(y) := f\left(\frac{\alpha y + \gamma}{\beta y + \delta}\right),$$

where $g = \left(\begin{smallmatrix} \alpha & \beta \\ \gamma & \delta \end{smallmatrix}\right) \in \mathrm{SL}(2, \mathbb{R})$. In $\mathrm{SL}(2, \mathbb{R})$ we find three one dimensional subgroups $g_j\colon \mathbb{R} \to \mathrm{SL}(2, \mathbb{R})$,

$$g_1(t) = \begin{pmatrix} 1 & 0 \\ t & 1 \end{pmatrix}, \quad g_2(t) = \begin{pmatrix} e^{t/2} & 0 \\ 0 & e^{-t/2} \end{pmatrix}, \quad g_3(t) = \begin{pmatrix} 1 & -t \\ 0 & 1 \end{pmatrix},$$

which induce the Lie algebra $\mathrm{sl}(2, \mathbb{R})$ with the induced basis (see (1.2))

$$X_1 = \frac{d}{dy}, \quad X_2 = y\frac{d}{dy}, \quad X_3 = y^2\frac{d}{dy}.$$

The equation of Lie's theorem (see (1.4)) has the form

$$z_1(t) + z_2(t)x(t) + z_3(t)x(t)^2 = \dot{x}(t),$$

with given initial condition $x(0) \in \mathbb{R}$. Equation (1.3) has the form

$$x(t) = \frac{\alpha(t)x^{(0)} + \gamma(t)}{\beta(t)x^{(0)} + \delta(t)}. \tag{2.1}$$

Rewriting for any three solutions $x^{(1)}$, $x^{(2)}$, $x^{(3)}$ ("any three" means that they are different, i.e. for initial conditions we have $x^{(j_1)}(0) \neq x^{(j_2)}(0)$ for $j_1 \neq j_2$) we get

$$x^{(1)}(t) = \frac{\alpha(t)x^{(1)}(0) + \gamma(t)}{\beta(t)x^{(1)}(0) + \delta(t)},$$
$$x^{(2)}(t) = \frac{\alpha(t)x^{(2)}(0) + \gamma(t)}{\beta(t)x^{(2)}(0) + \delta(t)},$$
$$x^{(3)}(t) = \frac{\alpha(t)x^{(3)}(0) + \gamma(t)}{\beta(t)x^{(3)}(0) + \delta(t)}.$$

This leads to a system of three linear equations

$$\gamma(t) + x^{(1)}(0)\alpha(t) - x^{(1)}(0)x^{(1)}(t)\beta(t) = x^{(1)}(t)\delta(t),$$
$$\gamma(t) + x^{(2)}(0)\alpha(t) - x^{(2)}(0)x^{(2)}(t)\beta(t) = x^{(2)}(t)\delta(t),$$
$$\gamma(t) + x^{(3)}(0)\alpha(t) - x^{(3)}(0)x^{(3)}(t)\beta(t) = x^{(3)}(t)\delta(t).$$

Solving for unknowns $\alpha(t)$, $\beta(t)$, $\gamma(t)$ and substituting into (2.1) we obtain the superposition formula in the form

$$x(t) = \frac{x^{(2)}(t)(x^{(1)}(t) - x^{(3)}(t))x(0) - x^{(3)}(t)(x^{(1)}(t) - x^{(2)}(t))}{(x^{(1)}(t) - x^{(3)}(t))x(0) - (x^{(1)}(t) - x^{(2)}(t))}.$$

3. Example 2: SL(2, ℝ) and Two Variables

Let us consider two nonequivalent realizations of sl(2, ℝ) on the space of functions of two real variables y_1, y_2 (we denote $\partial_j \equiv \partial/(\partial y_j)$):

$$X_1 = \partial_1, \quad X_2 = y_1\partial_1 + y_2\partial_2, \quad X_3 = y_1^2\partial_1 + (2y_1y_2 + Ky_2^2)\partial_2,$$

where $K \in \{0, 1\}$. The differential equation (1.4) has the form

$$\begin{aligned} \dot{x}_1(t) &= z_1(t) + z_2(t)x_1(t) + z_3(t)x_1^2(t), \\ \dot{x}_2(t) &= z_2(t)x_2(t) + z_3(t)\big(2x_1(t)x_2(t) + Kx_2^2(t)\big). \end{aligned} \tag{3.1}$$

The representation of the group on the space F is given by formula

$$(T_gf)(y_1, y_2) = f\left(\frac{\alpha y_1 + \gamma}{\beta y_1 + \delta}, \frac{y_2}{(\beta y_1 + \delta)(\beta(y_1 + Ky_2) + \delta)}\right).$$

To determine the matrix elements α, β, γ, δ we need two solutions $x^{(1)} \equiv (v_1, v_2)$, $x^{(2)} \equiv (w_1, w_2)$. The corresponding equations have the form:

$$v_1(t) = \frac{\alpha(t)v_1(0) + \gamma(t)}{\beta(t)v_1(0) + \delta(t)}, \tag{3.2}$$

$$v_2(t) = \frac{v_2(0)}{(\beta(t)v_1(0) + \delta(t))(\beta(t)(v_1(0) + Kv_2(0)) + \delta(t))}, \tag{3.3}$$

$$w_1(t) = \frac{\alpha(t)w_1(0) + \gamma(t)}{\beta(t)w_1(0) + \delta(t)}, \tag{3.4}$$

$$w_2(t) = \frac{w_2(0)}{(\beta(t)w_1(0) + \delta(t))(\beta(t)(w_1(0) + Kw_2(0)) + \delta(t))}. \tag{3.5}$$

In the case $K = 0$ Eqs. (3.2) and (3.4) give

$$\beta(t)v_1(0) + \delta(t) = \varepsilon\sqrt{\frac{v_2(0)}{v_2(t)}}, \quad \beta(t)w_1(0) + \delta(t) = \varepsilon'\sqrt{\frac{w_2(0)}{w_2(t)}},$$

where ε, $\varepsilon' = \pm 1$. A solution of these two linear equations exists iff $v_1(0) \neq w_1(0)$. Since

$$\begin{pmatrix} \alpha(t) & \beta(t) \\ \gamma(t) & \delta(t) \end{pmatrix} \in \mathrm{SL}(2, \mathbb{R}),$$

we must take into account the condition

$$\alpha(t)\delta(t) - \beta(t)\gamma(t) = 1. \tag{3.6}$$

Examining (3.6) we find that it is fulfilled iff $v_2(0)w_2(0) \neq 0$.

Thus the system of ODEs (3.1) for $K = 0$ has a superposition formula giving the general solution in terms of two solutions (v_1, v_2), (w_1, w_2) with initial conditions fulfilling $v_1(0) \neq w_1(0)$, $v_2(0)w_2(0) \neq 0$. Furthermore the components of the solutions are not independent:

$$R \equiv \frac{(v_1(t) - w_1(t))^2}{v_2(t)w_2(t)} = \text{const.} \tag{3.7}$$

For $K = 1$ the situation is similar. We get three requirements on the initial conditions, namely

$$v_1(0) \neq w_1(0), \quad v_2(0)w_2(0) \neq 0, \quad \widetilde{R} \neq 0,$$

where

$$\widetilde{R} \equiv \frac{(v_1(t) - w_1(t))(v_1(t) - w_1(t) + v_2(t) - w_2(t))}{v_2(t)w_2(t)} = \text{const.} \tag{3.8}$$

So, in distinction with the preceeding Example 1, the superposition formula connects the general solution with (two) other solutions which, however, are not quite freely chosen. Their initial conditions fulfil some relations and this is a common feature of the studied problem.

4. General Case

We consider a realization of $\mathrm{sl}(n+1, \mathbb{R})$ of the form

$$E_{ij} = y_i\partial_j + e_{ij}, \quad E_{n+1,i} = -\partial_i, \quad E_{i,n+1} = y_i\Big(\sum_j y_j\partial_j\Big) + \sum_j y_j e_{ij}, \tag{4.1}$$

where e_{ij} are such that they satisfy commutation relations of $\mathrm{sl}(n, \mathbb{R})$, $\mathrm{lin}\{e_{ij} \mid i, j = 1, \ldots, n\} = \mathrm{sl}(n, \mathbb{R})$ and $[e_{ij}, y_k] = [e_{ij}, \partial_k] = 0$.

Namely we consider $e_{\alpha\beta} = y_{\alpha+n}\partial_{\beta+n}$, $e_{n,\alpha} = -\partial_{\alpha+n}$, $e_{\alpha,n} = y_{\alpha+n}(y_{\beta+n}\partial_{\beta+n})$, α, $\beta = 1, \ldots, n-1$. (The case $e_{ij} = 0$ was considered earlier [**4**].)

For $n + 1 = 3$, i.e. $\mathrm{sl}(3, \mathbb{R})$ we have

$$\begin{aligned} &E_{31} = \partial_1, \quad E_{32} = \partial_2, \quad E_{21} = y_2\partial_1 + \partial_3, \\ &E_{11} = y_1\partial_1 + y_3\partial_3, \quad E_{22} = y_2\partial_2 - y_3\partial_3, \quad E_{12} = y_1\partial_2 - y_3^2\partial_3, \\ &E_{13} = y_1(y_1\partial_1 + y_2\partial_2) - (y_2y_3^2 - y_1y_3)\partial_3, \\ &E_{23} = y_2(y_1\partial_1 + y_2\partial_2) - (y_2y_3 - y_1)\partial_3. \end{aligned} \tag{4.2}$$

The system of ODEs has the form

$$
(4.3)\quad
\begin{aligned}
\dot x_j &= z_{n+1,j} + \sum_{i=1}^{n} z_{ij}x_i + x_j\sum_{i=1}^{n} z_{i,n+1}x_i, \quad 1\le j\le n,\\
\dot x_{n+\alpha} &= z_{n+\alpha}\sum_{\beta=1}^{n-1} z_{\beta\alpha}x_{n+\beta} - z_{nn}x_{n+\alpha} - x_{n+\alpha}\sum_{\beta=1}^{n-1} z_{\beta n}x_{n+\beta}\\
&\quad +(x_\alpha - x_{n+\alpha}x_n)\sum_{\beta=1}^{n-1}(z_{\beta,n-1}x_{n+\beta}+z_{n,n+1}),\ 1\le\alpha\le n-1.
\end{aligned}
$$

For $n+1=3$ it yields

$$
(4.4)\quad
\begin{aligned}
\dot x_1 &= z_{31}+z_{11}x_1+z_{21}x_2+z_{13}x_1^2+z_{23}x_1x_2,\\
\dot x_2 &= z_{32}+z_{12}x_1+z_{22}x_2+z_{13}x_1x_2+z_{23}x_2^2,\\
\dot x_3 &= z_{21}+(z_{11}-z_{22})x_3-z_{12}x_3^2-z_{13}x_3(x_3x_2-x_1)-z_{23}(x_3x_2-x_1).
\end{aligned}
$$

Finite transformations are computed using the theory of induced representations [**3, 9**].

For $g\in \mathrm{sl}(n+1,\mathbb{R})$ we have

$$
(4.5)\quad
\begin{aligned}
&T_g f(y_1,\dots,y_{2n-1}) = f(\widetilde y_1,\dots,\widetilde y_{2n-1}), \text{ where}\\
&\widetilde y_i = \frac{\sum_{j=1}^{n} g_{ji}y_j + g_{n+1,i}}{\sum_{j=1}^{n} g_{j,n+1}y_j + g_{n+1,n+1}}, \qquad i=1,\dots,n,\\
&\widetilde y_{\alpha+n} = \frac{F(\alpha)}{F(n)}, \quad \alpha=1,\dots,n-1,\\
&F(\alpha) = \sum_{\beta=1}^{n-2}\sum_{\gamma=1}^{n-1-\beta}(y_\beta y_{\gamma+\beta+n} - y_{\gamma+\beta}y_{\beta+n})A_{\beta,\alpha,\beta+\gamma,n+1}\\
&\qquad\qquad + \sum_{\beta=1}^{n-1}(y_\beta - y_n y_{\beta+n})A_{\beta,\alpha,n,n+1}\\
&\qquad\qquad + \sum_{\beta=1}^{n-1}(-y_{\beta+n})A_{\beta,\alpha,n+1,n+1} - A_{n,\alpha,n+1,n+1},
\end{aligned}
$$

where $A_{ijkl} = \begin{vmatrix} g_{ij} & g_{il}\\ g_{kj} & g_{kl}\end{vmatrix}$. For $n+1=3$ we have

$$
g = \begin{pmatrix} g_{11} & g_{12} & g_{13}\\ g_{21} & g_{22} & g_{23}\\ g_{31} & g_{32} & g_{33}\end{pmatrix}, \quad \det g = 1, \quad T_g f(y_1,y_2;y_3) = f(\widetilde y_1,\widetilde y_2,\widetilde y_3), \text{ where}
$$

$$
\begin{aligned}
\widetilde y_1 &= \frac{g_{11}y_1+g_{21}y_2+g_{31}}{g_{13}y_1+g_{23}y_2+g_{33}},\\
\widetilde y_2 &= \frac{g_{12}y_1+g_{22}y_2+g_{32}}{g_{13}y_1+g_{23}y_2+g_{33}},\\
\widetilde y_3 &= \frac{(-y_1+y_2y_3)A_{1123}+y_3A_{1133}+A_{2133}}{(-y_1+y_2y_3)A_{1223}+y_3A_{1233}+A_{2233}}.
\end{aligned}
$$

Rewriting Eq. (4.6) for 3 solutions $x^{(1)}\equiv(u_1,u_2,u_3)$, $x^{(2)}\equiv(v_1,v_2,v_3)$, $x^{(3)}\equiv(w_1,w_2,w_3)$ and solving the system of equations for the unknowns g_{ij} we find

(i) A superposition formula exists if and only if we have

(a) $\begin{vmatrix} u_1(0)-w_1(0) & v_1(0)-w_1(0)\\ u_2(0)-w_2(0) & v_2(0)-w_2(0)\end{vmatrix} \neq 0.$

(b) At least two of the products

$$UV(0)\,UW(0),\ VW(0)\,VU(0),\ WU(0)\,WV(0),$$

where

$$UV(t) := u_3(t)\big(u_2(t) - v_2(t)\big) - u_1(t) + v_1(t),$$

(and similarly for UW, $VU, \dots$), are nonzero.

(ii) The components of solutions u_j, v_j, w_j are not independent:

$$\frac{UW(t)WV(t)VU(t)}{UV(t)VW(t)WU(t)} = \text{const.}$$

For the case $n+1=4$ we have similar conclusions: dim $\mathrm{sl}(4,\mathbb{R}) = 15$, the number of components of a solution is 5, we need $m=3$ solutions for reconstruction. There is no relation among components. The general case $\mathrm{sl}(n+1,\mathbb{R})$ can be summarized in the following theorem:

THEOREM. (i) *The number m of particularly chosen solutions necessary for reconstruction of g_{ik} is $m = k+2$ when $n = 2k$ and $m = k+1$ when $n = 2k-1$.*

(ii) *The number of constraints is $\tau = 3k-2$ when $n=2k$ and $\tau = k-2$ when $n = 2k-1$.*

The reconstruction is linear in the sense that we only need to solve linear algebraic relations (once m solutions are known).

5. Conclusions

Bäcklund transformations for integrable nonlinear partial differential equations (PDE) involving two independent variables x and t typically have the form

$$u_x = F(x,t,u,v,v_x,v_t), \quad u_t = G(x,t,u,v,v_x,v_t), \tag{5.1}$$

where u and v a some scalar, vector, or matrix functions of x and t. These two equations are in general incompatible. The compatibility condition requires that $u(x,t)$ and $v(x,t)$ satisfy some nonlinear PDEs. If they satisfy the same PDE, we have an "auto-Bäcklund transformation". The functions F and G are themselves specified by the equations satisfied by u and v. They always have the form (1.4), i.e. they can be viewed as ordinary differential equations with superposition formulas. For equations of the AKNS series [**1**] the underlying Lie algebra is $\mathrm{sl}(2,\mathbb{R})$, or $\mathrm{sl}(2,\mathbb{C})$. The Bäcklund transformations are equivalent to Riccati equations (independently in x and in t).

The equations presented here, based on an imprimitive action of $\mathrm{SL}(N,\mathbb{R})$ should occur as Bäcklund transformations for integrable systems on flag manifolds.

References

1. M. J. Ablowitz and P. A. Clarkson, *Solitons, nonlinear evolution equations and inverse scattering*, Cambridge University Press, London, 1991.
2. M. Adler, L. Haine, and P. van Moerbeke, *Limit matrices for the Toda flow and periodic flags for loop groups*, Math. Ann. **296** (1993), 1–33.
3. A. O. Barut and R. Raczka, *Theory of group representations and applications*, 2nd edition, World Scientific Publ. Co., Singapore, 1986.
4. J. Harnad, P. Winternitz, and R. L. Anderson, *Superposition principles for matrix Riccati equations*, J. Math. Phys. **24** (1983), 1062–1072.

5. M. Havlíček and W. Lassner, *Canonical realizations of the Lie algebras* gl$(n, \mathbb{R})$ *and* SL$(n, \mathbb{R})$. I. *Formulae and classification*, Rep. Mathematical Phys. **8** (1975), no. 3, 391–399.
6. M. Havlíček, S. Pošta, and P. Winternitz, *Nonlinear superposition formulas based on imprimitive group action*, J. Math. Phys. **40** (1999), 3104–3122.
7. S. Lie, *Vorlesungen über continuirliche Gruppen mit geometrischen und anderen Anwendungen*, Bearbeitet und herausgegeben von Georg Scheffers. Nachdruck der Auflage des Jahres 1893. Chelsea Publishing Co., Bronx, N.Y., 1971.
8. ———, *Theorie der Transformationsgruppen*. III, Unter Mitwirkung von F. Engel, Teubner, Leipzig, 1893.
9. G. W. Mackey, *The theory of unitary group representations*, University of Chicago Press, Chicago, 1976.

Department of Mathematics and Doppler Institute, FNSPE, CTU, Trojanova 13, CZ-120 00 Prague 2, Czech Republic.
E-mail address, M. Havlíček: havlicek@km1.fjfi.cvut.cz
E-mail address, S. Pošta: posta@km1.fjfi.cvut.cz

Centre de Recherches Mathématiques, Univ. de Montréal, C.P. 6128, succ. Centre-ville, Montréal, Québec H3C 3J7, Canada.
E-mail address: wintern@crm.umontreal.ca

Centre de Recherches Mathématiques
CRM Proceedings and Lecture Notes
Volume **29**, 2001

Symmetries of Differential Difference Equations and Lie Algebra Contractions

R. Hernandez Heredero, *D. Levi*, M. A. Rodríguez, and P. Winternitz

ABSTRACT. We consider the problem of constructing point symmetries for equations on a lattice. We look for infinitesimal symmetry generators which depend on several points on the lattice and which in the continuum limit go over to dilation symmetries. We obtain such symmetries in the case of the Volterra equation and show that in the continuum limit we have a Lie algebra contraction in which a subset of the infinite dimensional algebra of generalized symmetries goes over to a finite dimensional algebra of point symmetries.

1. Introduction

The results presented here are part of a program, the aim of which is to use Lie theory to study symmetries of discrete equations. The equations are difference equations on a regular lattice. Previous studies have shown [**10, 13–15, 22**] that for difference equations the class of point symmetries is somewhat restricted. Many interesting symmetries are obtained only if one considers group actions on a set of points on the lattice [**4, 8, 12, 19**] rather than just at one point. Here we present the infinite dimensional Lie algebra of point symmetries and higher symmetries of the Volterra differential difference equation [**9**]

$$\dot{a}_n = a_n(a_{n-1} - a_{n+1}). \tag{1.1}$$

Point symmetries and higher symmetries of the Volterra hierarchy were obtained [**11**] using the Lax pair, i.e. the integrability properties of the hierarchy. In particular we show that the continuous limit from a difference equation to a differential one corresponds to a Lie algebra contraction. Lie algebra contractions were first introduced by Inonu and Wigner [**6**] to study the relation between relativistic and nonrelativistic theories when the velocity of light c satisfies $c \to \infty$. The Lorentz group then "contracts" to the Galilei one. A large body of literature now exists on Lie algebra contractions (see Refs. [**7,18,21**] and references therein). In our case the

---

2000 *Mathematics Subject Classification.* Primary: 37K10; Secondary: 17B80.

We acknowledge support from DGES (PB98-0821) and NATO (0717/96). This work was carried out during a stay of D.L. at Universidad Complutense of Madrid in the framework of the Universidad Complutense–Università di Roma Tre exchange agreement. The research of P.W. was partially supported by NSERC of Canada and FCAR du Québec.

This is the final form of the paper.

contraction parameter is the level spacing h. The infinite dimensional Lie algebra S of all symmetries of the difference equation, contracts to an infinite dimensional algebra $L(D)$ of symmetries of the differential equation. The crucial new element is that a subset S of elements of $L(\Delta)$ contracts to a subalgebra L_p of $L(D)$. Moreover, L_p is a Lie algebra of point symmetries, while the subset $S \subset L(\Delta)$ does not form a Lie algebra. It is not closed under commutation and it contains both point symmetries and generalized symmetries of the discrete equation. In Section 2 we give an overview of the problems associated with the construction of symmetries for difference equations. In Section 3 we present the symmetries of the Volterra equation in a form adapted to our needs. We reproduce some known results [**11**] to make this article self-contained. Section 4 is devoted to the structure of the Lie algebra $L(\Delta)$. To obtain the commutation relations we make use of the integrability properties of the hierarchy [**2, 3, 17**]. We use the spectral transform to relate the studied nonlinear evolution equations with linear equations for the reflection coefficient. Symmetries acting in the solution space of the evolution equation are transformed into symmetries acting in the space of the reflection coefficient. These turn out to be much simpler to deal with. In Section 5 we introduce the limiting procedure, taking the Volterra equation into the Korteweg–de Vries equation. The Lie algebras are subjected to the same limiting procedure. Explicitly we present the continuous limits of those symmetries that go into point ones.

2. Symmetries for Difference Equations

The power of Lie group theory in the treatment of differential equations is well known [**1, 20, 23**]. Part of the results obtained from Lie theory for differential equations can be carried on to the case of differential difference equations. To show it, we have to review briefly, at first, the case of differential equations. Lie group theory for differential equations deals with transformations which are analytical with respect to the group parameters and differentiable with respect to the dependent and independent variables appearing in the equation. When the dependent and independent variables resulting from the transformation depend only on the dependent and independent variables themselves, we say that we are considering Lie point symmetries. These transformations form a group which can be used to obtain new solutions from the old ones, to reduce the order of the equation or the number of independent variables, to classify equations into equivalent classes, etc. Consider a differential equation of order n

$$F(x, u, u_{x_i}, u_{x_i x_j}, \ldots, u_{x_{i_1} x_{i_2} \cdots x_{i_n}}) = 0, \tag{2.1}$$

where $u \in R^q$, $x \in R^p$ and by u_{x_i} we denote the partial derivative of u with respect to the x_i variable ($i = 1, \ldots, p$). A point symmetry is given by a transformation

$$u' = \Omega_g(x, u), \quad x' = \Lambda_g(x, u), \tag{2.2}$$

with g denoting a set of group parameters. Lie showed that, whenever Ω and Λ are functions that depend analytically on g and are infinitely differentiable with respect to x and u, the transformation is completely characterized by its infinitesimal generators

$$\widehat{X} = \sum_{i=1}^{p} \xi_i(x, u)\partial_{x_i} + \sum_{j=1}^{q} \phi_j(x, u)\partial_{u_j}, \tag{2.3}$$

where ξ_i and ϕ_j are the first order coefficients in the Taylor expansion of Λ and Ω in terms of the group parameters. We can extend the class of point symmetries by considering transformations, where the new dependent and independent variables depend also on the derivatives, up to some order m. In such a case

$$(2.4) \qquad \xi_i = \xi_i(x, u, u_{x_i}, \dots, u_{x_{i_1} x_{i_2} \cdots x_{im}}), \quad \phi_j = \phi_j(x, u, u_{x_i}, \dots, u_{x_{i_1} x_{i_2} \cdots x_{im}}),$$

and we say that we are considering generalized symmetries of order m. Usually the value of m depends on the equation under study and can be obtained only by successive trials. Moreover not every equation has generalized symmetries. Generically generalized symmetries are associated with integrable differential equations, i.e. equations which are either linear or, being nonlinear, are linearizable in a non trivial way via the solution of a linear spectral problem. Generalized symmetries usually form infinite dimensional Lie algebras. Once a generalized symmetry is obtained, the composition rule of symmetries together with the point symmetries will generate new generalized symmetries introducing higher order derivatives of the dependent variables. The coefficients ξ_i and ϕ_j of the generators of the Lie point symmetries are derivable algorithmically using programs written out in various symbolic manipulation languages [**5**], now easily available. Moreover one can construct, starting from the symmetry algebra the group transformations given by Ω and Λ by integrating the system of differential equations

$$(2.5) \qquad \frac{dx'_i}{d\lambda} = \xi_i(x', u'), \quad \frac{du'_j}{d\lambda} = \phi_j(x', u'),$$

with

$$(2.6) \qquad x'_i(\lambda = 0) = x_i, \quad u'_j(\lambda = 0) = u_j.$$

An alternative, equivalent, way to get the group transformation is by exponentiating the symmetry vector $\widehat{X}$:

$$(2.7) \qquad x'_i = e^{\lambda \widehat{X}} x_i, \quad u'_j = e^{\lambda \widehat{X}} u_j.$$

In the case of point symmetries both ways are available and provide the same. In the case of generalized symmetries the differential equations relating the Lie algebra and the Lie group are partial differential equations and, being often nonlinear in the dependent variables, are no longer easily integrable. However, by assuming the existence of the solution, expressed as a Taylor series expansion in the group parameter (by assumption, the group transformation is an analytic function of the group parameters) we get a recurrence relation whose solution is formally given by the exponential formula when we substitute $\widehat{X}$ by its prolongation [**1, 20**]. We can use generalized symmetries to reduce the equation and to get new solutions.

A first step in the derivation of symmetries for equations depending on discrete variables has been given by Maeda [**15, 16**], who introduced similarity methods for purely difference equations in the context of Hamiltonian and Lagrangian formulations of discrete mechanics. The procedure introduced by Maeda has been shown to be applicable also in the case of differential difference equations [**13, 14, 22**]. The obtained symmetries of difference equations have the same properties as those of differential equations, namely

(1) One can present an efficient algorithm for finding the symmetry group. Infinitesimal methods are applied for calculating the connected subgroups of the symmetry group.

(2) One can construct the group transformations.
(3) Symmetry reduction (to equations with fewer variables) can be performed using both continuous and discrete subgroups of the symmetry group.
(4) The concept of *conditional symmetry*, i.e. transformations which leave only a subclass of the solutions invariant, can be extended and applied to get new solutions.
(5) It is possible to construct difference equations with prechosen symmetry groups.
(6) One can perform a group classification of given classes of difference equations.

The Maeda approach to Lie point symmetries for difference equations consists in looking for transformations where the discrete independent variable is left invariant. This variable will, however, figure in the transformation of the continuous variables. For an equation depending essentially on p—continuous independent variables $x \in R^p$, M—discrete independent variables $n \in Z^M$ and q—dependent variables $u \in R^q$, we will be looking for local transformations of the form:

$$u'(x', n') = \Omega_g\big(x, n, u(x, n)\big), \tag{2.8}$$

$$x' = \Lambda_g\big(x, u(x, n)\big), \tag{2.9}$$

$$n' = n, \tag{2.10}$$

where u and u' are two a priori different solutions of the same equation and g, as before, denotes the set of group parameters. These group transformations generate a Lie algebra realized in terms of vector fields of the form:

$$\widehat{X} = \sum_{i=1}^{p} \xi_i\big(x, u(x, n)\big)\partial_{x_i} + \sum_{j=1}^{q} \phi_j\big(x, n, u(x, n)\big)\partial_{u_j(x,n)}, \tag{2.11}$$

or equivalently, in the evolutionary formalism [**20**]:

$$\widehat{X}_e = \sum_{j=1}^{q}\Big[\phi_j\big(x, n, u(x, n)\big) - \sum_{i=1}^{p} \xi_i\big(x, u(x, n)\big)u_{x_i}(x, n)\Big]\partial_{u_j(x,n)}. \tag{2.12}$$

In order to be an infinitesimal generator of a symmetry, the prolongation of the vector field, either $\widehat{X}$ or $\widehat{X}_e$, must annihilate the equation on its solution surface. If

$$\Delta\big(x, u(x, n), u_x(x, n), u(x, n+1), u(x, n-1), \ldots\big) = 0, \tag{2.13}$$

is a differential-difference equation, then

$$\mathrm{pr}\,\widehat{X}\Delta|_{\Delta=0} = \mathrm{pr}\,\widehat{X}_e\Delta|_{\Delta=0} = 0, \tag{2.14}$$

is the invariance condition which will define the infinitesimal generators of the symmetry. The symmetries we obtain in this way will be called *intrinsic Lie point symmetries*.

This class of transformations has been applied with success to given differential difference equations of interest, providing symmetries like translations of various forms or Galilean boosts, but it cannot provide, by construction, any dilation with respect to the discrete variable. This prompted an investigation on possible extensions of the intrinsic Lie point symmetries to a case in which more general symmetries exist which, for example, in the continuum limit reduce to dilations, rotations or Lorentz transformations. Prompted by the results obtained in the linear case

[4,19] using the standard operator approach of quantum mechanics, the formalism has been extended by considering coefficients of the infinitesimal generators which depend on u at various points of the lattice [12]. The infinitesimal generators in the evolutionary formalism read:

$$\widehat{X}_e = \sum_{j=1}^{q} Q_j\big(x, n, \{u_k(x, n+m)\}_{m=-a_{jk}}^{b_{jk}}, \{u_{k,x_s}(x, n+m)\}_{m=-c_{jks}}^{d_{jks}}\big)\partial_{u_j(x,n)}, \tag{2.15}$$

where Q_j is a function linear in $\{u_{k,x_s}(x, n+m)\}_{m=-a_{jks}}^{b_{jks}}$ and the curly brackets represent sets of variables with indices varying in the interval defined by the set of positive integer numbers $(a_{jk}, b_{jk}, c_{jks}, d_{jks})$, to be determined together with the functions Q_j. In this way one is able to get, in the nonlinear case, new symmetries, often very complicated at first glance, which, however, when written in terms of a continuous n—variable, correspond just to an n—translation [12]. This suggests that these extended symmetries are closer to generalized symmetries than to point ones.

3. The Volterra Equation and its Symmetries

The Volterra Eq. (1.1) is associated to the discrete Schrödinger spectral problem

$$\psi(n-1, t; \lambda) + a_n \psi(n+1, t; \lambda) = \lambda \psi(n, t; \lambda). \tag{3.1}$$

The Volterra Eq. (1.1) is a member of the the Volterra hierarchy [9]:

$$\dot{a}_n = g_1(\mathcal{L}, t)\big[a_n(a_{n-1} - a_{n+1})\big], \tag{3.2}$$

where the recursive operator $\mathcal{L}$ is given by

$$\mathcal{L}p_n = a_n[p_n + p_{n+1} + a_{n-1}s_{n-1} - a_{n+1}s_{n+1}], \tag{3.3}$$

where s_n is a solution of the inhomogeneous first order equation

$$s_{n+1} = s_n - p_n/a_n. \tag{3.4}$$

For any equation of the hierarchy (3.2) we can write an explicit evolution equation for the function $\psi(n, t; \lambda)$ [9] such that λ does not evolve in time. We impose the following boundary conditions

$$\lim_{|n|\to\infty} a_n - 1 = \lim_{|n|\to\infty} s_n = 0, \tag{3.5}$$

on the fields a_n and s_n. We can then associate to Eq. (3.1) a spectrum defined in the complex plane of the function z $(\lambda = z + z^{-1})$:

$$\{R(z, t), z \in \mathbf{C}_1; z_j, c_j(t), |z_j| < 1, j = 1, 2, \ldots, N\}, \tag{3.6}$$

where $R(z, t)$ is the reflection coefficient, $\mathbf{C}_1$ is the unit circle in the complex z plane, z_j are isolated points inside the unit disk and c_j are some complex constants related to the residues of $R(z, t)$ at the poles z_j. When a_n and s_n satisfy the boundary conditions (3.5), the spectral data define the potentials in a unique way. Thus, there is a one-to-one correspondence between the evolution of the potential a_n of the discrete Schrödinger spectral problem (3.1), given by Eq. (3.2) and that of the reflection coefficient $R(\lambda, t)$, given by

$$\frac{\mathrm{d}R(z, t)}{\mathrm{d}t} = \mu\lambda g_1(\lambda^2, t)R(z, t), \quad \mu = z^{-1} - z. \tag{3.7}$$

In Eq. (3.7) and below, $\mathrm{d}/\mathrm{d}y$ denotes the total derivative with respect to y. The Volterra equation is obtained from Eq. (3.2) by choosing $g_1(\lambda^2, t) = 1$ and thus the evolution equation of the reflection coefficient is given by

$$\frac{\mathrm{d}R(z,t)}{\mathrm{d}t} = \mu\lambda R(z,t). \tag{3.8}$$

The symmetries for the Volterra equation (1.1) are provided by all flows commuting with the equation itself [**11**]. An infinite number of such symmetries is given by the equations

$$a_{n,\varepsilon_k} = (\mathcal{L})^k \{a_n[a_{n-1} - a_{n+1}]\}. \tag{3.9}$$

Here k is any positive integer and ε_k is a group parameter. From the point of view of the spectral problem (3.1) the above equations correspond to isospectral deformations, i.e. we have $\lambda_{\varepsilon_k} = 0$. For any ε_k, the solution of the Cauchy problem for Eq. (3.9), provides a solution of the Volterra equation $a_n(\varepsilon_k)$ in terms of the initial condition $a_n(\varepsilon_k = 0)$. The group transformation corresponding to the group parameter ε_k can usually be written explicitly only for few values of k. In all other cases one can just use the symmetries (3.9) to do, for example, symmetry reduction and to reduce the equation under consideration to an ordinary differential equation, or possibly a functional one.

The proof of the validity of Eq. (3.9) is easily given by taking into account the one-to-one correspondence between the equation and the spectrum (3.6), under the asymptotic conditions (3.5). In fact, in such a case we can biunivocally associate to both the Eq. (1.1) and the symmetries (3.9) an evolution of the reflection coefficient. In the case of the symmetries (3.9), we have:

$$\frac{\mathrm{d}R}{\mathrm{d}\varepsilon_k} = \mu\lambda^{2k+1} R. \tag{3.10}$$

It is easy to prove that the flows (3.8) and (3.10) commute and hence the same must be true for the corresponding equations for the fields (1.1) and (3.9).

We can extend the class of symmetries by considering nonisospectral deformations of the spectral problem (3.1) [**9**]. Thus for the Volterra equation we obtain

$$\begin{aligned} a_{n,\varepsilon} = \mathcal{L}^k a_n\big[a_n - (n-1)a_{n-1} + (n+2)a_{n+1} - 4\big] \\ + t\mathcal{L}^k(\mathcal{L} - 2)\{a_n[a_{n-1} - a_{n+1}]\}. \end{aligned} \tag{3.11}$$

In correspondence with Eq. (3.11) we have the evolution of the reflection coefficient (3.6), given by

$$\frac{\mathrm{d}R}{\mathrm{d}\varepsilon_k} = \mu\lambda^{2k+1}(\lambda^2 - 2)tR, \quad \lambda_{\varepsilon_k} = \frac{1}{2}\mu^2\lambda^{2k+1}. \tag{3.12}$$

As in the case of isospectral symmetries (3.9), we must prove that the nonisospectral flows (3.11) commute with the corresponding evolution Eq. (1.1). This is reduced to an easier task, namely showing that the flows (3.12) in the space of the reflection coefficients, commute with the evolution Eq. (3.8). In addition to the above two hierarchies of symmetries (3.9) and (3.11) for the Volterra equation, we can construct one further case, which, however, does not satisfy the asymptotic boundary conditions (3.5). It is:

$$a_{n,\varepsilon} = t\dot{a}_n + a_n. \tag{3.13}$$

As the exceptional symmetry does not satisfy the asymptotic boundary conditions (3.5), we cannot write a corresponding evolution equation for the reflection coefficient (3.6).

4. Commutation Relations

To define the structure of the symmetry algebra for the Volterra equation we need to compute the commutation relations between the symmetries, i.e. the flows commuting with the equation. Apart from the exceptional case, the proof that the symmetry flows commute with the equations has been carried out using the one-to-one correspondence between the integrable equations and the evolution equations for the reflection coefficients. Without using this correspondence it would be extremely difficult to prove the existence of an infinite denumerable number of symmetries. In this section we extend the approach using the reflection coefficients to calculate the commutation relations between the symmetries in order to be able to analyze the structure of the obtained infinite dimensional Lie algebra.

Let us consider the commutation of the isospectral symmetry generators (3.9). We can write the generators for the isospectral symmetries (3.9) as:

(4.1) $$\widehat{X}_k = \mathcal{L}^k\big[a_n(a_{n-1} - a_{n+1})\big]\partial_{a_n}.$$

The fact, proven in Section 3, that Eq. (3.9) provides symmetries for the Volterra equation, implies

(4.2) $$[\widehat{X}_k, \widehat{X}_m] = 0.$$

A natural way of representing the fact that $(\mathrm{d}^2R)/(\mathrm{d}\varepsilon_k \mathrm{d}\varepsilon_m) = (\mathrm{d}^2R)/(\mathrm{d}\varepsilon_m \mathrm{d}\varepsilon_k)$, is to introduce symmetry generators in the space of the reflection coefficients. These generators are written as

(4.3) $$\widehat{\mathcal{X}}_k = \mu\lambda^{2k+1}R\partial_R.$$

In terms of the vector fields $\widehat{\mathcal{X}}_k$, this result is written as

(4.4) $$[\widehat{\mathcal{X}}_k, \widehat{\mathcal{X}}_m] = [\mu\lambda^k R\partial_R, \mu\lambda^m R\partial_R] = 0.$$

So far, the use of the vector fields in the reflection coefficient space has just reexpressed a known result. Taking into account Eq. (3.12), we can define the symmetry generators $\widehat{Y}_k$ corresponding to (3.11) in the reflection coefficient space as:

(4.5) $$\widehat{\mathcal{Y}}_k = \mu\lambda^{2k+1}t[\lambda^2 - 2]R\partial_R + \frac{1}{2}\mu^2\lambda^{2k+1}\partial_\lambda.$$

Commuting $\widehat{\mathcal{Y}}_k$ with $\widehat{\mathcal{Y}}_m$ we have:

(4.6) $$[\widehat{\mathcal{Y}}_k, \widehat{\mathcal{Y}}_m] = (m-k)[\widehat{\mathcal{Y}}_{k+m+1} - 4\widehat{\mathcal{Y}}_{k+m}].$$

From the isomorphism between the spectral space and the space of the solutions, we conclude that the vector fields representing the symmetries of the studied evolution equations, satisfy the same commutation relations. Hence we have

(4.7) $$[\widehat{Y}_k, \widehat{Y}_m] = (m-k)[\widehat{Y}_{k+m+1} - 4\widehat{Y}_{k+m}].$$

In a similar manner we can work out the commutation relations between the $\widehat{Y}_k$ and $\widehat{X}_m$ symmetry generators. We get:

(4.8) $$[\widehat{\mathcal{X}}_k, \widehat{\mathcal{Y}}_m] = -(1+k)\widehat{\mathcal{X}}_{k+m+1} + 2(2k+1)\widehat{\mathcal{X}}_{k+m},$$

and consequently

$$[\widehat{X}_k, \widehat{Y}_m] = -(1+k)\widehat{X}_{k+m+1} + 2(2k+1)\widehat{X}_{k+m}. \tag{4.9}$$

Let us now consider the commutation relations involving the exceptional symmetry (3.13) $\widehat{Z}$. As mentioned in Section 3, these symmetries do not satisfy the asymptotic conditions (3.5). Hence we cannot write them in the space of the reflection coefficient and cannot commute them simultaneously with all the other symmetries. Commuting $\widehat{Z}$ explicitly with the lowest symmetries in the $\widehat{X}$ and $\widehat{Y}$ series, we obtain:

$$[\widehat{Z}, \widehat{X}_0] = X_0, \quad [\widehat{Z}, \widehat{X}_1] = 2(\widehat{X}_0 + \widehat{X}_1), \quad [\widehat{Z}, \widehat{Y}_0] = \widehat{Y}_0 + 4\widehat{Z}. \tag{4.10}$$

Moreover it is easy to prove that the commutator of $\widehat{Z}$ and $\widehat{Y}_k$ with $\widehat{X}_j$ will give rise to a combination of $\widehat{X}_l$. For the Volterra equation $\widehat{X}_0$ and $\widehat{Z}$ are point symmetries. All the others are higher ones. The structure of the Lie algebra is $L = L_0 \niplus L_1$ with $L_0 = \{\widehat{Z}, \widehat{Y}_0, \widehat{Y}_1, \widehat{Y}_2, \dots\}$ and $L_1 = \{\widehat{X}_0, \widehat{X}_1, \dots\}$. Moreover L_0 is a perfect Lie algebra, i.e. we have $[L_0, L_0] = L_0$ and $\{\widehat{Z}, \widehat{X}_0, \widehat{X}_1, \dots, \widehat{X}_k\}$ is, for any finite k a finite dimensional subalgebra of L.

5. Contraction of the Symmetry Algebras in the Continuous Limit

It is well known [**10, 11**] that the Volterra equation has the Korteweg–de Vries equation as continuous limit.

By setting

$$u_n(t) = 1 + h^2 q(x,\tau), \tag{5.1}$$

$$x = (n-2t)h, \tag{5.2}$$

$$\tau = -\frac{1}{3}h^3 t, \tag{5.3}$$

we can write Eq. (1.1) as

$$q_{,\tau} = q_{,xxx} + 6qq_{,x} + \mathcal{O}(h^2), \tag{5.4}$$

i.e. the Korteweg–de Vries equation. Let us now rewrite the symmetry generators in the new coordinate system defined by (5.1)–(5.3) and develop them in Taylor series for small h. We have:

$$\widehat{X}_0 = \left\{-2hq_{,x}(x,\tau) - \frac{1}{3}h^3 q_{,\tau}(x,\tau)\right\}\partial_q, \tag{5.5}$$

$$\widehat{X}_1 = \left\{-8hq_{,x}(x,\tau) - \frac{10}{3}h^3 q_{,\tau}(x,\tau) + \mathcal{O}(h^5)\right\}\partial_q, \tag{5.6}$$

$$\widehat{Y}_0 = \{2[2q(x,\tau) + xq_{,x}(x,\tau) + 3\tau q_{,\tau}(x,\tau)] + \mathcal{O}(h)\}\partial_q, \tag{5.7}$$

$$\widehat{Z} = \left\{\frac{1}{h^2}[1 + 6\tau q_{,x}(x,\tau)] + \mathcal{O}(1)\right\}\partial_q. \tag{5.8}$$

The symmetry generators, written in the evolutionary form, for the Korteweg–de Vries Eq. (5.4) read:

$$\widehat{P}_0 = q_{,\tau}(x,\tau)\partial_q, \tag{5.9}$$

$$\widehat{P}_1 = q_{,x}(x,\tau)\partial_q, \tag{5.10}$$

$$\widehat{B} = \big[1 + 6\tau q_{,x}(x,\tau)\big]\partial_q, \tag{5.11}$$

$$\widehat{D} = \big[2q(x,\tau) + xq_{,x}(x,\tau) + 3\tau q_{,\tau}(x,\tau)\big]\partial_q, \tag{5.12}$$

and their commutation table is:

(5.13)

	$\widehat{P}_0$	$\widehat{P}_1$	$\widehat{B}$	$\widehat{D}$
$\widehat{P}_0$	0	0	$-6\widehat{P}_1$	$-3\widehat{P}_0$
$\widehat{P}_1$		0	0	$-\widehat{P}_1$
$\widehat{B}$			0	$2\widehat{B}$
$\widehat{D}$				0

We can write down some new symmetry generators for the Volterra equation, obtained as linear combinations of the generators (5.5), (5.6), (5.7), (5.8), such that in the continuous limit they go over to the generators of the Korteweg–de Vries equation:

$$\widetilde{P}_0 = \frac{1}{2h^3}(4\widehat{X}_0 - \widehat{X}_1), \tag{5.14}$$

$$\widetilde{P}_1 = \frac{1}{12h}(\widehat{X}_1 - 10\widehat{X}_0), \tag{5.15}$$

$$\widetilde{D} = \frac{1}{2}\widehat{Y}_0, \tag{5.16}$$

$$\widetilde{B} = h^2\widehat{Z}. \tag{5.17}$$

Taking into account the commutation table between the generators $\widehat{X}_0$, $\widehat{X}_1$, $\widehat{Z}$ and $\widehat{Y}_0$, (4.9), (4.10) and the fact that the continuous limit of $\widehat{X}_2$ is given by:

$$\widehat{X}_2 = \left[-32hq_{,x}(x,\tau) - \frac{64}{3}h^3 q_{,\tau}(x,\tau) + \mathcal{O}(h^5)\right]\partial_q, \tag{5.18}$$

we get:

(5.19)

	$\widetilde{P}_0$	$\widetilde{P}_1$	$\widetilde{B}$	$\widetilde{D}$
$\widetilde{P}_0$	0	$\mathcal{O}(h^4)$	$-6\widetilde{P}_1 + \mathcal{O}(h^2)$	$-3\widetilde{P}_0 + \mathcal{O}(h^2)$
$\widetilde{P}_1$		0	$\mathcal{O}(h^4)$	$-\widetilde{P}_1 + \mathcal{O}(h^2)$
$\widetilde{B}$			0	$2\widetilde{B} + \mathcal{O}(h^2)$
$\widetilde{D}$				0

Comparing the commutation tables (5.19) and (5.13) we see that the infinite dimensional Lie algebra generated by $\widehat{X}_0$, $\widehat{X}_1$, $\widehat{Z}$ and $\widehat{Y}_0$, reduces, in the continuous limit, when h goes to 0, to the Lie algebra of the point symmetries of the Korteweg–de Vries equation.

References

1. G. W. Bluman and S. Kumei, *Symmetries and differential equations*, Springer, New York, 1989.
2. H. Flaschka, *On the Toda lattice.* II: *Inverse scattering solutions*, Progr. Theoret. Phys. **51** (1974), 703–716.
3. ———, *The Toda lattice.* II: *Existence of integrals*, Phys. Rev. B **9** (1974), 1924–1925.
4. R. Floreanini and L. Vinet, *Lie symmetries of finite-difference equations*, J. Math. Phys. **36** (1995), 7024–7042.
5. W. Hereman, *Lie symmetry analysis with symbolic software*, Encyclopedia Math., Supplement , vol. I, Kluwer, 1997, pp. 351–355.
6. E. Inonu and E. P. Wigner, *On the contraction of groups and their representations*, Proc. Nat. Ac. Sci. U.S.A **39** (1953), 510–524.
7. A. A. Izmest'ev, G. S. Pogosyan, A. N. Sissakian, and P. Winternitz, *Contraction of Lie algebras and separation of variables. The n-dimensional sphere*, J. Math. Phys. **40** (1999), 1549–1573.
8. D. Levi, R. H. Heredero, and P. Winternitz, *Symmetries of the discrete Burgers equation*, J. Phys. **32** (1999), 2685–2695.
9. D. Levi and O. Ragnisco, *Nonlinear differential-difference equations with N-dependent coefficients.* II, J. Phys. A **12** (1979), L163–L167.
10. D. Levi and M. A. Rodríguez, *Symmetry group of partial differential equations and of differential difference equations*: *The Toda lattice versus the Korteweg–de Vries equation*, J. Phys. A **25** (1992), L975–979.
11. ———, *Lie symmetries for integrable evolution equations on the lattice*, J. Phys. A **32** (1999), 8303–8316.
12. D. Levi, L. Vinet, and P. Winternitz, *Lie group formalism for difference equations*, J. Phys. A **30** (1997), 633–649.
13. D. Levi and P. Winternitz, *Continuous symmetries of discrete equations*, Phys. Lett. A **152** (1991), 335–338.
14. ———, *Symmetries and conditional symmetries of differential-difference equations*, J. Math. Phys. **34** (1993), 3713–3730.
15. S. Maeda, *The similarity method for difference equations*, IMA J. Appl. Math. **38** (1987), 129–134.
16. ———, *Canonical structure and symmetries for discrete systems*, Math. Japon. **25** (1980), 405–420.
17. S. V. Manakov, *Complete integrability and stochastization of discrete dynamical systems*, Soviet Phys. JETP **40** (1975), 269–274.
18. R. V. Moody and J. Patera, *Discrete and continuous graded contractions of representation of Lie algebras*, J. Phys. A **24** (1991), 2227–2258.
19. J. Negro, L. M. Nieto, R. Floreanini, and L. Vinet, *Symmetries of the heat equation on the lattice*, Lett. Math. Phys. **36** (1996), 351–355.
20. P. J. Olver, *Applications of Lie groups to differential equations*, 2nd ed., Springer, New York, 1986.
21. J. Patera, G. S. Pogosyan, and P. Winternitz, *Graded contractions of the Lie algebra* $e(2,1)$, J. Phys. A **32** (1999) 805–826.
22. G. R. W. Quispel, H. W. Capel, and R. Sahadevan, *Continuous symmetries of differential-difference equations*: *The Kac–van Moerbeke equation and Painlevé reduction*, Phys. Lett. A **170** (1992), 379–383.
23. P. Winternitz, *Lie groups and solutions of nonlinear partial differential equations*, Integrable Systems, Quantum Groups and Quantum Field Theory, Kluwer, 1993, pp. 429–495.

Dept. de Física Teórica, Universidad Complutense, E-28040 Madrid, Spain.
Current address: Mathematics Department 379 Applied Sciences, University of California at Santa Cruz, 1156 High Street, Santa Cruz, CA 95064, USA.
E-mail address: rafahh@eucmos.sim.ucm.es

Dipartimento di Fisica, Università di Roma Tre and INFN-Sezione di Roma Tre, Via della Vasca Navale 84, I-00146 Roma, Italy.
E-mail address: levi@amaldi.fis.uniroma3.it

Dept. de Física Teórica, Universidad Complutense, E-28040 Madrid, Spain.
E-mail address: rodrigue@eucmos.sim.ucm.es

Centre de Recherches Mathématiques, Univ. de Montréal, C.P. 6128, succ. Centre-ville, Montréal, Québec H3C 3J7, Canada.
E-mail address: wintern@crm.umontreal.ca

Centre de Recherches Mathématiques
CRM Proceedings and Lecture Notes
Volume **29**, 2001

Bäcklund Transformations from the Bilinear Viewpoint

Jarmo Hietarinta

ABSTRACT. We discuss Bäcklund transformations within Hirota's bilinear formalism. In particular we comment on the "expansion method" used in generating Bäcklund transformations and the verification of the self-consistency of the obtained equations.

1. Introduction

Nonlinear PDE's are notoriously difficult to solve. We may say that one requirement for integrability is the existence of some methods by which we can nevertheless find solutions that are general enough. One of the topics of this workshop is the Bäcklund transformations, which is such a method: Given a seed solution one can construct new solutions (with free parameters) by solving a simpler set of equations [**12**]. In Hirota's bilinear formalism the original equation is "bilinear" in the dependent variables, but the Bäcklund transformation equations are truly bilinear, i.e., linear both in the new and in the old variables. Thus in the bilinear formalism one only needs to solve a set of *linear* equations in order to get new solutions.

2. Hirota's Method

Hirota's bilinear formalism has turned out to be a very fruitful method in the study of integrable systems. Hirota proposed it in 1971 as a convenient tool for obtaining soliton solutions directly, after a change of dependent variables [**8**]. Subsequently the new dependent variables (the τ-functions) were found to have interesting mathematical properties and this led to further deep mathematical developments by Sato and his followers (the Kyoto school) [**10**]. Before discussing the use of Hirota's method for Bäcklund transformations it is useful to illustrate it in its original application, in generating soliton solutions.

The first (and the only nonalgorithmic) step in this formalism is to make a transformation from the original nonlinear form to the "bilinear form". As an example let us take the Korteweg–de Vries equation

$$u_{xxx} + 6uu_x + u_t = 0.$$

2000 *Mathematics Subject Classification.* Primary: 35Q53; Secondary: 35N05.
This is the final form of the paper.

Hirota proposed [**8**] the transformation

$$u = 2\partial_x^2 \log f,$$

which yields (after one x-integration):

$$f_{xxxx}f - 4f_{xxx}f_x + 3f_{xx}^2 + f_{xt}f - f_xf_t = 0.$$

This may look more complicated than the original equation, but it can be written in a condensed form using Hirota's D operator:

$$(D_x^4 + D_xD_t)f \cdot f = 0,$$

where $D_x \equiv \partial_{x_1} - \partial_{x_2}$ acts on a pair of functions (the 'dot product') anti-symmetrically:

$$D_x^n f \cdot g = (\partial_{x_1} - \partial_{x_2})^n f(x_1)g(x_2)\big|_{x_2=x_1=x}.$$

The best motivation for using f instead of u is that soliton solutions look particularly simple in terms of f, as will be shown below. Indeed, the antisymmetry of the D-operator is essential and implies in particular that $P(D)e^{x\cdot p} \cdot e^{x\cdot p} = P(0)e^{2x\cdot p}$. From this it follows that power series expansions in $e^{x\cdot p}$ can truncate, because the leading term vanishes when $P(0) = 0$.

Let us now use the above observations and construct soliton solutions for any PDE's that can be transformed into the form

$$P(D_x, D_y, \dots)f \cdot f = 0,$$

with constant coefficients and with the property $P(0, 0, \dots) = 0$. Direct calculations now show that the one- *and* two-soliton solutions look very simple in terms of f:

1SS: $f = 1 + e^{\eta}$, where $\eta = px + qy + \dots$, and the parameters $p, q, \dots$ satisfy the dispersion relation $P(p, q, \dots) = 0$.

2SS: $f = 1 + e^{\eta_1} + e^{\eta_2} + A_{12}e^{\eta_1+\eta_2}$ with the same dispersion relations $P(p_i, q_i, \dots) = 0$, and with a phase factor

$$A_{ij} = -\frac{P(p_i - p_j, q_i - q_j, \dots)}{P(p_i + p_j, q_i + q_j, \dots)}.$$

It is clear that in terms of u these solutions look much more complicated.

The above constructions work for any $P(D_x, D_y, \dots)$ without a constant term, but this does *not* imply that one can similarly construct three-soliton solutions: they are possible only for integrable PDE's.

Indeed, in the integrable case the 2SS extends to a 3SS without any new freedom:

$$f = 1 + e^{\eta_1} + e^{\eta_2} + e^{\eta_3} \\ + A_{12}e^{\eta_1+\eta_2} + A_{13}e^{\eta_1+\eta_3} + A_{23}e^{\eta_2+\eta_3} + A_{12}A_{13}A_{23}e^{\eta_1+\eta_2+\eta_3}.$$

Here the coefficients are determined uniquely when we consider limits where one of the solitons is far away, i.e., when $\eta_k \to \pm\infty$, and insist that the remaining part is given by the 2SS above, up to shifts in the η_i. In general $\big(A_{ij} = e^{\varphi(i,j)}\big)$:

$$f = \sum_{\substack{\mu_i=0,1\\ 1\le i\le N}} \exp\Bigg(\sum_{1\le i<j\le N} \varphi(i,j)\mu_i\mu_j + \sum_{i=1}^{N} \mu_i\eta_i\Bigg).$$

The fact that three soliton solutions of the above form seem to be possible only for integrable PDE's can be used to search for such equations [**1–5**]. Since integrable nonlinear evolution equations are rather scarce one of the long term activities of the "integrable community" has been the search for new integrable systems. In order to start such a search one has to choose a definition of integrability, or a property closely associated with it. The property that is analyzed will then imply certain assumptions under which the search is sufficiently general yet possible in practice. In this conference P. Olver reported on a search based on symmetry properties and there the natural starting assumption is that the equation should have the form $u_t = f(u, u_x, u_{xx}, \dots)$ [**9, 11**]. This means that the x, t dependence is fixed but the nonlinearity is left completely open. It is interesting to observe that the starting assumptions in the search based on the existence of a three-soliton solution [**1–5**] are the opposite: There the nonlinearity is fixed (it follows from $P(D)f \cdot f = 0$ after the substitution $f = e^u$), while the x, $y, \dots$ dependence is left open.

3. The Kadomtsev–Petviashvili Equation

3.1. Obtaining the bilinear Bäcklund transformation using the expansion method. In 1974 Hirota showed how his bilinear method can be used to find Bäcklund transformations for integrable equations [**6**]. Suppose the bilinear equation is of the form

$$P(D)f \cdot f = 0.$$

Hirota's idea was to consider the combination

$$[P(D)f \cdot f]f'f' - [P(D)f' \cdot f']ff,$$

and write it in the terms some $A(D)f \cdot f'$ and $B(D)f \cdot f'$.

As an example let us consider the Kadomtsev–Petviashvili equation, which in the bilinear form reads

$$P(D)f \cdot f \equiv (D_x^4 + D_x D_y + 3D_z^2)f \cdot f = 0.$$

We have the following identity:

$$\begin{aligned}(3.1)\quad &[(D_x^4 + D_x D_y + 3D_z^2)f \cdot f]f'f' - [(D_x^4 + D_x D_y + 3D_z^2)f' \cdot f']ff \\ &\quad + 2D_x\big[(D_x^3 - 3D_x D_z + D_y - 3\lambda D_x - 3\mu D_z + \rho)f' \cdot f\big] \cdot (ff') \\ &\quad - 6D_x\big[(D_x^2 + D_z + \mu D_x + \lambda)f' \cdot f\big](D_x f' \cdot f) \\ &\quad + 6D_z\big[(D_x^2 + D_z + \mu D_x + \lambda)f' \cdot f\big] \cdot (f'f) = 0,\end{aligned}$$

and from this one concludes that the Bäcklund transformation for KP is given by the pair

$$(3.2)\quad \begin{aligned}A(D)f' \cdot f &\equiv (D_x^3 - 3D_x D_z + D_y - 3\lambda D_x - 3\mu D_z + \rho)f' \cdot f = 0, \\ B(D)f' \cdot f &\equiv (D_x^2 + D_z + \mu D_x + \lambda)f' \cdot f = 0.\end{aligned}$$

Here λ, μ and ρ are free parameters.

Indeed, on the basis of (3.1), (3.2) we can conclude that

if f solves $P(D)f \cdot f = 0$,
and if for this f one can solve f' from $A(D)f' \cdot f = 0$ and $B(D)f' \cdot f = 0$,
then f' solves $P(D)f' \cdot f' = 0$.

3.2. The consistency condition. Note open question about *consistency*: Given f, do equations $A(D)f' \cdot f = 0$ and $B(D)f' \cdot f = 0$ really have common solutions f'?

In order to study this question let us define the differential operators L_A and L_B by

$$\begin{aligned} L_A(g) &\equiv f^{-1}A(D)g \cdot f \\ &\equiv \big[\partial_x^3 - 3(f_x/f)\partial_x^2 - 3\partial_x\partial_z + (f_{xx}/f + f_z/f - \lambda)\partial_x + \partial_y \\ &\qquad + 3(f_x/f - \mu)\partial_z + f^{-1}(-3f_{xz} - f_{xxx} + 3\lambda f_x - f_y + 3\mu f_z) + \rho\big]g, \\ L_B(g) &\equiv f^{-1}B(D)g \cdot f \\ &\equiv \big[\partial_x^2 + (\mu - 2f_x/f)\partial_x + \partial_z + f^{-1}(f_{xx} - \mu f_x - f_z) + \lambda\big]g. \end{aligned}$$

One can show that for these operators

$$L_A L_B - L_B L_A = \big[\partial_x(f^{-2}P(D)f \cdot f)\big] - 3\big[\partial_x(f^{-2}D_x^2 f \cdot f)\big]L_B,$$

and the equations are indeed consistent, provided that $P(D)f \cdot f = 0$. [Note the appearance of a term linear in L_B, this is due to the *bilinear* form of (3.2): Instead of L_A we could consider the operator $\tilde{L}_A(g) := L_A(g) + 3f\partial_x\big(f^{-1}L_B(g)\big)$, which is independent of λ. Then $[\tilde{L}_A, L_B] = \big[\partial_x\big(f^{-2}P(D)f \cdot f\big)\big]$, but on the other hand $f\tilde{L}_A$ is no longer bilinear in f and g.]

The Bäcklund transformation (3.2) is therefore consistent. However, if variable coefficients are allowed in P, A and B it is not difficult to construct counterexamples, which satisfy some expansion identity similar to (3.1) but for which the resulting equations are inconsistent.

4. The sine-Gordon Equation

As the second example let us consider the sine-Gordon equation

$$u_{xy} = \sin(u),$$

Hirota proposed [**7**] the substitution

$$u = 4\arctan(g/f),$$

after which the equation becomes proportional to

$$(f^2 - g^2)\big[(D_xD_y - 1)g \cdot f\big] - fg\big[D_xD_y(f \cdot f - g \cdot g)\big] = 0. \tag{4.1}$$

Thus the bilinear equivalent of SG is

$$\begin{cases} A(D)g \cdot f \equiv (D_xD_y - 1)g \cdot f = 0, \\ B(D)(f \cdot f - g \cdot g) \equiv D_xD_y(f \cdot f - g \cdot g) = 0. \end{cases} \tag{4.2}$$

4.1. Expansion method. The proposed Bäcklund transformation for the sine-Gordon is [**6**]

$$W_1 \equiv (D_x + \lambda)(f \cdot f' - g \cdot g') = 0, \tag{4.3}$$

$$W_2 \equiv (D_x + \lambda + \beta)(f \cdot g' + g \cdot f') = 0, \tag{4.4}$$

$$W_3 \equiv (D_y + \mu - 1/\beta)(f \cdot f' + g \cdot g') = 0, \tag{4.5}$$

$$W_4 \equiv (D_y + \mu)(f \cdot g' - g \cdot f') = 0. \tag{4.6}$$

Indeed, one can derive the following identities

$$\begin{aligned}(4.7)\quad & \big[A(D)g\cdot f\big]f'g' - \big[A(D)f'\cdot g'\big]fg \\ & \qquad + \frac{1}{2}D_yW_1\cdot(ff'-gg') - \frac{1}{2}D_xW_3\cdot(ff'+gg') = 0,\end{aligned}$$

$$\begin{aligned}(4.8)\quad & \big[A(D)g\cdot f\big](f'^2-g'^2) - \big[B(D)(f'\cdot f'-g'\cdot g')\big]fg \\ & -\frac{1}{2}\big[(D_y+1/\beta)W_1\cdot(fg'+gf') + (D_y-1/\beta)W_2\cdot(ff'-gg') \\ & \qquad -(D_x-\beta)W_3\cdot(fg'-gf') - (D_x+\beta)W_4\cdot(ff'+gg')\big] = 0,\end{aligned}$$

$$\begin{aligned}(4.9)\quad & \big[B(D)(f\cdot f-g\cdot g)\big](f'^2-g'^2) - \big[B(D)(f'\cdot f'-g'\cdot g')\big](f^2-g^2) \\ & \qquad -2D_yW_1\cdot(ff'-gg') + 2D_xW_4\cdot(fg'-gf') = 0,\end{aligned}$$

$$\begin{aligned}(4.10)\quad & \big[B(D)(f\cdot f-g\cdot g)\big]f'g' - \big[A(D)f'\cdot g'\big](f^2-g^2) \\ & -\frac{1}{2}\big[(D_y-1/\beta)W_1\cdot(fg'+gf') + (D_y+1/\beta)W_2\cdot(ff'-gg') \\ & \qquad +(D_x-\beta)W_3\cdot(fg'-gf') + (D_x+\beta)W_4\cdot(ff'+gg') = 0.\end{aligned}$$

$$\begin{aligned}(4.11)\quad & D_yW_1\cdot(ff'-gg') + D_yW_2\cdot(fg'+gf') \\ & \qquad = D_xW_3\cdot(ff'+gg') + D_xW_4\cdot(fg'-gf').\end{aligned}$$

If we just take the expansions (4.7), (4.8) we can make the following surprising statement:

> If f, g are such that they solve $A(D)g\cdot f=0$,
> and if for these f, g one can solve f', g' from (4.3)–(4.6)
> then these f', g' solve $A(D)g'\cdot f'=0$ *and* $B(D)(f'\cdot f'-g'\cdot g')=0$,

Thus we would ostentiously have a Bäcklund transformation from one equation to a pair of equations. This is because the expansion method does not say anything about the self consistency of (4.3)–(4.6).

4.2. Consistency. Resolving (4.3)–(4.6) for the primed quantities yields two matrix equations

$$(-M\partial_x + M_x + \lambda M + \beta L_1)F' = 0, \tag{4.12}$$

$$\left(-M^t\partial_y + M_y^t + \mu M^t - \frac{1}{\beta}L_2\right)F' = 0, \tag{4.13}$$

where

$$M = \begin{pmatrix} f & -g \\ g & f\end{pmatrix},\quad L_1 = \begin{pmatrix} 0 & 0 \\ g & f\end{pmatrix},\quad L_2 = \begin{pmatrix} f & g \\ 0 & 0\end{pmatrix},\quad F' = \begin{pmatrix} f' \\ g'\end{pmatrix}.$$

Equations (4.12), (4.13) can also be written as

$$(\partial_x - A)F' = \lambda F', \tag{4.14}$$

$$(\partial_y - B)F' = \mu F', \tag{4.15}$$

where

$$A = M^{-1}(M_x + \beta L_1),\quad B = (M^t)^{-1}\left(M_y^t - \frac{1}{\beta}L_2\right).$$

The standard consistency condition for (4.14), (4.15) is

$$A_y - B_x + [A, B] = 0,$$

and when this is calculated it yields a matrix proportional to (4.1). Thus, if we assume that f, g solve $A(D)g \cdot f = 0$, the consistency condition is that they should also solve $B(D)(f \cdot f - g \cdot g) = 0$.

5. Conclusion

In this contribution we have discussed Bäcklund transformations from the bilinear point of view. The expansion method used for generating Bäcklund transformations somewhat obscures the relationship between the equations. A safe approach is the following: *A set of equations is a Bäcklund transformation between those equations that arise as its consistency conditions.* That is:

Assume we have two sets of dependent variables, unprimed and primed ones.

Assume a set of bilinear equations, linear both in the unprimed and in the primed quantities. [These equations should contain enough free parameters to guarantee that the new solution is genuinely new.]

The consistency condition for solving the primed quantities from the above equations leads to an equation (or equations) for the unprimed ones, and vice versa. [These equations may or may not be bilinear.]

The set of equations in 2 is a Bäcklund transformation between the equations obtained in 3.

Thus the expansion method may be useful for finding a Bäcklund transformation candidate, but the consistency condition must be verified before it can be called a true Bäcklund transformation. One should also make sure that the Bäcklund transformation really adds some freedom into the solution.

References

1. J. Hietarinta, *Equations that pass Hirota's three-soliton condition and other tests of integrability*, Nonlinear Evolution Equations and Dynamical Systems (S. Carillo and O. Ragnisco, eds.), Springer, 1990, pp. 46–50.
2. ———, *A search of bilinear equations passing Hirota's three-soliton condition.* I. KdV*-type bilinear equations*, J. Math. Phys. **28** (1987), no. 8, 1732–1742.
3. ———, *A search for bilinear equations passing Hirota's three-soliton condition.* II. mKdV*-type bilinear equations*, J. Math. Phys. **28** (1987), no. 9, 2094–2101.
4. ———, *A search for bilinear equations passing Hirota's three-soliton condition.* III. *Sine-Gordon-type bilinear equations*, J. Math. Phys. **28** (1987), no. 11, 2586–2592.
5. ———, *A search for bilinear equations passing Hirota's three-soliton condition.* IV. *Complex bilinear equations*, J. Math. Phys. **29** (1988), no. 3, 628–635.
6. R. Hirota, *A new form of Bäcklund transformations and its relation to the inverse scattering problem*, Progr. Theoret Phys. **52**, (1974), 1498–1512.
7. ———, *Exact solution of the Sine–Gordon equation for multiple collisions of solitons*, J. Phys. Soc. Japan **33** (1972), 1459–1463.
8. ———, *Exact solution of the Korteweg–de Vries equation for multiple collisions of solitons*, Phys. Rev. Lett. **27** (1971), 1192–1194.
9. A. V. Mikhailov, A. B. Shabat, and R. I. Yamilov, *The symmetry approach to the classification of nonlinear equations. Complete lists of integrable systems*, Uspekhi Mat. Nauk **42** (1987), no. 4(256), 3–53 (Russian).
10. T. Miwa, M. Jimbo, and E. Date, *Mathematics of solitons*, Cambridge University Press, Cambridge, 1999.

11. P. J. Olver and V. V. Sokolov, *Integrable evolution equations on associative algebras*, Comm. Math. Phys. **193** (1998) 245–268.
12. C. Rogers and W. F. Shadwick, *Bäcklund transformations and their applications*, Mathematics in Science and Engineering, vol. 161, Academic Press Inc., New York-London, 1982.

Department of Physics, University of Turku, FIN-20014 Turku, Finland.
E-mail address: hietarin@utu.fi

Centre de Recherches Mathématiques
CRM Proceedings and Lecture Notes
Volume **29**, 2001

Towards the Lax Formulation of SU(2) Principal Models with Nonconstant Metric

L. Hlavatý

ABSTRACT. The equations that define the Lax pairs for generalized principal chiral models can be solved for any constant nondegenerate bilinear form on su(2). The solution is dependent on one free variable that can serve as the spectral parameter. Necessary conditions for the nonconstant metric on SU(2) that define the integrable models are given.

1. Introduction

In the paper [**6**], analogues of the Bäcklund transformation for the Sine–Gordon equation were used to derive the Lax pair for the evolution equations of the of the chiral $\mathrm{SU}(2) \times \mathrm{SU}(2)$ fields (see also [**5**]). The Lax formulation of the evolution equations for the related principal chiral models

$$\partial_\nu (g^{-1} \partial^\nu g) = 0, \tag{1.1}$$

where $g = g(x_0, x_1)$ are elements of a Lie group G, were given in [**9**].

Both the Lax pair and the Bäcklund transformation for a generalization of the equations (1.1)

$$\partial_\nu (\Lambda^{\nu\mu} A_\mu) = 0, \tag{1.2}$$

where $A_\mu := -i(g^{-1}\partial_\mu g)$ and Λ is a (x_0, x_1) dependent matrix 2×2, were found in [**3**].

In this paper we aim at the Lax formulation of a different generalization of the principal chiral models. It may provide us with the Bäcklund transformations for the generalized models by a procedure similar to [**3**].

2000 *Mathematics Subject Classification.* Primary: 35L10, 35L15; Secondary: 34A55.

Key words and phrases. Principal models, chiral models, sigma models, Lax pair, integrable models, SU(2).

The author gladly acknowledge assistance of the systems Reduce 3.6 [**4**] and Mathematica 3.0 [**8**] by the presented calculations. This work was done with the support of the Grant No. 828/1999 of Ministry of Education of the Czech Republic and the Grant K1019601 of Academy of Sciences of the Czech Republic.

This is the final form of the paper.

2. Formulation of Generalized Principal Chiral Models

The evolution equations of the principal models (1.1) can be derived from the action

$$I[g] = -\int d^2x \eta^{\mu\nu} L(A_\mu, A_\nu), \tag{2.1}$$

where $\eta := \mathrm{diag}(1, -1)$ and L is the Killing form on the Lie algebra $\mathcal{L}(G)$. They are integrable by the inverse scattering method [**9**].

An immediate generalization of the principal chiral models is obtained when in the action (2.1) one considers a general bilinear form instead of the Killing [**7**]. Such a type of models were introduced e.g. as a quasiclassical limit of the Baxter quantum XYZ model [**2**]. Next step of generalization (2.1) is introducing G-dependent symmetric bilinear forms into the action. In the coordinate dependent version the generalized principal chiral models are defined by the action

$$I[g] = \int d^2x L_{ab}(g) \eta^{\mu\nu} (g^{-1}\partial_\mu g)^a (g^{-1}\partial_\nu g)^b, \tag{2.2}$$

where $L_{ab}(g)$ is matrix $\dim G \times \dim G$ corresponding to the G-dependent bilinear form $L(g)$

$$L_{ab}(g) := L(g)(t_a \otimes t_b), \tag{2.3}$$

and t_j are elements of a basis in the Lie algebra of the left-invariant fields. It is useful to consider the bilinear form $L(g)$ as a metric on the group manifold. Lie products of elements of the basis define the structure coefficients

$$[t_a, t_b] = i f_{ab}^c t_c, \tag{2.4}$$

and in the same basis we define the coordinates of the field A

$$iA_\nu = (g^{-1}\partial_\nu g) = (g^{-1}\partial_\nu g)^b t_b = iA_\nu^b t_b. \tag{2.5}$$

Varying the action (2.2) w.r.t. $\eta := g^{-1}\delta g$ we obtain equations of motion for the generalized principal chiral models

$$\partial_\mu A^{\mu,a} + \Gamma_{bc}^a A_\mu^b A^{\mu,c} = 0, \tag{2.6}$$

where

$$\Gamma_{bc}^a := S_{bc}^a + \gamma_{bc}^a, \tag{2.7}$$

and S_{bc}^a is the so-called flat connection given by the structure coefficients

$$S_{bc}^a := -\frac{1}{2}(F_{bc}^a + F_{cb}^a), \quad F_{bc}^a := (L^{-1})^{ap} f_{pb}^q L_{qc}, \tag{2.8}$$

while γ_{bc}^a is the Christoffel symbol for the metric $L_{ab}(g)$ on the group manifold

$$\gamma_{bc}^a := \frac{1}{2}(L^{-1})^{ad}(U_b L_{cd} + U_c L_{bd} - U_d L_{bc}). \tag{2.9}$$

The vector fields U_a in (2.9) are defined in the local group coordinates θ_i as

$$U_a := U_a^i(\theta)\frac{\partial}{\partial \theta_i}, \tag{2.10}$$

where the matrix U is inverse to the matrix V of vielbein coordinates

$$U_a^i(\theta) := (V^{-1})_a^i(\theta), \quad V_i^a(\theta) := -i\left(g^{-1}\frac{\partial g}{\partial \theta_i}\right)^a. \tag{2.11}$$

Note that the connection (2.7) is symmetric in the lower indices

$$\Gamma^a_{bc} = \Gamma^a_{cb}. \tag{2.12}$$

3. The Lax Pairs

In the paper [**7**], ansatz for the Lax formulation of the generalized chiral models was taken in the form

$$[i\partial_0 + P_{ab}A^b_0 t_a + Q_{ab}A^b_1 t_a, i\partial_1 + P_{ab}A^b_1 t_a + Q_{ab}A^b_0 t_a] = 0, \tag{3.1}$$

where P, Q are two auxiliary $\dim G \times \dim G$ matrices. The ansatz (3.1) is a generalization of the Lax pair for L equal to the Killing form $L_{ab} = \mathrm{Tr}(t_a t_b)$ (where Q and P are multiples of the unit matrix) and for the anisotropic SU(2) model where $L_{ab} := L_a \delta_{ab}$ (no summation) with $L_a = \text{const}$, $L_1 = L_2 \neq L_3$. A less general ansatz used in [**1**] leads just to the symmetric spaces, i.e. $L_{ab} := \delta_{ab}$ for SU(2). Necessary conditions that the operators in (3.1) form the Lax pair for the equations of motion (2.6) are

$$P_{ab} f^b_{pq} + (P_{bp}P_{cq} - Q_{bp}Q_{cq}) f^a_{bc} = 0, \tag{3.2}$$

$$\frac{1}{2} f^a_{cd}(P_{cp}Q_{dq} + P_{cq}Q_{dp}) = Q_{ab}\Gamma^b_{pq}. \tag{3.3}$$

If Q is invertible, that we shall assume in the following, then these conditions are also sufficient. Note that the first condition is independent of the bilinear form L so that one can start with solving the equation (3.2) and then look for the bilinear forms L that admit solution of the equation (3.3).

3.1. Solution of the equation (3.2). The structure coefficients for su(2) can be chosen in terms of the totally antisymmetric Levi–Civita tensor

$$f^c_{ab} = \varepsilon_{abc}. \tag{3.4}$$

In this case the equation (3.2) can be rewritten to the form

$$(\mathrm{Adj}\, Q)_{ab} = (\mathrm{Adj}\, P)_{ab} + P_{ba}, \tag{3.5}$$

and solved by

$$Q = \pm (P^T + \mathrm{Adj}\, P)^{-1} / \sqrt{\det(P^T + \mathrm{Adj}\, P)}, \tag{3.6}$$

where P^T is the transpose of P and elements $(\mathrm{Adj}\, N)_{ab}$ of the adjoint matrix to N are obtained as determinants of matrix N with dropped b-th row and a-th column multiplied by $(-)^{a+b}$. If we assume that Q is invertible then the solution (3.6) is unique up to the sign.

Inserting (3.6) into the conditions (3.3) that remain to be solved we obtain rather complicated set of equations of the form

$$G^b_{pq}(P) := \frac{1}{2} R_{ba}\varepsilon_{cda}(P_{cp}R^{-1}_{dq} + P_{cq}R^{-1}_{dp}) = \Gamma^b_{pq}(L), \tag{3.7}$$

where $R = P^T + \mathrm{Adj}\, P$. As we shall see, the fact that the left-hand side is expressed only in terms of elements of the matrix P while the right-hand side only in terms of elements of the metric L imposes restrictions on the metric for which the above given form of Lax pair exists. On the other hand, solvability of these equations for L imposes conditions for P independent of L.

3.2. Conditions for the metric. As the matrix L_{ab} is symmetric we can diagonalize it by orthogonal transformations and the structure coefficients (3.4) remain invariant. It means that for the SU(2) models we can assume without loss of generality that L is diagonal

$$L = \operatorname{diag}\big(L_1(g), L_2(g), L_3(g)\big). \tag{3.8}$$

The right-hand sides of (3.7), i.e. elements of the connection $\Gamma(L)$ have rather special and simple form in this case

$$\Gamma^a_{bc} = \Gamma^a_{cb} = \frac{L_b - L_c}{2L_a}, \quad \text{for } a \neq b,\ a \neq c,\ c \neq b. \tag{3.9}$$

$$\Gamma^a_{bb} = -\frac{U_a L_b}{2L_a}, \quad \text{for } a \neq b, \text{ no sums.} \tag{3.10}$$

$$\Gamma^a_{ab} = \Gamma^a_{ba} = \frac{U_b L_a}{2L_a}, \quad \text{no sums.} \tag{3.11}$$

On the other hand, the left-hand sides of (3.7) satisfy two important identities, namely

$$G^a_{ab} = 0, \quad G^a_{bb} = 0, \tag{3.12}$$

that hold for arbitrary P. Indeed, using the antisymmetry of Levi–Civita symbol ε and (3.7) one gets

$$G^b_{bq} = \frac{1}{2} R_{ba}\varepsilon_{cda} P_{cb} R^{-1}_{dq} = \frac{1}{2}(\operatorname{Adj} P)_{ba}\varepsilon_{cda} P_{cb} R^{-1}_{dq} = \delta_{ca}\varepsilon_{cda} R^{-1}_{dq} \det P = 0.$$

Proof of the second identity in (3.12) can be done e.g. by expressing both inverse and adjoint matrix via Levi–Civita symbol ε but it is very tedious. The easiest way is to check it by computer.

From (3.7), (3.10) and (3.12) we get

$$U_1(L_1 - L_2 - L_3) = 0 \text{ and cyclic permutations of } (1, 2, 3), \tag{3.13}$$

from where we find that

$$L_1(g) = f_2(g) + f_3(g) \text{ and cyclic permutations of } (1, 2, 3), \tag{3.14}$$

where the functions f_1, f_2, f_3 are invariant with respect to the fields U_1, U_2, U_3, respectively. From (3.7), (3.11) and (3.12) we get

$$\sum_{b=1}^{3} U_a \log L_b = 0 \quad \forall a \Rightarrow \det L = L_1 L_2 L_3 = \text{const}, \tag{3.15}$$

because the fields U_1, U_2, U_3 form a basis in T_gG.

3.3. Conditions for the matrix *P*. Comparing (3.10) and (3.11) we immediately see that the following identities hold for arbitrary diagonal metric and $a \neq b$

$$L_a \Gamma^a_{bb} + L_b {\Gamma^b}_{ba} = 0, \quad \text{no sums.} \tag{3.16}$$

Using the equation (3.7), we can replace Γ^a_{bc} by G^a_{bc} and obtain a set of six linear equations for L_j. Their solvability then yields three algebraic equations for elements of the matrix P

$$G^a_{bb} G^b_{aa} = G^b_{ba} G^a_{ba}, \quad \text{no sums.} \tag{3.17}$$

Another set of conditions for G^a_{bc} can be obtained from (3.7) for $a \neq b$, $a \neq c$, $c \neq b$ because in these cases we get three linear equations for L_j

$$2L_a G^a_{cb} = L_b - L_c \text{ no sums,} \tag{3.18}$$

due to (3.9). The solvability condition for these equations reads

$$G^1_{23} G^2_{13} G^3_{12} + G^1_{23} + G^2_{13} + G^3_{12} = 0. \tag{3.19}$$

Note that the equations (3.17) and (3.19) are independent of L. Unfortunately, they are highly nonlinear in elements of P and it seems impossible to solve them without an ansatz.

4. Solutions for General Constant Metric

As it was mentioned above, without loss of generality we can assume that the metric L is diagonal. We shall prove that we can satisfy the conditions (3.2), (3.3) for any constant diagonal L by matrices P, Q containing one free ("spectral") parameter. It means that any generalized principal SU(2) model with constant metric has a Lax pair.

4.1. Diagonal ansatz. The most natural extensions of the results obtained e.g. in [**7**] and [**9**] is the diagonal ansatz for P

$$P = \operatorname{diag}(P_1, P_2, P_3). \tag{4.1}$$

It is rather easy to check that in this case the conditions (3.19) and (3.17) for the matrix P are satisfied identically.

Inserting (4.1) into (3.6) one can immediately see that the matrix Q is diagonal as well, namely $Q = \operatorname{diag}(Q_1, Q_2, Q_3)$ where

$$Q_1 = \pm\sqrt{\frac{R_{22}R_{33}}{R_{11}}}, \quad Q_2 = \frac{R_{33}}{Q_1}, \quad Q_3 = \frac{R_{22}}{Q_1}, \tag{4.2}$$

and

$$R_{11} = P_2P_3 + P_1, \quad R_{22} = P_3P_1 + P_2, \quad R_{33} = P_1P_2 + P_3. \tag{4.3}$$

Inserting (4.1), (4.2) and (4.3) into the left-hand side of (3.7) for $p = q$ and $p = b$ we get zero and using (3.11), (3.10) we find that the metric must be constant in this case.

The equations (3.7) for diagonal P reduce to three nonlinear nonhomogeneous equations for P_j

$$E_1 := P_1P_2\big[\sigma_3(P_3^2 - 1) + P_2^2 - P_1^2\big] + P_3\big[(P_1^2 + P_2^2)\sigma_3 + P_2^2 - P_1^2\big] = 0, \tag{4.4}$$

$$E_2 := P_2P_3\big[\sigma_1(P_1^2 - 1) + P_3^2 - P_2^2\big] + P_1\big[(P_2^2 + P_3^2)\sigma_1 + P_3^2 - P_2^2\big] = 0, \tag{4.5}$$

$$E_3 := P_3P_1\big[\sigma_2(P_2^2 - 1) + P_1^2 - P_3^2\big] + P_2\big[(P_3^2 + P_1^2)\sigma_2 + P_1^2 - P_3^2\big] = 0. \tag{4.6}$$

These equations, where

$$\sigma_1 = \frac{L_2 - L_3}{L_1}, \quad \sigma_2 = \frac{L_3 - L_1}{L_2}, \quad \sigma_3 = \frac{L_1 - L_2}{L_3},$$

are not independent because the following relation holds identically

$$\begin{aligned} E_1P_1L_1(-L_1 + L_2 + L_3) + E_2P_2L_2(-L_2 + L_3 + L_1) \\ + E_3P_3L_3(-L_3 + L_1 + L_2) = 0, \end{aligned} \tag{4.7}$$

and that's why the variety of solutions of (4.4)–(4.6) has the dimension one. The solution curves can be written as

$$(4.8)\qquad \begin{gathered} P_1 = \kappa_1 \frac{\sqrt{\mu + L_2}\sqrt{\mu + L_3}}{\sqrt{L_2}\sqrt{L_3}}, \quad P_2 = \kappa_2 \frac{\sqrt{\mu + L_3}\sqrt{\mu + L_1}}{\sqrt{L_3}\sqrt{L_1}}, \\ P_3 = \kappa_3 \frac{\sqrt{\mu + L_1}\sqrt{\mu + L_2}}{\sqrt{L_1}\sqrt{L_2}}, \end{gathered}$$

where μ is a free parameter and

$$\kappa_1^2 = \kappa_2^2 = 1, \quad \kappa_3 = \kappa_1 \kappa_2.$$

Inserting (4.8) into (4.2) we get

$$(4.9)\qquad Q_1 = \omega_1 \frac{\sqrt{\mu}\sqrt{\mu + L_1}}{\sqrt{L_2}\sqrt{L_3}}, \; Q_2 = \omega_2 \frac{\sqrt{\mu}\sqrt{\mu + L_2}}{\sqrt{L_3}\sqrt{L_1}}, \; Q_3 = \omega_3 \frac{\sqrt{\mu}\sqrt{\mu + L_3}}{\sqrt{L_1}\sqrt{L_2}},$$

where

$$\omega_1 = \kappa_2 \omega_3, \quad \omega_2 = \kappa_1 \omega_3, \quad \omega_3^2 = 1.$$

The formulas (4.1), (4.2), (4.8), and (4.9) yield the solution of the equations (3.2), (3.3) for

$$(4.10)\qquad L_{ab} = L_a \delta_{ab}, \quad f_{ab}^c = \varepsilon_{abc},$$

and up to an eventual orthogonal transformation of the algebra basis, they define the Lax pair for the generalized principal SU(2) chiral model with the constant anisotropic metric. It is easy to check that for $L_1 = L_2 = L_3$ and $L_1 = L_2 \neq L_3$, the Lax pairs coincide with the previously known cases [**7**].

4.2. Block-diagonal ansatz. Our next goal is to find model with the G-dependent metric. As it follows from the preceding subsection, the only possibility is the nondiagonal matrix P. On the other hand the calculations with the general matrix seem hopelessly complicated so that we can try the block-diagonal form

$$(4.11)\qquad P = \begin{pmatrix} p_1 & b_1 & 0 \\ b_2 & p_2 & 0 \\ 0 & 0 & p_3 \end{pmatrix}.$$

However, as we shall see, this ansatz leads again to the constant metric.

Inserting (4.11) into (3.6) we find that the matrix Q has the same block-diagonal form as P. The conditions (3.17) are satisfied identically for block-diagonal P while the equation (3.19) now reads

$$(4.12)\qquad \begin{aligned} 0 = {} & p_3(b_1 p_1 + b_2 p_2)(b_1^2 - b_2^2 - p_1^2 + p_2^2) \\ & \times \left(2p_1 p_2 - 2b_1 b_2 + p_3(1 + b_1^2 + b_2^2 + p_1^2 + p_2^2 - p_3^2)\right). \end{aligned}$$

On the other hand, from (3.7), (3.10) and (3.11) we find that

$$(4.13)\qquad \begin{aligned} U_3 L_k = {} & (-)^k L_k p_3 (b_1 p_1 + b_2 p_2) \\ & \times \left(2p_1 p_2 - 2b_1 b_2 + p_3(1 + b_1^2 + b_2^2 + p_1^2 + p_2^2 - p_3^2)\right), \end{aligned}$$

for $k = 1, 2$ and $U_a L_b = 0$ for all other combinations of indices. Comparing (4.12) and (4.13) we can see that the Lax pair for nonconstant metric can exist only if

$$(4.14)\qquad b_1^2 - b_2^2 = p_1^2 - p_2^2.$$

Unfortunately in this case $G_{12}^3 = 0$ so that from (3.7) and (3.9) we get $L_1 = L_2$ and from (4.13) we find that the metric must be constant.

5. Conclusions

The ansatz of the form (3.1) for the Lax formulation of the generalized principal chiral model equations (2.6) leads to a complicated set of quadratic equations (3.2), (3.3) for elements of the matrices P, Q and the metric L. For the group SU(2), the matrix Q can be solved in terms of P and one can derive conditions for the group dependence of the metric L under which the Lax pair for the generalized principal chiral model may exist.

It seems that there is no other way to solve the equations (3.2), (3.3) but using an ansatz. Using the diagonal ansatz for the matrix P we have found the explicit form of the Lax pair with the spectral parameter for the general anisotropic SU(2) model with constant metric. We have proved that the Lax pair for the nonconstant metric (if it exists in the form (3.1)) requires a more general forms of the matrix P than the block-diagonal (4.11). Unfortunately, all ansätze that we have tried (Jordan form and others) did not simplify the equations for P and L to a form that would be solvable.

References

1. M. Bordemann, M. Forger, J. Laartz, and U. Schaeper, *2-dimensional nonlinear sigma models and Poisson structure*, Harmonic Maps and Integrable Systems (John C. Wood and Allan P. Fordy, eds.), Aspects Math., vol. E23, Vieweg, Braunschweig, 1994, pp. 175–192.
2. I. V. Cherednik, *Relativistic-invariant quasiclassical limits of integrable two-dimensional quantum models*, Teoret. Mat. Fiz. **47** (1981), no. 2, 225–229 (in Russian).
3. M. Gürses and A. Karasu, *Integrable sigma models*, Internat. J. Modern Phys. A **6** (1991), no. 3, 487–500.
4. A. C. Hearn, *Reduce* 3.6, Konrad–Zuse–Zentrum, Berlin, 1996.
5. F. Lund, *Example of a relativistic, completely integrable, Hamiltonian systems*, Phys. Rev. Lett. **38** (1977), 1175–1178.
6. K. Pohlmayer, *Integrable Hamiltonian systems and interaction through quadratic constraints*, Comm. Math. Phys. **46** (1976), 207–221.
7. N. Sochen, *Integrable generalized principal chiral models*, Phys. Lett. B **391** (1997), 374–380.
8. S. Wolfram, *The mathematica book*, 3rd ed., Wolfram Media Inc., Champaign, IL; Cambridge University Press, Cambridge, 1996.
9. V. E. Zakharov and A. V. Mikhailov, *Relativistically invariant two-dimensional models in field theory integrable by the inverse problem technique*, J. Experiment. Theoret. Phys. **74** (1978), no. 6, 1953–1973 (in Russian).

Department of Physics, Faculty of Nuclear Sciences and Physical Engineering, Czech Technical University, Břehová 7, 115 19, Prague 1, Czech Republic.

E-mail address: `hlavaty@br.fjfi.cvut.cz`

Centre de Recherches Mathématiques
CRM Proceedings and Lecture Notes
Volume **29**, 2001

Transcendental Solutions of the sine-Gordon Equation

C. A. Hoenselaers and S. Miccichè

Abstract. We consider solutions of the sine-Gordon equation which are given in terms of Elliptic functions.

In the first part of the paper we devote our attention to solutions obtained through a Separation Ansatz. As a result, the equation decouples into two nonlinear equations for α and β, solutions of which turn out to be Jacobi Elliptic functions. The explicit form of the functions is determined by the zeroes of some quartic polynomials. According to the nature of their roots we shall distinguish several cases and provide a full classification of the solutions.

In the second part we consider a Bäcklund transformation to a seed which describes a steady progressing wave. This simplifies the two Riccati equations associated to the Bäcklund transformation considerably. As a result these equations can be solved exactly. As before, several cases are considered.

Computer generated 3-D plots of the relevant solutions are shown.

1. Introduction

Solutions of the sine-Gordon equation consisting of one or more solitons have been studied by many authors, cf. [**1,5**] for further references. The solutions contain trigonometric and hyperbolic functions of expressions linear in the coordinates.

In this paper we shall study solutions which contain Jacobi elliptic functions. One class of solutions will be derived by way of a Separation Ansatz. Such a Separation Ansatz has previously been used[1] to derive the two soliton solutions in an easy manner.

Another class of solutions will be obtained by applying a Bäcklund transformation to a seed describing a steady wave. The first to derive a solution describing the nonlinear superposition of a soliton and a wave was Wahlquist for the Korteweg–de Vries equation [**7**]. We believe, however, that his way of deriving the solution can be simplified considerably by the approach which we shall outline below for the sine-Gordon equation.

2000 *Mathematics Subject Classification.* Primary: 35B99; Secondary: 35C05.

This is the final form of the paper.

[1]The paper we refer to from memory was written about 20 years ago by three authors. C.A.H. only remembers that a Separation Ansatz was used to derive the two soliton solution. At the meeting R. Conte attracted our attention to [**6**], where a Separation Ansatz has also been used.

We shall give an exhaustive list of the solutions obtainable in both these two approaches. We shall also show plots of some typical cases.

2. The Separation Ansatz

We write the sine-Gordon equation in space-time coordinates x and t as

$$\phi_{xx} - \phi_{tt} = \sin\phi. \tag{2.1}$$

A subscript denotes differentiation with respect to the indicated variable. With the usual substitution

$$\phi = 4\arctan\delta, \tag{2.2}$$

the equation becomes

$$(\delta_{xx} - \delta_{tt})(1+\delta^2) - 2\delta(\delta_x^2 - \delta_t^2) + \delta(\delta - 1) = 0. \tag{2.3}$$

We now make the Ansatz that δ can be separated into a function of x only and a function of t only:

$$\delta = \frac{\alpha(x)}{\beta(t)}. \tag{2.4}$$

With the further substitutions $\alpha_x = \sqrt{A(\alpha)}$ and $\beta_t = \sqrt{B(\beta)}$, Eq. (2.3) can be rewritten as:

$$(\alpha^2+\beta^2)(\beta A_\alpha - \alpha B_\beta) - 2\alpha\beta(A+B) + \alpha\beta(\alpha^2+\beta^2) = 0.$$

After some calculations it can be shown that:

$$\alpha_x^2 = k_4\alpha^4 + k_2\alpha^2 + k_0, \tag{2.5a}$$

$$\beta_t^2 = -k_4\beta^4 + (k_2 - 1)\beta^2 - k_0. \tag{2.5b}$$

If either k_4 or k_0 vanishes, the solution becomes one of the two-soliton solutions. We shall thus assume that neither k_4 nor k_0 is zero. By an appropriate scaling of α and β we can set $k_4 = \varepsilon = \pm 1$. The solutions of these equations are Jacobi elliptic functions. The functional form of α and β is determined by the zeros of the right-hand sides of (2.5). As biquadratics the polynomials can have four real zeros, two real and two purely imaginary zeros, four purely imaginary zeros or four genuinely complex zeros. This leads to the classification in the next section. The explicit form of the two functions α and β in the various cases will be given in a forthcoming paper [**4**].

2.1. Some plots. Below we shall show some typical solutions. The quantity plotted is the Hamiltonian for the sine-Gordon equation, which for our case can be rewritten as

$$H = \frac{8}{(\alpha^2+\beta^2)^2}(\beta^2\alpha_x^2 + \alpha^2\beta_t^2 + \alpha^2\beta^2). \tag{2.6}$$

We have chosen the parameters in each plot so that they display the features we deem most interesting.

Case 1: Four real zeros for α_x^2.

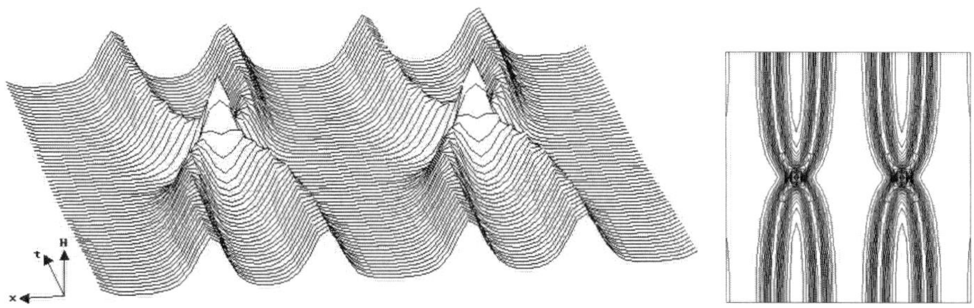

FIGURE 1. Separation Ansatz: Case 1-a. The picture on the right is the contour plot of the picture on the left.

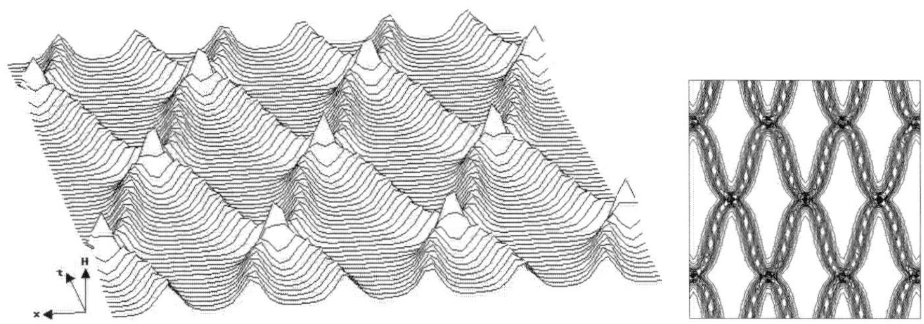

FIGURE 2. Separation Ansatz: Case 1-b. The picture on the right is the contour plot of the picture on the left.

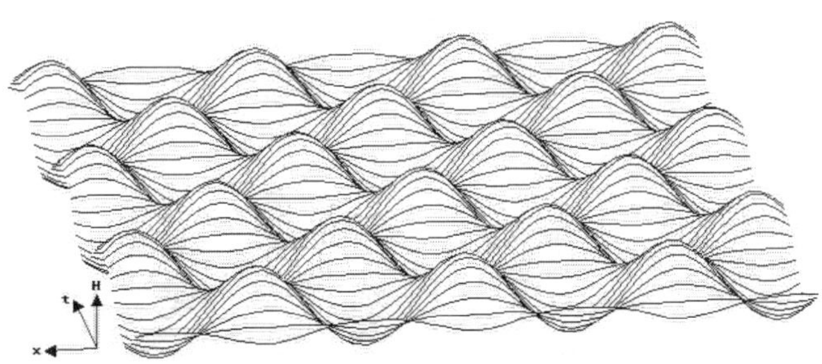

FIGURE 3. Separation Ansatz: Case 1-c.

According to the nature of the zeroes in (2.5b) there are 3 subcases to be considered: four real zeroes for β_t^2 (Case 1-a), four imaginary zeroes for β_t^2 (Case 1-b), four complex zeroes for β_t^2 (Case 1-c).

This is an example of Case 1-a. The Case 1-a describes the interaction of solitons and antisolitons. For particular parameters, the solution is akin to the marginally bound two-soliton solution; the soliton and antisoliton take a long time before interacting with their next neighbour.

This is Case 1-b. We see a train of solitons moving to the right while the anti-solitons move to the left—or vice versa; n.b. the Hamiltonian does not distinguish

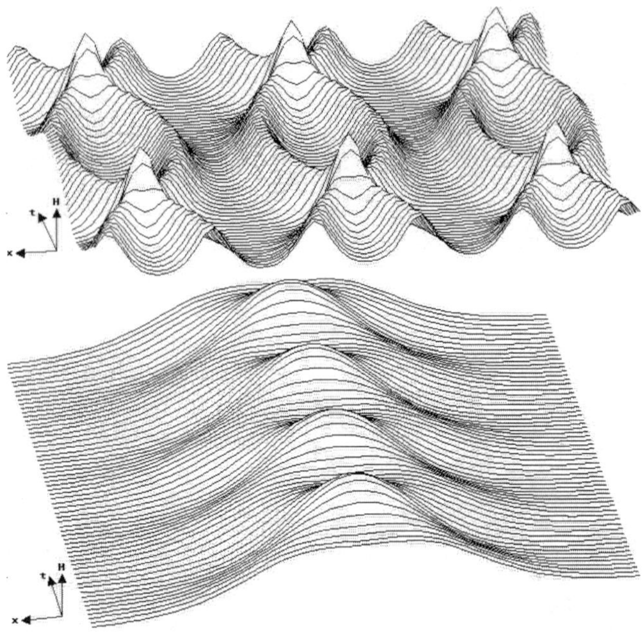

FIGURE 4. Separation Ansatz: Case 2.

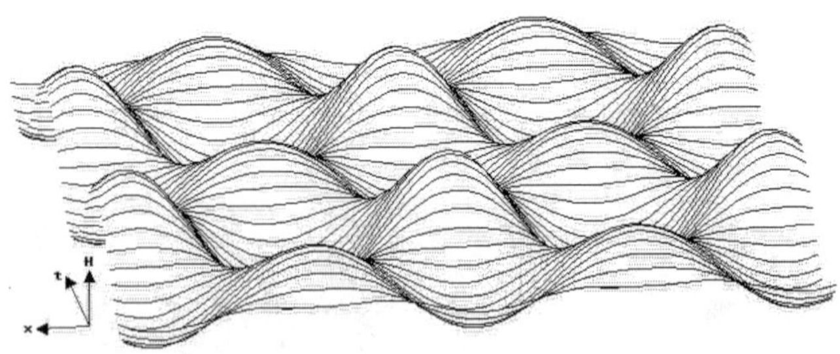

FIGURE 5. Separation Ansatz: Case 3.

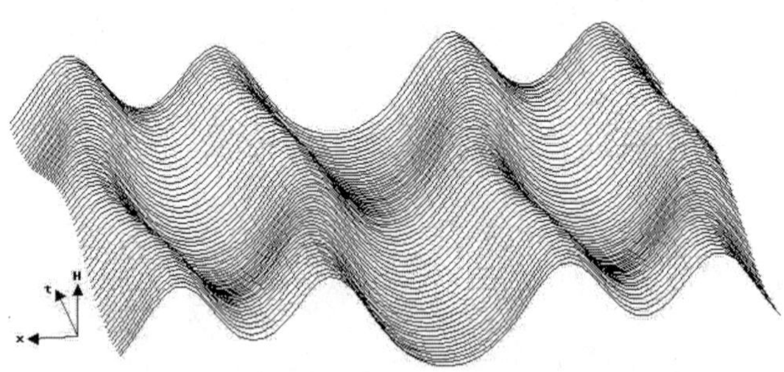

FIGURE 6. Separation Ansatz: Case 4.

between solitons and antisolitons. The contour plots for this and the above figure are shown below.

We have chosen this figure to illustrate a situation which occurs frequently. The parameters are chosen such that the actual paths of solitons and/or antisolitons are virtually indistinguishable and it becomes a moot point whether one is dealing with solitons or simply a doubly periodic solution of the sine-Gordon equation. Finally we mention that the case of β_t^2 having two real and two imaginary zeroes is ruled out.

Case 2: Two real and two imaginary roots for α_x^2 and $\varepsilon = -1$.

This example from Case 2 shows the archetypical breather solution, here, of course, extended periodically in x. With the appropriate choice of parameters the breathers can be rather closely bound. The figure is to be extended periodically to the right and to the left.

Case 3: Two real and two imaginary roots for α_x^2 and $\varepsilon = 1$.

This example from Case 3 is a rather peculiar one. Please note the different height of the bumps. Again it is up to the beholder whether one wants to interpret this solution in terms of (anti)solitons or as a doubly periodic solution of the equation.

Case 4: Four imaginary zeros for α_x^2 and $\varepsilon = 1$.

This solution consists only of solitons, say. We know from the two-soliton solutions that solitons repel each other and consequently the situation can be thought of as solitons being almost held in their place by neighbouring solitons. Note that, also in this case, the plot can be made to look like Fig. 3 by an appropriate choice of parameters.

3. Solitons and Waves

We are now looking for a solution describing a steady progressing wave, i.e. for a solution $\phi = \phi(x - vt)$. As the sine-Gordon equation is invariant under Lorentz-transformations, we can transform ϕ into a function of x only. This is tantamount to chosing a frame in which the wave is at rest. Here, we are neglecting waves moving at superluminal speed; for those, ϕ could be transformed into a function of t only. With $\phi = \phi(x)$ Eq. (2.1) becomes

$$\phi_{xx} = \sin\phi. \tag{3.1}$$

This equation admits a first integral

$$\mathcal{E} = \cos\phi + \frac{1}{2}\phi_x^2, \tag{3.2}$$

and has the solutions (in the notation for the Jacobi elliptic functions and the elliptic integrals we are following Ref. [**2**])

$$\phi_1 = 2\,\mathrm{am}\left(\frac{x}{k}, k\right) - \pi, \qquad \mathcal{E} = \frac{2}{k^2} - 1, \tag{3.3a}$$

$$\phi_2 = 2\arctan\left(k\frac{\mathrm{sn}(x,k)}{\mathrm{dn}(x,k)}\right), \quad \mathcal{E} = 2k^2 - 1. \tag{3.3b}$$

These are the seed solutions which we shall use in the subsequent sections. It should be noted that ϕ_2 becomes the standard soliton solution in the limit $k \to 1$. Throughout this paper we neglect translations along x and t.

3.1. The Bäcklund transformation. The standard Bäcklund transformation for the sine-Gordon equation is normally written in double null coordinates related to the space-time coordinates by $x = u + v, t = u - v$. It reads

$$\gamma_u = \phi_u + \lambda \sin\gamma, \quad \gamma_v = \lambda^{-1}\sin(\gamma - \phi).$$

The integrability condition $\gamma_{uv} = \gamma_{vu}$ is equivalent to the sine-Gordon equation in the form $\phi_{uv} = \sin\phi$ and a new solution is given by

$$\psi = \phi - 2\gamma.$$

These equations can be transformed into Riccati equations by

$$\gamma = 2\arctan\delta,$$

and read in space-time coordinates:

$$\delta_x = \frac{1}{4\lambda}\left[(\delta^2 - 1)\sin\phi + 2\delta(\cos\phi + \lambda^2) + \lambda(\delta^2 + 1)(\phi_x + \phi_t)\right], \tag{3.4a}$$

$$\delta_t = \frac{1}{4\lambda}\left[(1 - \delta^2)\sin\phi - 2\delta(\cos\phi - \lambda^2) + \lambda(\delta^2 + 1)(\phi_x + \phi_t)\right]. \tag{3.4b}$$

The new solution is given by

$$\psi = \phi - 4\arctan\delta.$$

Our seed solution is a function of x only. However, it involves Jacobi elliptic functions. Trying to determine the x-dependence of δ through the standard linearization of the Riccati Eq. (3.4a) would yield a linear second order equation for δ the coefficients of which are rather complicated functions of x. We prefer to solve (3.4b) first.

The substitution

$$\delta = \frac{1}{\Delta a}\Delta_t, \quad a(x) = \frac{1}{4}(\lambda^{-1}\sin\phi - \phi_x), \tag{3.5}$$

transforms Eq. (3.4b) into:

$$\Delta_{tt} + \frac{1}{2\lambda}(\cos\phi + \lambda^2)\Delta_t - \frac{1}{16}(\lambda^{-2}\sin^2\phi - \phi_x^2)\Delta = 0. \tag{3.6}$$

It should be stressed again that this equation determines the t dependence of Δ; the coefficients are thus to be treated as constants. An Ansatz $\Delta = e^{gt}$ is indicated with g being a function of x only. Upon inserting this Ansatz into the equation and solving the resulting quadratic for g the solution is

$$g = f \pm c, \quad f = \frac{1}{4}(\lambda - \lambda^{-1}\cos\phi), \tag{3.7a}$$

$$c = \frac{1}{4\lambda}\sqrt{1 - 2\mathcal{E}\lambda^2 + \lambda^4}. \tag{3.7b}$$

In deriving this result we have made use of (3.2) to replace terms in ϕ_x^2. c is a constant independent of x. The solutions of (3.4b) are given by

$$\delta = \frac{1}{a}\big(c\coth\big(ct + b(x)\big) + f\big) \quad \text{in Case A.} \tag{3.8a}$$

$$\delta = \frac{1}{a}\big(c\tanh\big(ct + b(x)\big) + f\big) \quad \text{in Case B.} \tag{3.8b}$$

$$\delta = \frac{1}{a}\big(c\tan\big(ct + b(x)\big) + f\big) \quad \text{in Case C.} \tag{3.8c}$$

$$\delta = \frac{1}{a}\left(f + \frac{1}{b(x) + t}\right) \quad \text{in Case D.} \tag{3.8d}$$

$$\delta = \frac{1}{a}f \quad \text{in Case E.} \tag{3.8e}$$

Cases C, D, E are to be considered if $\phi = \phi_1$ only. In fact, if we take ϕ_1 as the seed solution we have $\mathcal{E} > 1$ and there are values for λ for which c becomes imaginary. Moreover, $b(x)$ is proportional to c and the hyperbolic functions are to be replaced by the corresponding trigonometric functions. In this case—Case C—c is replaced by $|c|$. Cases D and E correspond to letting $c = 0$ which, again, is only possible for $\phi = \phi_1$. Finally, it may be noticed that case E is a trivial one. Indeed, the solution generated from this δ does not depend on time and, moreover, it can be shown by direct calculation that it has the same first integral $\mathcal{E}$, (3.2), than the seed solution. It can thus differ from the seed only by a translation in x and thus is, for the purpose of the present paper, the same solution.

The integration function $b(x)$ is determined by inserting δ into (3.4a). After some calculation it is found that

$$b_x = c\frac{\lambda\phi_x + \sin\phi}{\lambda\phi_x - \sin\phi}.$$

The integration can be performed [**3**] and, as an example, we quote the result for $\phi = \phi_1$ and real c:

$$b(x) = -cx + \frac{k}{2}\left[\left(\frac{n_1}{\lambda} - \lambda\right)\Pi\left(\operatorname{am}\left(\frac{x}{k}, k\right), \frac{1}{n_1}, k\right)\right. \tag{3.9a}$$
$$\left. - \left(\frac{n_2}{\lambda} - \lambda\right)\Pi\left(\operatorname{am}\left(\frac{x}{k}, k\right), \frac{1}{n_2}, k\right)\right] + \frac{1}{4}\log\left(\frac{\operatorname{sn}^2(x/k, k) - n_2}{\operatorname{sn}^2(x/k, k) - n_1}\right),$$

$$n_{1,2} = \frac{1}{2}(\lambda^2 + 1) \pm 2c\lambda. \tag{3.9b}$$

3.2. Some plots. The functional form of the solutions is complicated enough to defy interpretation without the aid of computer-generated plots. Below we shall show some typical instances. The quantity plotted is again the Hamiltonian, or energy density,

$$H = \frac{1}{2}(\psi_x^2 + \psi_t^2) + \cos\psi - 1, \tag{3.10}$$

or, in some cases, its logarithm.

Case I: $\phi = \phi_1 = 2\operatorname{am}(x/k, k) - \pi$.

- I-a $\delta = \big(c\coth(ct + b) + f\big)/a$.

For this solution we have $k = 0.98$ and $\lambda = 0.817$. The plot shows the interaction of a rather "small" soliton with the background waves. A phenomenon common to all solutions of this class we have seen is not really visible here due to the "smallness" of the soliton but becomes more and more pronounced the larger the soliton becomes.

Here, the parameters are $k = 0.5$ and $\lambda = 10$. One sees clearly that the soliton creates a visible depression among the waves with which it interacts. One is reminded of the interaction of solitons with solitons which repell each other. One might therefore tentatively describe this class of solutions as a soliton interacting with the background solitons.

- I-b $\delta = \big(c \tanh(ct + b) + f\big)/a$.

In the case shown here the parameters are $k = 0.9$ and $\lambda = 3$. The solution shows a markedly different behaviour from the ones in case A. Instead of creating a depression the soliton actually creates what may, with some stretch of imagination be called an elevation. This particular graph is reminiscent of the interaction of solitons and antisolitons which attract each other. The solutions of this class may be thought of as describing an antisoliton interacting with the background solitons.

This phenomenon persists also for larger solitons; here $k = 0.5$ and $\lambda = 10$.

- I-c $\delta = \big(f + 1/(b + t)\big)/a$.

This is the particular situation in which $c = 0$. It can be shown that k and λ are no longer independent. The values $k = 0.985$ and $\lambda = 1.191$ yield the small soliton shown here. It interacts with the background in the usual manner. For a large soliton, $k = 0.28$ and $\lambda = 7$, we encounter again the phenomenon of a considerable depression before and after the passage of the soliton through the waves. Note that for this phenomenon to be as visible as in the case here the soliton has to be large enough to necessitate a logarithmic scaling.

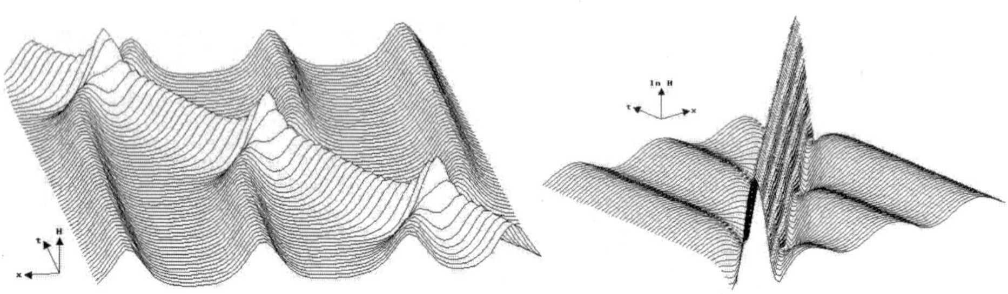

FIGURE 7. Bäcklund transformation: Case I-a.

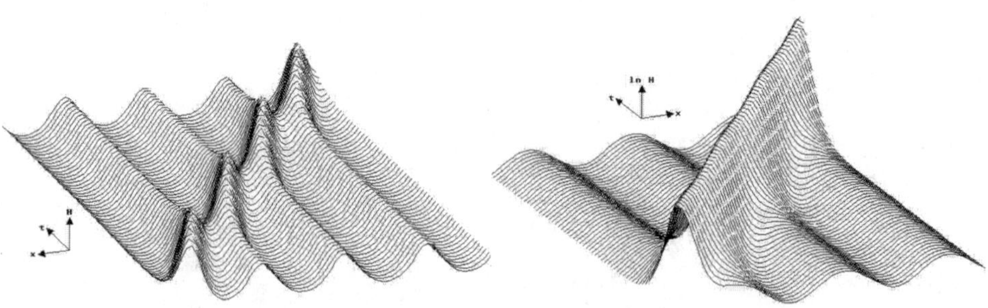

FIGURE 8. Bäcklund transformation: Case I-b.

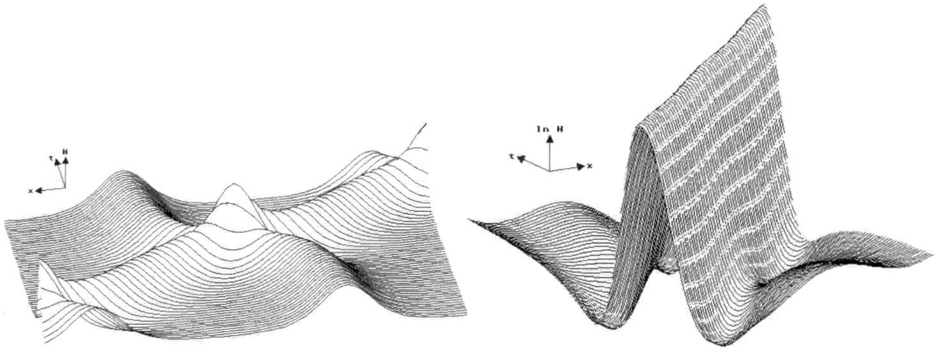

FIGURE 9. Bäcklund transformation: Case I-c.

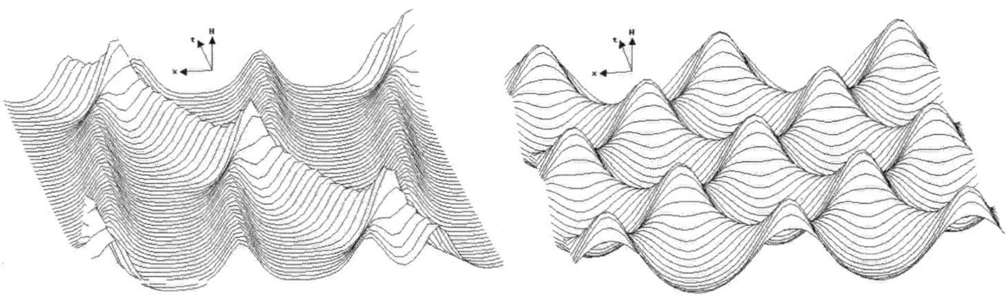

FIGURE 10. Bäcklund transformation: Case I-d.

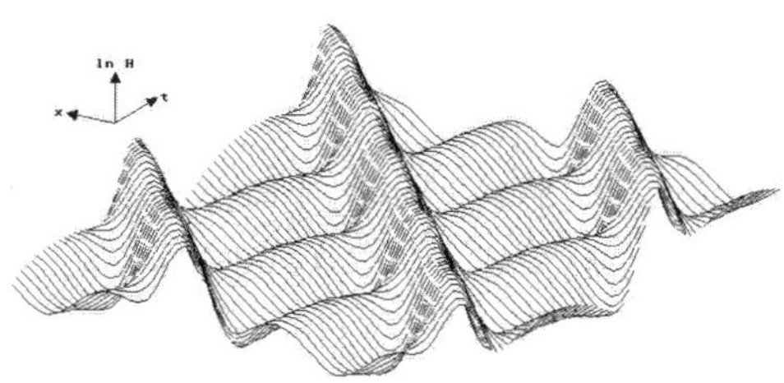

FIGURE 11. Bäcklund transformation: Case I-d.

• I-d $\delta = \big(c\tan(ct+b)+f\big)/a$.

The solutions of this class are multi-soliton solutions. Here we see a typical solution for $k=0.95$ and $\lambda=0.82$. These are rather small solitons and they interact with the waves in a manner one would have expected from the corresponding figures for the two-soliton solutions. This solution, however, for $k=0.5$ and $\lambda=1$ is rather akin to solutions obtained earlier by the Separation Ansatz. At this point we should like to stress that words like "solitons" and "background waves" are terms inspired by the graphs and are nor really applicable to solutions such as this one.

The solution shown here describes rather large solitons moving through the background waves. Here, we have $k=0.5$ and $\lambda=0.36$.

Case II: $\phi=\phi_2=2\arctan\big(\mathrm{sn}(x,k)/\,\mathrm{dn}(x,k)\big)$.

• II-a $\delta = \big(c\coth(ct+b)+f\big)/a$.

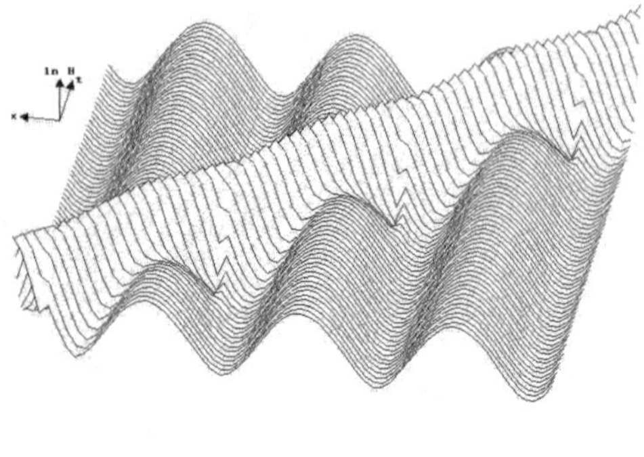

FIGURE 12. Bäcklund transformation: Case II-a.

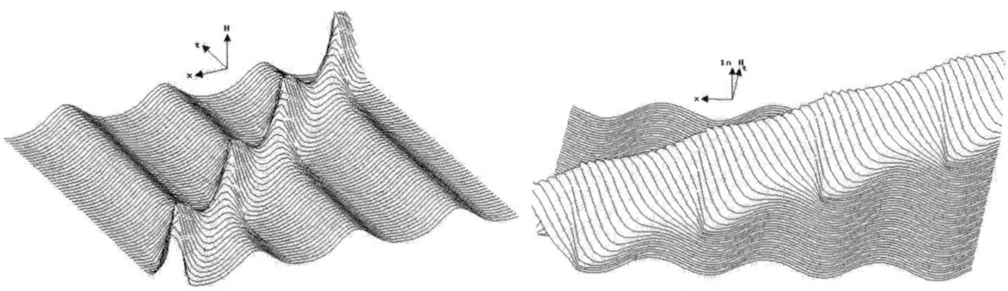

FIGURE 13. Bäcklund transformation: Case II-b.

Here, $k = 0.9$ and $\lambda = 4$. Note the strange feature at the place where the soliton meets the trailing flank of the wave—a similar indentation is present at the leading flank. This is due to rather close subsequent zeros in the denominator of b_x , cf. (3.1). The function b oscillates quickly between $\pm\infty$. For smaller values of λ—n.b. c attains its minimum at $\lambda = 1$—we have not been able to resolve these oscillations numerically which resulted in a jump in δ.

• II-b $\delta = \big(c\tanh(ct+b)+f\big)/a$.

This graph is similar to the one shown in Case I-b. The parameters are $k = 0.9$ and $\lambda = 3$. It almost looks like having imparted some momentum to the background waves in the past and the waves are now bouncing off each other.

In this case one sees again the phenomenon described above, namely the indentation at the flanks of the background waves. The cause is the same: close zeros in the denominator of b. The parameters are $k = 0.5$ and $\lambda = 4.8$.

References

1. R. K. Dodd, J. C. Eilbeck, J. D. Gibbon, and H. C. Morris, *Solitons and nonlinear wave equations*, Academic Press Inc., London-New York, 1982.
2. I. S. Gradshteyn and I. M. Ryzhik, *Table of integrals, series and products*, Academic Press, New York, 1965.
3. W. Gröbner and N. Hofreiter, *Integraltafel*, Springer-Verlag, Wien, 1975.
4. C. A. Hoenselaers and S. Miccichè, 1999, in preparation.
5. R. Meinel, G. Neugebauer, and H. Steudel, *Soliton*, Nonlinear structures, Akademie Verlag, Berlin, 1991 (German).
6. R. Steuerwald, *Über Ennepersche Flächen und Bäcklundsche Transformationen*, Abh. Bayerische Akadem. Wissenschaft **40** (1936), 1–105.

7. H. Wahlquist, *Bäcklund transformations of potentials of thc Korteweg–de Vries equation and the interation of solitons with cnoidal waves*, Bäcklund Transformations, the Inverse Scattering Method, Solitons, and Their Applications (A. Dokl and B. Eckmann, eds.) (Nashville, 1974), Lecture Notes in Math., vol. 515, Springer, Berlin, 1976, pp. 162–183.

DEPARTMENT OF MATHEMATICAL SCIENCES, LOUGHBOROUGH UNIVERSITY, LOUGHBOROUGH, LEICS. LE11 3TU, GB.

E-mail address, C. A. Hoenselaers: C.A.Hoenselaers@Lboro.ac.uk

E-mail address, S. Micciché: S.Micciche@Lboro.ac.uk

Centre de Recherches Mathématiques
CRM Proceedings and Lecture Notes
Volume **29**, 2001

Three Dimensional Skyrmions and Harmonic Maps

T. Ioannidou, B. Piette, and *W. J. Zakrzewski*

ABSTRACT. We review our recent work [**7,8**] in which we have shown how some harmonic maps from S^2 to CP^{N-1} can be used to find nontrivial spherically symmetric static solutions of the SU(N) Skyrme model in 3 dimensions and to generate some of its low energy field configurations.

1. Introduction

Solitons (or, in general, extended structures) arise as classical solutions of many models in $(1+1)$ dimensions. A good review of them can be found in the book by Ablowitz and Clarkson [**1**]. In fact there are two classes of solitons, those that arise in integrable models and those that are topological in nature. The first class of them has been studied in much detail. Many properties of solitons from this class can be proved by exploiting the integrability of the models.

In physical applications one often encounters solitons of the second class. Such solitons are based on topological considerations; the basic field takes values in a manifold with topologically nontrivial properties which are responsible for the existence of solitons. An example is given by the field which takes values in S^1. A field configuration whose energy density is finite must take the same value as $x \to \pm\infty$, hence it becomes a map of $S^1 \to S^1$. Such maps have a topological quantity which we can identify with the soliton number and the minimal field configurations are called "topological solitons".

In $(2+1)$ dimensions topological solitons are based on the maps $S^2 \to S^2$ and have been a subject of many investigations. Most physical applications, however, involve 3 spatial dimensions; here the solitonic applications involve magnetic monopoles, magnetic bubbles, skyrmions and many other topologically nontrivial field configurations. Here the topology is associated with the maps $S^3 \to S^3$.

In this talk we want to discuss some recent work on the Skyrme model. This model was initially proposed as a theory of strong interactions of hadrons [**11**], but recently, it was shown to be the low energy limit of QCD in the large N_c limit [**12**]. As such it presents an opportunity to understand nuclear physics as a low energy limit of quantum chromodynamics (QCD). This is based on the assumption that topologically nontrivial solutions of the Skyrme model, known as skyrmions,

2000 *Mathematics Subject Classification.* Primary: 35Q99; Secondary: 53Z05.
This is the final form of the paper.

can be identified with classical ground states of light nuclei. However a thorough understanding of the structure and dynamics of multi-skyrmion configurations is required before a more qualitative assessment of the validity of this application of the model can be made.

The SU(N) Skyrme model involves fields which take value in SU(N); i.e. are described by SU(N) valued functions of $\vec{x}$ and t. Its static solutions provide us with multiskyrmions and in this talk we will discuss some of our recent work [**7, 8**].

Static fields describing multiskyrmions are stationary points (either maxima or saddle points) of the static energy functional, which is given by

$$E = \int_{R^3} \left\{ -\frac{1}{2} \operatorname{tr} \left(\partial_i U U^{-1}\right)^2 - \frac{1}{16} \operatorname{tr} \left[\partial_i U U^{-1}, \partial_j U U^{-1}\right]^2 \right\} d^3\vec{x}, \tag{1.1}$$

where $U(\vec{x}) \in \mathrm{SU}(N)$.

The equations for multiskyrmions are therefore

$$\partial_i \left(\partial_i U U^{-1} - \frac{1}{4} \left[\partial_j U U^{-1}, \left[\partial_j U U^{-1}, \partial_i U U^{-1}\right]\right] \right) = 0. \tag{1.2}$$

Here, for simplicity, the mass terms have been set to zero. All the discussion can be easily generalised to include mass terms [**7**].

Finiteness of the energy functional requires that $U(\vec{x})$ approaches a constant matrix at spatial infinity, which can be reduced to the identity matrix by a global $U(N)$ transformation. So, without any loss of generality, we can impose that $U \to I$ as $|\vec{x}| \to \infty$.

Since $U \to I$ as $|\vec{x}| \to \infty$ is a mapping from $S^3 \to \mathrm{SU}(N)$, it can be classified by the third homotopy group $\pi_3(\mathrm{SU}(N)) \equiv Z$ or, equivalently, by the integer valued winding number

$$B(U) = \frac{1}{24\pi^2} \int_{R^3} \varepsilon_{ijk} \operatorname{tr} \left(\partial_i U U^{-1} \partial_j U U^{-1} \partial_k U U^{-1}\right) d^3\vec{x}, \tag{1.3}$$

which is a topological invariant. This winding number classifies the solitonic sectors in the model [**11**] and $B(U)$ may be identified with the baryon number of the field configuration.

So far most of the studies of the Skyrme model have concentrated on the SU(2) version of the model and its embeddings into SU(N). The simplest nontrivial classical solution involves a single skyrmion ($B = 1$) and has already been discussed by Skyrme [**11**]. Its energy density is radially symmetric and, as a result, using the so-called hedgehog ansatz one can reduce (1.2) to an ordinary differential equation, which then has to be solved numerically.

Many solutions with $B > 1$ of the SU(2) model have also been computed numerically and, in all cases, the solutions are very symmetrical (c.f. Battye et al. [**3**] and references therein). These studies have shown that the energy density of the two skyrmion solution forms a torus, while the energy density of the $B = 3$ solution has the symmetry of a tetrahedron. For larger B the solutions describe semi-radially symmetric structures in which skyrmions are split into connected parts which are all located on a spherical hollow shell and, as was shown by Houghton et al. in [**6**], the positions of these parts on S^2 are very symmetrical.

All this work has involved the SU(2) skyrmions. An interesting question then arises as to whether there are any finite energy solutions of the SU(N) ($N > 2$) model which are not embeddings of the SU(2) model and, if they exist, whether they have lower energies than their SU(2) counterparts.

Some such *nonembedding* solutions have already been found and studied [**2,10**] but it is generally believed that the SU(2) embeddings provide the lowest energy states. Recently [**7, 8**] we have generalised some very recent work of Houghton et al. [**6**] finding some new solutions of SU(N) Skyrme models (radially symmetric solutions) and also presenting a general procedure for finding low energy field configurations of a given SU(N) Skyrme model. In this talk we discuss the main ideas of our procedure.

2. Harmonic Maps

In a recent paper Houghton et al. [**6**] exploited the similarity of the energy densities of multi-skyrmion solutions with the SU(2) BPS monopoles and related them to some harmonic maps on the two-dimensional sphere S^2. Namely, they showed that the solutions of the Skyrme model can be well approximated by the expressions of the form

$$U(r,\theta,\phi) = \exp\big(ig(r)\hat{\boldsymbol{n}}\cdot\boldsymbol{\sigma}\big), \tag{2.1}$$

where (r,θ,ϕ) are the usual polar coordinates on $\mathbf{R}^3$, and

$$\hat{\boldsymbol{n}} = \frac{1}{1+|R|^2}\big(2\Re(R), 2\Im(R), 1-|R|^2\big), \tag{2.2}$$

where R are some rational functions of $\xi = \tan(\theta/2)e^{i\phi}$ and where $g(r)$ is a real function satisfying the boundary conditions: $g(0) = \pi$ and $g(\infty) = 0$. In other words, the configurations (2.1) factorise into products of a radial profile function $g(r)$ and a harmonic map from the two dimensional sphere, which can be identified with concentric spheres, centered at the origin, in $\mathbf{R}^3$, into an S^2 submanifold of $\mathrm{SU}(2) \equiv S^3$. Moreover, it is easy to check that the baryon number B is given by the degree of the harmonic map $\hat{\boldsymbol{n}}$.

The idea of Houghton et al. was generalised by us [**7**] to SU(N) by modifying their ansatz to

$$U(r,\theta,\phi) = e^{2ig(r)(P-I/N)} = e^{-2ig(r)/N}\left(I + (e^{2ig}-1)P\right), \tag{2.3}$$

where P is a $N\times N$ Hermitian projector which depends only on the angular variables (θ,ϕ). When $N=2$ this expression coincides with (2.1). Note that, the matrix P is a harmonic map from S^2 into CP^{N-1}. Hence it is convenient, rather than using the polar coordinates, to map the sphere onto the complex plane via a stereographic projection and, instead of θ and ϕ, use the complex coordinate ξ and its conjugate. Thus, P can be written as

$$P(V) = \frac{V\otimes V^\dagger}{|V|^2}, \tag{2.4}$$

where V is a N-component complex vector (dependent on ξ and $\bar{\xi}$).

In [**8**] we generalised the method of [**7**] further, by involving more projectors. In this we exploited some ideas from the theory of harmonic maps of $S^2 \to \mathrm{CP}^{N-1}$ [**5**].

To do this we recalled [**5**] that in N dimensional space there is a "natural" set of projectors constructed from $S^2 \to \mathrm{CP}^{N-1}$ maps. They are constructed as follows:

Write each projector P in the form (2.4). The variables ξ and $\bar{\xi}$ on which V depends can be expressed in terms of the familiar θ and φ as $\xi = \tan\theta/2\, e^{i\varphi}$. Now consider a series of such projectors P, changing V. Then for the first projector take $V = f(\xi)$, i.e. an analytic vector of ξ. For the other projectors we take V's which are obtained from the original V by differentiation and Gram Schmidt orthogonalisation. In fact it is sufficient to define an operator P_+ by its action on any vector $v \in \mathrm{C}^N$ [**5**] as

$$P_+ v = \partial_\xi v - \frac{v(v^\dagger \partial_\xi v)}{|v|^2}, \tag{2.5}$$

and then define further vectors $P_+^k v$ by induction: $P_+^k v = P_+(P_+^{k-1} v)$.

Then as our new projectors we take projectors given by (2.4) with V given by $P_+^k f$, where $f = f(\xi)$, i.e.

$$P_0 = P(f), \quad P_1 = P(P_+ f), \quad P_2 = P(P_+^2 f), \ldots, P_k = P(P_+^k f). \tag{2.6}$$

We note, that due to the orthogonality of P_i we have $\sum_{k=0}^{N-1} P_k = 1$.

The orthogonality properties of our projectors follow from the following properties of $P_+^k f$ when f is holomorphic:

$$(P_+^k f)^\dagger P_+^l f = 0, \quad k \neq l, \tag{2.7}$$

$$\partial_{\bar{\xi}} \left(P_+^k f\right) = -P_+^{k-1} f \frac{|P_+^k f|^2}{|P_+^{k-1} f|^2}, \quad \partial_\xi \left(\frac{P_+^{k-1} f}{|P_+^{k-1} f|^2}\right) = \frac{P_+^k f}{|P_+^{k-1} f|^2}. \tag{2.8}$$

Note that for SU(N) the last projector P_{N-1} in this sequence corresponds to an anti-analytic vector; (i.e. the components of $V = P_+^{N-2} f$ are functions of only $\bar{\xi}$).

As we have $\sum_{k=0}^{N-1} P_k = 1$ we see that for SU(N) only $N-1$ projectors are independent so we can choose the first $N-1$, i.e.: P_0, $P_1 \ldots P_{N-2}$.

Then our new ansatz involves the introduction of $N-1$ profile functions and so we consider the field of the form:

$$\begin{aligned} U &= \exp\left[-ig_0\left(\frac{I}{N} - P_0\right) - ig_1\left(\frac{I}{N} - P_1\right) - \ldots - ig_{N-2}\left(\frac{I}{N} - P_{N-2}\right)\right] \\ &= e^{-ig_0/N}(I + A_0 P_0) e^{-ig_1/N}(I + A_1 P_1) \ldots e^{-ig_{N-2}/N}(I + A_{N-1} P_{N-2}), \end{aligned} \tag{2.9}$$

where $g_k = g_k(r)$, for $k = 0, \ldots, N-2$, are the profile functions and $A_k = e^{ig_k} - 1$. The vector V in the projector P_i is given by $V = P_+^i f$. Our original generalisation corresponds to putting all the profile functions, but the first one, equal to zero.

What should we take for the original vector f? A little thought shows that if we take

$$N = 2, \quad f = (1, \xi)^t, \tag{2.10}$$

$$N = 3, \quad f = (1, \sqrt{2}\xi, \xi^2)^t, \tag{2.11}$$

$$N = 4, \quad f = (1, \sqrt{3}\xi, \sqrt{3}\xi^2, \xi^3)^t, \tag{2.12}$$

$$N = n, \quad f = (f_0, f_1, \ldots, f_{r-1})^t : f_k = \xi^k \sqrt{C_{k+1}^{n-1}}, \tag{2.13}$$

where C_{k+1}^{n-1} denotes the binomial coefficients then the modulus of the N-dimensional vector f is a power of $(1 + |\xi|^2)$. In fact, we have $|f|^2 = \left(1 + |\xi|^2\right)^{N-1}$. Moreover, it is also easy to prove [**8**] that

$$|P_+^k f|^2 = k!(N-1)(N-2)\ldots(N-k)\left(1 + |\xi|^2\right)^{N-2k-1}. \tag{2.14}$$

3. General Results

With our generalisations we have obtained several interesting results. All details can be found in [7] and [8]; here we briefly mention only some of them.

First of all, in [8] we showed that when we use $N-1$ projectors (for the SU(N) model) our procedure reduces the full equations to $N-1$ coupled and nonlinear ordinary equations for the profile functions $g_k(r)$. As for SU(N) $\sum_{k=0}^{N-1} P_k = 1$, we will use the first $N-1$ projectors but our expression can be rewritten using any other $N-1$ projectors (with appropriate redefinitions of profile functions). Then, when we solve, numerically, the equations for the profile functions, our procedure gives us exact solutions of SU(N) Skyrme models.

The key ingredient in this construction is the form of the initial analytic vectors f (2.13) and the property (2.14). Detailed calculations reported in [8] show that when we substitute (2.9) into (1.2) all the derivative terms of the projectors cancel and all the nonvanishing terms in the equation of motion are proportional to $1/N - P_i$ (moreover, all the residual dependence on ξ and $\bar{\xi}$ cancels). As there are $N-1$ such independent terms their coefficients provide us with the equations for the profile functions $g_i(r)$. Note that had we taken fewer projectors or different functions f our procedure would not have worked.

The equations for the profile functions take the form

$$
\begin{aligned}
(3.1)\qquad & -\frac{2l+1}{N}\sum_{i=0}^{N-2}\ddot{F}_i + 2\sum_{k=0}^{l}\sum_{i=k}^{N-2}\ddot{F}_i + \ddot{F}_l(l+1)(N-l-1)(1-\cos F_l) \\
& + \frac{1}{2}\dot{F}_l^2(l+1)(N-l-1)\sin F_l \\
& + \frac{2}{r}\left(-\frac{2(l+1)}{N}\sum_{i=0}^{N-2}\dot{F}_i + 2\sum_{k=0}^{l}\left(\sum_{i=k}^{N-2}\dot{F}_i\right) + \dot{F}_l D_{l+1}\right) \\
& - \frac{2}{r^2}(l+1)(N-l-1)\sin F_l + \frac{4}{r^2}(l+1)^2(N-l-1)^2(1-\cos F_l)\sin F_l \\
& - \frac{2}{r^2}(l+1)(N-l-1)\sin F_l\big[l(N-l)(1-\cos F_{l-1}) \\
& \qquad\qquad + (l+2)(N-l-2)(1-\cos F_{l+1})\big] = 0,
\end{aligned}
$$

where

$$F_k = g_k - g_{k+1}, \quad (k = 0, \dots, N-2),$$

and where, of course, $F_{N-2} = g_{N-2}$. These equations have to be solved numerically. To do this one has to specialise to the particular case of N and diagonalise the terms involving the second derivatives.

It is easy to calculate the topological charge B, (1.3), of our configurations. We find [8] that B is given by

$$B = \frac{1}{4\pi}\sum_{i=0}^{N-2}(i+1)(N-i-1)\left(F_i - \sin F_i\right)_{r=0}^{r=\infty} = -\frac{1}{4\pi}\sum_{i=0}^{N-2}(i+1)(N-i-1)F_i(0),$$

where we have used the fact that $g_i(\infty) = 0$ and so the only contributions to the topological charge come from $F_i(0)$.

Our $N-1$ Eqs. (3.1) for $N-1$ functions F_i and so g_i have many symmetries. These symmetries allow us to check whether there exist special solutions which involve a smaller number of functions. The main symmetry, which is relatively easy to spot [8] is the symmetry under the independent interchanges

$$F_0 \leftrightarrow F_{N-2}, \quad F_1 \leftrightarrow F_{N-3}, \quad F_k \leftrightarrow F_{N-k-2}, \ldots.$$

In the next section we restrict our attention to the SU(3) case. An interested reader who wants to see solutions of the SU(N) model for $N > 3$ can find them in [1] and [8].

4. SU(3) Case

• Exact solutions

In this case $N = 3$ and we have two functions: F_0 and F_1. Their equations are:

$$\frac{2}{3}(\ddot{F}_1 + \ddot{F}_0) + 2\ddot{F}_1(1 - \cos F_1) + \dot{F}_1^2 \sin F_1 + \frac{4}{r^2}\sin F_1 - \frac{8}{r^2}\sin(2F_1) + \frac{8}{r^2}\cos F_0 \sin F_1 = 0, \tag{4.1}$$

$$\frac{2}{3}(\ddot{F}_0 + \ddot{F}_1) + 2\ddot{F}_0(1 - \cos F_0) + \dot{F}_0^2 \sin F_0 + \frac{4}{r^2}\sin F_0 - \frac{8}{r^2}\sin(2F_0) + \frac{8}{r^2}\cos F_1 \sin F_0 = 0. \tag{4.2}$$

These equations can be solved for our two functions F_0 and F_1. Clearly, we cannot put either of them to zero but we can take, for instance, $F_0 = F_1 = F$ in which case both the equations become

$$2\ddot{F} + 2\ddot{F}(1 - \cos F) + \dot{F}^2 \sin F + \frac{4}{r^2}\sin F - \frac{4}{r^2}\sin(2F) = 0. \tag{4.3}$$

In addition, there is a further symmetry; we can put $F_0 = -F_1 = G$. In this case both equations reduce to

$$\frac{2}{3}\ddot{G} + 2\ddot{G}(1 - \cos G) + \dot{G}^2 \sin G + \frac{4}{r^2}\sin G - \frac{4}{r^2}\sin(2G) = 0. \tag{4.4}$$

This last case, as $F_0 = g_0 - g_1$ and $F_1 = g_1$ corresponds to the case of $g_0 = 0$ and so our solution involves only one projector, namely P_1. This solution was found and discussed in [7].

So we see that, in general, we have two profile functions but there are cases when we have just one function. Incidentally, the case of Balachandran et al. [2] corresponds to taking a case of two functions and choosing the boundary conditions $g_0(0) = \pi$ and $g_1(0) = 0$.

•Approximate solutions

If we now take an analytic vector f (more general than (2.14)) we will not get a solution of the model but only a low energy field configuration given by (2.3).

Its energy is

$$E = \frac{1}{3\pi}\int dr \left(A_N g_r^2 r^2 + 2\mathcal{N}\sin^2 g(1 + g_r^2) + \mathcal{I}\frac{\sin^4 g}{r^2} \right), \tag{4.5}$$

where $A_N = 2/N(N-1)$ and

$$\mathcal{N} = \frac{i}{2\pi}\int d\xi\, d\bar{\xi}\, \mathrm{tr}\left(|\partial_\xi P|^2\right), \tag{4.6}$$

$$\mathcal{I} = \frac{i}{4\pi}\int d\xi\, d\bar{\xi}\, (1 + |\xi|^2)^2\, \mathrm{tr}\left([\partial_\xi P, \partial_{\bar{\xi}} P]^2\right). \tag{4.7}$$

As the integrals $\mathcal{N}$ and $\mathcal{I}$ in (4.5) are independent of r, we can minimise (4.5) by first minimising $\mathcal{N}$ and $\mathcal{I}$ as functions of the parameters of P and then with respect to the profile function g.

The baryon number is given by

$$B = \frac{i}{\pi^2} \int d\xi \, d\bar{\xi} \, \mathrm{tr}\left(P[\partial_\xi P, \partial_{\bar{\xi}} P]\right) \int_0^\infty dr \sin^2 g \, g_r = \frac{i}{2\pi} \int d\xi \, d\bar{\xi} \, \mathrm{tr}\big(P[\partial_{\bar{\xi}} P, \partial_\xi P]\big), \tag{4.8}$$

which is the topological charge of the two-dimensional $\mathrm{CP}^{(N-1)}$ sigma model.

We have looked [**7**] at various field configurations for the case of SU(3) model. In all cases the lowest energies were quite close to the energies of SU(2) embeddings but in each case slightly higher. The field configurations had, in most cases, very different symmetries (as seen from their energy or baryon charge densities). So both, the embeddings and our configurations, could be used as starting points of possible minimisation programs for SU(3) and higher Skyrme models. Only when such work has been performed will we be able to judge whether our configurations are close to real solutions, how many such solutions there are and which of them are global maxima. Clearly, these are important questions which will be answered in the near future.

5. Conclusions

In this talk we have discussed some results of our recent papers [**7**] and [**8**]. We have shown how to construct radially symmetric solutions of the SU(N) Skyrme model. In the general case such solutions depend on $N-1$ profile functions which have to be determined numerically. In some cases we can exploit symmetries of our expressions and reduce the number of functions.

Note that our procedure uses solutions of the CP^{N-1} model in two dimensions, which is integrable, to construct solutions of the three dimensional Skyrme model which is not integrable. It is clear that the integrability of the CP^{N-1} model is not sufficient; only its very special solutions can be used in our construction. A little thought shows that the key property of the CP^{N-1} fields that make our procedure work is (2.14), i.e. that all ξ and $\bar{\xi}$ dependence of all $|P_+^k f|^2$ resides in $(1+|\xi|^2)^\alpha$.

It is interesting to observe that vectors $P_+^k f$ with this property correspond to conformally minimal immersions of S^2 into CP^{N-1} of constant curvature. In fact our sequence of vectors f, $P+f, \ldots P_+^{N-2} f$ is called the Veronese sequence in [**4**][1]. This observation suggests further avenues of exploration; like looking at immersions with curvatures of a given symmetry.

We have also shown that when we restrict our attention to a given SU(N) model, and take a more general vector f, our construction gives us nontrivial low energy field configurations. They are not solutions of the full equations of motion but, basing our expectations on similar results for the SU(2) model, we expect them to be close to true solutions of the model. As such, they would be good starting points of numerical relaxation schemes designed to find true solutions. Finally, let us add that the projector and multiprojector ideas have also recently been used in the context of finding expressions for the Higgs field in monopole systems [**9**].

[1]W.J.Z. thanks John Bolton for drawing his attention to [**4**].

References

1. M. J. Ablowitz and P. A. Clarkson, *Solitons, nonlinear evolution equations and inverse scattering*, Cambridge University Press, Cambridge, 1991.
2. A. P. Balachandran, A. Barducci, F. Lizzi, V. G. J. Rodgers, and A. Stern, *Doubly strange Diabaryon in the chiral model*, Phys. Rev. Lett. **52** (1984), 887–890.
3. R. A. Battye and P. M Sutcliffe, *Symmetric skyrmions*, Phys. Rev. Lett. **79** (1997), 363–366.
4. J. Bolton, G. R. Jensen, M. Rigoli, and L. M. Woodward, *On conformal minimal immersions of S^2 into* CP^n, Math. Ann. **279** (1988), no. 4, 599–620.
5. A. Din and W. J. Zakrzewski, *General classical solutions of the* CP^{N-1} *model*, Nuclear Phys. B **174** (1980), no. 2-3, 397–406.
6. C. J. Houghton, N. S. Manton, and P. M. Sutcliffe, *Rational maps, monopoles and Skyrmions*, Nuclear Phys. B **510** (1998), no. 3, 507–537.
7. T. Ioannidou, B. Piette, and W. J. Zakrzewski, *Low energy states in the* $\mathrm{SU}(N)$ *Skyrme model*, Proceedings of ISPM-98 (M. Eliashvili *et al.*, eds.), 1998, 91–123.
8. ———, *Spherically symmetric solutions of the* $\mathrm{SU}(N)$ *Skyrme models*, J. Math. Phys. **40** (1999), no. 12, 6223–6233.
9. T. Ioannidou and P. M. Sutcliffe, *Monopoles from rational maps*, Phys. Lett. B **457** (1999), 133–138.
10. V. B. Kopeliovich, B. E. Schwesinger, and B. E. Stern, SU(3) *Skyrmion with $B = 2$ and large strangeness content*, JETP Lett. **62** (1995), 185–90.
11. T. H. R. Skyrme, *A nonlinear field theory*, Proc. Roy. Soc. London Ser. A **260** (1961), 127–138.
12. E. Witten, *Global aspects of current algebra*, Nuclear Phys. B **223** (1983), no. 2, 422–432.
13. W. J. Zakrzewski, *Low-dimensional sigma models*, Adam Hilger Ltd., Bristol, 1989.

Institute of Mathematics, University of Kent at Canterbury, Canterbury CT2 7NF, UK.

E-mail address: T.Ioannidou@ukc.ac.uk

Department of Mathematical Sciences, University of Durham, Durham DH1 3LE, UK.

E-mail address, B. Piette: B.M.A.G.Piette@durham.ac.uk

E-mail address, W. J. Zakrzewski: W.Zakrzewski@durham.ac.uk

Centre de Recherches Mathématiques
CRM Proceedings and Lecture Notes
Volume **29**, 2001

Induced Surfaces and Their Integrable Deformations

B. G. Konopelchenko and G. Landolfi

Abstract. Generalized Weierstrass representations for surfaces immersed into $\mathbb{R}^4$, $\mathbb{R}^{2,2}$ and $\mathbb{R}^{3,1}$ are presented. Integrable deformations of surfaces in these spaces via the Davey–Stewartson hierarchy are discussed.

1. Introduction

Soliton Theory is an interdisciplinary area of research representing the confluence of interests from a wide variety of fields. One of its most intriguing subjects is the study of solitonic equations in connection with the Differential Geometry of surfaces. Mostly due to the development of the Inverse Spectral Transform method (I.S.T.), in the last three decades growing specialization and diversification have brought a lot of papers dealing with several problems and features of interrelations among the differential geometry of surfaces and the theory of integrable systems. Among others, results obtained within the "soliton surfaces approach" proposed by Sym [**10**] deserve to be mentioned (see also [**1**]). In this approach a surface is associated with a given solution of a $1+1$-dimensional integrable PDE that arises as the compatibility condition of two linear problems of the type $\psi_x = Q\psi$ and $\psi_t = P\psi$.

A different use of I.S.T. method in the framework of the differential geometry of surfaces has been proposed recently by one of us (B.G.K.) in connection with $2+1$-dimensional integrable PDE's [**2**]. Although it could be also considered as a direct generalization to "3d soliton hypersurfaces" of the Sym's approach, it is more adequate to consider the approach discussed in [**2**] as a way to generate surfaces and their solitonic evolutions (*integrable deformations*). In fact, surfaces are generated by means of 2D linear problems and features which are typical of $2+1$ soliton equations reflect themselves only in the dynamical evolutions (deformations) of these surfaces.

In this paper we present the generalized Weierstrass representations for surfaces immersed into four-dimensional pseudo-Euclidean spaces. These representations, recently introduced in [**4, 5**], are generated through linear problems of the Davey–Stewartson type. We also discuss integrable deformations of surfaces in four-dimensional pseudo-Euclidean spaces. We point out that once the generalized

2000 *Mathematics Subject Classification.* 53B21.
This is the final form of the paper.

Weierstrass representations are adopted, then integrable deformations are naturally defined by the Davey–Stewartson hierarchy of soliton equations. These equations have an infinite set of integrals of motion which can be interpreted as functionals over surfaces. Remarkably, the total squared mean curvature (Willmore functional) is the simplest of them.

The outline is as follows. In Section 2 we introduce the generalized Weierstrass representation for surfaces in $\mathbb{R}^4$. We also characterize representations for classes of surfaces of interest in differential geometry and discuss applications to the study of rigid string models. In Section 3 we define integrable deformations of surfaces in $\mathbb{R}^4$ via the Davey–Stewartson II hierarchy of integrable equations. Sections 4 and 5 are devoted, respectively, to space-like and time-like surfaces both in $\mathbb{R}^{2,2}$ and $\mathbb{R}^{3,1}$. The last section is a brief summary.

2. Generalized Weierstrass Representation for Surfaces in $\mathbb{R}^4$

The generalized Weierstrass representation for surfaces in $\mathbb{R}^4$ has been proposed recently in [**4, 5**]. It starts with the two two-dimensional Dirac equations

$$\begin{aligned} \psi_{1z} &= p\varphi_1, & \psi_{2z} &= \overline{p}\varphi_2, \\ \varphi_{1\bar{z}} &= -\overline{p}\psi_1, & \varphi_{2\bar{z}} &= -p\psi_2, \end{aligned} \tag{2.1}$$

where z is a complex variable, the bar denotes the complex conjugation, and $\psi_1(z,\bar{z})$, $\varphi_1(z,\bar{z})$, $\psi_2(z,\bar{z})$, $\varphi_2(z,\bar{z})$ and $p(z,\bar{z})$ are complex-valued functions. Given the potential p and solutions ψ_1, φ_1, ψ_2, φ_2 of the system (2.1), one defines four-real functions X^i $(i = 1,2,3,4)$ as follows

$$\begin{aligned} X^1 + iX^2 &= \int_\Gamma (-\varphi_1\varphi_2\, dz' + \psi_1\psi_2\, d\bar{z}'), \\ X^3 + iX^4 &= \int_\Gamma (\varphi_1\overline{\psi}_2 dz' + \psi_1\overline{\varphi}_2\, d\bar{z}'), \end{aligned} \tag{2.2}$$

where Γ is an arbitrary contour in $\mathbb{C}$. By virtue of (2.1) the integrals (2.2) do not depend on the curve of integration Γ. If one treats $X^i(z,\bar{z})$ as coordinates in $\mathbb{R}^4$, then the formulae (2.1)–(2.2) can be used to describe conformal immersions of surfaces S into $\mathbb{R}^4$ with the induced metric ds^2 and the Gaussian curvature K given by

$$ds^2 = u_1 u_2\, dz\, d\bar{z}, \tag{2.3}$$

$$K = -\frac{2}{u_1 u_2}\left[\log(u_1 u_2)\right]_{z\bar{z}}, \tag{2.4}$$

where $u_i = |\psi_i|^2 + |\varphi_i|^2$, $i = 1,2$. Moreover, the squared mean curvature is

$$\vec{H}^2 = 4\frac{|p|^2}{u_1 u_2}, \tag{2.5}$$

so that the Willmore functional $W \equiv \int \vec{H}^2[dS]$ takes the form $\big(x = \mathrm{Re}(z), y = \mathrm{Im}(z)\big)$

$$W = 4\int |p|^2 dx\, dy. \tag{2.6}$$

Since the geometric characteristics take a simple form, then the generalized Weierstrass representation (2.1)–(2.2) turns out to be a powerful tool to study surfaces in $\mathbb{R}^4$. In the following we give some examples of its applications to mathematics as well as to physics.

2.1 Particular classes of surfaces. One of the classical problems in differential geometry is to furnish detailed descriptions of surfaces in the presence of particular conditions (minimal surfaces, surfaces of revolution etc.). Here we characterize the generalized Weierstrass representations corresponding to some geometrically interesting cases.

1. *Minimal surfaces* are surfaces with vanishing mean curvature vector, $\vec{H} = 0$. By virtue of (2.5), they are associated with

$$p \equiv 0. \tag{2.7}$$

2. *Superminimal surfaces* form the subclass of minimal surfaces for which the constraint $(\partial\partial\vec{X})^2 = 0$ is also satisfied. In this case functions involved in the generalized Weierstrass representation (2.1)–(2.2) obey the constraints

$$p = 0 \quad \text{and} \quad \begin{cases} \varphi_1 & = a\overline{\psi}_2 \quad \text{or,} \\ \varphi_2 & = a\overline{\psi}_2. \end{cases} \tag{2.8}$$

3. *Developable surfaces* are surfaces with vanishing Gaussian curvature. From (2.4) it follows that they are given by the generalized Weierstrass representation (2.1)–(2.2) with the additional constraint

$$\left(|\psi_1|^2 + |\varphi_1|^2\right)\left(|\psi_2|^2 + |\varphi_2|^2\right) = \left|A(z)\right|^2, \tag{2.9}$$

where $A(z)$ is an arbitrary holomorphic function.

4. Finally, it is interesting to consider surfaces generated by *one-dimensional reductions* of the potential p in the linear problem (2.1). In particular, assuming

$$p = p(x), \tag{2.10}$$

where $x = \mathrm{Re}(z)$, the solutions of (2.1) take the form

$$\psi_\alpha = \chi_\alpha(x)\exp(\lambda_\alpha y), \quad \varphi_\alpha = \widetilde{\chi}_\alpha(x)\exp(\lambda_\alpha y), \tag{2.11}$$

where $y = \mathrm{Im}(z)$ and $\alpha = 1, 2$. For real λ_1 and λ_2, formulae (2.2) and (2.11) imply

$$\vec{X}(x, y) = \vec{Y}(x)\exp\left[(\lambda_1 + \lambda_2)y\right], \tag{2.12}$$

and

$$\vec{Y}(x) \cdot \vec{Y}(x) = \frac{1}{(\lambda_1 + \lambda_2)^2}. \tag{2.13}$$

We get surfaces obtained by homothety by the factor $\exp\left[(\lambda_1 + \lambda_2)y\right]$ of curves $\vec{Y}(x)$ lying on the sphere $\mathbb{S}^3$ of the radius $1/(\lambda_1 + \lambda_2)$. In the case of pure imaginary λ_1 and λ_2, we get surfaces of revolution in $\mathbb{R}^4$

$$ds^2 = A(x)\,dz\,d\bar{z} = \left(|\chi_1|^2 + |\widetilde{\chi}_1|^2\right)\left(|\chi_1|^2 + |\widetilde{\chi}_1|^2\right) dz\,d\bar{z}. \tag{2.14}$$

2.2 Applications to string theory. The generalized Weierstrass representation (2.1)–(2.2) turns out to be useful also in the study of models of physical interest. As an example, it can be used in the attempt to understand as much as possible the classical and quantum aspects of rigid string models described by the action [**9**]

$$S = S_{\rm NG} + S_{\rm extr}, \tag{2.15}$$

where $S_{\rm NG}$ and $S_{\rm extr}$ are the Nambu–Goto action and the extrinsic curvature action, i.e.

$$S_{\rm NG} = \mu_0 \int [d\Sigma], \quad S_P = \frac{1}{\alpha_0} \int \vec{H}^2 [d\Sigma], \tag{2.16}$$

where $d\Sigma$ denotes the infinitesimal element of the string worldsheet area. By virtue of (2.3) and (2.6), if the string worldsheets are immersed into the Euclidean space $\mathbb{R}^4$ we find

$$S_{\rm NG} = \mu_0 \int u_1 u_2 \, dx \, dy, \quad S_P = \frac{4}{\alpha_0} \int |p|^2 dx \, dy. \tag{2.17}$$

Special classical configurations, referred to as instantons and anti-instantons, play an important role in rigid string theory. They are solutions of the system [**9**]

$$t_{\pm}^{\mu\nu} = c_{\pm}^{\mu\nu}, \tag{2.18}$$

where $t_{\pm}^{\mu\nu} = t^{\mu\nu} \pm t^{*\mu\nu}$, $t^{\mu\nu} = (\varepsilon^{ab})/(\sqrt{\det g})(\partial X^\mu)(\partial \xi^a)(\partial X^\nu)(\partial \xi^b)$, $\det g$ is the determinant of the induced metric on string worldsheet, $t^{*\mu\nu} = 1/2 \varepsilon^{\mu\nu\rho\sigma} t^{\rho\sigma}$ and, finally, $c_{\pm}^{\mu\nu}$'s are constant tensors. In [**6**] equations (2.18) have been studied by means of the generalized Weierstrass representation (2.1)–(2.2). In particular, it has been pointed out that both instantons and anti-instantons of the string theory in $\mathbb{R}^4$ governed by the action (2.15) are given by superminimal immersions.

Generalized Weierstrass formulae (2.1)–(2.2) turn out to be useful also for a one-loop analysis of rigid strings in $\mathbb{R}^4$. Correlators of primary fields can be explicitly computed [**7**].

3. Integrable Deformations of Surfaces in $\mathbb{R}^4$

An advantage of the generalized Weierstrass representation (2.1)–(2.2) is that it naturally provides a way to construct *integrable deformations* of surfaces immersed into $\mathbb{R}^4$ as follows.

Let us assume that the functions ψ, φ and p in (2.1)–(2.2) depend on an evolution parameter t_n and suppose that the evolutions are described by

$$\psi_{t_n} = A_n \psi + B_n \varphi, \quad \varphi_{t_n} = C_n \psi + D_n \varphi, \tag{3.1}$$

where A_n, B_n, C_n, D_n are differential operators of nth order. The basic idea is to define deformations of surfaces by requiring that the evolutions (3.1) are compatible with the linear system (2.1). In the language of soliton theory, the linear problem (2.1) consists of two particular cases of the Davey–Stewartson II linear system (see e.g. [**3**]). Compatibility (2.1) and (3.1) gives rise to an infinite set of $2+1$ nonlinear soliton equations which belong to the Davey–Stewartson II hierarchy of soliton equations. In the special case $n = 2$ one has, in particular, the Davey–Stewartson II equation (see e.g. [**3**])

$$p_{t_2} = i(p_{zz} + p_{\bar{z}\bar{z}} + up), \quad u_{z\bar{z}} = 2|p|^2_{zz} + 2|p|^2_{\bar{z}\bar{z}}, \tag{3.2}$$

where

$$u = w_1 + w_2, \quad w_{1z} = 2|p|_{\bar{z}}, \quad w_{2\bar{z}} = -2|p|^2{}_z.$$

Corresponding operators in (3.1) are of the form

$$(3.3) \qquad \begin{aligned} A_2 &= i(\partial_{\bar{z}}^2 + w_1), \quad & B_2 &= i(p_z - p\partial_z), \\ C_2 &= i(\overline{p}_{\bar{z}} - \overline{p}\partial_{\bar{z}}), \quad & D_2 &= -i(\partial_z^2 + w_2), \end{aligned}$$

for ψ_1, φ_1 and

$$(3.4) \qquad \begin{aligned} A_2 &= -i(\partial_{\bar{z}}^2 + w_1), \quad & B_2 &= -i(\overline{p}_z - \overline{p}\partial_z), \\ C_2 &= -i(p_{\bar{z}} - p\partial_{\bar{z}}), \quad & D_2 &= i(\partial_z^2 + w_2), \end{aligned}$$

for ψ_2, φ_2. Explicit deformations of the coordinates can then be straightforwardly computed [**5**].

Equations of the Davey–Stewartson hierarchy II are integrable by the I.S.T. method. Since the inverse spectral transform method allows us to linearize the initial-value problem

$$(3.5) \qquad p(z, \bar{z}, t_n = 0) \to p(z, \bar{z}, t_n),$$

for soliton equations of the Davey–Stewartson hierarchy, the generalized Weierstrass formulae (2.1)–(2.2) allow us to linearize the initial-value problem for the corresponding deformations of surfaces in $\mathbb{R}^4$

$$(3.6) \qquad \vec{H}(z, \bar{z}, 0) \to \vec{H}(z, \bar{z}, t_n),$$

hence deformations of surfaces in $\mathbb{R}^4$ generated by the Davey–Stewartson II hierarchy can be referred to as *integrable* ones. Such integrable deformations of surfaces inherit all properties of equations of the Davey–Stewartson II hierarchy. In particular, integrals of motions of the soliton equations of the Davey–Stewartson II hierarchy are certain functionals on surfaces invariant under deformations generated by these equations. It must be pointed out that the simplest integral of motion takes the form

$$(3.7) \qquad C_1 = \int |p|^2 \, dx \, dy.$$

Therefore the Willmore functional for surfaces in $\mathbb{R}^4$ is invariant under deformations of the Davey–Stewartson II hierarchy ($W_{t_n} = 0$).

4. Generalized Weierstrass Representations and Integrable Deformations for Space-Like Surfaces in $\mathbb{R}^{2,2}$ and $\mathbb{R}^{3,1}$

The generalized Weierstrass representation for space-like surfaces in $\mathbb{R}^{2,2}$ is similar to that of the $\mathbb{R}^4$ case. In particular, if one starts from the linear systems

$$(4.1) \qquad \begin{aligned} \psi_{1z} &= p\varphi_1, \quad & \psi_{2z} &= \overline{p}\varphi_2, \\ \varphi_{1\bar{z}} &= \overline{p}\psi_1, \quad & \varphi_{2\bar{z}} &= p\psi_2, \end{aligned}$$

then the coordinates X^i $(i = 1, 2, 3, 4)$ of a surface in $\mathbb{R}^{2,2}$ can be defined via

$$(4.2) \qquad \begin{aligned} X^1 + iX^2 &= \int_\Gamma (\varphi_1\varphi_2 dz' + \psi_1\psi_2 d\bar{z}'), \\ X^3 + iX^4 &= i\int_\Gamma (\overline{\psi}_1\varphi_2 dz' + \overline{\varphi}_1\psi_2 d\bar{z}'). \end{aligned}$$

Therefore, the induced metric and Gaussian curvature are

$$ds^2 = v_1 v_2 \, dz \, d\bar{z}, \quad K = -\frac{2}{v_1 v_2} \left[\log(v_1 v_2)\right]_{z\,\bar{z}}, \tag{4.3}$$

where $v_i = |\psi_i|^2 - |\varphi_i|^2$ $(i = 1, 2)$. Further, squared mean curvature $\vec{H}^2$ and Willmore functional W take the form

$$\vec{H}^2 = -4 \frac{|p|^2}{v_1 v_2}, \quad W = -4 \int |p|^2 \, dx \, dy. \tag{4.4}$$

Integrable deformations of surfaces in $\mathbb{R}^{2,2}$ can be defined according to the scheme discussed in the previous section. They are generated via the Davey–Stewartson II hierarchy associated with (4.1) and the Willmore functional turns out to be an integral of motion for the whole hierarchy.

Formulae describing conformal immersions of space-like surfaces in $\mathbb{R}^{3,1}$ are based on the linear systems

$$\psi_{\alpha z} = p \varphi_\alpha, \quad \varphi_{\alpha \bar{z}} = q \psi_\alpha, \qquad \alpha = 1, 2, \tag{4.5}$$

where q and p are real-valued functions. Coordinates of surfaces are then defined via the integral formulae

$$\begin{aligned} X^1 + iX^2 &= \int_\Gamma (\varphi_1 \overline{\psi}_2 \, dz' + \psi_1 \overline{\varphi}_2 \, d\bar{z}'), \\ X^3 + X^4 &= \int_\Gamma (\overline{\psi}_1 \varphi_1 \, dz' + \psi_1 \overline{\varphi}_1 \, d\bar{z}'), \\ X^3 - X^4 &= -\int_\Gamma (\overline{\psi}_2 \varphi_2 \, dz' + \psi_2 \overline{\varphi}_2 \, d\bar{z}'). \end{aligned} \tag{4.6}$$

From (4.6) we get

$$ds^2 = |\psi_1 \varphi_2 - \varphi_1 \psi_2|^2 \, dz \, d\bar{z}, \tag{4.7}$$

for the induced metric, and

$$K = -\frac{2}{|\psi_1 \varphi_2 - \psi_2 \varphi_1|^2} \left[\log(|\psi_1 \varphi_2 - \varphi_1 \psi_2|)\right]_{z\,\bar{z}}, \tag{4.8}$$

for the Gaussian curvature. Finally, the squared mean curvature $\vec{H}^2$ and the Willmore functional W are given by

$$\vec{H}^2 = -\frac{4qp}{|\psi_1 \varphi_2 - \psi_2 \varphi_1|^2}, \quad W = -4 \int qp \, dx \, dy. \tag{4.9}$$

The generalized Weierstrass representation (4.5)–(4.6) is based on a Davey–Stewartson II linear system with real p, q. So integrable deformations of surfaces in $\mathbb{R}^{3,1}$ are generated via the corresponding Davey–Stewartson II hierarchy. The Willmore functional is an integral of motion for the whole hierarchy also in this case.

5. Generalized Weierstrass Representations and Integrable Deformations for Time-Like Surfaces in $\mathbb{R}^{2,2}$ and $\mathbb{R}^{3,1}$

In this section we present the generalized Weierstrass representations for time-like surfaces in $\mathbb{R}^{2,2}$ and $\mathbb{R}^{3,1}$. They are based on Davey–Stewartson linear systems too. However, in contrast with space-like cases considered in the previous section, isothermal coordinates on surfaces are now real. As a consequence, linear problems are of the Davey–Stewartson I type.

To start with, let us consider the Weierstrass type formulae

$$
\begin{aligned}
X^1 &= \frac{1}{2}\int_\Gamma \big[(\varphi_1\widetilde{\varphi}_2 + \widetilde{\varphi}_1\varphi_2)\,d\xi' + (\psi_1\widetilde{\psi}_2 + \widetilde{\psi}_1\psi_2)\,d\eta'\big],\\
X^2 &= \frac{1}{2}\int_\Gamma \big[(\varphi_1\widetilde{\varphi}_1 - \widetilde{\varphi}_2\varphi_2)\,d\xi' + (\psi_1\widetilde{\psi}_1 - \psi_2\widetilde{\psi}_2)\,d\eta'\big],\\
X^3 &= \frac{1}{2}\int_\Gamma \big[(\widetilde{\varphi}_1\varphi_2 - \varphi_1\widetilde{\varphi}_2)\,d\xi' + (\widetilde{\psi}_1\psi_2 - \psi_1\widetilde{\psi}_2)\,d\eta'\big],\\
X^4 &= \frac{1}{2}\int_\Gamma \big[(\varphi_1\widetilde{\varphi}_1 + \varphi_2\widetilde{\varphi}_2)\,d\xi' + (\psi_1\widetilde{\psi}_1 + \widetilde{\psi}_2\psi_2)\,d\eta'\big],
\end{aligned}
\tag{5.1}
$$

where

$$
\begin{aligned}
\psi_{\alpha\xi} &= p\varphi_\alpha, & \widetilde{\psi}_{\alpha\xi} &= q\widetilde{\varphi}_\alpha,\\
\varphi_{\alpha\eta} &= q\psi_\alpha, & \widetilde{\varphi}_{\alpha\eta} &= p\widetilde{\psi}_\alpha,
\end{aligned}
\qquad (\alpha = 1, 2),
\tag{5.2}
$$

ψ_α, φ_α, $\widetilde{\psi}_\alpha$, $\widetilde{\varphi}_\alpha$ ($\alpha = 1, 2$) and p, q are real-valued functions, Γ is a contour of integration in $\mathbb{R}^2$. Formulae (5.1)–(5.2) define a conformal immersion into $\mathbb{R}^{2,2}$ of a surface parametrized by the real minimal lines ξ and η. The induced metric is

$$
ds^2 = (\psi_1\varphi_2 - \psi_2\varphi_1)(\widetilde{\varphi}_1\widetilde{\psi}_2 - \widetilde{\psi}_1\widetilde{\varphi}_2)\,d\xi\,d\eta,
\tag{5.3}
$$

while the Gaussian and the squared mean curvatures are

$$
K = -\frac{2}{(\psi_1\varphi_2 - \psi_2\varphi_1)(\widetilde{\varphi}_1\widetilde{\psi}_2 - \widetilde{\psi}_1\widetilde{\varphi}_2)} \times \big\{\log\big[(\psi_1\varphi_2 - \psi_2\varphi_1)(\widetilde{\varphi}_1\widetilde{\psi}_2 - \widetilde{\psi}_1\widetilde{\varphi}_2)\big]\big\}_{\xi\eta},
\tag{5.4}
$$

$$
\vec{H}^2 = \frac{4qp}{(\psi_1\varphi_2 - \psi_2\varphi_1)(\widetilde{\psi}_1\widetilde{\varphi}_2 - \widetilde{\psi}_2\widetilde{\varphi}_1)},
\tag{5.5}
$$

and the Willmore functional is given by

$$
W = -2\varepsilon\int qp\,d\xi\,d\eta,
\tag{5.6}
$$

where $\varepsilon = \operatorname{sign}\big[(\psi_1\varphi_2 - \psi_2\varphi_1)(\widetilde{\varphi}_1\widetilde{\psi}_2 - \widetilde{\psi}_1\widetilde{\varphi}_2)\big]$.

Finally, let us consider immersions into the Minkowski space $\mathbb{R}^{3,1}$ of generic time-like surfaces parametrized by minimal lines (ξ, η). They can be described by the formulae

$$X^1 = \frac{1}{2}\int_\Gamma \left[(\overline{\varphi}_1\varphi_2 + \varphi_1\overline{\varphi}_2)\,d\xi' + (\overline{\psi}_1\psi_2 + \psi_1\overline{\psi}_2)\,d\eta'\right],$$
$$X^2 = \frac{i}{2}\int_\Gamma \left[(\overline{\varphi}_1\varphi_2 - \varphi_1\overline{\varphi}_2)\,d\xi' + (\overline{\psi}_1\psi_2 - \psi_1\overline{\psi}_2)\,d\eta'\right],$$
$$X^3 = \frac{1}{2}\int_\Gamma \left[(\varphi_1\overline{\varphi}_1 - \varphi_2\overline{\varphi}_2)\,d\xi' + (\psi_1\overline{\psi}_1 - \psi_2\overline{\psi}_2)\,d\eta'\right],$$
$$X^4 = \frac{1}{2}\int_\Gamma \left[(\varphi_1\overline{\varphi}_1 + \varphi_2\overline{\varphi}_2)\,d\xi' + (\psi_1\overline{\psi}_1 + \psi_2\overline{\psi}_2)d\eta'\right], \tag{5.7}$$

where Γ is a contour of integration in $\mathbb{R}^2$ and the functions ψ_α, φ_α $(\alpha = 1, 2)$ satisfy the linear system

$$\psi_{\alpha\xi} = p\varphi_\alpha, \quad \varphi_{\alpha\eta} = \overline{p}\psi_\alpha, \qquad \alpha = 1, 2. \tag{5.8}$$

We get

$$ds^2 = -|\varphi_1\psi_2 - \psi_1\varphi_2|^2\,d\eta\,d\xi, \tag{5.9}$$

for the induced metric, and

$$K = \frac{2}{|\varphi_1\psi_2 - \psi_1\varphi_2|^2}\left[\log(|\varphi_1\psi_2 - \psi_1\varphi_2|)\right]_{\xi\eta}, \tag{5.10}$$

for the Gaussian curvature. Moreover, the squared mean curvature $\vec{H}^2$ and the Willmore functional W are given by

$$\vec{H}^2 = 4\frac{|p|^2}{|\varphi_1\psi_2 - \psi_1\varphi_2|^2}, \quad W = 2\int |p|^2 d\eta\,d\xi. \tag{5.11}$$

Since their generalized Weierstrass representations are based on the linear problems (5.1) and (5.8) which are of the Davey–Stewartson I type, the integrable deformations of time-like surfaces in $\mathbb{R}^{2,2}$ as well as in $\mathbb{R}^{3,1}$ are generated via hierarchies of the Davey–Stewartson I type. In both cases the Willmore functionals are integrals of motion for the corresponding hierarchies.

6. Summary

Generalized Weierstrass representations (2.1)–(2.2), (4.1)–(4.2), (4.5)–(4.6), (5.1)–(5.2) and (5.7)–(5.8) are generated by particular cases of the Davey–Stewartson linear system

$$\psi_{u_1} = p\varphi, \quad \varphi_{u_2} = q\psi. \tag{6.1}$$

In particular, space-like surfaces are associated with the so-called Davey–Stewartson II linear problem, i.e. $(u_1, u_2) \equiv (z, \bar{z})$ with $z \in \mathbb{C}$, while time-like surfaces are associated with the Davey–Stewartson I linear problem, i.e. $(u_1, u_2) \equiv (\xi, \eta)$ with ξ, $\eta \in \mathbb{R}^2$. In all the cases Willmore functionals are proportional to

$$C_1 = \int qp[d^2u], \tag{6.2}$$

(see Eqs. (2.6), (4.4), (4.9), (5.6), and (5.11)).

Since generalized Weierstrass representations for surfaces in four-dimensional pseudo-Euclidean spaces are based on the linear system (6.1), they easily provide a way to construct special deformations of surfaces. These deformations are generated by the compatibility condition of (6.1) and equations of evolution of the type (3.1). So they are generated by equations of the Davey–Stewartson hierarchy of $2+1$ soliton equations. Since these equations are integrable by the inverse spectral transform method (see e.g. [**3**]), one gets evolutions of surfaces which can be referred as *integrable* ones. Specifically, once representations of generalized Weierstrass type are adopted, then integrable deformations of space-like surfaces immersed into $\mathbb{R}^4$, $\mathbb{R}^{3,1}$ and $\mathbb{R}^{2,2}$ can be defined in terms of hierarchies of the Davey–Stewartson II type, whereas integrable deformations of time-like surfaces in $\mathbb{R}^{3,1}$ and $\mathbb{R}^{2,2}$ can be defined in terms of hierarchies of the Davey–Stewartson I type.

It should be pointed out that in all the cases the Willmore functionals W are invariant under the corresponding deformations ($\partial_{t_n} W = 0$). Indeed, the quantity C_1 in (6.2) is nothing but the simplest integral of motion associated with the $2+1$-dimensional soliton equations of the Davey–Stewartson hierarchy which follows from the compatibility of (6.1) and evolution equations.

As a final comment, we would like to mention that above results hold also in the case of surfaces immersed into conformally pseudo-Euclidean spaces [**5**].

References

1. A. I. Bobenko, *All constant mean curvature tori in* $\mathbb{R}^3$, $\mathbb{S}^3$ *and* $\mathbb{H}^3$ *in terms of theta-functions*, Math. Ann. **290** (1990), 209–245.
2. B. G. Konopelchenko, *Induced surfaces and their integrable dynamics*, Stud. Appl. Math. **96** (1996), 9–51.
3. ———, *Solitons in multidimension*, World Scientific, Singapore, 1993.
4. ———, *Weierstrass representation for surfaces in* 4D *spaces and their integrable deformations via the* DS *hierarchy*, Ann. Global Anal. Geometry **18** (2000), no. 1, 61–74.
5. B. G. Konopelchenko and G. Landolfi, *Induced surfaces and their integrable dynamics.* II. *Generalized Weierstrass representations in* 4D *spaces and deformations via* DS *hierarchy*, Stud. Appl. Math. **104** (2000), no. 2, 129–169.
6. ———, *On rigid string instantons in four-dimensions*, Phys. Lett. B **459**, 522–529.
7. ———, *One-loop analysis of* QCD *strings world-sheets in* $\mathbb{R}^4$ *through a Weierstrass-type representation*, Internat. J. Modern Physics A (to appear).
8. ———, *Quantum effects for extrinsic geometry of strings via the generalized Weierstrass representation*, Phys. Lett. B **444** (1998), 299–308.
9. A. M. Polyakov, *Fine structure of strings*, Nuclear Phys. B **268** (1986), 406–412.
10. A. Sym, *Soliton surfaces and their applications*, Lecture Notes in Physics, vol. 239, Springer-Verlag, Berlin, 1985, pp. 154–231.

Dipartimento di Fisica, Università Degli Studi di Lecce, Via Arnesano, 73100 Lecce, C.P. 193, Italy.

Email address, B. G. Konopelchenko: `Konopel@le.infn.it`

Email address, G. Landolfi: `Landolfi@le.infn.it`

Centre de Recherches Mathématiques
CRM Proceedings and Lecture Notes
Volume **29**, 2001

Properties of a Class of Slowly Decaying Oscillatory Solutions of KdV

M. Kovalyov

ABSTRACT. In this report we discuss a class of slowly decaying oscillatory solutions of KdV.

Recall that the Korteweg–de Vries equation

$$u_t - 6uu_x + u_{xxx} = 0, \qquad \lim_{x\to\pm\infty} u(t,x) = 0, \tag{KdV}$$

is obtained by adding a nonlinear correction term $-6uu_x$ to the so-called linear KdV

$$u_t + u_{xxx} = 0. \tag{ℓKdV}$$

In applications u is small and thus the correction term is of higher degree of smallness compared to the linear terms. That suggests that, at least, some solutions of the KdV may be viewed as nonlinear perturbations of solutions of ℓKdV due to addition of the nonlinear term. Since the exponential functions $e^{ik(x-k^2t)}$ are solutions of ℓKdV, it is only natural to ask how they are transformed into solutions of the KdV by addition of the nonlinear term. We can construct more complicated solutions of ℓKdV as superpositions of the exponents $e^{ik(x-k^2t)}$ in the form $\int_{-\infty}^{\infty} e^{ik(x-k^2t)}\rho(k)\,dk$. How, then, addition of the nonlinear term perturbs these solutions of ℓKdV into solutions of KdV? The last, but probably, the most interesting question is how the addition of the nonlinear term perturbs the very process of integration or, in other words, how do we construct *nonlinear* superposition for KdV? Attempts to answer these questions lead to the work whose summary is provided here. This report contains only a review of the main results, for details and discussions the reader is referred to [**Kov1, Kov2, Kov3, Kov4**] and references therein. Some of the results have been also independently obtained in [**Mat1, Mat2, Mat3**], [**Ark1**] and [**Nov1**].

2000 *Mathematics Subject Classification.* Primary: 37K10; Secondary: 74J30.
This is the final form of the paper.

In the class of sufficiently smooth and sufficiently fast decaying at infinity solutions, the Cauchy problem for KdV is usually "solved" by means of the direct scattering/inverse scattering method as follows:

(1) First the direct scattering problem is solved for $L = -\partial^2/\partial x^2 + u$ for $t = 0$, i.e. we find the solutions of $L\psi = k^2\psi$, $\psi(x,k) \sim e^{ikx}$, as $x \to +\infty$ and determine scattering coefficients $a(k,0)$, $b(k,0)$ and $r(k,0) = b(k,0)/a(k,0)$ in the asymptotic $\psi(x,k) \sim a(k,0)e^{ikx} - b(-k,0)e^{-ikx}$ as $x \to -\infty$. In general, this step can be done in theory only.

(2) A function

$$F(x,t) = \sum_{\substack{\text{discrete}\\ \text{spectrum of } L}} \frac{ib(k,t)}{a'(k)} e^{ikx+8ik^3t} + \frac{1}{2\pi}\int_{-\infty}^{+\infty} r(k,t)e^{ikx+8ik^3t}\,dk,$$

is then constructed and the integral equation $K(x,y,t) + F(x+y,t) + \int_x^{+\infty} K(x,z,t)F(y+x,t)\,dz = 0$ is solved for $K(x,y,t)$ time evolution of the corresponding solution of KdV is given by $u(x,t) = -2(d/dx)K(x,x,t)$.

In the case of $r(k) \equiv 0$,

$$F(x,t) = \sum_{\substack{\text{discrete}\\ \text{spectrum of } L}} \frac{ib(k,t)}{a'(k)} e^{ikx},$$

gives us N-soliton solutions which can be viewed as nonlinear superpositions of N solitons, each soliton viewed as a simplest component [**Gar1**]. In the more interesting solitonless case of $F(z) = \int_{-\infty}^{+\infty} r(k)e^{ikz}\,dk$, the simplest solutions of KdV are obtained by taking $F(z)$ generated by

$$r(k) = \lim_{\varepsilon\to 0}\begin{cases} e^{i\lambda\gamma+p\varepsilon+8i(\lambda+i\varepsilon)^3t}, & |k-\lambda| < \varepsilon,\\ e^{-i\lambda\gamma+p\varepsilon+8i(-\lambda+i\varepsilon)^3t}, & |k+\lambda| < \varepsilon,\\ 0, & \text{otherwise.}\end{cases} \tag{1}$$

Nonlinear superposition of N such components is obtained by taking

$$r(k) = \lim_{\varepsilon\to 0}\begin{cases} e^{i\lambda_n\gamma_n+p_n\varepsilon+8i(\lambda_n+i\varepsilon)^3t}, & |k-\lambda_n| < \varepsilon\\ e^{-i\lambda_n\gamma_n+p_n\varepsilon+8i(-\lambda_n+i\varepsilon)^3t}, & |k+\lambda_n| < \varepsilon\,, \quad n = 1,2,\ldots,N,\\ 0, & \text{otherwise}\end{cases} \tag{2}$$

which results in

$$\begin{gathered} u(x,t) = -2\frac{d^2}{dx^2}\ln\det\mathcal{D},\\ \mathcal{D}_{nm} = \begin{cases} \frac{\sin(\Gamma_n-\Gamma_m)}{\lambda_n-\lambda_m} - \frac{\sin(\Gamma_n+\Gamma_m)}{\lambda_n+\lambda_m}, & n \neq m,\\ \tau_n = p_n - 12\lambda_n^2t - x - \frac{\sin 2\Gamma_n}{2\lambda_n}, & n = m,\end{cases}\end{gathered} \tag{3}$$

where $1 \le n,\, m \le N$ and $\Gamma_n = \lambda_n(\gamma_n - 4\lambda_n^2t - x)$.

For $N = 1$ we obtain the nonlinear analogues of the exponents we set out to compute at the very beginning of the paper. Possibly, the most interesting property of these solutions is a nonlinear analogue of interference. What it means is that if $\lambda_N \to \lambda_{N-1}$, while keeping $\lambda_N\gamma_N = \lambda_{N-1}\gamma_{N-1} + 2\pi m$, $m \in \mathbb{N}$, then

$$(4)\quad \det\begin{pmatrix} \tau_1 & \mathcal{D}_{12} & \dots & \mathcal{D}_{1N-1} & \mathcal{D}_{1N} \\ \mathcal{D}_{12} & \tau_2 & \dots & \mathcal{D}_{2N-1} & \mathcal{D}_{2N} \\ \vdots & & \vdots & & \vdots \\ \mathcal{D}_{1N-1} & \mathcal{D}_{2N-1} & \dots & \tau_{N-1} & \mathcal{D}_{N-1\,N} \\ \mathcal{D}_{1\,N} & \mathcal{D}_{2\,N} & \dots & \mathcal{D}_{N-1\,N} & \tau_N \end{pmatrix} \to \det\begin{pmatrix} \tau_1 & \mathcal{D}_{12} & \dots & \mathcal{D}_{1\,N-1} \\ \mathcal{D}_{12} & \tau_2 & \dots & \mathcal{D}_{2\,N-1} \\ \vdots & & \vdots & \vdots \\ \mathcal{D}_{1\,N-1} & \mathcal{D}_{2\,N-1} & \dots & \widetilde{\tau}_{N-1} \end{pmatrix},$$

where

$$\tau_{N-1} = \widetilde{p}_{N-1} - 12\lambda_{N-1}^2 t - x - \frac{\sin 2\lambda_{N-1}(\gamma_{N-1} - 4\lambda_{N-1}^2 t - x)}{2\lambda_{N-1}},$$

and $1/\widetilde{p}_{N-1} = 1/p_{N-1} + 1/p_N$. When all p_n's are large and x is restricted to a region whose size is small compare to the values of the p_n's, (4) becomes

$$(5)\qquad u \approx -4\sum_1^N \frac{\lambda_n}{p_n}\sin 2\Gamma_n.$$

Nonlinear interference then is reduced to a trivial statement that any two terms in the sum with the same frequency can be combined together by adding up corresponding coefficients λ_n/p_n. The argument shows close similarity of (3) to the Fourier sums and Fourier integrals suggesting a possibility of using (3) to modulate miscellaneous wave-packets and profiles. This point is addressed in [**Kov2**] where a number of pictures are produced to illustrate that.

Schrödinger operators with potentials given by (4) have neither continuous nor discrete spectra due to singularity of the potential. Yet they do possess somewhat analogous concepts which we will refer to as "continuous" and "discrete" spectra.

Decaying at infinity eigenfunctions of $L = -\partial^2/\partial x^2 + u$ corresponding to the "discrete" spectrum are obtained from the following algebraic system:

$$(6)\qquad \sum_{m=1}^N \mathcal{D}_{mn}\psi_m(x) = \sin\Gamma_n, \quad n = 1, 2, \dots, N,$$

whereas increasing at infinity eigenfunctions corresponding to the "discrete" spectrum are

$$(7)\qquad \begin{aligned} \phi_m = {} & 2\sum_{\substack{n=1\\ n\neq m}}^N \frac{\psi_n}{\lambda_n^2 - \lambda_m^2}\big[(x - p_m - 12\lambda_m^2 t)(\lambda_n\cos\Gamma_m\cos\Gamma_n \\ & \qquad + \lambda_m\sin\Gamma_m\sin\Gamma_n) + \cos\Gamma_m\sin\Gamma_n\big] \\ & + 4\lambda_m\sum_{\substack{n=1\\ n\neq m}}^N \frac{\psi_n}{(\lambda_n^2 - \lambda_m^2)^2}\big[\lambda_m\cos\Gamma_m\sin\Gamma_n + \lambda_n\sin\Gamma_m\cos\Gamma_n\big] \\ & + \frac{\psi_m}{\lambda_m}\big[(x - p_m - 12\lambda_m^2 t)\cos^2\Gamma_m - 2\tau_m\big] \\ & \qquad + (x - p_m - 12\lambda_m^2 t)\cos\Gamma_m. \end{aligned}$$

Eigenfunctions corresponding to the "continuous" spectrum are of the form

$$\psi(k,t,x) = \left[1 - 2\sum_{n=1}^{N} \frac{1}{k^2-\lambda_n^2}(\lambda_n \cos\Gamma_n + ik\sin\Gamma_n)\psi_n\right]e^{ikx}. \tag{8}$$

If all $p_n - 12\lambda_n^2 t - x$'s are large and of the same order of magnitude, we obtain the following asymptotics:

$$u = \sum_n \frac{-4\lambda_n}{p_n - 12\lambda_n^2 t - x}\sin 2\lambda_n(\gamma_n - 4\lambda_n^2 t - x) + o\left(\frac{1}{p_n - 12\lambda_n^2 t - x}\right), \tag{9a}$$

$$\psi_n = \frac{1}{p_n - 12\lambda_n^2 t - x}\sin \lambda_n(\gamma_n - 4\lambda_n^2 t - x) + o\left(\frac{1}{p_n - 12\lambda_n^2 t - x}\right), \tag{9b}$$

$$\frac{\phi_n}{p_n} = \frac{p_n - 12\lambda_n^2 t - x}{p_n}\cos \lambda_n(\gamma_n - 4\lambda_n^2 t - x) + o\left(\frac{1}{p_n - 12\lambda_n^2 t - x}\right). \tag{9c}$$

Let us assume that $x \in (-X, X)$, $X > 0$, $|t| + 1 \ll X \ll \min_n |p_n|$. Then the eigenfunctions $\eta_n = \left[\phi_n/(p_n - X) - i(p_n - X)\psi_n\right]e^{i\lambda_n(\gamma_n - 4\lambda_n^2 t)}$ behave for $x \sim X$ as $\eta \sim e^{i\lambda_n x}$, and for $x \sim -X$ as

$$\eta \sim \frac{1}{2}\frac{p_n^2 + X^2}{p_n^2 - X^2}e^{i\lambda_n x} + \frac{1}{2}\frac{2p_n X}{p_n^2 - X^2}e^{2i\lambda_n(\gamma_n - 4\lambda_n^2 t) - i\lambda_n x} = a(\lambda_n)e^{i\lambda_n x} - b(-\lambda_n)e^{-i\lambda_n x},$$

where the numbers $a(\lambda_n) = (p_n^2 + X^2)/(p_n^2 - X^2)$ and $b(\lambda_n) = (2p_n X)/(X^2 - p_n^2)e^{-2i\lambda_n(\gamma_n - 4\lambda_n^2 t)}$ satisfy $|a_n|^2 - |b_n|^2 = 1$ and can be viewed as scattering coefficients on a large interval.

To obtain conservation laws let us consider

$$\begin{aligned}
\chi(x,k) &= \frac{1}{\psi e^{-ikx}}\frac{d\psi e^{-ikx}}{dx} \\
&= -2\frac{\sum_{n=1}^{N} \lambda_n/(k^2 - \lambda_n^2)(\lambda_n \sin\Gamma_n - ik\cos\Gamma_n)\psi_n}{[1 - 2\sum_{n=1}^{N} 1/(k^2 - \lambda_n^2)(\lambda_n\cos\Gamma_n + ik\sin\Gamma_n)\psi_n]} \\
&\qquad - 2\frac{\sum_{n=1}^{N} 1/(k^2 - \lambda_n^2)(\lambda_n\cos\Gamma_n + ik\sin\Gamma_n)(d\psi_n)/(dx)}{[1 - 2\sum_{n=1}^{N} 1/(k^2 - \lambda_n^2)(\lambda_n\cos\Gamma_n + ik\sin\Gamma_n)\psi_n]} \\
&= \sum_{n=1}^{N}\sum_{m=1}^{\infty}\left[\frac{\lambda_n^{2m-1}}{k^{2m}}\frac{\partial}{\partial x}\left(\frac{\sin 2\Gamma_n}{p_n - 12\lambda_n^2 t - x}\right)\right. \\
&\qquad \left. + i\frac{\lambda_n^{2m-2}}{k^{2m-1}}\frac{\partial}{\partial x}\left(\frac{1 - \cos 2\Gamma_n}{p_n - 12\lambda_n^2 t - x}\right)\right] + o\left(\frac{1}{x^2}\right),^{1}
\end{aligned} \tag{10}$$

where we used (9) to obtain the last equality. Eq. (10) yields conserved densities in the form

$$\chi_m(x) = \begin{cases} (2i)^m \sum_{n=1}^{N} \lambda_n^{m-1}\frac{\partial}{\partial x}\left(\frac{\sin 2\Gamma_n}{p_n - 12\lambda_n^2 t - x}\right) + o\left(\frac{1}{x^2}\right)^1, & m \text{ is even}, \\ i(2i)^m \sum_{n=1}^{N} \lambda_n^{m-1}\frac{\partial}{\partial x}\left(\frac{1 - \cos 2\Gamma_n}{p_n - 12\lambda_n^2 t - x}\right) + o\left(\frac{1}{x^2}\right)^1, & m \text{ is odd}. \end{cases} \tag{11}$$

[1]Here $o(1/x^2)$ is understood in the sense of distributions.

The conservation laws for (3) on $\mathbb{R}$ take the form of:

$$\lim_{A\to+\infty}\frac{A}{2}\int_A^{2A}\chi_m(x)dx=\begin{cases}0, & m\text{ is even},\\ 2^m(-1)^{(m+3)/2}\sum_{n=1}^N\lambda_n^{m-1}, & m\text{ is odd},\end{cases}\tag{12}$$

where we used that

$$\lim_{x\to\pm\infty}\frac{x\sin 2\Gamma_n}{p_n-12\lambda_n^2t-x}=\lim_{x\to\pm\infty}\frac{x\cos 2\Gamma_n}{p_n-12\lambda_n^2t-x}=0,$$

in the sense of distributions.

Another type of conservation laws can also be derived by integrating (12) under the assumptions $x\in(-X,X)$, $X>0$, $|t|+1\ll X\ll \min_n|p_n|$:

$$\int_{-X}^{X}\chi_m(x)dx=\begin{cases}0, & m\text{ is even},\\ 2^m(-1)^{(m+3)/(2)}\sum_{n=1}^N\lambda_n^{m-1}\frac{2X}{p_n^2-X^2}, & m\text{ is odd},\end{cases}$$
$$+\text{ lower order terms.}$$

Defining $\varepsilon_n=\pi/X$ and by virtue of

$$\frac{2X}{p_n^2-X^2}\approx\frac{2}{p_n^2}\approx\frac{1}{\pi}\varepsilon_n\frac{2X^2}{p_n^2}\approx\frac{1}{\pi}\varepsilon_n\ln a_n,$$

we obtain

$$\int_{-X}^{X}\chi_m(x)dx=\begin{cases}0, & m\text{ is even},\\ 2^{m+1}(-1)^{(m+1)/2}\frac{1}{\pi}\sum_{n=1}^N\lambda_n^{m-1}\varepsilon_n\ln a_n, & m\text{ is odd},\end{cases}+o\left(\frac{1}{X}\right).$$

It may appear that solutions of type (3) should exist for all integrable systems. In particular, the argument in [**Fad1**, Ch. III, §8] seems to suggest existence of such solutions for self-focusing NLS, yet our computations yielded none. However, solutions of type (3) do exist for defocusing NLS as shown in [**Bar1**].

Apendix: A Picture's Worth a Thousand Words

Below are a few pictures illustrating the main point of this article.

Figure 1 illustrates a typical simplest solution (3) obtained when $N=1$. Figure 2 illustrates a solution (3) obtained by taking $N=2$ and $\lambda_1-\lambda_2$ small compare to $\lambda_1\approx\lambda_2$. Figures 3 and 4 illustrate (3) obtained with $N=50$, and $\lambda_n=0.15n$, $\gamma_n=\frac{\pi}{4}4$, $p_n=-10n^2$ whereas Figures 5 and 6 illustrate (3) obtained with $N=50$, and $\lambda_n=0.15n$, $\gamma_n=0$, $p_n=-10n^2$. Figures 4 and 6 are just the central parts of Figures 3 and 5 shown in higher resolution.

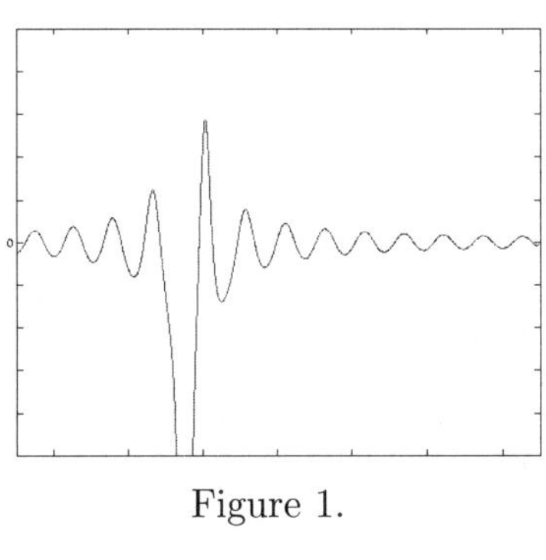

Figure 1.

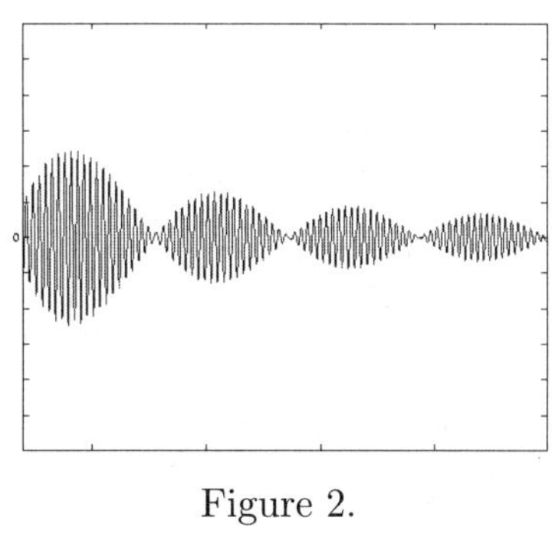

Figure 2.

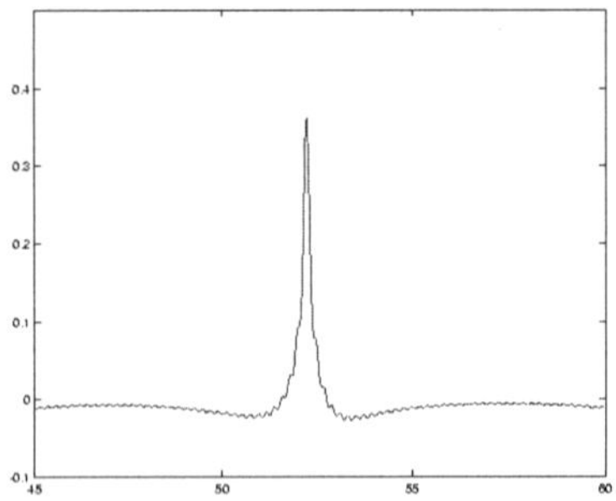

Figure 3.

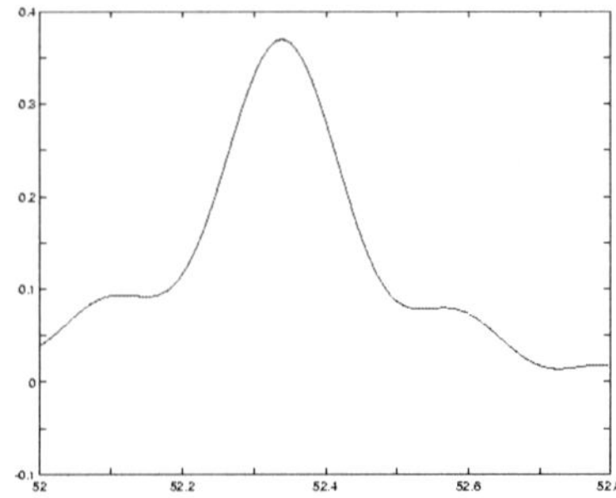

Figure 4.

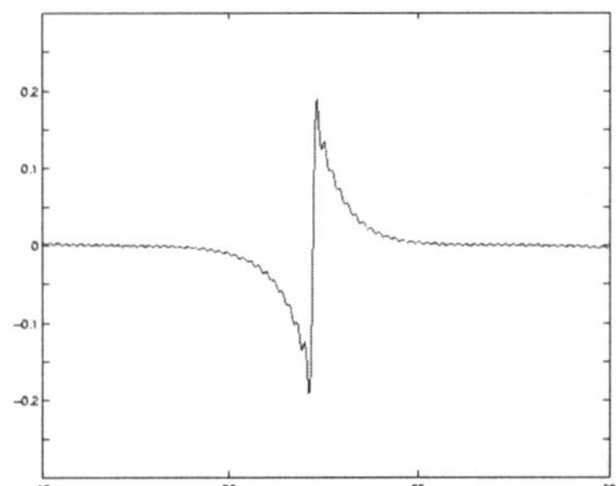

Figure 5.

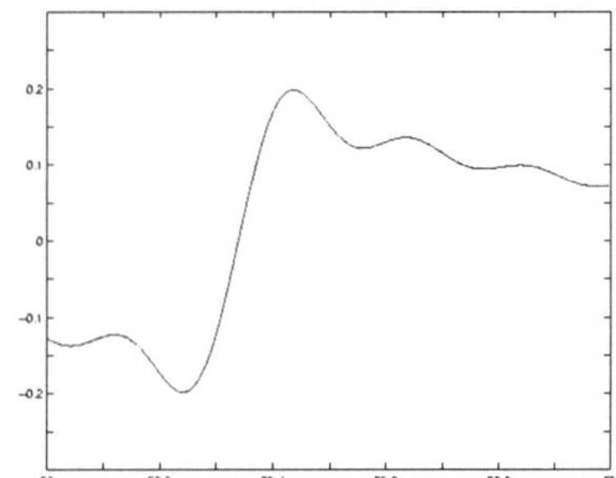

Figure 6.

References

[Ark1] V. A. Arkad'ev, A. K. Pogrebkov, and M. K. Polivanov, *The method of the inverse scattering problem in application to singular solutions of nonlinear equations.* I, Theoret. and Math. Phys. **53** (1982), no. 2, 1051–1062.

[Bar1] Ş. Barran and M. Kovalyov, *A note on slowly decaying solutions of the defocusing nonlinear Schrödinger equation*, J. Phys. A **32** (1999), no. 34, 6121–6125.

[Beu1] R. Beutler, *Positon solutions of the sine-Gordon equation*, J. Math. Phys. **34** (1993), no. 7, 3098–3109.

[Fad1] L. Fadeev and L. A. Takhtajan, *Hamiltonian methods in the theory of solitons*, Springer-Verlag, Berlin-New York; Russian transl., 1987.

[Gar1] C. S. Gardner, J. M. Greene, M. D. Kruskal, and R. Miura, *Korteweg–de Vries equation and generalizations.* IV. *Methods for exact solutions*, Comm. Pure Appl. Math. **27** (1974), 97–133.

[Kov1] M. Kovalyov, *Basic motions of the Korteweg–de Vries equation*, Nonlinear Anal. **31** (1998), no. 5/6, 599–619.

[Kov2] ———, *Modulating properties of harmonic breather solutions of* KdV *equation*, J. Phys. A **31** (1998), 5117–5128.

[Kov3] ———, *Slowly decaying solutions of* KdV, Nonlinear PDE's, Dynamics and Continuum Physics (South Hadley, 1998), Contemp. Math., vol. 255, Amer. Math. Soc., Providence, RI, 2000, pp. 163–180.

[Kov4] M. Kovalyov and Hosseini Ali Abadi Mohammad, *An explicit formula for a class of solutions of* KdV, Phys. Lett. A **254** (1999), no. 1-2, 47–52.

[Mat1] V. B. Matveev, *Asymptotics of the multipositon-soliton τ-function of the Korteweg–de Vries equation and the supertransparency*, J. Math. Phys. **35** (1994), no. 6, 2955–2970.

[Mat2] ———, *Positon-positon and soliton-positon collisions*: KdV *case*, Phys. Lett. A **166** (1992), no. 3-4, 209–212.

[Mat3] ———, *Trivial S-matrices, Wigner-Von Neumann resonances and positon solutions of the integrable systems*, Algebraic and Geometric Methods in Mathematical Physics (Kaciveli, 1993) (A. Boutet de Monvel and V. Marchenko, eds.), Math. Phys. Stud., vol. 19, Kluwer Acad. Publ., Dordrecht, pp. 343–355.

[Mat4] ———, *Oscillating and rational potentials with trivial S-matrix and solutions of* KdV *equation*, Proceedings of All Union Conference on PDE in honor of the 75th birthday of academician I. G. Petrovskiy (P. S. Alexandrov and O. A. Oleinik, eds.), Moscow University Press, 1978, pp. 378–379. (Russian)

[Nov1] R. Novikov and G. Khenkin, *Oscillating weakly localized solutions of the Korteweg–de Vries equation*, Teoret. Mat. Fiz. **61** (1984), no. 2, 199–213. (Russian)

Depart. of Mathematics, University of Alberta, Edmonton, Alberta, T6G 2G1, Canada.

E-mail address: mkovalyo@ualberta.ca

Centre de Recherches Mathématiques
CRM Proceedings and Lecture Notes
Volume **29**, 2001

Blending Two Discrete Integrability Criteria: Singularity Confinement and Algebraic Entropy

S. Lafortune, A. Ramani, B. Grammaticos,
Y. Ohta, and K. M. Tamizhmani

ABSTRACT. We confront two integrability criteria for rational mappings. The first is the singularity confinement based on the requirement that every singularity, spontaneously appearing during the iteration of a mapping, disappear after some steps. The second, recently proposed, is the algebraic entropy criterion associated to the growth of the degree of the iterates. The algebraic entropy results confirm the previous findings of singularity confinement on discrete Painlevé equations. The degree-growth methods are also applied to linearisable systems. The result is that systems integrable through linearisation have a slower growth than systems integrable through isospectral methods. This may provide a valuable detector of not just integrability but also of the precise integration method. We propose an extension of the Gambier mapping in N dimensions. Finally a dual strategy for the investigation of the integrability of discrete systems is proposed based on both singularity confinement and the low growth requirement.

1. Introduction

The detection of discrete integrability has for the past few years relied on the criterion of singularity confinement [**12**]. This criterion is based on the observation that for integrable discrete systems any singularity that appears spontaneously disappears after some iterations. Let us give one example. In the mapping

$$x_{n+1} + x_{n-1} = \frac{a}{x_n} + \frac{1}{x_n^2}, \tag{1.1}$$

where a is constant, we assume that, at some iteration, x_{n-1} vanishes while x_{n-2} is finite. This has as consequence a diverging x_n, and iterating once more a vanishing

2000 *Mathematics Subject Classification.* Primary: 39A13; Secondary: 37J35.

S. Lafortune was supported by FCAR du Québec for his Ph.D. and by "Programme de Soutien aux Cotutelles de Thèse de Doctorat du Gouvernement du Québec" for his stay in Paris.

The authors are grateful to several anonymous referees who, by insisting that the singularity confinement results are wrong in the light of the Hietarinta and Viallet findings, pushed us to learn the techniques of degree-growth which allowed us to *confirm the validity of our previous results.*

This is the final form of the paper.

x_{n+1}. If we try to compute x_{n+2} we obtain an indeterminate form, $\infty - \infty$. In order to lift the indeterminacy we introduce a small parameter ε and assume that $x_{n-1} = \varepsilon$. Iterating we obtain successively, expanding in ε, $x_n = 1/\varepsilon^2 + a/\varepsilon - x_{n-2} + \mathcal{O}(\varepsilon)$, $x_{n+1} = -\varepsilon + a\varepsilon^2 + \mathcal{O}(\varepsilon^3)$, $x_{n+2} = x_{n-2} + 2(ax_{n-2}+1)\varepsilon + \mathcal{O}(\varepsilon^2)$. By taking the appropriate limit $\varepsilon \to 0$ we find that x_{n+2} is finite, and moreover contains the information on the initial data, i.e. x_{n-2}. The subsequent x_n's are indeed finite and the singularity is confined to the sequence $\{0, \infty^2, 0\}$, the square in the infinite sign being a reminder of the $1/\varepsilon^2$. Singularity confinement is tailored for rational mappings although this should not be considered as an absolute restriction. What singularity confinement *cannot* treat are mappings which do *not* possess singularities. For example a mapping of the form $x_{n+1} = \lambda x_n(1 - x_n)$ lies beyond the reach of the confinement method [**4**]. Our tacit conjecture has always been that polynomial mappings, like the logistic one, are *not* integrable, with only exception the linear one [**26**].

While confinement has turned out to be a most useful integrability detector, its application was based on the unwarranted assumption that its necessary character will be constraining enough to make it sufficient. Still, from the outset it was clear that confinement was not a sufficient criterion. Let us recall the example presented already in [**22**]. If we examine the singularity structure of the mapping $x_{n+1} = f(x_n, n)$ where the f is rational in x_n and analytic in n we find that all f's of the form $f = \sum_k \alpha_k/(x_n + \beta_k)^{\nu_k}$, with $\nu_k \in \mathbb{Z}$, lead to confinement. Clearly not all these mappings can be integrable and, indeed, only the homographic one, $x_{n+1} = \alpha + \lambda/(x_n + \beta)$, is. The non sufficient character of singularity confinement was put to a wider perspective by Hietarinta and Viallet [**14**] who verified that the mapping:

$$x_{n+1} + x_{n-1} = x_n + \frac{1}{x_n^2}, \tag{1.2}$$

is confining, the singularity sequence being $\{0, \infty^2, \infty^2, 0\}$, and at the same time exhibits chaotic behaviour. Moreover this is not an isolated example and one can easily construct mappings which are confining without being integrable.

Clearly a more stringent integrability criterion was needed and the authors of [**14**] (see also [**3**]) have proposed one based on the ideas of Arnold and Veselov. According to Arnold [**1**] the complexity (in the case of mappings of the plane) is the number of intersection points of a fixed curve with the image of a second curve obtained under the mapping at hand. While the complexity grows exponentially with the iteration for generic mappings, it can be shown [**27**] to grow only polynomially for a large class of integrable mappings. As Veselov points out, "integrability has an essential correlation with the weak growth of certain characteristics". Thus the authors of [**14**] proposed to directly test the degree of the successive iterates and introduced the notion of algebraic entropy. It is defined as $E = \lim_{n\to\infty} \log(d_n)/n$ where d_n is the degree of the nth iterate of some initial data under the action of the mapping. Since a generic nonintegrable mapping exhibits exponential degree-growth, a nonzero algebraic entropy indicates nonintegrability. Integrable mappings must have zero algebraic entropy, associated to slower-than-exponential (typically polynomial) degree-growth. It is expected that the requirement of zero algebraic entropy is strong enough to be a sufficient integrability condition.

In order to make the ideas clearer we present here our method for the study of the degree-growth. It presents a few differences with respect to the one of [**14**] which, we feel, make its practical implementation simpler. We start by introducing homogeneous variables through the appropriate choice of initial data. Typically, in a three-point mapping we choose to introduce $x_0 = r$, $x_1 = p/q$ and thus the weight of r is 0 while that of p, q are taken equal to 1. We then compute the homogeneous degree in p, q in the denominator and numerator of x_n at every iteration. Let us follow the first few iterations of mapping (1.1) in order to understand the mechanism of the degree-growth. We find

$$x_2 = \frac{q^2 + apq - rp^2}{p^2}, \qquad x_3 = \frac{pP_4}{q(q^2 + apq - rp^2)^2},$$
$$x_4 = \frac{(q^2 + apq - rp^2)P_6}{P_4^2}, \quad x_5 = \frac{P_4 P_9}{qP_6^2},$$

where the P_k's are homogeneous polynomials in p, q of degree k (remember that r is of zero homogeneous degree in our convention). A pattern becomes apparent. Whenever a new polynomial appears in the numerator of x_n, its square will appear in the denominator of x_{n+1} and it will appear one last time as a factor of the numerator of x_{n+2}, after which it disappears due to factorisations. The singularities we are working with in the singularity confinement approach correspond to the zeros of any of these polynomials, which explains the pattern $\{0, \infty^2, 0\}$. The singularity confinement is intimately related to this factorisation which plays a crucial role in the algebraic entropy approach. If we calculate the degree of the iterates, we obtain: 0, 1, 2, 5, 8, 13, 18, 25, 32, 41, ... Clearly the degree-growth is polynomial: we have $d_{2m} = 2m^2$ and $d_{2m+1} = 2m^2 + 2m + 1$. (A remark is necessary at this point. In order to obtain a closed-form expression for the degrees of the iterates, we start by computing a sufficient number of them. Once the expression of the degree has been heuristically established we compute the next few ones and check that they agree with the analytical expression predicted). Thus the algebraic entropy of this mapping is zero in agreement with its integrable character [**20**].

On the other hand, if we iterate mapping (1.2) we obtain the sequence

$$x_2 = \frac{p^3 - p^2qr + q^3}{p^2q}, \qquad x_3 = \frac{P_8}{p^2(p^3 - p^2qr + q^3)^2},$$
$$x_4 = \frac{pP_{22}}{q(p^3 - p^2qr + q^3)^2{P_8}^2}, \quad x_5 = \frac{(p^3 - p^2qr + q^3)P_{58}}{q{P_8}^2{P_{22}}^2}.$$

Again some factorisations and simplifications do occur which explain why this mapping has the confinement property. However the degree of the new terms appearing at every iteration grows too rapidly and thus the simplifications cannot curb the exponential growth. The degree sequence now is 0, 1, 3, 8, 23, 61, 160, 421, ..., i.e. the degrees obey the relation $d_{n+1} - 3d_n + d_{n-1} = 2/3(1 + j^{n+1} + j^{2(n+1)})$ where j is a complex cubic root of 1 leading to an exponential growth with asymptotic ratio $(3+\sqrt{5})/2$ and algebraic entropy $\log\big((3+\sqrt{5})/2\big)$. The situation is even worse in the case of nonconfining mappings where no simplifications occur and the growth rate is maximal.

Since a considerable volume of results on integrable discrete systems were obtained with the method of singularity confinement it is natural to question their integrability in the light of the findings concerning the non sufficiency of the singularity confinement criterion. One of the aims of this review is to provide the confirmation of these results based on the more stringent criterion of algebraic entropy. In particular we shall examine several discrete Painlevé equations that we have obtained over the years and obtain their growth properties. The main result of this analysis is that, in every case studied, the singularity confinement criterion *when used for deautonomisation of an integrable mapping*, turns out to be sufficient [**18**]. This means that an equation, the autonomous form of which is integrable and which has been deautonomised following the constraints provided by singularity confinement, leads to exactly the same degree-growth as the initial, autonomous, one. Thus we expect the deautonomised equation to be as integrable as its autonomous limit. This finding is of capital importance since it validates the results previously obtained with the singularity confinement approach. Moreover it sets the frame for a new, composite, approach. Whenever we wish to obtain the integrable cases of an equation containing a certain number of parameters we can perform a first, exploratory, study using the confinement method. Once the confinement constraints have been used in order to limit the freedom, the algebraic entropy, or low-growth criterion, can be implemented in order to pin down the truly integrable cases. Examples of this dual approach can be found in [**25**].

2. Discrete Painlevé Equations

Let us first recall what has always been our approach to the derivation of discrete Painlevé equations. We start from an autonomous system the integrability of which has been independently established. In the case of discrete Painlevé equations, this system is the QRT mapping [**20**]:

$$f^{(1)}(x_n) - (x_{n+1} + x_{n-1})f^{(2)}(x_n) + x_{n+1}x_{n-1}f^{(3)}(x_n) = 0. \tag{2.1}$$

When the $f^{(i)}$'s are quartic functions, satisfying specific constraints, the mapping (2.1) is integrable in terms of elliptic functions. Since the elliptic functions are the autonomous limits of the Painlevé transcendents, the mapping (2.1) is the appropriate starting point for the construction of the nonautonomous discrete systems which are the analogues of the Painlevé equations. The procedure we used, often referred to as 'deautonomisation', consists in finding the dependence of the coefficients of the quartic polynomials appearing in (2.1) with respect to the independent variable n, which is compatible with the singularity confinement property. Namely, the n-dependence is obtained by asking that the singularities are indeed confined. The reason why this procedure can be justified is the following. Since the autonomous starting point is integrable, it is expected that the growth of the degree of the iterates is polynomial. Now it turns out that the application of the singularity confinement deautonomisation corresponds to the requirement that the nonautonomous mappings lead to the same factorizations and subsequent simplifications and have precisely the same growth properties as the autonomous ones. These considerations will be made more transparent thanks to the examples we present in what follows.

Let us start with a simple case. We consider the mapping:

$$x_{n+1} + x_{n-1} = \frac{a}{x_n} + \frac{1}{x_n^2}, \tag{2.2}$$

where a depends on n. The singularity confinement result is that a must satisfy the conditions $a_{n+1} - 2a_n + a_{n-1} = 0$, i.e. a is linear in n. Assuming now that a is an arbitrary function of n we compute the iterates of (2.2). We obtain the sequence:

$$x_2 = \frac{q^2 + a_1 pq - p^2 r}{p^2}, \qquad x_3 = \frac{pQ_4}{q(q^2 + a_1 pq - p^2 r)^2},$$
$$x_4 = \frac{(q^2 + a_1 pq - p^2 r)Q_7}{pQ_4^2}, \quad x_5 = \frac{pQ_4 Q_{12}}{q(q^2 + a_1 pq - p^2 r)Q_7^2},$$

where the Q_k's are homogeneous polynomials in q, r of degree k. The simplifications that do occur are insufficient to curb the asymptotic growth. As a matter of fact, if we follow a particular factor we can check that it keeps appearing either in the numerator or the denominator (where its degree is alternatively 1 and 2). This corresponds to the unconfined singularity pattern $\{0, \infty^2, 0, \infty, 0, \infty^2, 0, \infty, \dots\}$. The confinement condition $a_{n+1} - 2a_n + a_{n-1} = 0$ is the condition for q to divide exactly Q_7, for both q and $r^2 + a_1 qr - pq^2$ to divide exactly Q_{12}, etc. Let us now turn to the computation of the degrees of x_n. We obtain successively 0, 1, 2, 5, 10, 21, 42, 85,... The growth is exponential, the degrees behaving like $d_{2m-1} = (2^{2m}-1)/3$ and $d_{2m} = 2d_{2m-1}$, a clear indication that the mapping is not integrable in general. Already at the fourth iteration the degrees differ in the autonomous and nonautonomous cases. Our approach consists in requiring that the degree in the nonautonomous case be *identical* to the one obtained in the autonomous one. If we implement the requirement that d_4 be 8 instead of 10 we find the condition $a_{n+1} - 2a_n + a_{n-1} = 0$, i.e. precisely the one obtained through singularity confinement. Moreover, once this condition is satisfied, the subsequent degrees of the nonautonomous case coincide with that of the autonomous one. Thus this mapping, leading to polynomial growth, should be integrable, and, in fact, it is. As we have shown in [**20**], where we presented its Lax pair, equation (2.2) with $a_n = \alpha n + \beta$ is a discrete form of the Painlevé I equation. In the examples that follow, we shall show that in all cases the nonautonomous form of an integrable mapping obtained through singularity confinement leads to exactly the same degrees of the iterates as the autonomous one.

We now turn to what is known as the "standard" discrete Painlevé equations [**21**] and compare the results of singularity confinement to those of the algebraic entropy approach. We start with d-P_I in the form:

$$x_{n+1} + x_n + x_{n-1} = a + \frac{b}{x_n}. \tag{2.3}$$

The degrees of the iterates of the autonomous mapping are 0, 1, 2, 3, 6, 9, 12, 17, 22,..., i.e. a quadratic growth with $d_{3m+k} = 3m^2 + (2m+1)k$, for $k = 0, 1, 2$, while those of the generic nonautonomous one are 0, 1, 2, 3, 6, 11,... Requiring two extra factorisations at that level (so as to bring d_5 down to 9) we find the following conditions $a_{n+1} = a_n$, so a_n must be a constant, and $b_{n+2} - b_{n+1} - b_n + b_{n-1} = 0$, i.e. b_n is of the form $b_n = \alpha n + \beta + \gamma(-1)^n$ which are exactly the result of singularity confinement. Implementing these conditions we find that the autonomous and nonautonomous mappings have the same (polynomial) growth [**18**]. Both are integrable, the Lax pair of the nonautonomous one, namely d-P_I having been given in [**6, 15, 19**].

For the discrete P_II equation we have

$$x_{n+1} + x_{n-1} = \frac{ax_n + b}{x_n^2 - 1}. \tag{2.4}$$

The degrees of the iterates in the autonomous case are $d_n = 0, 1, 2, 4, 6, 9, 12, 16, 20, \ldots$, (i.e. $d_{2m-1} = m^2$, $d_{2m} = m^2 + m$) while in the generic nonautonomous case we find the first discrepancy for d_4 which is now 8. To bring it down to 6 we find two conditions, $a_{n+1} - 2a_n + a_{n-1}=0$ and $b_{n+1} = b_{n-1}$. This means that a is linear in n and b is an even/odd constant, as predicted by singularity confinement. Once we implement these constraints, the degrees of the nonautonomous and autonomous cases coincide. The Lax pair of equation (2.4) in the nonautonomous form, i.e. d–$\mathrm{P_{II}}$, has been presented in [**5, 17, 19**].

The q-$\mathrm{P_{III}}$ equation was obtained from the deautonomisation of the mapping:

$$x_{n+1}x_{n-1} = \frac{(x_n - a)(x_n - b)}{(1 - cx_n)(1 - x_n/c)}. \tag{2.5}$$

In the autonomous case we obtain the degrees $d_n = 0, 1, 2, 5, 8, 13, 18, \ldots$, just like for equation (2.2), while in the generic nonautonomous case we have 0, 1, 2, 5, 12,... For d_4 to be 8 instead of 12, one needs four factors to cancel out. The conditions are $c_{n+1} = c_{n-1}$ and $a_{n+1}b_{n-1} = a_{n-1}b_{n+1} = a_n b_n$. Thus c is a constant up to an even/odd dependence, while a and b are proportional to λ^n for some λ, with an extra even/odd dependence, just as predicted by singularity confinement in [**21**]. The Lax pair for q-$\mathrm{P_{III}}$ has been presented in [**16, 19**].

For the remaining three discrete Painlevé equations the Lax pairs are not known yet. It is thus important to have one more check of their integrability provided by the algebraic entropy approach. We start with d-$\mathrm{P_{IV}}$ in the form:

$$(x_{n+1} + x_n)(x_{n-1} + x_n) = \frac{(x_n^2 - a^2)(x_n^2 - b^2)}{(x_n + z_n)^2 - c^2}, \tag{2.6}$$

where a, b and c are constants. If z_n is constant we obtain for the degrees of the successive iterates $d_n = 0, 1, 3, 6, 11, 17, 24, \ldots$ The general expression of the growth is $d_n = 6m^2$ if $n = 3m$, $d_n = 6m^2 + 4m + 1$ if $n = 3m + 1$ and $d_n = 6m^2 + 8m + 3$ if $n = 3m + 2$. This polynomial (quadratic) growth is expected since in the autonomous case this equation is integrable, its solution being given in terms of elliptic functions. For a generic z_n we obtain the sequence $d_n = 0, 1, 3, 6, 13, \ldots$ The condition for the extra factorizations to occur in the last case, bringing down the degree d_4 to 11, is for z to be linear in n. We can check that the subsequent degrees coincide with those of the autonomous case.

For the discrete Painlevé V equation we start from:

$$(x_{n+1}x_n - 1)(x_{n-1}x_n - 1) = \frac{(x_n^2 + ax_n + 1)(x_n^2 + bx_n + 1)}{(1 - z_n c x_n)(1 - z_n d x_n)}, \tag{2.7}$$

where a, b, c and d are constants. If moreover z is also a constant, we obtain exactly the same sequence of degrees $d_n = 0, 1, 3, 6, 11, 17, 24, \ldots$, as in the d-$\mathrm{P_{IV}}$ case. Again, this polynomial (quadratic) growth is expected since this mapping is also integrable in terms of elliptic functions. For the generic nonautonomous case we again find the sequence $d_n = 0, 1, 3, 6, 13, \ldots$ Once more we require a factorization bringing down d_4 to 11. It turns out that this entails a z which is exponential in n, which then generates the same sequence of degrees as the autonomous case. In both the d-$\mathrm{P_{IV}}$ and q-$\mathrm{P_V}$ cases we find the n-dependence already obtained through singularity confinement. Since this results to a vanishing algebraic entropy we expect both equations to be integrable.

The final system we shall study is the one related to the discrete P_{VI} equation:

$$\frac{(x_{n+1}x_n - z_{n+1}z_n)(x_{n-1}x_n - z_{n-1}z_n)}{(x_{n+1}x_n - 1)(x_{n-1}x_n - 1)} = \frac{(x_n^2 + az_nx_n + z_n^2)(x_n^2 + bz_nx_n + z_n^2)}{(x_n^2 + cx_n + 1)(x_n^2 + dx_n + 1)}, \tag{2.8}$$

where a, b, c and d are constants. In fact the generic symmetric QRT mapping can be brought to the autonomous (z_n constant) form of equation (2.8) through the appropriate homographic transformation. In the autonomous case, we obtain the degree sequence $d_n = 0, 1, 4, 9, 16, 25, \ldots$, i.e. $d_n = n^2$. Since mapping (2.8) is rather complicated we cannot investigate its full freedom. Still we were able to perform two interesting calculations. First, assume that in the rhs instead of the function z_n a different function ζ_n appears. In this case the degrees grow like 0, 1, $5, \ldots$, and the condition to have $d_2 = 4$ instead of 5 is $z_{n+1}z_{n-1}z_n^2 = \zeta_n^4$. Assuming this is true, we compute the degree d_3 of the next iterate and find $d_3 = 13$ instead of 9. To bring down d_3 to the value 9 we need $z_n^2 = \zeta_n^2$, which up to a redefinition of a and b means $z_n = \zeta_n$. This implies $z_{n+1}z_{n-1} = z_n^2$, and z_n is thus an exponential function of n, $z_n = z_0\lambda^n$ (which is in agreement with the results of [**7**]). Then a quartic factor drops out and d_3 is just 9. One can then check that the next degree is 16, just as in the autonomous case. Thus the q-P_{VI} equation leads to the same growth as the generic symmetric QRT mapping and is thus expected to be integrable. As a matter of fact we were able to show that the generic *asymmetric* QRT mapping leads to the same growth $d_n = n^2$ as the symmetric one. This is not surprising, given the integrability of this mapping. What is interesting is that the growth of the generic symmetric and asymmetric QRT mappings are the same. Thus $d_n = n^2$ is the maximal growth one can obtain for the QRT mapping in the homogeneous variables we are using. As a matter of fact we have also checked that the asymmetric nonautonomous q-P_{VI} equation, introduced in [**7**] led to exactly the same degree-growth $d_n = n^2$.

The singularity confinement results have been confronted to the algebraic entropy approach for several other discrete Painlevé equations. In every case examined the deautonomisation obtained has turned out to be the right one, it was the condition for the degree-growth to be identical to the one of the autonomous system. Thus despite the non sufficiency of the singularity confinement the integrability predictions for discrete Painlevé equations based on this criterion are confirmed.

3. Linearisable Equations

In this section we shall examine this particular class of mappings which are linearisable and study their growth properties. Most of these systems were obtained using the singularity confinement criterion and thus a study of the growth of the degree of the iterates would be an interesting complementary information. Moreover, as we will show, the linearisable systems do possess particular growth properties which set them apart from the other integrable discrete systems.

The first mapping we are going to treat is a two-point mapping of the form $x_{n+1} = f(x_n, n)$ where f is rational in x_n and analytical in n. As explained in the introduction for all f's of the form $\sum_k \alpha_k/(x_n + \beta_k)^{\nu_k}$ the singularity confinement requirement is satisfied. However only the discrete Riccati, $x_{n+1} = \alpha + \lambda/(x_n + \beta)$, is expected to be integrable. Our argument in [**13**], for the rejection of these confining but nonintegrable cases, was based on the proliferation of the preimages of a given point. If we solve the mapping for x_n in terms of x_{n+1} we do not find a uniquely defined x_n and, iterating, the number of x_{n-k} grows exponentially. In what follows we shall analyse this two-point mapping in the light of the algebraic entropy approach. We start from the simplest case which we expect to be nonintegrable,

$$x_{n+1} = \alpha + \frac{\lambda}{x_n + \beta} + \frac{\mu}{x_n + \gamma}. \tag{3.1}$$

The initial condition we are going to iterate is $x_0 = p/q$ and the degree we calculate is the homogeneous degree in p and q of the numerator (or the denominator) of the iterate. We obtain readily the following degree sequence $d_n = 1, 2, 4, 8, 16, \ldots$, i.e. $d_n = 2^n$. Thus the algebraic entropy of the mapping is $\log(2) > 0$, an indication that the mapping cannot be integrable. Now we ask how can one curb the growth and make it nonexponential. It turns out that the only possibilities are $\lambda\mu = 0$ or $\beta = \gamma$. In either case mapping (3.1) becomes a homography. The degree in this case is simply $d_n = 1$ for all n. This is an interesting result, clearly due to the fact that the homographic mapping is linearisable through a simple Cole–Hopf transformation.

The second mapping we shall examine is one due to Bellon and collaborators [**2**]

$$x_{n+1} = \frac{x_n + y_n - 2x_n y_n^2}{y_n(x_n - y_n)}, \qquad y_{n+1} = \frac{x_n + y_n - 2x_n^2 y_n}{x_n(y_n - x_n)}. \tag{3.2}$$

The degree-growth in this case is studied starting from $x_0 = r$, $y_0 = p/q$ and again we calculate the homogeneous degree of the iterate in p and q, i.e. we set the degree of r to zero. (Other choices could have been possible but the conclusion would not depend on these details.) We obtain the degrees $d_{x_n} = 0, 2, 2, 4, 4, 6, 6, \ldots$ and $d_{y_n} = 1, 1, 3, 3, 5, 5, \ldots$, i.e. a linear degree-growth. This is in perfect agreement with the integrable character of the mapping. As was shown in [**10**] it does satisfy the unique preimage requirement and possesses a constant of motion $k = (1 - x_n y_n)/(y_n - x_n)$, the use of which reduces it to a homographic mapping for x_n or y_n.

The third mapping we are going to study is the one proposed in [**13**]

$$x_{n+1} = \frac{x_n(x_n - y_n - a)}{x_n^2 - y_n}, \qquad y_{n+1} = \frac{(x_n - y_n)(x_n - y_n - a)}{x_n^2 - y_n}, \tag{3.3}$$

where a was taken constant. We start by assuming that a is an arbitrary function of n and compute the growth of the degree. We find $d_{x_n} = 0, 1, 2, 3, 4, 5, 6, 7, 8, \ldots$ and $d_{y_n} = 1, 2, 3, 4, 5, 6, 7, 8, 9, \ldots$, i.e. again a linear growth. This is an indication that (3.3) is integrable for arbitrary a_n and indeed it is. Dividing the two equations we obtain $y_{n+1}/x_{n+1} = 1 - y_n/x_n$, i.e. $y_n/x_n = 1/2 + k(-1)^n$ whereupon (3.3) is reduced to a homographic mapping for x. Thus in this case the degree-growth has successfully predicted integrability.

A picture starts emerging at this point. While in our study of discrete Painlevé equations and the QRT mapping we found quadratic growth of the degree of the iterate, linearisable second-order mappings seem to lead to slower growth. In order to investigate this property in detail we shall analyse the three-point mapping we have studied in [**22, 24**] from the point of view of integrability in general and linearisability in particular. The generic mapping studied in [**24**] was one trilinear in x_n, x_{n+1}, x_{n-1}. Several cases were considered. Our starting point is the mapping,

$$\begin{aligned}(3.4)\quad x_{n+1}x_nx_{n-1} + \beta x_nx_{n+1} + \zeta\eta x_{n+1}x_{n-1} + \gamma x_nx_{n-1} \\ + \beta\gamma x_n + \eta x_{n-1} + \zeta x_{n+1} + 1 = 0.\end{aligned}$$

We start with the initial conditions $x_0 = r$, $x_1 = p/q$ and compute the homogeneous degree in p, q at every n. We find $d_n = 0, 1, 1, 2, 3, 5, 8, 13, \ldots$, i.e. a Fibonacci sequence $d_{n+1} = d_n + d_{n-1}$ leading to exponential growth of d_n with asymptotic ratio $(1+\sqrt{5})/2$. Thus mapping (3.4) is not expected to be integrable in general. However, as shown in [**24**] integrable subcases do exist. We start by requiring that the degree-growth be less rapid and as a drastic decrease in the degree we demand that $d_3 = 1$ instead of 2. We find that this is possible when either $\beta = \zeta = 0$ in which case the mapping reduces to:

$$(3.5)\qquad x_{n+1} = -\gamma - \frac{\eta}{x_n} - \frac{1}{x_nx_{n-1}},$$

or $\gamma = \eta = 0$, giving a mapping identical to (3.5) after $x \to 1/x$. In this case the degree is $d_n = 1$ for $n > 0$. Equation (3.5) is the generic projective three-point mapping, written in canonical form. Its linearisation can be obtained [**24**] in terms of a system of three linear equations, a fact which explains the constancy of the degree.

Non generic subcases of (3.4) some of which are integrable do exist. They have been studied in [**23, 24**].

Linearisability through the reduction to a projective system is not the only possibility. Other possibilities do exist. Let us start by considering the generic three-point mapping that can be considered as the discrete derivative of a (discrete) Riccati equation. Let us start from the general homographic mapping which we can write as

$$(3.6)\qquad Ax_nx_{n+1} + Bx_n + Cx_{n+1} + D = 0,$$

where A, B, C, D are linear in some constant quantity κ. In order to take the discrete derivative we extract the constant κ and rewrite (3.6) as:

$$(3.7)\qquad \kappa = \frac{\alpha x_nx_{n+1} + \beta x_n + \gamma x_{n+1} + \delta}{\varepsilon x_nx_{n+1} + \zeta x_n + \eta x_n up + \theta}.$$

Using the fact that κ is a constant, it is now easy to obtain the discrete derivative by downshifting (3.7) and subtracting it form (3.7) above. Instead of examining this most general case we concentrate on the forms proposed in [**8**]. They correspond to the reduction of (3.7) to the two cases:

$$(3.8)\qquad \kappa = x_{n+1} + a + \frac{b}{x_n},$$

$$(3.9)\qquad \kappa = \frac{x_{n+1}(x_n + a)}{x_n + b}.$$

Next we compute the discrete derivatives of (3.8) and (3.9). We find:

$$(3.10)\qquad x_{n+1} = x_n + a_{n-1} - a_n - \frac{b_n}{x_n} + \frac{b_{n-1}}{x_{n-1}},$$

and

$$x_{n+1} = x_n \frac{x_{n-1} + a_{n-1}}{x_n + a_n} \frac{x_n + b_n}{x_{n-1} + b_{n-1}}. \tag{3.11}$$

The study of the degree of growth of (3.10) and (3.11) can be performed in a straightforward way. For both mappings we find the sequence $d_n = 0, 1, 2, 3, 4, 5, 6, \ldots$, i.e. a linear growth just as in the cases of mappings (3.2) and (3.3). If we substitute b_{n-1} by c_{n-1} in the last term of the rhs of (3.10) or the denominator of (3.11) we find $d_n = 0, 1, 2, 4, 8, 16, \ldots$, i.e. $d_n = 2^n$ for $n > 0$ unless $c = b$. Investigating all the possible ways to curb the growth we find for both (3.10) and (3.11) that $c = 0$ is also a possibility to bring d_3 down to 3. However a detailed analysis of this case shows that for $c = 0$ we have $d_n = 0, 1, 2, 3, 5, 8, 13, 21, \ldots$, i.e. a Fibonacci sequence with slower, but still exponential, growth (i.e. ratio $(1+\sqrt{5})/2$ instead of 2).

A bonus study: The N-dimensional Gambier mapping. While most of the results included in this paper were presented in previous publications of ours, in this section we shall present a study which appear for the first time in these proceedings. It concerns the extension of the Gambier mapping to N dimensions.

The two-dimensional Gambier mapping was introduced in [**9, 11**] as a discretisation of the second order differential equation discovered by Gambier, in his study of second order ODE's having the Painlevé property. The Gambier equation is in fact a system of two Riccati's in cascade. The latter means that the system consists in one Riccati for one variable and a second one (for the second variable) with coefficients depending (linearly) on the variable of the first one. The discretisation of the Gambier system is straightforward: just substitute homographic mapping, i.e. discrete Riccati, in the previous sentence. Thus the general second-order Gambier mapping is given by:

$$y_{n+1} = \frac{ay_n + b}{cy_n + d}, \tag{3.12a}$$

$$x_{n+1} = \frac{(ey_n + f)x_n + (gy_n + h)}{(jy_n + k)x_n + (ly_n + m)}, \tag{3.12b}$$

where $a, b, \ldots, m$ are functions of n. In [**11**] we have studied this mapping in detail from the point of view of the singularity structure. This was motivated by the fact that we aimed at being able to express the solution as an infinite product of matrices, *even across singularities*. On the other hand if we are not interested in this fine point, the linearisability of (3.12) can be obtained through a Cole–Hopf transformation for each variable.

The study of the degree-growth of (3.12) is straightforward. We start from $x_0 = r$, $y_0 = p/q$ and compute the homogeneous in p, q degree of (4.1a) and (4.1b). Since (3.12a) is a Riccati its degree does not grow, i.e. we have $d_{y_n} = 1$. Given the structure of (3.12b) we have $d_{x_{n+1}} = d_{x_n} + d_{y_n}$ and thus $d_{x_n} = n$. What is interesting here is that the Gambier mapping exhibits a linear degree-growth independently of the precise values of $a, b, \ldots, m$. The fact that it can be reduced to Riccati's in cascade is enough to guarantee its integrability.

The generalisation of (3.12) to $N+1$ dimensions is straightforward. We find the system:

$$
\begin{aligned}
x_{n+1}^{(0)} &= \frac{a^{(0)}x_n^{(0)} + b^{(0)}}{c^{(0)}x_n^{(0)} + d^{(0)}}, \\
x_{n+1}^{(i)} &= \frac{(e^{(i)}x_n^{(i-1)} + f^{(i)})x_n^{(i)} + (g^{(i)}x_n^{(i-1)} + h^{(i)})}{(j^{(i)}x_n^{(i-1)} + k^{(i)})x_n^{(i)} + (l^{(i)}x_n^{(i-1)} + m^{(i)})}, \qquad i = 1,\ldots,N.
\end{aligned} \tag{3.13}
$$

The study of the degree-growth of system (3.13) can be performed along the $N = 1$ case. From (3.13) we have the recurrence: $dx_{n+1}^{(k)} = dx_n^{(k)} + dx_n^{(k-1)}$. We obtain formally $dx_{n+1}^{(k)} = \sum_{p=0}^{n} dx_p^{(k-1)}$. Thus starting from $dx_n^{(0)} = 1$ we find $dx_n^{(N)} \propto n^N$. Thus the $(N+1)$-dimensional Gambier mapping has polynomial growth for every dimension. The linearisation is obtained through a Cole–Hopf transformation.

Given the structure of (3.13) it is clear that we can solve successively for each variable and express finally (3.13) as a single $(N+2)$-point mapping. This leads us to another integrable discretisation of linearisable mappings with cascade structure. Let us consider the second order system

$$
y_{n+1} = \frac{ay_n + b}{cy_n + d}, \qquad x_{n+1} = \frac{f_1(y_n)x_n - f_2(y_n)}{f_4(y_n) - f_3(y_n)x_n}, \tag{3.14}
$$

where the f_i's are polynomial in y_n. From the structure of (3.14) it is clear that it can be linearised, independently of the precise form of the f_i. Similarly the degree-growth of x_n is always linear. On the other hand it is not, in general, possible to express (3.14) as a single three-point mapping. Extension of (3.14) to N-dimensions can be obtained along the lines of the N-dimensional discrete Gambier system.

In this section we have studied the growth properties of various linearisable systems identified through the singularity confinement criterion. In every case the singularity confinement results were confirmed. Moreover it turned out that the linearisable systems lead to slower growth than systems which are integrable by other methods. This property could be used for the classification of integrable systems and be a valuable indication as to the precise method of their integration.

4. Conclusion

In this work we have presented a comparative review of results obtained with the singularity confinement and the algebraic entropy methods. While the former approach is not a sufficient criterion of integrability, the confirmation of its results (through the second, more stringent, criterion) in the domain of discrete Painlevé equations leads to interesting new insights. In every case examined, the singularity confinement, when used for the deautonomisation of an integrable autonomous mapping turned out to give sufficient constraints for the degree-growth to be non exponential. Thus we can propose a new strategy for the detection of integrable discrete systems. Given a mapping which contains several parameters the singularity confinement necessary criterion can be used in order to screen it for possible integrable discrete cases. Once the research domain is reduced the growth properties can be studied leading to better integrability candidates.

Another interesting result of our studies concerns the degree-growth of linearisable systems. We found that, while for second-order mappings the generic integrable case is associated to quadratic growth, the linearisable mappings lead to zero or linear growth. The growth exponent is of course a property which depends on the dimension. We surmise that the generic integrable Nth order mapping will lead to growth n^N, while the linearisable mappings of the same dimension will lead to slower growth. Our study of the Nth order Gambier system shows that the growth is n^{N-1}, and we expect that the projective system of order N to lead to zero growth. Thus the detailed study of the growth properties can become a precious indication as to the precise method of integration of a given discrete system.

References

1. V. I. Arnold, *Dynamics of complexity of intersections*, Bol. Soc. Brasil Mat. (N.S.) **21** (1990), 1–10.
2. M. P. Bellon, J.-M. Maillard, and C.-M. Viallet, *Rational mappings, arborescent iterations, and the symmetries of integrability*, Phys. Rev. Lett. **67** (1991), 1373–1376.
3. M. P. Bellon and C.-M. Viallet, *Algebraic entropy*, Comm. Math. Phys. **204** (1999), 425–437.
4. R. Conte and M. Musette, *A new method to test discrete Painlevé equations*, Phys. Lett. A **223** (1996), 439–448.
5. B. Grammaticos, F. W. Nijhoff, V. G. Papageorgiou, A. Ramani, and J. Satsuma, *Linearization and solutions of the discrete Painlevé* III *equation*, Phys. Lett. A **185** (1994), 446–452.
6. B. Grammaticos and A. Ramani, *From continuous Painlevé* IV *to the asymmetric discrete Painlevé* I, J. Phys. A **31** (1998), 5787–5798.
7. ———, *On a novel q-discrete analogue of the Painlevé* IV *equation*, Phys. Lett. A **257** (1999), 288–292.
8. ———, *Retracing the Painlevé–Gambier classification for discrete systems*, Methods Appl. Anal. **4** (1997), 196–211.
9. ———, *The Gambier mapping*, Phys. A **223** (1995), 125–136.
10. ———, *Investigating the integrability of discrete systems*, Internat. J. Modern Phys. B **7** (1993), 3551–3565.
11. B. Grammaticos, A. Ramani, and S. Lafortune, *The Gambier mapping, revisited*, Phys. A **253** (1998), 260–270.
12. B. Grammaticos, A. Ramani, and V. G. Papageorgiou, *Do integrable mappings have the Painlevé property?*, Phys. Rev. Lett. **67** (1991), 1825–1826.
13. B. Grammaticos, A. Ramani, and K. M. Tamizhmani, *Nonproliferation of pre-images in integrable mappings*, J. Phys. A **27** (1994), 559–566.
14. J. Hietarinta and C.-M. Viallet, *Singularity confinement and chaos in discrete systems*, Phys. Rev. Lett. **81** (1998), 325–328.
15. A. R. Its, A. V. Kitaev, and A. S. Fokas, *An isomonodromy approach to the theory of two-dimensional quantum gravity*, Uspekhi Mat. Nauk **45** (1990), 135–136; English transl. Russian Math. Surveys **45** (1990), 155–157.
16. M. Jimbo and H. Sakai, *A q-analog of the sixth Painlevé equation*, Lett. Math. Phys. **38** (1996), 145–154.
17. N. Joshi, D. Burtonclay, and R. G. Halburd, *Nonlinear nonautonomous discrete dynamical systems from a general discrete isomonodromy problem*, Lett. Math. Phys. **26** (1992), 123–131.
18. Y. Ohta, K. M. Tamizhmani, B. Grammaticos, and A. Ramani, *Singularity confinement and algebraic entropy: The case of the discrete Painlevé equations*, Phys. Lett. A **262** (1999), 152–156.
19. V. G. Papageorgiou, F. W. Nijhoff, B. Grammaticos, and A. Ramani, *Isomonodromic deformation problems for discrete analogues of Painlevé equations*, Phys. Lett. A **164** (1992), 57–64.
20. G. R. W. Quispel, J. A. G. Roberts, and C. J. Thompson, *Integrable mappings and soliton equations* II, Phys. D **34** (1989), 183–192.
21. A. Ramani, B. Grammaticos, and J. Hietarinta, *Discrete versions of the Painlevé equations*, Phys. Rev. Lett. **67** (1991), 1829–1832.
22. A. Ramani, B. Grammaticos, and G. Karra, *Linearizable mappings*, Phys. A **180** (1992), 115–127.

23. A. Ramani, B. Grammaticos, S. Lafortune, and Y. Ohta, *Linearizable mappings and the low growth criterion*, J. Phys. A **33** (2000), 288–292.
24. A. Ramani, B. Grammaticos, K. M. Tamizhmani, and S. Lafortune, *Again, linearizable mappings*, Phys. A **252** (1998), 138–150.
25. K. M. Tamizhmani, B. Grammaticos, A. Ramani, and Y. Ohta, *Integrability criteria for differential-difference systems: A comparison of singularity confinement and low-growth requirements*, J. Phys. A **32** (1999), 6679–6685.
26. A. P. Veselov, *What is an integrable mapping?*, What is integrability? (V. E. Zakharov, ed.), Springer Ser. Nonlinear Dynamics, Springer, Berlin, 1991, pp. 251–272.
27. ______, *Growth and integrability in the dynamics of mappings*, Comm. Math. Phys. **145** (1992), 181–193.

LPTMC et GMPIB, Univ. Paris VII, Tour 24-14, 5ᵉÉtage, 75251 Paris, France.
Current address: CRM, Université de Montréal, Montréal, C.P. 6128, succ. Centre-ville, Montréal, Québec H3C 3J7, Canada.
E-mail address: `lafortus@crm.umontreal.ca`

CPT, Ecole Polytechnique, CNRS, UMR 7644, 91128 Palaiseau, France.
E-mail address: `ramani@orphee.polytechnique.fr`

GMPIB, Univ. Paris VII, Tour 24-14, 5ᵉÉtage, Case 7021, 75251 Paris, France.
E-mail address: `grammati@paris7.jussieu.fr`

Department of Applied Mathematics, Faculty of Engineering, Hiroshima University, 1-4-1 Kagamiyama, Higashi-Hiroshima, 739-8527 Japan.
E-mail address: `ohta@kurims.kyoto-u.ac.jp`

Department of Mathematics, Pondicherry University, Kalapet, Pondicherry, 605014 India.
E-mail address: `tamizh@yahoo.com`

Centre de Recherches Mathématiques
CRM Proceedings and Lecture Notes
Volume **29**, 2001

Bäcklund Transformations of Soliton Systems from Symmetry Constraints

Wen-Xiu Ma and Xianguo Geng

ABSTRACT. Binary symmetry constraints are applied to constructing Bäcklund transformations of soliton systems, both continuous and discrete. Construction of solutions to soliton systems is split into finding solutions to lower-dimensional Liouville integrable systems, which also paves a way for separation of variables and exhibits integrability by quadratures for soliton systems. Illustrative examples are provided for the KdV equation, the AKNS system of nonlinear Schrödinger equations, the Toda lattice, and the Langmuir lattice.

1. Introduction

Symmetry constraints [**7, 12**] play an important role in showing integrability by quadratures for soliton systems, both continuous and discrete. They help to generate finite-dimensional integrable systems [**4, 5, 12**] and integrable symplectic mappings [**14**], and further provide a way of constructing finite-gap solutions to soliton systems by means of Riemann-theta functions [**2**]. Based on Lax pairs, symmetries themselves can also be applied to constructing Bäcklund transformations of soliton systems [**15**]. However, there is little work showing the importance of symmetry constraints in the study of Bäcklund transformations. In this paper we focus on the construction of Bäcklund transformations by using symmetry constraints, and show in certain cases that symmetry constraints can break up soliton systems into lower-dimensional Liouville integrable systems.

Let us recall some fundamental concepts. A system of continuous equations $u_t = K(u, u_x, \dots)$ is said to have a continuous Lax pair

$$\phi_x = U(u,\lambda)\phi, \quad \phi_t = V(u, u_x, \dots; \lambda)\phi, \tag{1.1}$$

if it is equivalent to the compatibility condition $U_t - V_x + [U, V] = 0$ of (1.1) under the isospectral condition $\lambda_t = 0$, and a system of discrete equations $u_t =$

2000 *Mathematics Subject Classification.* Primary: 37K35, 35Q58; Secondary: 37K05, 39A10.

The work was supported by a strategic research grant from City University of Hong Kong (Project No. 7001041) and a competitive earmarked research grant from the Research Grants Council of Hong Kong (Project No. 9040466).

This is the final form of the paper.

$K(u, E^{-1}u, Eu, \dots)$ (where E is the shift operator, i.e., $(Eu)(n) = u(n+1)$) is said to have a discrete Lax pair

$$E\phi = U(u,\lambda)\phi, \quad \phi_t = V(u, E^{-1}u, Eu, \dots; \lambda)\phi, \tag{1.2}$$

if it is equivalent to the compatibility condition $U_t-(EV)U+UV = 0$ of (1.2) under the isospectral condition $\lambda_t = 0$. The corresponding adjoint Lax pairs presenting the same compatibility conditions read as

$$\psi_x = -U^T(u,\lambda)\psi, \qquad \psi_t = -V^T(u, u_x, \dots; \lambda)\psi, \tag{1.3}$$

$$E^{-1}\psi = \big(E^{-1}U^T(u,\lambda)\big)\psi, \quad \psi_t = -V^T(u, E^{-1}u, Eu, \dots; \lambda)\psi, \tag{1.4}$$

where $(\cdot)^T$ denotes matrix transpose. Adjoint Lax pairs can help us determine the variational derivative of the spectral parameter with respect to the potential u (see, for example, [**6, 10**] for the continuous case).

A soliton hierarchy of continuous or discrete systems

$$u_{t_n} = K_n = \Phi^n K_0 = JG_n = J\frac{\delta \widetilde{H}_n}{\delta u}, \quad n \geq 0, \tag{1.5}$$

can be generated through the isospectral ($\lambda_{t_n} = 0$) compatibility conditions of continuous Lax pairs

$$\phi_x = U(u,\lambda)\phi, \quad \phi_{t_n} = V^{(n)}(u, u_x, \dots; \lambda)\phi, \tag{1.6}$$

or discrete Lax pairs

$$E\phi = U(u,\lambda)\phi, \quad \phi_{t_n} = V^{(n)}(u, E^{-1}u, Eu, \dots; \lambda)\phi, \tag{1.7}$$

where $V^{(n)}$ are Laurent polynomials in λ, Φ is a hereditary recursion operator to map symmetries into symmetries, and J is a Hamiltonian operator to map conserved covariants to symmetries.

In this paper, we would like to show that symmetry constraints can be applied to constructing Bäcklund transformations of soliton systems from Lax pairs. The resulting Bäcklund transformations separate each soliton system (in a hierarchy) into two lower-dimensional integrable systems. Thus, symmetry constraints are shown to be very useful in exposing integrability by quadratures for soliton systems. Illustrative examples will be given in both continuous and discrete cases.

2. Symmetry Constraints

Let us consider the space parts and the time parts of Lax pairs and adjoint Lax pairs:

$$\begin{cases} \phi_x = U(u,\lambda)\phi, \\ \psi_x = -U^T(u,\lambda)\psi, \end{cases} \qquad \begin{cases} \phi_{t_n} = V^{(n)}(u, u_x, \dots; \lambda)\phi, \\ \psi_{t_n} = -V^{(n)T}(u, u_x, \dots; \lambda)\psi, \end{cases}$$

or

$$\begin{cases} E\phi = U(u,\lambda)\phi, \\ E^{-1}\psi = (E^{-1}U^T(u,\lambda))\psi, \end{cases} \qquad \begin{cases} \phi_{t_n} = V^{(n)}(u, E^{-1}u, Eu, \dots; \lambda)\phi, \\ \psi_{t_n} = -V^{(n)T}(u, E^{-1}u, Eu, \dots; \lambda)\psi. \end{cases}$$

By using the space parts, we can work out

$$\frac{\delta\lambda}{\delta u} = \alpha\psi^T\frac{\partial U(u,\lambda)}{\partial u}\phi, \quad \text{or} \quad \frac{\delta\lambda}{\delta u} = \beta(E\psi^T)\frac{\partial U(u,\lambda)}{\partial u}\phi, \tag{2.1}$$

where α and β are two constants. Note that the Lie homomorphism $J\delta/(\delta u)$ transforms conserved functionals to symmetries. Therefore, $J\big((\delta\lambda)/(\delta u)\big)$ is a symmetry of each system $u_{t_n} = K_n$, since λ is a conserved functional, i.e., $\lambda_{t_n} = \big(\lambda(u)\big)_{t_n} = 0$ when $u_{t_n} = K_n$.

Now for all $m_0 \geq 0$, we can make symmetry constraints

$$(2.2)\qquad \begin{aligned} K_{m_0} &= \frac{1}{\alpha} J \frac{\delta\lambda}{\delta u} = J\psi^T \frac{\partial U(u,\lambda)}{\partial u}\phi, \quad \text{or} \\ K_{m_0} &= \frac{1}{\beta} J \frac{\delta\lambda}{\delta u} = J(E\psi^T) \frac{\partial U(u,\lambda)}{\partial u}\phi. \end{aligned}$$

If we take distinct eigenvalues $\lambda_1, \dots, \lambda_N$, and suppose that

$$(2.3)\qquad \phi_x^{(j)} = U(u,\lambda_j)\phi^{(j)}, \quad \psi_x^{(j)} = -U^T(u,\lambda_j)\psi^{(j)},$$

or

$$(2.4)\qquad E\phi^{(j)} = U(u,\lambda_j)\phi^{(j)}, \quad E^{-1}\psi^{(j)} = \big(E^{-1}U^T(u,\lambda_j)\big)\psi^{(j)},$$

where

$$\phi^{(j)} = (\phi_{1j}, \dots, \phi_{sj})^T, \quad \psi^{(j)} = (\psi_{1j}, \dots, \psi_{sj})^T,$$

we can make more systematical symmetry constraints

$$(2.5)\qquad K_{m_0} = \sum_{j=1}^N \frac{1}{\alpha_j} J \frac{\delta\lambda_j}{\delta u}, \quad \text{or} \quad K_{m_0} = \sum_{j=1}^N \frac{1}{\beta_j} J \frac{\delta\lambda_j}{\delta u},$$

namely,

$$(2.6)\qquad \begin{aligned} K_{m_0} &= \sum_{j=1}^N J\psi^{(j)T} \frac{\partial U(u,\lambda_j)}{\partial u}\phi^{(j)}, \quad \text{or} \\ K_{m_0} &= \sum_{j=1}^N J(E\psi^{(j)T}) \frac{\partial U(u,\lambda_j)}{\partial u}\phi^{(j)}, \end{aligned}$$

where α_j and β_j are the constants defined as in (2.1). These symmetry constraints suggest

$$(2.7)\qquad G_{m_0} = \sum_{j=1}^N \psi^{(j)T} \frac{\partial U(u,\lambda_j)}{\partial u}\phi^{(j)}, \quad \text{or} \quad G_{m_0} = \sum_{j=1}^N (E\psi^{(j)T}) \frac{\partial U(u,\lambda_j)}{\partial u}\phi^{(j)}.$$

Among those constraints between the potential, u, and the eigenfunctions and adjoint eigenfunctions, $\phi^{(j)}$ and $\psi^{(j)}$, the Bargmann constraint will be applied to constructing Bäcklund transformations between soliton systems and lower-dimensional integrable systems.

3. Bäcklund Transformations

Let us take the Bargmann constraint, i.e., the constraint (2.7) with $G_{m_0} = G_{m_0}(u)$ not involving any $\partial^i u$ $\big(\partial = \partial/(\partial x)\big)$, $i > 0$, or any $E^i u$, $i \neq 0$. Note that the discrete constraint defined by (2.7) can be rewritten as

$$(3.1)\qquad G_{m_0}(u) = \frac{\delta \widetilde{H}_{m_0}(u)}{\delta u} = \sum_{j=1}^N \frac{1}{\beta_j}\frac{\delta\lambda_j}{\delta u} = \sum_{j=1}^N \psi^{(j)T} U^{-1}(u,\lambda_j) \frac{\partial U(u,\lambda_j)}{\partial u}\phi^{(j)},$$

and therefore, in each of continuous and discrete cases, the Bargmann constraint defined by (2.7) is an algebraic equation on u, $\phi^{(j)}$, and $\psi^{(j)}$. Assume that solving the corresponding algebraic equation for u gives rise to an explicit expression of u:

$$u = f(\phi^{(1)}, \phi^{(2)}, \dots, \phi^{(N)}; \psi^{(1)}, \psi^{(2)}, \dots, \psi^{(N)}). \tag{3.2}$$

Substituting this expression of u into Lax pairs and adjoint Lax pairs leads to two systems, called binary constrained Lax pairs. Binary constrained continuous Lax pairs read as

$$\begin{cases} \phi_x^{(j)} = U(f, \lambda_j)\phi^{(j)}, & 1 \le j \le N, \\ \psi_x^{(j)} = -U^T(f, \lambda_j)\psi^{(j)}, & 1 \le j \le N, \end{cases} \tag{3.3}$$

$$\begin{cases} \phi_{t_n}^{(j)} = V^{(n)}(f, f_x, \dots; \lambda_j)\phi^{(j)}, & 1 \le j \le N, \\ \psi_{t_n}^{(j)} = -V^{(n)T}(f, f_x, \dots; \lambda_j)\psi^{(j)}, & 1 \le j \le N, \end{cases} \tag{3.4}$$

the first of which is a system of ordinary differential equations, but the second of which is a system of partial differential equations since $V^{(n)}$ contains some derivatives of u with respect to x. Binary constrained discrete Lax pairs read as

$$\begin{cases} E\phi^{(j)} = U(f, \lambda_j)\phi^{(j)}, & 1 \le j \le N, \\ E^{-1}\psi^{(j)} = (E^{-1}U^T(f, \lambda_j))\psi^{(j)}, & 1 \le j \le N, \end{cases} \tag{3.5}$$

$$\begin{cases} \phi_{t_n}^{(j)} = V^{(n)}(f, E^{-1}f, Ef, \dots; \lambda_j)\phi^{(j)}, & 1 \le j \le N, \\ \psi_{t_n}^{(j)} = -V^{(n)T}(f, E^{-1}f, Ef, \dots; \lambda_j)\psi^{(j)}, & 1 \le j \le N, \end{cases} \tag{3.6}$$

the first of which is a system of difference equations, but the second of which is a system of difference-differential equations since $V^{(n)}$ contains some of $E^i u$, $i \neq 0$. However, the second systems can be transformed into systems of ordinary differential equations by using the first systems. Furthermore, it can be shown by r-matrix formulation that all binary constrained Lax pairs, both continuous and discrete, are integrable in the Liouville sense [**1, 3**].

Therefore, (3.2) provides Bäcklund transformations between soliton systems and integrable binary constrained Lax pairs, and construction of solutions $u = f(\phi^{(j)}, \psi^{(j)})$ to soliton systems is split into finding solutions $\phi^{(j)}$ and $\psi^{(j)}$ to two lower-dimensional integrable systems.

4. Examples of Continuous Systems

EXAMPLE 4.1. Let us consider the KdV Equation

$$u_{t_1} = \frac{1}{4}u_{xxx} + \frac{3}{2}uu_x = J\frac{\delta \widetilde{H}_1}{\delta u}, \quad J = \partial, \quad \widetilde{H}_1 = \int \left(\frac{1}{8}uu_{xx} + \frac{3}{12}u^3\right) dx, \tag{4.1}$$

which can be written as $U_t - V_x + [U, V] = 0$ with

$$U = \begin{pmatrix} 0 & 1 \\ \lambda - u & 0 \end{pmatrix}, \quad V = \begin{pmatrix} -\frac{1}{4}u_x & \lambda + \frac{1}{2}u \\ \lambda^2 - \frac{1}{2}u\lambda - \frac{1}{4}u_{xx} - \frac{1}{2}u^2 & \frac{1}{4}u_x \end{pmatrix}. \tag{4.2}$$

Take the Bargmann symmetry constraint

$$K_{m_0} = 2\partial_x \sum_{j=1}^N \psi^{(j)T} \frac{\partial U(u, \lambda_j)}{\partial u} \phi^{(j)}, \text{ where } K_{m_0} = K_0 = u_x, \tag{4.3}$$

which implies the following equation

$$u_x = 2\partial_x \sum_{j=1}^{N} (\psi_{1j}, \psi_{2j}) \begin{pmatrix} 0 & 0 \\ -1 & 0 \end{pmatrix} \begin{pmatrix} \phi_{1j} \\ \phi_{2j} \end{pmatrix} = 2\partial_x \sum_{j=1}^{N} \phi_{1j}\psi_{2j}.$$

Integrating this equation with respect to x yields a Bäcklund transformation

$$u = f(\phi_{ij}; \psi_{ij}) = 2\sum_{j=1}^{N} \phi_{1j}\psi_{2j} = 2\langle P_1, Q_2\rangle, \tag{4.4}$$

where $\langle \cdot, \cdot \rangle$ denotes the standard inner product of $\mathbb{R}^N$, and P_i and Q_i are defined by

$$P_i = (\phi_{i1}, \phi_{i2}, \dots, \phi_{iN})^T, \quad Q_i = (\psi_{i1}, \psi_{i2}, \dots, \psi_{iN})^T, \qquad i = 1, 2.$$

A general Bäcklund transformation $u = 2\langle P_1, Q_2\rangle + c$ with an arbitrary constant c can also result from the above symmetry constraint, but it will not generate essentially new integrable systems from Lax pair of the KdV equation and so we omit to discuss it.

Keeping (4.4) in mind, the corresponding constrained Lax pairs (3.3) and (3.4), where two matrices U and $V^{(n)} = V$ are defined by (4.2), can simultaneously be rewritten as

$$P_{ix} = -\frac{\partial H^c}{\partial Q_i}, \quad Q_{ix} = \frac{\partial H^c}{\partial P_i}, \quad H^c = -F_3, \qquad i = 1, 2, \tag{4.5}$$

$$P_{it} = -\frac{\partial H_1^c}{\partial Q_i}, \quad Q_{it} = \frac{\partial H_1^c}{\partial P_i}, \quad H_1^c = -F_4, \qquad i = 1, 2, \tag{4.6}$$

where the functions F_m are given by

$$F_0 = 0, \quad F_1 = 1, \quad F_2 = 0,$$

$$F_3 = \langle P_2, Q_1\rangle - \langle P_1, Q_2\rangle^2 + \langle AP_1, Q_2\rangle,$$

$$F_m = \sum_{i=0}^{m-4} (\bar{a}_i\bar{a}_{m-i-4} + \bar{b}_i\bar{c}_{m-i-4}) + \bar{c}_{m-3} + \bar{b}_{m-2} - \frac{1}{2}f\bar{b}_{m-3}, \quad m \geq 4,$$

$$\bar{a}_i = \langle A^iP_1, Q_1\rangle - \langle A^iP_2, Q_2\rangle, \quad \bar{b}_i = \langle A^iP_1, Q_2\rangle, \quad \bar{c}_i = \langle A^iP_2, Q_1\rangle,$$

through $F = \det V|_{u=f} = \sum_{m=0}^{\infty} F_m\lambda^{-m}$, $V = \sum_{i=0}^{\infty} V_i\lambda^{-i}$ satisfying $V_x = [U, V]$. Note that we always accept

$$A = \mathrm{diag}(\lambda_1, \dots, \lambda_N).$$

These two systems (4.5) and (4.6) are Liouville integrable [**1**], since they have involutive integrals of motion F_m, $m \geq 0$, and

$$\overline{F}_j = \phi_{1j}\psi_{1j} + \phi_{2j}\psi_{2j}, \quad 1 \leq j \leq N, \tag{4.7}$$

among which $F_3, F_4, \dots, F_{N+2}$ and $\overline{F}_1, \dots, \overline{F}_N$ are functionally independent. All functions

$$u(x, t_1) = 2\big\langle P_1(x, t_1), Q_2(x, t_1)\big\rangle = \langle g_H^x g_{H_1}^{t_1} P_{10}, g_H^x g_{H_1}^{t_1} Q_{20}\rangle, \tag{4.8}$$

where g_H^x and $g_{H_1}^{t_1}$ are the Hamiltonian flows associated with (4.5) and (4.6) respectively, and P_{10} and Q_{20} are two arbitrary constant vectors, will determine solutions to the KdV equation (4.1).

EXAMPLE 4.2. Let us now consider the coupled nonlinear Schrödinger system:

$$u_{t_2} = \begin{pmatrix} q \\ r \end{pmatrix}_{t_2} = \begin{pmatrix} -\frac{1}{2}q_{xx} + q^2 r \\ \frac{1}{2}r_{xx} - qr^2 \end{pmatrix} = J\frac{\delta \widetilde{H}_2}{\delta u}, \tag{4.9}$$

with the Hamiltonian operator J and the Hamiltonian functional $\widetilde{H}_2$:

$$J = \begin{pmatrix} 0 & -2 \\ 2 & 0 \end{pmatrix}, \quad \widetilde{H}_2 = \frac{1}{12}\int (qr_{xx} - q_x r_x + q_{xx} r - 3q^2 r^2)\, dx. \tag{4.10}$$

It has a Lax pair

$$U = \begin{pmatrix} -\lambda & q \\ r & \lambda \end{pmatrix}, \quad V = \begin{pmatrix} -\lambda^2 + \frac{1}{2}qr & q\lambda - \frac{1}{2}q_x \\ r\lambda + \frac{1}{2}r_x & \lambda^2 - \frac{1}{2}qr \end{pmatrix}. \tag{4.11}$$

The Bargmann symmetry constraint (2.6) reads as

$$K_{m_0} = J\sum_{j=1}^{N} \psi^{(j)T}\frac{\partial U(u,\lambda_j)}{\partial u}\phi^{(j)}, \text{ where } K_{m_0} = K_0 = J\begin{pmatrix} r \\ q \end{pmatrix}. \tag{4.12}$$

This implies the following equation

$$J\begin{pmatrix} r \\ q \end{pmatrix} = J\sum_{j=1}^{N} \begin{pmatrix} (\psi_{1j},\psi_{2j})\begin{pmatrix} 0 & 1 \\ 0 & 0 \end{pmatrix}\begin{pmatrix} \phi_{1j} \\ \phi_{2j} \end{pmatrix} \\ (\psi_{1j},\psi_{2j})\begin{pmatrix} 0 & 0 \\ 1 & 0 \end{pmatrix}\begin{pmatrix} \phi_{1j} \\ \phi_{2j} \end{pmatrix} \end{pmatrix} = J\sum_{j=1}^{N}\begin{pmatrix} \phi_{2j}\psi_{1j} \\ \phi_{1j}\psi_{2j} \end{pmatrix}, \tag{4.13}$$

which equivalently engenders a Bäcklund transformation

$$u = \begin{pmatrix} q \\ r \end{pmatrix} = f(\phi_{ij};\psi_{ij}) = \sum_{j=1}^{N}\begin{pmatrix} \phi_{1j}\psi_{2j} \\ \phi_{2j}\psi_{1j} \end{pmatrix} = \begin{pmatrix} \langle P_1, Q_2\rangle \\ \langle P_2, Q_1\rangle \end{pmatrix}. \tag{4.14}$$

Keeping (4.14) in mind, the corresponding constrained Lax pairs (3.3) and (3.4), where two matrices U and $V^{(n)} = V$ are defined by (4.11), can simultaneously be transformed into the following

$$P_{ix} = -\frac{\partial H^c}{\partial Q_i}, \quad Q_{ix} = \frac{\partial H^c}{\partial P_i}, \quad H^c = F_2 - \frac{1}{4}F_1^2, \qquad i = 1,2, \tag{4.15}$$

$$P_{it_2} = -\frac{\partial H_2^c}{\partial Q_i}, \quad Q_{it_2} = \frac{\partial H_2^c}{\partial P_i}, \quad H_2^c = F_3 - \frac{1}{2}F_1F_2 + \frac{3}{24}F_1^3, \qquad i = 1,2, \tag{4.16}$$

where the functions F_m are given by

$$\begin{aligned} F_1 &= \langle P_2, Q_2\rangle - \langle P_1, Q_1\rangle, \\ F_m &= \sum_{i=0}^{m-2}\Big[\frac{1}{4}(\langle A^i P_1, Q_1\rangle - \langle A^i P_2, Q_2\rangle) \\ &\qquad (\langle A^{m-i-2}P_1, Q_1\rangle - \langle A^{m-i-2}P_2, Q_2\rangle) + \langle A^i P_1, Q_2\rangle\langle A^{m-i-2}P_2, Q_1\rangle\Big] \\ &\quad + \langle A^{m-1}P_2, Q_2\rangle - \langle A^{m-1}P_1, Q_1\rangle, \quad m \ge 2. \end{aligned}$$

These two systems are completely integrable in the Liouville sense [**1**], since they have involutive integrals of motion F_m, $m \ge 0$, defined above, and $\overline{F}_j$, $1 \le j \le N$, defined by (4.7), among which $F_1, F_2, \dots, F_N$ and $\overline{F}_1, \dots, \overline{F}_N$ are functionally independent. The Bäcklund transformation (4.14) determines solutions to the coupled nonlinear Schrödinger system (4.9):

$$
(4.17) \qquad \begin{cases} q(x,t_2) = \langle P_1(x,t_2), Q_2(x,t_2)\rangle = \sum_{j=1}^N \phi_{1j}(x,t_2)\psi_{2j}(x,t_2), \\ r(x,t_2) = \langle P_2(x,t_2), Q_1(x,t_2)\rangle = \sum_{j=1}^N \phi_{2j}(x,t_2)\psi_{1j}(x,t_2), \end{cases}
$$

if $\phi_{ij}(x,t_2)$ and $\psi_{ij}(x,t_2)$ simultaneously solve two integrable finite-dimensional Hamiltonian systems (4.15) and (4.16).

5. Examples of Discrete Systems

EXAMPLE 5.1. Let us consider the Toda lattice [**13**]:

$$
(5.1) \qquad a_t(n,t) = a(n,t)\big(b(n+1,t) - b(n,t)\big), \quad b_t(n,t) = a(n,t) - a(n-1,t),
$$

which associates with the discrete spectral problem

$$
(5.2) \qquad E\phi = U(u,\lambda)\phi, \quad U(u,\lambda) = \begin{pmatrix} 0 & a \\ -1 & \lambda - b \end{pmatrix}, \quad \phi = \begin{pmatrix} \phi_1 \\ \phi_2 \end{pmatrix},
$$

where $u = (a,b)^T$ and λ is a spectral parameter. In order to derive a hierarchy of lattice equations associated with (5.2), we first solve the stationary discrete zero-curvature equation:

$$
(5.3) \qquad (EV)U - UV = 0, \quad V = (V_{ij})_{2\times 2},
$$

by assuming

$$
V_{11} = aB + (b-\lambda)C, \quad V_{12} = E^{-1}aC, \quad V_{21} = -C, \quad V_{22} = E^{-1}aB,
$$

where

$$
B = \sum_{i\geq 0} B_{i-1}\lambda^{-i}, \quad C = \sum_{i\geq 0} C_{i-1}\lambda^{-i}.
$$

The discrete zero-curvature equation (5.3) requires

$$
(5.4) \qquad JG_{-1} = 0, \quad MG_{n-1} = JG_n, \qquad n \geq 0,
$$

where $G_n = (B_n, C_n)^T$, and J, M are two skew-symmetric operators:

$$
(5.5) \qquad J = \begin{pmatrix} 0 & a\Delta \\ -\Delta^* a & 0 \end{pmatrix}, \quad M = \begin{pmatrix} a(\Delta - \Delta^*)a & a\Delta b \\ -b\Delta^* a & a\Delta - \Delta^* a \end{pmatrix}.
$$

We choose $G_{-1} = (0,1)^T$, and assume that all terms of G_n, $n \geq 0$, do not belong to $\ker J = \{G_{-1}, G_{-2}\}$ where $G_{-2} = (a^{-1}, 0)^T$, when we uniquely determine all G_n, $n \geq 0$. For instance, the second member has to be $G_0 = (1,b)^T$. This requirement also means that we just choose the key vector fields to form systems of lattice equations.

Let $\lambda_1, \dots, \lambda_N$ be distinct eigenvalues. Then we have

$$
(5.6) \qquad (E\phi_{1j}, E\phi_{2j}) = (\phi_{1j}, \phi_{2j})U(u,\lambda_j)^T, \quad 1 \leq j \leq N,
$$

$$
(5.7) \qquad (E\psi_{1j}, E\psi_{2j}) = (\psi_{1j}, \psi_{2j})U(u,\lambda_j)^{-1}, \quad 1 \leq j \leq N.
$$

It is easy to see that

$$
(5.8) \qquad M\frac{\delta\lambda_j}{\delta u} = \lambda_j J \frac{\delta\lambda_j}{\delta u},
$$

where $(\delta\lambda_j)/(\delta u)$ is determined by

$$(5.9)\qquad \frac{\delta\lambda_j}{\delta u}=\begin{pmatrix}\frac{\delta\lambda_j}{\delta a}\\ \frac{\delta\lambda_j}{\delta b}\end{pmatrix}=\frac{\beta_j}{a}\begin{pmatrix}(\lambda_j-b)\phi_{2j}\psi_{1j}+\phi_{2j}\psi_{2j}\\ a\phi_{2j}\psi_{1j}\end{pmatrix},\quad \beta_j=\text{const.}$$

Now the Bargmann constraint $G_0=\sum_{j=1}^N\beta_j^{-1}(\delta\lambda_j)/(\delta u)$ leads to a Bäcklund transformation

$$(5.10)\qquad a=\langle AP_2,P_1\rangle+\langle P_2,Q_2\rangle-\langle P_2,Q_1\rangle^2,\quad b=\langle P_2,Q_1\rangle,$$

where $A=\text{diag}(\lambda_1,\dots,\lambda_N)$, $P_i=(\phi_{i1},\dots,\phi_{iN})^T$, $Q_i=(\psi_{i1},\dots,\psi_{iN})^T$, and $\langle\cdot,\cdot\rangle$ is the standard inner product of $\mathbb{R}^N$, as defined before.

Substituting (5.10) into (5.6) and (5.7) yields a discrete Bargmann system

$$(5.11)\qquad \begin{cases}EP_1=(\langle AP_2,Q_1\rangle+\langle P_2,Q_2\rangle-\langle P_2,Q_1\rangle^2)P_2,\\ EP_2=-P_1-\langle P_2,Q_1\rangle P_2+AP_2,\\ EQ_1=\frac{Q_2-\langle P_2,Q_1\rangle Q_1+AQ_1}{\langle AP_2,Q_1\rangle+\langle P_2,Q_2\rangle-\langle P_2,Q_1\rangle^2},\\ EQ_2=-Q_1,\end{cases}$$

which determines a symplectic mapping H:

$$(5.12)\qquad (EP_1,EP_2,EQ_1,EQ_2)=H(P_1,P_2,Q_1,Q_2),$$

since we have by a direct calculation

$$\sum_{j=1}^N d(E\phi^{(j)})\wedge d(E\psi^{(j)})=\sum_{j=1}^N d\phi^{(j)}\wedge d\psi^{(j)}.$$

The generating function $\mathcal{F}_\lambda=\det V|_{u=f}$:

$$\mathcal{F}_\lambda=-Q_\lambda(AP_1,Q_1)-Q_\lambda(P_1,Q_2)+\langle P_1,Q_1\rangle Q_\lambda(P_2,Q_1)\\ +\begin{vmatrix}Q_\lambda(P_1,Q_1) & Q_\lambda(P_1,Q_2)\\ Q_\lambda(P_2,Q_1) & Q_\lambda(P_2,Q_2)\end{vmatrix}=\sum_{m\ge0}F_m\lambda^{-m-1},$$

where

$$Q_\lambda(\xi,\eta)=\sum_{j=1}^N\frac{\xi_j\eta_j}{\lambda-\lambda_j}=\sum_{m\ge0}\langle A^m\xi,\eta\rangle\lambda^{-m-1},$$

generates a hierarchy of invariants of (5.11):

$$F_0=\langle AP_1,Q_1\rangle-\langle P_1,Q_2\rangle+\langle P_1,Q_1\rangle\langle P_2,Q_1\rangle,$$
$$F_m=-\langle A^{m+1}P_1,Q_1\rangle-\langle A^mP_1,Q_2\rangle+\langle P_1,Q_1\rangle\langle A^mP_2,Q_1\rangle\\ +\sum_{i=1}^m\begin{vmatrix}\langle A^{i-1}P_1,Q_1\rangle & \langle A^{m-i}P_2,Q_1\rangle\\ \langle A^{i-1}P_1,Q_2\rangle & \langle A^{m-i}P_2,Q_2\rangle\end{vmatrix},\quad m\ge1.$$

A direct computation can show the involutivity

$$\{F_m,\bar F_l\}=0,\quad m,l\ge0,$$

where the variants $\bar F_j=\phi_{1j}\psi_{1j}+\phi_{2j}\psi_{2j}$, $1\le j\le N$, defined as before. Now we can easily see that the symplectic mapping (5.11) is Liouville integrable [**3**].

Introduce an initial-value problem

$$(5.13)\qquad P_{it}=\frac{\partial F_0}{\partial Q_i},\quad Q_{it}=-\frac{\partial F_0}{\partial P_i},\ (P_i(t),Q_i(t))\big|_{t=0}=(P_{i0},Q_{i0}),\qquad i=1,2,$$

where P_{i0} and Q_{i0}, $i = 1, 2$, are arbitrary constant vectors. Let $\big(P_i(t), Q_i(t)\big)$, $i = 1, 2$, be a solution to the initial-value problems (5.13), and further define

$$\big(P_1(n,t), P_2(n,t), Q_1(n,t), Q_2(n,t)\big) = H^n\big(P_1(t), P_2(t), Q_1(t), Q_2(t)\big). \tag{5.14}$$

Then $a(n,t)$ and $b(n,t)$ determined by the Bäcklund transformation (5.10) solves the Toda lattice (5.1).

EXAMPLE 5.2. Let us now consider the Langmuir lattice [**16**]:

$$a_t(n,t) = a(n,t)\big(a(n+1,t) - a(n-1,t)\big), \tag{5.15}$$

which associates with a reduction of the discrete spectral problem (5.2):

$$E\phi = U(a,\lambda)\phi, \quad U(a,\lambda) = \begin{pmatrix} 0 & a \\ -1 & \lambda \end{pmatrix}, \quad \phi = \begin{pmatrix} \phi_1 \\ \phi_2 \end{pmatrix}. \tag{5.16}$$

Assume that its discrete zero-curvature equation

$$(EV)U - UV = 0, \quad V = (V_{ij})_{2\times 2}, \tag{5.17}$$

has a solution

$$V_{11} = aB - \lambda^2 C, \quad V_{12} = \lambda E^{-1} aC, \quad V_{21} = -\lambda C, \quad V_{22} = E^{-1} aB,$$

where

$$B = \sum_{i\geq 0} B_i \lambda^{-2i}, \quad C = \sum_{i\geq 0} C_i \lambda^{-2i}.$$

Then upon choosing $B_0 = C_0 = 1$, we can easily find that (5.17) is equivalent to

$$\begin{aligned} &MB_{n-1} = JB_n, \quad B_0 = 1, \quad ej \geq 1, \\ &C_n = (\Delta^* a - a\Delta)^{-1}\Delta^* aB_n, \quad n \geq 1, \end{aligned} \tag{5.18}$$

where two skew-symmetric operators J and M read as

$$M = a(\Delta - \Delta^*)a, \quad J = a\Delta(\Delta^* a - a\Delta)^{-1}\Delta^* a. \tag{5.19}$$

Let $\lambda_1, \ldots, \lambda_N$ be distinct eigenvalues, then we have

$$\begin{cases} (E\phi_{1j}, E\phi_{2j}) = (\phi_{1j}, \phi_{2j})U(a,\lambda_j)^T, & 1 \leq j \leq N, \\ (E\psi_{1j}, E\psi_{2j}) = (\psi_{1j}, \psi_{2j})U(a,\lambda_j)^{-1}, & 1 \leq j \leq N, \end{cases} \tag{5.20}$$

and

$$M\frac{\delta\lambda_j}{\delta a} = \lambda_j^2 J\frac{\delta\lambda_j}{\delta a}, \text{ where } \frac{\delta\lambda_j}{\delta a} = \frac{\beta_j}{a}(\lambda_j\phi_{2j}\psi_{1j} + \phi_{2j}\psi_{2j}), \quad \beta_j = \text{const}. \tag{5.21}$$

Now similarly, the Bargmann constraint $G_0 = B_0 = \sum_{j=1}^N \beta_j^{-1}(\delta\lambda_j)/(\delta a)$ leads to a Bäcklund transformation

$$a = \langle AP_2, Q_1\rangle + \langle P_2, Q_2\rangle. \tag{5.22}$$

Substituting (5.22) into (5.20), we obtain another discrete Bargmann system

$$\begin{cases} EP_1 &= \big(\langle AP_2, Q_1\rangle + \langle P_2, Q_2\rangle\big)P_2, \\ EP_2 &= -P_1 + AP_2, \\ EQ_1 &= \frac{Q_2 + AQ_1}{\langle AP_2, Q_1\rangle + \langle P_2, Q_2\rangle}, \\ EQ_2 &= -Q_1, \end{cases} \tag{5.23}$$

which determines a symplectic mapping H:

$$(EP_1, EP_2, EQ_1, EQ_2) = H(P_1, P_2, Q_1, Q_2). \tag{5.24}$$

This symplectic mapping is Liouville integrable [**3**], since we have involutive invariants: $\overline{F}_j = \phi_{1j}\psi_{1j} + \phi_{2j}\psi_{2j}$, $1 \le j \le N$, and F_m, $m \ge 0$, defined by

$$F_0 = -\langle A^2P_1, Q_1\rangle - \langle AP_1, Q_2\rangle + \langle P_1, Q_1\rangle\big(\langle AP_2, Q_1\rangle + \langle P_2, Q_2\rangle\big),$$

$$F_m = -\langle A^{2m+2}P_1, Q_1\rangle - \langle A^{2m+1}P_1, Q_2\rangle + \langle P_1, Q_1\rangle\langle A^{2m+1}P_2, Q_1\rangle + \langle P_2, Q_2\rangle\langle A^{2m}P_1, Q_1\rangle + \sum_{i=1}^{m} \begin{vmatrix} \langle A^{2i-2}P_1, Q_1\rangle & \langle A^{2m-2i+1}P_2, Q_1\rangle \\ \langle A^{2i-1}P_1, Q_2\rangle & \langle A^{2m-2i+2}P_2, Q_2\rangle \end{vmatrix}.$$

Let $\big(P_i(t), Q_i(t)\big)$, $i = 1, 2$, be a solution to an initial-value problem:

$$P_{it} = \frac{\partial F_0}{\partial Q_i}, \quad Q_{it} = -\frac{\partial F_0}{\partial P_i}, \quad \big(P_i(t), Q_i(t)\big)\big|_{t=0} = (P_{i0}, Q_{i0}), \qquad i = 1, 2, \tag{5.25}$$

where P_{i0} and Q_{i0}, $i = 1, 2$, are arbitrary, and similarly define

$$\big(P_1(n,t), P_2(n,t), Q_1(n,t), Q_2(n,t)\big) = H^n\big(P_1(t), P_2(t), Q_1(t), Q_2(t)\big). \tag{5.26}$$

Then $a(n,t)$ determined by the Bäcklund transformation (5.22):

$$a(n,t) = \big\langle AP_2(n,t), Q_1(n,t)\big\rangle + \big\langle P_2(n,t), Q_2(n,t)\big\rangle,$$

provides a solution to the Langmuir lattice (5.15).

6. Concluding Remarks

It has been shown that solving symmetry constraints for u can give rise to Bäcklund transformations between soliton systems and lower-dimensional Liouville integrable systems, which supplements the study of binary nonlinearization of Lax pairs [**8–10, 12**]. Construction of solutions to soliton systems is split into finding solutions to the space and time parts of integrable constrained Lax pairs, which gives a way to separate variables for soliton systems and exhibits integrability by quadratures for soliton systems. Upon solving the Riemann–Jacobi inversion problems for constrained Lax pairs, the resulting Bäcklund transformations can generate finite-gap solutions to soliton systems in terms of Riemann-theta functions.

We remark that all symmetry constraints defined by (2.7) can put Lax pairs into integrable symplectic mappings and/or integrable finite-dimensional Hamiltonian systems. The corresponding constrained Lax pairs may have some specific properties, e.g., bi-Hamiltonian and quasi-bi-Hamiltonian structures. Therefore, symmetry constraints are very powerful in constructing lower-dimensional integrable systems from Lax pairs of soliton systems. Nevertheless, there exist symmetry constraints which do not force Lax pairs into integrable systems with constant coefficient symplectic forms [**11**], and the problem of integrability has not been solved for the time parts of the original constrained Lax pairs, i.e., systems of partial differential equations

$$\begin{cases} \phi^{(j)}_{t_n} = V^{(n)}(f, f_x, \dots; \lambda_j)\phi^{(j)}, & 1 \le j \le N, \\ \psi^{(j)}_{t_n} = -V^{(n)T}(f, f_x, \dots; \lambda_j)\psi^{(j)}, & 1 \le j \le N, \end{cases} \tag{6.1}$$

and systems of difference-differential equations

$$\begin{cases} \phi^{(j)}_{t_n} = V^{(n)}(f, Ef, E^{-1}f, \dots; \lambda_j)\phi^{(j)}, & 1 \le j \le N, \\ \psi^{(j)}_{t_n} = -V^{(n)T}(f, Ef, E^{-1}f, \dots; \lambda_j)\psi^{(j)}, & 1 \le j \le N. \end{cases} \tag{6.2}$$

We are curious to known whether they are good candidates for integrable systems.

References

1. V. I. Arnold, *Mathematical methods of classical mechanics*, Grad. Texts in Math., vol. 60, Springer-Verlag, New York, 1989.
2. E. D. Belokolos, A. I. Bobenko, V. Z. Enol'skii, A. R. Its, and V. B. Matveev, *Algebra-geometric approach to nonlinear integrable equations*, Springer-Verlag, Berlin, 1994.
3. M. Bruschi, O. Ragnisco, P. M. Santini, and G. Z. Tu, *Integrable symplectic maps*, Phys. D **49** (1991), 273–294.
4. C. W. Cao, *Nonlinearization of the Lax system for* AKNS *hierarchy*, Sci. China Ser. A **33** (1990), 528–536.
5. C. W. Cao and X. G. Geng, *Classical integrable systems generated through nonlinearization of eigenvalue problems*, Nonlinear Physics (Shanghai, 1989), Res. Rep. Phys., Springer, Berlin, 1990, pp. 68–78.
6. A. S. Fokas and R. L. Anderson, *On the use of isospectral eigenvalue problems for obtaining hereditary symmetries for Hamiltonian systems*, J. Math. Phys. **23** (1982), 1066–1073.
7. B. G. Konopelchenko, J. Sidorenko, and W. Strampp, $(1+1)$*-dimensional integrable systems as symmetry constraints of* $(2+1)$*-dimensional systems*, Phys. Lett. A **157** (1991), 17–21.
8. W. X. Ma, *New finite-dimensional integrable systems by symmetry constraint of the* KdV *equations*, J. Phys. Soc. Japan **64** (1995), 1085–1091.
9. W. X. Ma and B. Fuchssteiner, *Binary nonlinearization of Lax pairs*, Nonlinear Physics: Theory and Experiment (Lecce, 1995), World Sci. Publishing, River Edge, NJ, 1996, pp. 217–224.
10. W. X. Ma, B. Fuchssteiner, and W. Oevel, *A* 3×3 *matrix spectral problem for* AKNS *hierarchy and its binary nonlinearization*, Phys. A **233** (1996), 331–354.
11. W. X. Ma and Y. S. Li, *Do symmetry constraints nonlinearize spectral problems into Hamiltonian systems?*, Phys. Lett. A **268** (2000).
12. W. X. Ma and W. Strampp, *An explicit symmetry constraint for the Lax pairs and the adjoint Lax pairs of* AKNS *systems*, Phys. Lett. A **185** (1994), 277–286.
13. M. Toda, *Nonlinear waves and solitons*, Mathematics and its Applications (Japanese Series), vol. 5, Kluwer Academic Publishers, Dordrecht, 1989.
14. Y. T. Wu and X. G. Geng, *A new integrable symplectic map associated with lattice soliton equations*, J. Math. Phys. **37** (1996), 2338–2345.
15. X. F. Yang and R. Schmid, *Bäcklund transformations induced by symmetries*, Phys. Lett. A **195** (1994), 63–73.
16. V. E. Zakharov, S. V. Manakov, S. P. Novikov, and L. P. Pitaevski, *Theory of solitons*, Nauka, Moscow, 1980.

Department of Mathematics, City University of Hong Kong, Kowloon, Hong Kong, P. R. of China.

E-mail address: `mawx@math.cityu.edu.hk`

Department of Mathematics, City University of Hong Kong, Tat Chee Avenue 83, Kowloon, Hong Kong.

E-mail address: `gengxg@public2.zz.ha.cn`

Centre de Recherches Mathématiques
CRM Proceedings and Lecture Notes
Volume **29**, 2001

Open Problems for the Super KdV Equations

P. Mathieu

ABSTRACT. After a review of the basic results concerning the $N = 1, 2$ supersymmetric extensions of the Korteweg–de Vries equation, with a pedagogical presentation of the superspace techniques, we discuss some basic open problems mainly in relation with the $N = 2$ extensions.

1. Introduction

Supersymmetry offers a powerful tool for widening the scope of integrability. The field of supersymmetric integrable systems turns out to be remarkably rich in addition to further displaying novel features such as conserved nonlocal "Poisson square roots" of local conservation laws.[1] Not surprisingly, it started with the extension of the Korteweg–de Vries (KdV) equation although by now many other equations have been supersymmetrized.

The $N = 1, 2$ (where N refers to the number of supersymmetries) integrable supersymmetric versions of the KdV equation have been found about 10 years ago [**16, 17, 20–22, 27**].[2] The key points of this development were

(1) the realization that supersymmetrization could be restricted to the space variable only and

2000 *Mathematics Subject Classification.* Primary: 37K10; Secondary: 35Q53.

Work supported by NSERC (Canada) and FCAR (Québec).

I would like to thank Z. Popowicz for his helpful comments on the recent literature and the organisers of the workshop for their kind invitation to present this work.

This is the final form of the paper.

[1]Some physical motivations for considering supersymmetric integrable systems are scattered in the footnotes 2, 4 and 6, while geometrical implications are alluded to in the conclusion.

[2]The present restriction to $N \leq 2$ is partly motivated by the limitations of my own works. However there is also a physical motivation: these systems have found a remarquable application, albeit in their quantum formulation, in perturbed conformal field theory [**15,24,25**], in the context of which $N = 2$ is the maximal number of supersymmetries that is of real interest. Nevertheless, in some context, the $N = 3$ and 4 extensions may be physically relevant and the corresponding extensions of KdV have been considered; see for instance [**7, 12, 34**]. Our discussion will also be restricted to supersymmetric KdV extensions having an even Poisson brackets. A new $N = 2$ super KdV equation with an odd Poisson structure has been found recently in [**30**] but the resulting equations are somewhat less interesting in that the bosonic fields do not couple to the fermions in their time evolution. It would be of interest to see whether there are other "odd" integrable extensions displaying fermionic interactions in the bosonic evolution equations.

(2) that the crucial KdV structure whose core needs to be preserved is the KdV second Hamiltonian structure.[3]

In Fourier components, the underlying Hamiltonian operator is the Poisson bracket formulation of the Virasoro algebra [**10, 14**], for which supersymmetric extensions were already known and could then be used to construct the appropriate supersymmetric extension of KdV.

A number of developments have occured in the following years but recently the focus has moved toward the construction of extended super KdV hierarchies[4] (in the way the Boussinesq equation generalizes KdV, that is, via higher order Lax operators) [**3,6,28,29,35**]. However a certain number of problems associated to the $N = 1$, 2 KdV systems have remained unsolved and the goal of this presentation is to identify some of them. For this it is necessary to present a brief review of the supersymmetric formulation of the KdV equation and, for the benefit of those readers unfamiliar with super technologies, some manipulations will be worked out in some detail.

2. Supersymmetrization of the KdV Equation: $N = 1$

We will formulate the supersymmetric extension of the KdV equation in the *superspace formalism.* That amounts to extend the x variable to a doublet (x, θ) where θ is a Grassmannian (or *anticommuting*) variable: $\theta^2 = 0$. Ordinary (i.e. commuting) fields $f(x)$ (functions of x and t in fact but the time dependence will generally be suppressed) will be replaced by *superfields* $F(x, \theta)$. Given that $\theta^2 = 0$, these superfields have a very simple Taylor expansion in terms of θ:

$$F(x, \theta) = f(x) + \theta\gamma(x), \tag{2.1}$$

f and γ are called the *component fields.* γ is said to be the super-partner of f and vice-versa. In the present case, $F(x, \theta)$ is a *bosonic* superfield: it has the same "statistics" (i.e. commuting or anticommuting character) as the field appearing in the θ independent term (here f); on the other hand, γ is anticommuting, i.e. it is a *fermionic field.* In particular, $\gamma(x)\gamma(y) = -\gamma(y)\gamma(x)$ so that $\gamma(x)^2 = 0$; also for instance, $\theta\gamma = -\gamma\theta$. The final ingredient that we need is the *superderivative*

$$D = \theta\partial + \partial_\theta, \tag{2.2}$$

whose square is the usual space derivative: $D^2 = \partial$.

A supersymmetry transformation is nothing but a translation in superspace. Such a translation takes the form: $x \to x - \eta\theta$ and $\theta \to \theta + \eta$, where η is a constant

---

[3]Here are some comments on the literature concerning the $N = 1$ case. Already in 1984, Kupershmidt [**13**] has presented a simple fermionic (but not supersymmetric—see below) extension of the KdV equation by placing the emphasis on its Hamiltonian formulation (this system has actually two local Hamiltonian structures). This was an important step toward the formulation of the right supersymmetric extension given that both Hamiltonian operators were supersymmetric invariant (the Hamiltonian themselves were not) and could then serve in the formulation of a genuine supersymmetric system. The supersymmetric KdV equation was initially found independently of the work of Manin–Radul [**20**] on super KP hierarchy. It was realized afterwards that this general system has indeed a reduction to the supersymmetric KdV system.

[4]The most important physical application of these constructions concerns conformal field theory: the corresponding Poisson structures yield classical versions of super W algebras whose quantum form can be obtained via the quantization of the modified fields obtained through the Miura transformation (see e.g. [**23**]).

anticommuting parameter, supposed, in the following, to be very small. Consider then the effect of the translation in the superfield:

$$\begin{aligned} F(x,\theta) \to F(x-\eta\theta, \theta+\eta) &= F(x,\theta) - \eta\theta\partial F(x,\theta) + \eta\partial_\theta F(x,\theta), \\ &\equiv F(x,\theta) + \delta_\eta F(x,\theta), \\ &= f(x) + \theta\gamma(x) + \delta_\eta f(x) + \theta\delta_\eta\gamma(x). \end{aligned} \tag{2.3}$$

The second equality shows that δ_η is bosonic so that it commutes with θ. We read off the component-field transformations to be

$$\delta_\eta f = \eta\gamma, \quad \delta_\eta\gamma = \eta f_x. \tag{2.4}$$

This is called a *supersymmetry transformation*; it relates a bosonic field to a fermionic field and vice-versa. It has the remarquable virtue of linking a field transformation to a (super)space translation. Note that two successive supersymmetry transformations lead to

$$\delta_\eta\delta_{\eta'} f = \eta'\eta f_x, \quad \delta_\eta\delta_{\eta'}\gamma = \eta'\eta\gamma_x. \tag{2.5}$$

In other words, a translation in superspace, hence a supersymmetry transformation, is a sort of square root of an ordinary translation. Every local expression in the superfields and the superderivatives is manifestly supersymmetric invariant.

To supersymmetrize the KdV equation

$$u_t = -u_{xxx} + 6uu_x, \tag{2.6}$$

one should then start by extending the u field to a superfield. There are two ways of doing this: either as a fermonic superfield

$$u(x) \to \phi(x,\theta) = \theta u(x) + \xi(x), \tag{2.7}$$

or as a bosonic superfield

$$u(x) \to U(x,\theta) = u(x) + \theta\lambda(x). \tag{2.8}$$

It turns out that the first choice is the one that gives the interesting extension.[5] The KdV equation is homogeneous with respect to the scaling gradation: in the normalization where $\deg\partial = 1$, one finds that $\deg u = 2$. The identity $D^2 = \partial$ implies that $\deg D = 1/2$, so that $\deg\theta = -1/2$. For the superfield to be homogeneous, ξ must have degree $3/2$.

Let us then proceed with a direct extension of the KdV equation, multiplying each term by θ and rewriting the result in terms of superfields:

$$u_t \to \phi_t, \quad u_{xxx} \to \phi_{xxx}, \quad 3uu_x \to c\phi D\phi_x + (6-c)\phi_x(D\phi), \tag{2.9}$$

where c is a free constant. We thus observe that the nonlinear term does not have a unique extension in terms of superfields. Therefore, this direct extension leaves us with a supersymmetric version of the KdV equation containing a free parameter:

$$\phi_t = \phi_{xxx} + c(\phi D\phi)_x + (6-2c)\phi_x(D\phi). \tag{2.10}$$

[5]The other possibility would not have a second Hamiltonian structure associated to the super Virasoro algebra as a simple dimensional (i.e. degree counting) analysis shows (it requires the introduction of an anticommuting field of degree 3/2 as ξ and not of degree 5/2 as λ). The degree counting is explained below. Notice that to a large extend, we try to reserve Greek letters for anticommuting variables or fields.

It turns out that this equation is integrable only if $c = 3$ [**21, 22**][6]. We call the resulting equation the *super* KdV *equation*, or sKdV for short. Its component version reads[7]

$$u_t = -u_{xxx} + 6uu_x - 3\xi\xi_{xx}, \quad \xi_t = -\xi_{xxx} + 3(u\xi)_x. \tag{2.11}$$

It is not difficult to verify that the system (2.11) is invariant under the supersymmetry transformation $\delta_\eta u = \eta\xi_x$ and $\delta_\eta \xi = \eta u$. This is not so for the integrable fermionic extension proposed by Kupershmidt [**13**]:

$$u_t = -u_{xxx} + 6uu_x - 3\xi\xi_{xx}, \quad \xi_t = -4\xi_{xxx} + 6u\xi_x + 3u_x\xi. \tag{2.12}$$

With (2.11) being called the super KdV equation, it would be appropriate to call (2.12) the *Kuper*-KdV *equation.*

The integrability of (2.10) can be established in various ways. The most direct argument is that it has a Lax representation:[8]

$$L_t = [-4L_+^{3/2}, L], \quad L = \partial^2 - \phi D, \tag{2.13}$$

and the conservation laws are obtained as follows (the subscript gives the degree):

$$H_{2k+1} = \int dx\, d\theta\, \text{sRes}\, L^{(2k+1)/2}. \tag{2.14}$$

For super pseudodifferential operators, the + projection and the super residue sRes are defined as follows

$$\Lambda = \sum_{k=-\infty}^{N} \alpha_i D^i, \quad \Lambda_+ = \sum_{k=0}^{N} \alpha_i D^i, \qquad \text{sRes}\,\Lambda = \alpha_{-1}. \tag{2.15}$$

In the above expression for the conservation laws, we have also introduced the superintegration. The integration over the θ variable is defined as follows:

$$\int d\theta\, 1 = 0, \quad \int d\theta\, \theta = 1. \tag{2.16}$$

The integration over θ is thus essentially equivalent to the differentiation with respect to θ. With these rules, the superintegration of a superderivative vanishes (with the usual rule that the ordinary integral of a total derivative vanishes):

$$\int dx\, d\theta \big[D\phi(x,\theta)\big] = \int dx\, d\theta (\theta\xi_x + u) = \int dx\, \xi_x = 0. \tag{2.17}$$

---

[6]Actually the case $c = 0$ is also integrable but its leads to a somewhat trivial system in which the fermionic fields decouple from the bosonic equation which reduces then to the usual KdV equation. Nevertheless, this equation happens to be relevant in supersymmetric extensions of matrix models that describes superstrings in $d < 3/2$ dimensions, or equivalently, conformal field theories coupled to gravity [**2**].

[7]Notice that the product $\xi\xi_{xx}$ acts as a bosonic field: quite generally, a product of two fermions is a boson. It can be seen easily that passing an anticommuting variable in front of it does not induce an overall minus sign, e.g. $\xi\xi_{xx}\theta = -\xi\theta\xi_{xx} = \theta\xi\xi_{xx}$. Notice moreover that this term is a total derivative: $\xi\xi_{xx} = (\xi\xi_x)_x$ since the extra resulting term is $\xi_x\xi_x = 0$.

[8]Actually, the Lax operator is not unique: the choice $L = \partial^2 + \phi D - (D\phi)$ (the formal adjoint of $\partial^2 - \phi D$) leads to completely equivalent results.

For instance, the second conservation law is[9]

$$H_3 = \int dx\, d\theta (\phi D\phi) = \int dx(u^2 - \xi\xi_x). \tag{2.18}$$

Another way of establishing the integrability is to supersymmetrize the Gardner transformation [**9**]. This extension is unique:

$$\phi = \chi + \varepsilon\chi_x + \varepsilon^2 \chi D\chi, \tag{2.19}$$

with $\chi = \theta w + \sigma$. It maps a solution of the super Gardner equation[10]

$$\chi_t = -\chi_{xxx} + 3(\chi D\chi)_x + \varepsilon^2 3(D\chi)(\chi D\chi)_x, \tag{2.20}$$

into a solution of the sKdV equation. Since χ_t is a total superderivative, e.g.

$$(D\chi)(\chi D\chi)_x = \frac{1}{6} D[(D\chi)^3] + \frac{1}{2}[\chi(D\chi)^2]_x, \tag{2.21}$$

$\int dx\, d\theta\, \chi$ is conserved and by inverting the super Gardner transformation (2.19), we recover an infinite number of conservation laws:

$$\chi = \sum_{n=0}^{\infty} \varepsilon^n h_n[\phi] \Longrightarrow \frac{d}{dt} \int dx\, d\theta \sum_{n=0}^{\infty} \varepsilon^n h_n[\phi] = 0, \tag{2.22}$$

(where $h_n[\phi]$ stands for a differential polynomial in ϕ). Now the crucial point is that the sKdV equation is independent of ε so that each separate power of ε must be separately conserved. This produces an infinite number of conservation laws, half of which can be shown to be nontrivial, having a leading term $\phi(D\phi)^k$; these are bound to be the H_{2k+1} above [**21, 22**]. Note that these are all bosonic (χ is fermionic but the measure $dx\, d\theta$ is also fermionic).

Finally, we point out that the sKdV equation is biHamiltonian, the two Hamiltonian operators being[11] [**8, 21, 22, 26, 33**],

$$P_1 = \partial[D^3 - \phi]^{-1}\partial, \quad P_2 = -D^5 + 3\phi\partial + (D\phi)D + 2\phi_x. \tag{2.23}$$

Notice that P_1 is a very complicated nonlocal Hamiltonian operator, being essentially an infinite series: $[D^3 - \phi]^{-1} = D^{-3}[1 - D^{-3}\phi]^{-1}$. P_2 is the direct supersymmetrization of the KdV second Hamiltonian structure: $-\partial^3 + 4u\partial + 2u_x$.

There is a remarquable feature of the super case that is not present for the usual KdV equation which is the presence of *fermionic nonlocal conservation laws*

[9]Notice that for a fermionic variable, $\xi\xi_x$ is *not* a total derivative: a partial integration of $\int dx\, \xi\xi_x$ leads to $-\int dx\, \xi_x\xi$ and the interchanges of the two terms generates another minus sign so that the original expression is recovered.

[10]The component form of this superfield equation reads:

$$\begin{aligned} w_t &= -w_{xxx} + 6ww_x - \sigma\sigma_{xx} + \varepsilon^2[6w^2 w_x - 3(\sigma\sigma_x w)_x], \\ \sigma_t &= -\sigma_{xxx} + 3(\sigma w)_x + \varepsilon^2[3w(w\sigma)_x]. \end{aligned}$$

Notice that σ_t is not a total derivative. The usual Gardner equation is recovered by setting the fermionic field $\sigma = 0$ and the sKdV equation is the limiting case where $\varepsilon = 0$.

[11]The second Hamiltonian structure has been found in [**21, 22, 33**] and the first one in [**8, 26**]. Here and below, the action of the derivatives is always delimited by parentheses, e.g., $D\phi = (D\phi) - \phi D$.

[**5, 11**]. The first few of them are

$$
(2.24)\quad
\begin{aligned}
J_{1/2} &= \int dx\, d\theta (D^{-1}\phi) = \int dx\, \xi, \\
J_{3/2} &= \int dx\, d\theta (D^{-1}\phi)^2 = \int dx\, u(\partial^{-1}\xi), \\
J_{5/2} &= \int dx\, d\theta \big[(D^{-1}\phi)^3 - 6\partial^{-1}(\phi D\phi)\big] \\
&\qquad\qquad = \int dx\, \big[3\xi(\partial^{-1}u)^2 - 6\partial^{-1}(u^2 - \xi\xi_x)\big].
\end{aligned}
$$

They Poisson commute with the local bosonic conservation laws H_n but not among themselves:

$$
(2.25)\quad
\begin{aligned}
&\{J_{(4n+i)/2}, J_{(4m+i)/2}\} = H_{2(n+m)+i} \quad \text{with } i = 1, 3, \\
&\{J_{(4n+1)/2}, J_{(4m+3)/2}\} = 0.
\end{aligned}
$$

The J_i are thus some sort of Poisson square roots of the usual conservation laws. The first fermionic nonlocal conservation law is actually local in terms of the component fields. It signals the presence of the supersymmetry invariance (and in particular ξ is not a conserved density for the Kuper-KdV equation). The infinite sequence can be generated from the first two conservation laws by the application of the recursion operator $P_1^{-1}P_2$. But there is a more spectacular way of expressing them that makes manifest their supersymmetric origin: in superspace, ∂^2 not only has a square root but it also has a *fourth root*: $(\partial^2)^{1/4} = D$. The fermionic nonlocal conservation laws are thus related to the super residues of the fourth root of odd powers of the Lax operators as

$$
(2.26)\quad J_{k/2} = \int dx\, d\theta\, \mathrm{sRes}\, L^{k/4}, \quad (k \text{ odd}).
$$

The sKdV equation represents the first example of an integrable system for which nonlocal conservation laws arise in such a clean form. In that respect, we introduce the first open problem (**OP**):

OP-1. Find an integrable deformation that reproduces the fermionic nonlocal conservation laws.

3. Supersymmetrization of the KdV Equation: $N = 2$

The next extension to be considered is the addition of an extra supersymmetry which amounts to add an extra anticommuting space dimension. We thus extend x to a triplet (x, θ_1, θ_2), with $\theta_1^2 = \theta_2^2 = 0$, $\theta_1\theta_2 = -\theta_2\theta_1$ and introduce two super derivatives:

$$
(3.1)\quad D_1 = \theta_1\partial + \partial_{\theta_1}, \quad D_2 = \theta_2\partial + \partial_{\theta_2}, \quad D_1^2 = D_2^2 = \partial, \quad D_1D_2 = -D_2D_1.
$$

The superfields are now functions of (x, θ_1, θ_2) (as well as t) and their Taylor expansion in terms of the anticommuting variables contain four terms. For instance the $N = 2$ KdV superfield will be written as:

$$
(3.2)\quad \Phi(x, \theta_1, \theta_2) = \theta_2\theta_1 u(x) + \theta_1\xi_1(x) + \theta_2\xi_2(x) + v(x),
$$

ξ_1 and ξ_2 are two fermionic fields and v is a new bosonic field of degree 1 (in a supersymmetric theory, the number of bosonic and fermionic fields must be the same). Notice that Φ is a bosonic superfield.

One could then proceed to the direct supersymmetrization of the KdV equation and get a multiparameter $N = 2$ extension. However, a sounder approach, that reduces substantially the number of such free parameters, is to formulate the equation directly in terms of the $N = 2$ supersymmetric version of the second Hamiltonian structure (which is expected to be the core structure underlying integrability of nontrivial KdV extensions):

$$P_2 = D_1 D_2 \partial + 2\Phi\partial - (D_1\Phi)D_1 - (D_2\Phi)D_2 + 2\Phi_x, \tag{3.3}$$

e.g. as

$$\Phi_t = P_2 \frac{\delta}{\delta\Phi} \int dx\, d\theta_1\, d\theta_2 \big[\Phi(D_1 D_2 \Phi) + a\Phi^3\big]. \tag{3.4}$$

This Hamiltonian is the direct generalisation of the KdV Hamiltonian $\int dx\, u^2$ and of the $N = 1$ version $\int dx\, d\theta\, \phi(D\phi)$. However, the $N = 2$ generalisation is not unique and this introduces a free parameter a. The resulting equation is

$$\Phi_t = -\Phi_{xxx} + 3(\Phi D_1 D_2 \Phi)_x + \frac{(a-1)}{2}(D_1 D_2 \Phi^2)_x + 3a\Phi^2\Phi_x. \tag{3.5}$$

This is called the SKdV$_a$ equation (the capital S is used for $N = 2$). This system is integrable for exactly three values of a [**16, 17**]: $a = -2, 1, 4$. For $a = -2, 4$, the Lax representation is standard: $L_t = [-4L_+^{3/2}, L]$ with

$$L_{a=4} = -(D_1 D_2 + \Phi)^2, \quad L_{a=-2} = -\partial^2 + \sum_{i,j=1,2} \varepsilon_{ij} D_i (D_1 D_2 + \Phi) D_j, \tag{3.6}$$

(with $\varepsilon_{12} = -\varepsilon_{21} = 1$) while for $a = 1$ it is nonstandard [**27**]: $L_t = [-4L_{\geq 1}^3, L]$ with

$$L_{a=1} = \partial - \partial^{-1}\big[(D_1 D_2 \Phi) - (D_2\Phi)D_1 - (D_1\phi)D_2 + \Phi D_1 D_2\big]. \tag{3.7}$$

In all cases, there is an infinite number of conservation laws given by

$$H_{2k+1} = \int d d\theta_1\, d\theta_2\, \mathrm{SRes}\, L^{(2k+1)/2}, \tag{3.8}$$

where the $N = 2$ version of the residue of a pseudodifferential operator is the coefficient of $D_1 D_2 \partial^{-1}$.

Although there exists a Miura transformation, there are no known integrable (i.e., Gardner-type) deformations of it. This leads us to:

OP-2. Find an integrable deformation for SKdV$_{-2,1,4}$ that reproduces their conservations laws.

For $a = -2, 4$, there are two independent towers of fermionic nonlocal conservations laws [**5**]: in each case, the first one is

$$\int dx\, d\theta_1\, d\theta_2 (D_i^{-1}\Phi) = \int dx\, \xi_i, \quad (i = 1, 2). \tag{3.9}$$

However, although the Lax operator has two distinct fourth roots, these have not yet been related to these fermionic conservation laws:

OP-3. For SKdV$_{-2,4}$, find the relation between the fermionic nonlocal conservation laws and the Lax operator.

The existence of these fermionic conservation laws is natural in that there are two supersymmetries. On the other hand, for the $a = 1$ case, the first fermionic laws (3.9) do not generate infinite towers: they are isolated.

OP-4. Why there are no infinite towers of fermionic conservation laws for $SKdV_1$?

The $SKdV_{-2,4}$ equations are both biHamiltonian : their first Hamiltonian operator is [**26**]

$$P_1^{(a=4)} = \partial,\ P_1^{(a=-2)} = (D_1 D_2 \partial^{-1} - D_1^{-1}\Phi D_1^{-1} - D_2^{-1}\Phi D_2^{-1}) D_1^{-1}\Phi D_1^{-1}. \tag{3.10}$$

In that respect, the $SKdV_1$ stands as one of the rare example of a classical integrable system which is not (known to be) biHamiltonian.

OP-5. Is $SKdV_1$ biHamiltonian?

Another very natural question is:

OP-6. Why is there exactly three integrable $N = 2$ super KdV extensions?[12] Is there an underlying Lie algebraic interpretation for this threefold way (i.e., is this related to the existence of the three classical algebras)?

4. Concluding Questions

There are further general questions that could be formulated in relation with super integrable systems. As stressed here, these equations are naturally formulated in superspace. Over the years, it became clear that a lot of structure is contained in the Painlevé test: for instance its truncation leads to Backlünd transformations and the Lax operator [**31, 32**]. Yet, no Painlevé analysis has been done directly in superspace.

OP-7. Is it possible to formulate the Painlevé analysis in superspace?

Moreover, little is known concerning the solutions of super integrable systems. The Darboux transformation in the $N = 1$ case has been worked out in [**18, 19**] but nothing has been done at this point concerning the $N = 2$ cases.[13]

Finally, unraveling the deep relations between geometry and soliton theory has been an important theme of this workshop; little is known on the super version of this connection. In particular, the super KdV equations have super Sine–Gordon relatives and these should lead to very interesting geometrical structures.

References

1. M. A. Ayari, V. Hussin, and P. Winternitz, *Group invariant solutions for the $N = 2$ super Korteweg–de Vries equation*, J. Math. Phys. **40** (1999), no. 4, 1951–1965.
2. K. Becker and M. Becker, *Nonperturbative solution of the super-Virasoro constraints*, Modern Phys. Lett. A **8** (1993), no. 13, 1205–1214.
3. S. Bellucci, E. Ivanov, S. Krivonos, and A. Pichugin, $N = 2$ *super Boussinesq hierarchy: Lax pairs and conservation laws*, Phys. Lett. B **312** (1993), no. 4, 463–470.
4. L. Bonora, S. Krivonos, and A. Sorin, *Towards the construction of $N = 2$ supersymmetric integrable hierarchies*, Nuclear Phys. B **477** (1996), no. 3, 835–854.
5. P. Dargis and P. Mathieu, *Nonlocal conservation laws for supersymmetric* KdV *equations*, Phys. Lett. A **176** (1993), 67–74.
6. F. Delduc and L. Gallot, $N = 2$ KP *and* KdV *hierarchies in extended superspace*, Comm. Math. Phys. **190** (1997), no. 2, 395–410.

[12] There are in fact also three distinct SKdV hierarchies but they generalize (in the Lax sense) the $SKdV_{2,4}$ equations; the $SKdV_1$ equation appears as a sort of isolated point. Interesting technical observations in relation with the bosonic truncation of the Lax operators are presented in [**4**].

[13] Some solutions for the $SKdV_a$ equations have been reported in [**1**].

7. F. Delduc, L. Gallot, and E. Ivanov, *New super* KdV *system with* $N = 4$ SCA *as the Hamiltonian structure*, Phys. Lett. B **396** (1997), 122–132.
8. J. M. Figueroa–O'Farrill, J. Mas, and E. Ramos, *BiHamiltonian structure of the supersymmetric* SKdV *hierarchy*, Rev. Modern Phys. **3** (1991), 479.
9. C. S. Gardner, *Korteweg–de Vries equation and generalizations.* IV. *The Korteweg–de Vries equation as a Hamiltonian system*, J. Math Phys. **12** (1971), 1548–1551.
10. J. L. Gervais, *Infinite family of polynomial functions of the Virasoro generators with vanishing Poisson brackets*, Phys. Lett. B **160** (1985), 277–279.
11. P. H. M. Kersten, *Higher order supersymmetries and fermionic conservation laws of the supersymmetric extension of the* KdV *and the* mKdV *equation*, Phys. Lett. A **134** (1988), 25–30.
12. S. Krivonos, A. Pashnev, and Z. Popowicz, *Lax pairs for* $N = 2, 3$ SUSY KdV *equations and their extensions*, Modern Phys. Lett. A **13** (1998), no. 18, 1435–1443.
13. B. A. Kupershmidt, *A super Korteweg–de Vries equation: An integrable system*, Phys. Lett. A **102** (1984), 213–215.
14. ———, *Super Korteweg–de Vries equations associated to super extensions of the Virasoro algebra*, Phys. Lett. A **109** (1985), 417–423.
15. B. A. Kupershmidt and P. Mathieu, *Quantum Korteveg–de Vries like equations and perturbed conformal field theories*, Phys. Lett. B **227** (1989), 245–250.
16. P. Labelle and P. Mathieu, *A new* $N = 2$ *supersymmetric Korteweg–de Vries equation*, J. Math. Phys. **32** (1991), no. 4, 923–927.
17. C. A. Laberge and P. Mathieu, $N = 2$ *superconformal algebra and* O(2) *integrable fermionic extensions of the Korteweg–de Vries equation*, Phys. Lett. B **215** (1988), no. 4, 718–722.
18. Q. P. Liu, *Darboux transformations for supersymmetric Korteweg de Vries equations*, Lett. Math. Phys. **35** (1995), no. 2, 115–122.
19. Q. P. Liu and M. Manas, *Darboux transformations for* SUSY *integrable systems*, Supersymmetry and Integrable Models (H. Aratyn *et al.*, eds.), Lecture Notes in Phys. **35**, Springer-Verlag, 1995, pp. 269–281.
20. Yu. I. Manin and A. O. Radul, *A supersymmetric extension of the Kadomstev–Petviashvili hierarchy*, Comm. Math. Phys. **98** (1985), 65–77.
21. P. Mathieu, *Supersymmetric extension of the Korteweg–de Vries equation*, J. Math. Phys. **29** (1988), no. 11, 2499–2506.
22. ———, *Superconformal algebra and supersymmetric Korteweg–de Vries equation*, Phys. Lett. B **203** (1988), no. 3, 287–291.
23. ———, *Extended classical conformal algebras and the second Hamiltonian structure of Lax equations*, Phys. Lett. B **208** (1988), 101–106.
24. ———, *Integrability of perturbed superconformal minimal models*, Nuclear Phys. B **336** (1990), no. 2, 338–348.
25. P. Mathieu and M. A. Walton, *Integrable perturbations of* $N = 2$ *superconformal minimal models*, Phys. Lett. B **254** (1991), 106–110.
26. W. Oevel and Z. Popowicz, *The bi-Hamiltonian structure of the fully supersymmetric Korteweg–de Vries equation*, Comm. Math. Phys. **139** (1991), no. 3, 441–460.
27. Z. Popowicz, *The Lax formulation of the "new"* $N = 2$ SUSY KdV *equation*, Phys. Lett. A **174** (1993), 411–415.
28. ———, *Lax formulation of the* $N = 2$ *super Boussinesq equation*, Phys. Lett. B **319** (1993), no. 4, 478–484.
29. ———, *Extensions of the* $N = 2$ SUSY $a = -2$ *Boussinesq hierarchy*, Phys. Lett. A **236** (1997), 455–461.
30. ———, *Odd bi-Hamiltonian structures of new supersymmetric* $N = 2$*,* 4 *Korteweg–de Vries equation and odd* SUSY *Virasoro-like algebra*, Phys. Lett. B **459** (1999), 150–158.
31. J. Weiss, *The singular manifold method*, Painlevé Transcendents (Sainte-Adéle, 1990) (D. Levi and P. Winternitz, eds.), NATO Adv. Sci. Inst. Ser. B Phys. **278**, Plenum, New York, 1992, 225–247; and references therein.
32. J. Weiss, M. Tabor, and G. Carnavale, *The Painlevé property for partial differential equations*, J. Math. Phys. **24** (1983), no. 3, 522–526.
33. I. Yamanaka and R. Sasaki, *Super Virasoro algebra and solvable supersymmetric quantum field theories*, Progr. Theoret. Phys. **79** (1988), no. 5, 1167–1184.

34. C. M. Yung, *The $N = 3$ supersymmetric* KdV *hierarchy*, Modern Phys. Lett. A **8** (1993), no. 12, 1161–1169.
35. ———, *The $N = 2$ supersymmetric Boussinesq hierarchies*, Phys. Lett. B **309** (1993), 75–84.

Département de Physique, Université Laval, Québec, G1K 7P4, Canada.
E-mail address: `pmathieu@phy.ulaval.ca`

Centre de Recherches Mathématiques
CRM Proceedings and Lecture Notes
Volume **29**, 2001

Combinatorial Aspects of the Darboux Transformation

R. Milson

ABSTRACT. The present article undertakes a formal, combinatorial analysis of the Darboux transformation for the one-dimensional Schrödinger equation. To this end the article introduces a new combinatorial technique: the spectral residue sequence. This is an infinite sequence of numbers that represents each given potential function, and is related in a nonlinear, invertible manner to the power-series coefficients of the latter. The spectral residue sequence possesses a number of remarkable properties; one of these is the fact that exact-solvability manifests as the factorizability of the spectral residues with respect to the potential coupling constants. Another property is the fact that a Darboux transformation of the potential function—a nonlinear mapping—corresponds to a simple negation of the spectral residues. This fact gives a useful criterion for establishing whether a given potential is shape invariant, and permits, in principle, the creation of new shape-invariant potentials.

1. Introduction

The Darboux transformation is an indespensible tool in the study of exact solutions to the time-independent Schrödinger equation of quantum mechanics. The importance of this technique is two-fold. First, it is a general method for constructing new exactly-solvable potentials from potentials already known to be exactly-solvable. Second, utilizing the closely related paradigm of shape-invariance, one is able to construct new exactly-solvable potentials. Indeed, these shape-invariant examples form the core of the known 1-dimensional quantum-mechanical models with exact solutions [**2**].

The present paper aims to investigate shape-invariance from a formal, combinatorial point of view. One can, of course, proceed by representing potentials by a power series; however, this is not the approach pursued here. The key tool in the present investigation is the representation of a given potential function by means of a new combinatorial entity: the spectral residue sequence. Just as each function determines a corresponding infinite sequence of power series coefficients, each potential function can also be represented by a corresponding residue sequence. However, in the case of spectral residues, the mapping from potential to infinite sequence is not linear. This mapping also possesses some novel, and useful features.

2000 *Mathematics Subject Classification.* Primary: 05A19; Secondary: 81Q05.

Supported by a Dalhousie University Grant.

This is the final form of the paper.

Indeed, it will be shown in Theorem 4.1 below that a Darboux transformation of the original potential is equivalent to the negation of the corresponding residue sequence. This fact makes it possible to check for shape-invariance, and indeed allows one to construct, at least formally, new shape invariant potentials.

The spectral residue representation has another interesting property: exact-solvability, in a suitable sense, of a parameterized potential causes the factorization of the corresponding residue sequence. This interesting feature will be indicated in an example, but will not be discussed in detail. One should note that the factorizability of the spectral residue sequence is strongly reminiscent of the factorization property of the recently discovered Bender–Dunne polynomials associated to quasi-exactly-solvable potentials [**1**,**5**]. This connection certainly warrants closer scrutiny, however, the present paper merely aims to introduce the spectral residue sequence, and to indicate how this technique can be applied to the study of shape invariance. A more detailed investigation of factorizability, and of other combinatorial aspects of the residue sequence will have to wait for a subsequent publication.

2. Preliminaries

The present section introduces the spectral residue sequence and describes some of its important features. First, it is necessary to fix some notation and to discuss some related combinatorial notions.

An l-part composition of a natural number n is an *ordered* list of l positive integers $\mathbf{p} = (p_1, p_2, \ldots, p_l)$ such that $p_1 + \cdots + p_l = n$. Sums indexed by compositions arise naturally when one evaluates recursive expressions of the form

$$f_n = W_{0n} f_0 + W_{1n} f_1 + \cdots + W_{n-1,n} f_{n-1}, \tag{2.1}$$

where W_{ij}, $0 \le i < j$ is a given array of constants. Indeed, with the initial condition $f_0 = 1$, one has

$$f_n = \sum_{\mathbf{p}} W_{0 s_1} W_{s_1 s_2} W_{s_2 s_3} \ldots W_{s_{l-1} n}, \tag{2.2}$$

where the sum is taken over all compositions of n, and where s_j denotes the jth left partial sum, $p_1 + \cdots + p_j$ of a given composition $\mathbf{p}$.

The quantities f_n defined by (2.1) have a natural combinatorial interpretation: they are a weighted enumeration of uni-directional walks on a 1-dimensional lattice. Indeed, if W_{ij} with $i < j$ is the cost of a step from node i to node j, then f_n is the weighted enumeration of all uni-directional walks on the lattice from node 0 to node n.

Composition sums need to be employed when one considers formal solutions $f(z) = 1 + f_1 z + f_2 z^2 + \ldots$ of second order equations of the form

$$z^2 f_{zz} + (1-\nu) z f_z + U(z) f = 0, \quad f(0) = 1, \tag{2.3}$$

where the potential function $U(z) = U_1 z + U_2 z^2 + \ldots$ is a formal power series with zero constant term, and where ν is a spectral parameter. The expression in the left-hand side of the above equation defines a degree-raising operator, and consequently the coefficients of $f(z)$ satisfy the recursive relations

$$n(\nu - n) f_n = \sum_{k=0}^{n-1} U_k f_{n-k}. \tag{2.4}$$

The coefficients of the solution can therefore be described by the following composition sum:

$$f_n = \sum_{\mathbf{p}} \frac{U_{p_1}}{s_1(\nu - s_1)} \times \frac{U_{p_2}}{s_2(\nu - s_2)} \times \cdots \times \frac{U_{p_l}}{n(\nu - n)}, \tag{2.5}$$

where the index ranges over all compositions $(p_1, \ldots, p_l)$ of n and where s_j denotes the jth partial sum $p_1 + \cdots + p_j$.

It makes sense to regard $U(z)$ as a potential, because equation (2.3) is "nearly" self-adjoint in the following sense. If $f(z)$ is a formal solution of (2.3) then $\hat{f}(z) = z^{\nu/2} f(z)$ is a formal solution of the self-adjoint equation

$$z^2 \hat{f}_{zz} + \hat{f}_z + U(z)\hat{f} = \frac{\nu^2}{4}\hat{f}.$$

Thus, the "nearly" self-adjoint equation (2.3) leads to the same power series as the corresponding self-adjoint formulation; however it is better suited for specifying the relations that govern the coefficients of the solution. The preceding remarks also explain why it makes sense to regard ν as a spectral parameter.

For reasons that will be indicated below, it is useful to consider the residues of the coefficients f_n rather than the coefficients themselves. These residues can be most efficiently described in terms of a composition sum. For a composition $\mathbf{p} = (p_1, \ldots, p_l)$ of a natural number n, set

$$\pi(\mathbf{p}) = s_1 s_2 \ldots s_l,$$

where as before s_j is the jth left partial sum. Thus, if one interprets $\mathbf{p}$ as a walk from 0 to n with steps $p_1, \ldots, p_l$, then $\pi(\mathbf{p})$ is the product of numbers visited on that walk. Let $\mathbf{p}'$ denote the reversed composition $(p_l, \ldots, p_2, p_1)$. In this way $\pi(\mathbf{p}')$ will denote the product of visited numbers when one interprets $\mathbf{p}$ as a reverse walk from n down to 0.

DEFINITION 2.1. *Let $U(z) = U_1 z + U_2 z^2 + \ldots$ be a formal power series with a vanishing constant term. For each natural number $n \geq 1$ set*

$$\rho_n(U) = n \sum_{\mathbf{p}} \frac{U_{p_1} U_{p_2} \ldots U_{p_l}}{\pi(\mathbf{p})\pi(\mathbf{p}')}, \tag{2.6}$$

where the sum is taken over all compositions (of all lengths) $(p_1, \ldots, p_l)$ of n. The sequence $\rho_1(U)$, $\rho_2(U)$, ... will be called the spectral residue sequence corresponding to $U(z)$.

The reason for this name is made clear by the following. Let $U(z) = U_1 z + U_2 z^2 + \ldots$ be a formal power series, and let $f(z) = 1 + f_1 z + f_2 z^2 + \ldots$ be a formal solution of equation (2.3).

PROPOSITION 2.2. *The residue of f_n taken at $\nu = n$ is equal to $\rho_n(U)$.*

PROOF. This follows immediately from the composition sum representation of f_n shown in (2.5). □

To motivate the above definition it is necessary to consider the effect of gauge transformations on a second-order linear differential equation. Representing such an equation as $\mathcal{H}(f) = 0$, where $\mathcal{H}$ is a second-order differential operator, a gauge transformation amounts to the conjugation of $\mathcal{H}$ by a gauge factor $e^{\sigma(z)}$. At the level of solutions, $f(z)$ is a solution of the equation $\mathcal{H}(f) = 0$ if and only if $\hat{f} = e^{\sigma} f$

is a solution of $\widehat{\mathcal{H}}(\hat{f}) = 0$, where $\widehat{\mathcal{H}} = \mathrm{e}^{\sigma}\mathcal{H}\mathrm{e}^{-\sigma}$ is the conjugated second-order operator. More specifically, for

$$\mathcal{H} = z^2\partial_{zz} + (1-\nu)z\partial_z + U(z),$$

one has

$$\widehat{\mathcal{H}} = z^2\partial_{zz} + (1-\nu)z\partial_z - 2\sigma_z z^2\partial_z + U + z^2\big(-\sigma_{zz} + (\sigma_z)^2\big) + \nu z\sigma_z.$$

At this point, two observations are crucial. First, writing out the gauge factor as a power series one has

$$\hat{f}(z) = (1 + \sigma + \sigma^2/2! + \dots)f(z).$$

If one assumes that $\sigma = \sigma_1 z + \sigma_2 z^2 + \dots$, i.e. there is no constant term, then

$$\hat{f}_n = f_n + \text{ a linear combination of } f_0, \dots f_{n-1}.$$

However, looking back to the recursive definition of f_n given in (2.4) one sees that f_n contains a factor of $1/(\nu - n)$ whereas $f_0, f_1, \dots f_{n-1}$ do not. Consequently, even though $\hat{f}_n$ and f_n are not in general equal, there is equality of the corresponding residues taken at $\nu = n$. The above discussion can be summarized as the following generalization of Proposition 2.2.

PROPOSITION 2.3. *Let $U(z)$, $\sigma(z)$, $\widehat{\mathcal{H}}$, and $\hat{f}(z)$ be as above. The residue of $\hat{f}_n$ taken at $\nu = n$ is equal to $\rho_n(U)$. In other words, the spectral residue sequence is invariant with respect to gauge transformations.*

The two crucial features of the spectral residue sequence is the above-mentioned gauge-invariance, and the symmetry of the defining relation (2.6) with respect to $\mathbf{p}$ and $\mathbf{p}'$. The consequence of the first feature is the factorizability of the residue sequence for the class of exactly-solvable potentials. This will be illustrated in an example in the following section. The consequence of the second feature is the nice behaviour of the residue sequence with respect to Darboux transformations. This will be discussed in the section after next.

3. A Motivating Example

The present section considers the spectral residue sequence corresponding to an exactly-solvable family of potentials that were first described by Eckart [**3**]. These potentials arise naturally if one starts with the usual hypergeometric equation and adjusts it so that the corresponding second-order operator is self-adjoint. After this adjustment, the 3 parameters of the hypergeometric equation become the energy and 2 additional coupling constants. The elements of the spectral residue sequence turn out to be polynomials of these coupling constants.

The eigenfunctions of the Eckart Hamiltonian can be described in terms of the hypergeometric function; the bound states correspond to the Jacobi polynomials. As will be shown below, the existence of polynomial solutions to the hypergeometric equation causes the polynomials that comprise the residue sequence to factorize.

The Eckart potentials are functions of the form

$$U = \frac{u\mathrm{e}^x}{1-\mathrm{e}^x} + \frac{v\mathrm{e}^x}{(1-\mathrm{e}^x)^2},$$

where x is the physical distance variable, and u and v are coupling constants. Rewriting the Hamiltonian, $\mathcal{H}$, in terms of $z = \mathrm{e}^x$ one gets

$$\mathcal{H} = z^2\partial_{zz} + (1-\nu)z\partial_z + \frac{uz}{1-z} + \frac{vz}{(1-z)^2},$$

where the spectral parameter ν has been encoded into the first-order part of the operator (see the preceding section for a discussion of this matter). Expanding the potential into a series one obtains

$$U = \sum_k (u+kv)z^k.$$

The corresponding residue sequence is therefore given by

$$\rho_n(u,v) = n\sum_{\mathbf{p}} \frac{(u+vp_1)(u+vp_2)\dots(u+vp_l)}{\pi(\mathbf{p})\pi(\mathbf{p}')}, \tag{3.1}$$

where the index $\mathbf{p}$ ranges over all compositions of all lengths of n.

Computation of the first few terms of this residue sequence points to a remarkable factorization.

PROPOSITION 3.1. *In the even case, $n = 2m$, one has*

$$\rho_n(u,v) = K\prod_{i=0}^{m-1}\left[(u+v+a_i)(u+v+a_{i+1}) + b_i^2 u\right],$$

where $K = 1/n!(n-1)!$, $a_i = i(n-i)$ and $b_i = n-1-2i$. In the odd case, $n = 2m+1$, one has

$$\rho_n(u,v) = K(u+v+a_m)\prod_{i=0}^{m-1}\left[(u+v+a_i)(u+v+a_{i+1}) + b_i^2 u\right].$$

The proof of this proposition will be given shortly. First it is worth remarking that certain specializations of the potential parameters lead to interesting combinatorial identities.

COROLLARY 3.2. *The coefficient of v^k in*

$$v(v+1\cdot 2)(v+2\cdot 3)\dots\big(v+(n-1)n\big),$$

is given by

$$\sum_{\mathbf{p}} \frac{(n!)^2}{\pi(\mathbf{p})\pi(\mathbf{p}')}\prod_{j=1}^{k} p_j,$$

where the sum is taken over all k-part compositions of n.

PROOF. An examination of (3.1) will show $\rho_n(0,v)$ is a polynomial whose kth degree coefficient is given, up to a constant, by the composition sum above. Using Proposition 3.1 to factor $\rho_n(0,v)$, and noting that

$$(v+a_i)(v+a_{i+1}) + b_i^2 v = \big[v + i(i+1)\big]\big[v + (n-i-1)(n-i)\big],$$

one obtains

$$n!(n-1)!\rho_n(0,v) = v(v+1\cdot 2)(v+2\cdot 3)\dots\big(v+(n-1)n\big). \quad \square$$

In order to understand why the above factorization is occurring it is necessary to recall the connection between the Eckart Hamiltonian and the hypergeometric operator. The hypergeometric function [**4**],

$$F(\alpha, \beta, \gamma; z) = F_0 + F_1 z + F_2 z^2 + \dots,$$

can be defined as the solution of the hypergeometric differential equation:

$$\mathcal{G}(F) = 0, \quad F(0) = 1, \tag{3.2}$$

where $\mathcal{G}$ is the second order differential operator given by

$$\mathcal{G} = z(1-z)\partial_{zz} + \big(\gamma - (\alpha+\beta+1)z\big)\partial_z - \alpha\beta.$$

It is evident that $\mathcal{G}$ is a "non-raising" operator, i.e. $\mathcal{G}(z^n)$ is a polynomial of degree n or less. However, one can easily convert $\mathcal{G}$ to a raising operator. To wit, set

$$\begin{aligned}
\mathcal{G}_1 &= \frac{z}{1-z}\mathcal{G}, \\
&= z^2\partial_{zz} + (1-\nu)z\partial_z - \frac{z}{1-z}\big[(\alpha+\beta+\nu)z\partial_z + \alpha\beta\big], \\
&= z^2\partial_{zz} + (1-\nu)z\partial_z - \sum_{n=1}^{\infty} z^n\big[(\alpha+\beta+\nu)z\partial_z + \alpha\beta\big],
\end{aligned}$$

where $\nu = 1-\gamma$ is employed in place of γ to streamline subsequent formulas. In addition to the usual hypergeometric equation, one can equally well specify the hypergeometric function F as the solution of

$$\mathcal{G}_1(F) = 0, \quad F(0) = 1.$$

The operator $\mathcal{G}_1$ is related to the Eckart Hamiltonian by a gauge transformation. Indeed,

$$\mathcal{H} = \mathrm{e}^{\sigma} \cdot \mathcal{G}_1 \cdot \mathrm{e}^{-\sigma},$$

where

$$\begin{aligned}
\sigma(z) &= -\frac{1}{2}(\alpha+\beta+\nu)\log(1-z), & \mathrm{e}^{\sigma(z)} &= (1-z)^{-(\alpha+\beta+\nu)/2}, \\
u &= \frac{1}{4}(\alpha+\beta+\nu)(2-\alpha-\beta-\nu), & v &= \frac{1}{4}(\alpha-\beta-\nu)(\alpha-\beta+\nu).
\end{aligned} \tag{3.3}$$

It is therefore evident that eigenfunctions of the Eckart Hamiltonian are the product of a power of $(1-z)$ and of a hypergeometric function. According to Proposition 2.3, the spectral residue sequence can just as well be calculated by taking residues of the coefficients of the hypergeometric function F. These coefficients are given by the following well-known formula:

$$F_n = \frac{(\alpha)^n(\beta)^n}{n!(\gamma)^n}, \tag{3.4}$$

where $(x)^n$ denotes the rising factorial $x(x+1)\dots(x+n-1)$, and $\gamma = 1-\nu$. Consequently, the residue at $\nu = n$ is given by

$$(-1)^n\frac{(\alpha)^n(\beta)^n}{n!(n-1)!}. \tag{3.5}$$

Starting with the relations in (3.3), an elementary calculation shows that for $i = 0, 1, \ldots, n-1$

$$(3.6)\quad (\alpha+i)(\beta+i)(\alpha+n-1-i)(\beta+n-1-i) = (\hat{u}+\hat{v}+a_i)(\hat{u}+\hat{v}+a_{i+1})+b_i^2\hat{u},$$

where $\hat{u}$ and $\hat{v}$ are obtained from u and v, respectively, by setting $\nu = n$. Another calculation will show that if n is odd, say $n = 2m+1$, then

$$-(\alpha+m)(\beta+m) = \hat{u}+\hat{v}+a_m.$$

This observation explains why the $(-1)^n$ factor in (3.5) can be discarded. Indeed, the factorization described in Proposition 3.1 now follows immediately.

There is an alternate explanation of the above factorization phenomenon that deserves to be mentioned. It is clear from looking at (3.4) that negative integer values of the parameters α, β cause the hypergeometric series to truncate. Equation (3.6) indicates that

$$(3.7)\qquad (u+v+a_i)(u+v+a_{i+1})+b_i^2 u = 0,$$

for some positive integers $i < n$ if and only if one of α or β is a negative integer. Now the spectral residue sequence is gauge invariant, and hence can be obtained just as well by taking spectral residues of the coefficients of the hypergeometric function. Consequently, if u and v are such that (3.7) holds for some i and n, then $\rho_N(u, v)$ will have to vanish for all $N \geq n$. In summary, the presence of polynomial solutions to the hypergeometric equation corresponds to the existence of roots of the elements of the spectral sequence, and this in turn causes the factorization. It seems to the author that this point of view is closest in spirit to the factorization property of the Bender–Dunne polynomials [**1**]. At the moment there is no obvious way to construct spectral residue sequences for quasi-exactly-solvable potentials. It is to be hoped that a generalization can be achieved by searching for methodological commonalities between the present theory and the Bender–Dunne approach.

4. Darboux Transformations

This section considers the effect of a Darboux transformation of a potential $U(z)$ on the corresponding spectral residue sequence. The point of view is formal, in the sense that the potential is just a formal power series $U_1 z + U_2 z^2 + \ldots$; the spectral residue sequence approach requires that the constant term be zero.

The Darboux transformation is most easily described in terms of factorization of a self-adjoint second-order operator:

$$\partial_{xx} + U = (\partial_x + W)(\partial_x - W).$$

The function W is commonly referred to as the pre-potential. It is related to the potential by:

$$U = -W_x - W^2.$$

Reversing the factors in the above factorization one obtains a new self-adjoint operator:

$$\partial_{xx} + \widetilde{U} = (\partial_x - W)(\partial_x + W),$$

where

$$\widetilde{U} = W_x - W^2.$$

The potential $\widetilde{U}$ is called the Darboux transform of U. The preceding discussion demonstrates that at the level of the pre-potential W, the Darboux transformation simply corresponds to a negation.

Working with the independent variable $z = e^x$ one has the following relation between potential and pre-potential:

$$U = -zW_z - W^2.$$

If $U(z)$ is a formal power series with zero constant term, the above relation can be solved for W, to obtain a unique formal series solution that also has a zero constant term. Since the transformed potential is given by

$$\widetilde{U} = zW_z - W^2,$$

there exists a formal, nonlinear correspondence between the coefficients of U and $\widetilde{U}$ that encodes the Darboux transformation. Here are the first few relations that describe the formal Darboux transformation:

$$\widetilde{U}_1 = -U_1, \quad \widetilde{U}_2 = -2U_1^2 - U_2, \quad \widetilde{U}_3 = -2U_1^3 - 2U_1U_2 - U_3.$$

At this point it is natural to enquire about the effects of the Darboux transformation on the spectral residue sequence.

THEOREM 4.1. *Let U and $\widetilde{U}$ be formal powers series with zero constant term. If U and $\widetilde{U}$ are related by a Darboux transformation, then the corresponding residue sequence of one is the negative of the other, i.e. $\rho_n(U) = -\rho_n(\widetilde{U})$ for all n.*

PROOF. Since the Darboux transformation corresponds to the negation of the pre-potential $W(z) = W_1 z + W_2 z^2 + \dots$, the above result would be easy to explain if the dependence on the spectral residue sequence on the pre-potential were linear. It is not. However, it will be demonstrated shortly that each ρ_n, when considered as a function of the pre-potential, is given by a sum each of whose terms is a scalar times the product of an odd number of W_i's. For this reason, the negation of W also negates ρ_n.

To elucidate the relationship between the pre-potential and the spectral residue sequence, note that the Hamiltonian

$$\mathcal{H} = z^2\partial_{zz} + (1-\nu)z\partial_z + U,$$

is gauge equivalent to the following operator

$$\mathcal{H}_1 = z^2\partial_{zz} + (1-\nu)z\partial_z z + 2Wz\partial_z - \nu W.$$

The gauge factor that takes one to the other is just $e^{\int (W/z)\,dz}$. Hence, the coefficients of the formal series solution of

$$\mathcal{H}_1(f) = 0, \quad f(0) = 1,$$

are determined by the following recursive relations

$$n(\nu - n)f_n = \sum_{k=0}^{n-1} (2n - 2k - \nu)W_k f_{n-k},$$

and can therefore be evaluated by means of a composition sum. Recall that by Proposition 2.3 the spectral residue sequence can be obtained by taking the residue of f_n at $\nu = n$. Doing so yields the following:

$$\rho_n = n \sum_{\mathbf{p}} \frac{(n-2s_1)(n-2s_2)\dots(n-2s_l)}{\pi(\mathbf{p})\pi(\mathbf{p}')} W_{p_1} W_{p_2} \dots W_{p_l},$$

where the index $\mathbf{p}$ runs over all compositions of all lengths of n. Next, let $r_k = p_l + p_{l-1} + \dots + p_{l-k+1}$ denote the kth right partial sum of a partition $(p_1, \dots, p_l)$. Note that $r_k = n - s_{l-k}$ and hence

$$\rho_n = \sum_{\mathbf{p}} \frac{(r_l - s_0)(r_{l-1} - s_1)(r_{l-2} - s_2)\dots(r_0 - s_l)}{\pi(\mathbf{p})\pi(\mathbf{p}')} W_{p_1} W_{p_2} \dots W_{p_l},$$

where it is convenient to set $s_0 = r_0 = 0$. Hence, if l is even, then the term in the composition sum corresponding to any given $\mathbf{p}$ is the negative of the term corresponding to $\mathbf{p}'$. Therefore, the terms corresponding to compositions of even length contribute nothing to the sum. □

5. Shape Invariance

Let $U(x; \mathbf{a})$ be a parameterized potential function with $\mathbf{a} \in \mathbb{R}^n$ denoting a list of one or more potential parameters. The corresponding family of potentials is called shape-invariant if it is closed with respect to the Darboux transformation, i.e. if for every $\mathbf{a} \in \mathbb{R}^n$ there exists an $\mathbf{a}'$ such that $\widetilde{U(\mathbf{a})} = U(\mathbf{a}')$. Theorem 4.1 gives a formal criterion that can be applied to construct and recognize shape invariant potentials. As an example, consider the following questions. Is the family of Eckart potentials shape-invariant? If not, does there exist a sub-family that is shape-invariant?

Proposition 3.1 gives the answer to the first question: the Eckart potentials are not shape-invariant. In other words given a certain u and a v there do not exists, in general, a u' and a v' such that $\rho_n(u', v') = -\rho(u, v)$ for all n. In regards to the second question one has the following.

PROPOSITION 5.1. *Let $U(u, v) = uz/(1-z) + vz/(1-z)^2$ be a Eckart potential for some fixed values of u and v. The Darboux transform of U is another Eckart potential if and only if $v = (u+v)^2$. If this condition holds then $\widetilde{U} = U(-u-2v, v)$.*

The proof follows more or less directly from Proposition 3.1, and will not be given here.

To construct a shape invariant potential, one needs to begin with a parameterized residue sequence, $\rho_n(\mathbf{a})$, such that for every value of the parameters, $\mathbf{a}$, there exists a $\mathbf{a}'$ such that $\rho_n(\mathbf{a}') = -\rho_n(\mathbf{a})$ for all n. Having such a sequence one can reconstruct the potential function as follows. For each $n = 1, 2, \dots$ set

$$U_n = n \sum_{\mathbf{p}} \frac{(-1)^{l-1} \rho_{p_1} \rho_{p_2} \cdots \rho_{p_l}}{(p_1 + p_2)(p_2 + p_3) \dots (p_{l-1} + p_l)}, \tag{5.1}$$

where the sum is taken over all compositions $\mathbf{p} = (p_1, \dots, p_l)$ of n.

PROPOSITION 5.2. *Let ρ_n be a given sequence, let U_n be defined according to the above formula, and set $U(z) = U_1 z + U_2 z^2 + \dots$. Then, the spectral residue sequence corresponding to U is precisely $\rho_1, \rho_2, \dots$. In other words the transformations defined by equations (2.1) and (5.1) are inverses.*

The proof of the above fact lies beyond the scope of the present article. It may be better, in conclusion, to note that the constant sequence $\rho_n = t$, where t is a parameter, satisfies the above criterion of shape invariance (the transformation $t \mapsto -t$ negates the sequence.) Consequently the series

$$\sum_{1 \le l \le n} c_{ln} t^l z^n,$$

where

$$c_{ln} = \sum_{\mathbf{p}} \frac{(-1)^{l-1} n}{(p_1 + p_2)(p_2 + p_3) \dots (p_{l-1} + p_l)},$$

and $\mathbf{p}$ ranges over all l-part compositions of n, defines a shape-invariant potential with respect to the z variable. The Darboux transformation of this potential corresponds to the negation of t. The study of the analytic properties of this potential, as well as other shape-invariant potentials constructed by the method of spectral residue sequences will be done in a subsequent publication.

References

1. C. M. Bender and G. V. Dunne, *Quasi-exactly solvable systems and orthogonal polynomials*, J. Math. Phys. **37** (1996), no. 1, 6–11.
2. F. Cooper, A. Khare, and U. Sukhatme, *Supersymmetry and quantum mechanics*, Phys. Rep. **251** (1995), no. 5-6, 267–385.
3. C. Eckart, *The penetration of a potential barrier by electrons*, Phys. Rev. **35** (1930), 1303–1309.
4. A. Erdélyi and H. Bateman, Table errata: *Higher transcendental functions*, vol. I, McGraw-Hill, New York, 1953.
5. A. Krajewska, A. Ushveridze, and Z. Walczak, *Bender–Dunne orthogonal polynomials. General theory*, Modern Phys. Lett. A **12** (1997), no. 16, 1131–1144.

Dept. of Mathematics, Dalhousie University, Halifax, B3H 3J5, Canada.
E-mail address: milson@mscs.dal.ca

Centre de Recherches Mathématiques
CRM Proceedings and Lecture Notes
Volume **29**, 2001

Bäcklund Transformation and Nonlinear Superposition Formula of the Kaup–Kupershmidt and Tzitzéica Equations

M. Musette, *R. Conte*, and C. Verhoeven

ABSTRACT. The G5 and G25 nonlinear differential equations are the only ones in the Gambier list to be linearizable into a third order equation. Therefore, the x-part of the Bäcklund transformation of partial differential equations (PDEs) with a third order Lax pair can only be G5 or G25. We consider three such PDEs, and represent the Lax pair as a coupled two-component Riccati system. The linearizing transformation of G5 allows one to recover the Weiss truncation for the Sawada–Kotera equation, and that of G25 allows one to extend it to the Kaup–Kupershmidt (KK) equation, thus yielding the Bäcklund transformation of the latter. For the Tzitzéica equation, a truncation is defined which admits a unique solution; by elimination of one Riccati component, its Bäcklund transformation is obtained in the G5 class. The nonlinear superposition formula for KK is a second-order, second-degree nonlinear equation of the Appell type, and we give its general solution when one starts from the vacuum solution.

1. Introduction

This paper is a self-contained summary, with some extensions, of the recent derivation [**9, 38**] of the Bäcklund transformation (BT) of two nonlinear partial differential equations (PDEs). This was made possible by an extension of the *singular manifold method* of Weiss *et al.* (WTC) [**51, 54**], whose original version could not derive these BTs, despite successive improvements, see references in [**38**].

The first PDE belongs to the class of fifth order equations

$$u_t = \left(u_{xxxx} + (8a - 2b)uu_{xx} - 2(a+b)u_x^2 - \frac{20}{3}abu^3\right)_x, \tag{1.1}$$

integrable for the only three values of the ratio $b/a = (-1, -6, -16)$, corresponding respectively to the Sawada–Kotera (SK), KdV_5 and Kaup–Kupershmidt (KK) equations, which possess N-soliton solutions. Its reduction $(x,t) \mapsto \xi = x - ct$,

2000 *Mathematics Subject Classification.* Primary: 35Q51, 35Q58.

M.M. and R.C. acknowledge the financial support of the Tournesol grants numbers T95/004 and T99/040. R.C. acknowledges the partial financial support of the organizers. M.M. acknowledges the financial support of the IUAP Contract No. P4/08 funded by the Belgian government and the support of CEA.

This is the final form of the paper.

$u(x,t) = q_1(\xi)$ is equivalent to the Hénon–Heiles system [**15, 24**]:

$$(1.2)\quad q_1'' = bq_1^2 - aq_2^2 + \frac{c}{4a},\quad q_2'' = -2aq_1q_2 - \frac{\Lambda^2}{4a^2}q_2^{-3},\qquad \Lambda \text{ arbitrary constant,}$$

which is integrable for the same three values of the ratio b/a.

The KdV$_5$ equation belongs to the same hierarchy than KdV and therefore shares its properties of integrability (second order scattering problem, Riccati-type Bäcklund transformation (BT), classical Darboux transformation (DT)) which are well-known since many years. As for the SK and KK equations, they are dual equations linked to each other by a BT [**16, 21**], and they both possess a scattering problem of third order [**28**]. However, if the auto-BT for SK has been known since 1977 [**14, 43**] and can be easily recovered [**36**] by singularity approach [**54**], the one previously determined for KK [**41**] did not contain any Bäcklund parameter, a crucial quantity to build the nonlinear superposition formula. Last but not least, there still remains a challenge for the Hirota method to provide for KK a bilinear BT able to be linearized into the Lax pair.

The second PDE is the Tzitzéica equation

$$(1.3)\qquad E(u) \equiv u_{xt} + \alpha e^{u} + a_2 e^{-2u} = 0,\qquad \alpha a_2 \neq 0,\quad (\alpha, a_2) \text{ constant,}$$

originally found by Georges Tzitzéica [**46**] who looked for surfaces on which the total curvature at each point is proportional to the fourth power of the distance from a fixed point to the tangent plane. Many of its integrability results were established by Tzitzéica from purely geometric considerations, and they are summarized in his book [**49**]. For instance, he gave the linear system [**47**]

$$(1.4)\qquad -\tau_{xx} + U_x\tau_x + \sqrt{a_2}\lambda e^{-U}\tau_t = 0,$$

$$(1.5)\qquad -\tau_{tt} + U_t\tau_t + \sqrt{a_2}\lambda^{-1}e^{-U}\tau_x = 0,$$

$$(1.6)\qquad -\tau_{xt} - \alpha e^{U}\tau = 0,$$

with λ an arbitrary nonzero complex constant, in which the three commutators vanish whenever these three equations hold and U satisfies (1.3). Mikhailov [**32**] later gave a more conventional Lax pair of third order in the zero curvature representation. The Darboux transformation follows from the invariance of the system (1.4)–(1.6) under the involution [**17, 48**]

$$(1.7)\qquad (\tau, e^U, \lambda) \to \left(\frac{1}{\tau}, -e^U - \frac{2}{\alpha}\frac{\tau_x\tau_t}{\tau^2}, -\lambda\right),$$

and it can be written either as

$$(1.8)\qquad e^u = -e^U - \frac{2}{\alpha}\frac{\tau_x\tau_t}{\tau^2},$$

or as

$$(1.9)\qquad e^u = \frac{2}{\alpha}\partial_x\partial_t \operatorname{Log}\tau + e^U,$$

in which u and U are two different solutions of (1.3). Concerning the Tzitzéica equation, the problem which was to be solved by the singular manifold method was to define a truncation admitting a unique solution providing all the integrability results (Lax pair, Darboux transformation and Bäcklund transformation).

The results of this paper are established by singularity analysis only, they are the following:

(i) the BT of KK with a nonzero Bäcklund parameter,
(ii) the nonlinear superposition formula (NLSF) of KK (allowing the algebraic construction of some classes of solutions to the PDE),
(iii) a truncation of Tzitzéica equation providing the Lax pair and the Darboux transformation,
(iv) a BT of Tzitzéica equation obeying the definition recalled below.

The method for finding the BT of SK and KK is based on an *a priori* property of the x-part of the BT, which identifies it with one of the only two nonlinear ordinary differential equations (ODE) of the Gambier classification [**18**] (G5 and G25) which are linearizable into third order. Their respective linearizing transformations and involutions allowed us [**38**] to solve the equations of the Weiss truncation [**51, 54**], providing the Lax pair and the DT, then yielding after an elimination process the BT of SK and KK.

For KK equation, the NLSF we have established [**39, 50**] is new and provides an alternative to a perturbation scheme [**19**] for building the N-soliton solution.

2. The Help from Gambier Classification

Among the numerous definitions of the *integrability* of a PDE, one of the most widely accepted is the existence of a BT [**3, 42**]. This transformation between two given PDEs, taken for simplicity in two independent variables x, t,

$$E_1(u,x,t)=0, \quad E_2(U,X,T)=0, \tag{2.1}$$

is by definition ([**13**] vol. III, Chap. XII) a pair of relations

$$F_j(u,u_x,u_t,\dots,U,U_x,U_t,\dots,x,t,X,T)=0, \quad j=1,2, \tag{2.2}$$

with a transformation law between (x,t) and (X,T), such that the elimination of u (resp. U) between (F_1, F_2) implies $E_2(U,X,T)=0$ (resp. $E_1(u,x,t)=0$). In case the two PDEs are the same, the BT is called auto-BT.

In the case of $1+1$-dimensional PDEs, let us consider the ODE obtained by eliminating ∂_t between the two equations defining the (yet unknown) BT. This nonlinear ODE for $Y=u-U$, with coefficients depending on U and an arbitrary constant λ, has two properties. Firstly, it is linearizable since it results from the Lax pair, a linear system, and the DT [**31**] by an elimination process [**5**]. Secondly, it has the Painlevé property (PP) since it is linearizable. Therefore, if its order is small (at most three), it belongs to the appropriate finite list established by the Painlevé school between 1900 and 1910.

These very special nonlinear ODEs provide a link to both the Lax pair, *via* their linearizing transformation, and the Darboux transformation, *via* an involution which leaves them invariant.

The only ODE of first order and first degree with the PP is the Riccati equation, linearizable into a second order linear equation and defining a unique choice (3.25) for describing scattering problems which have order two.

Next, let us consider the ODEs of order two and degree one, inequivalent under the homographic group of transformations, which are linearizable into a third order equation, which we define as

$$\psi_{xxx}-a\psi_x-b\psi=0, \tag{2.3}$$

where a and b are functions of (x,t). Let us derive the constraints, if any, between (a,b). Introducing the two components

$$Y_1 = \frac{\psi_x}{\psi}, \quad Y_2 = \frac{\psi_{xx}}{\psi} + K\left(\frac{\psi_x}{\psi}\right)^2 + \alpha, \tag{2.4}$$

where K is a constant and α a function of (x,t), Eq. (2.3) is equivalent to

$$Y_{1,x} = Y_2 - (K+1)Y_1^2 - \alpha, \tag{2.5}$$

$$Y_{2,x} = \big(a - \alpha(2K-1)\big)Y_1 + (2K-1)Y_1Y_2 - K(2K+1)Y_1^3 + b + \alpha_x. \tag{2.6}$$

This first order system defines a birational transformation between Y_1 and Y_2 iff

$$K(2K+1) = 0, \tag{2.7}$$

a constraint which is assumed from now on. The elimination of Y_2 yields the second order ODE for Y_1

$$Y_{1,xx} + 3Y_1Y_{1,x} + Y_1^3 - aY_1 - b = 0, \tag{2.8}$$

i.e. the G5 equation of the Gambier classification, while the elimination of Y_1 gives:

$$\begin{aligned} W_{xx} &- \left(1 + \frac{K+1}{2K-1}\right)W^{-1}W_x^2 + \left(1 - \frac{2(K+1)}{2K-1}\right)\left(\frac{a_x}{2K-1} + b\right)WW_x \\ &- \frac{K+1}{2K-1}\left(\frac{a_x}{2K-1} + b\right)^2 W^3 + \left(\frac{a_{xx}}{2K-1} + b_x\right)W^2 - aW + 2K - 1 = 0, \end{aligned} \tag{2.9}$$

$$W = \left(Y_2 + \frac{a}{2K-1} - \alpha\right)^{-1}. \tag{2.10}$$

In the case $K = 0$, the equations (2.5)–(2.6) represent a coupled Riccati system of projective type [**1**], therefore the equation for W is necessarily homographically equivalent to the G5 equation.

In the case $K = -1/2$, making the transformation $W = Z/(b - a_x/2)$ in (2.9), the equation for Z possesses the PP iff [**4**]

$$b - \frac{a_x}{2} \equiv \lambda(t) \neq 0, \tag{2.11}$$

which allows us to identify it with the G25 equation of the Gambier classification

$$Z_{xx} - \frac{3}{4}Z^{-1}Z_x^2 + \frac{3}{2}ZZ_x + \frac{1}{4}Z^3 - aZ - 2\lambda(t) = 0, \tag{2.12}$$

$$Z = \frac{\tau_x}{\tau} = \frac{\lambda(t)\psi^2}{\psi_{xx}\psi - (1/2)\psi_x^2 - (a/2)\psi^2} \equiv \partial_x \operatorname{Log}\left(\psi_{xx}\psi - \frac{1}{2}\psi_x^2 - \frac{a}{2}\psi^2\right), \tag{2.13}$$

taking account of (2.3) and the constraint (2.11).

In conclusion, the only two generic choices among the ODEs of second order and first degree for describing scattering problems of third order like (2.3) are the equations (2.8) and (2.12), equivalent under the birational transformation

$$Z = \frac{\lambda(t)}{Y_{1,x} + (1/2)Y_1^2 - (a/2)}, \quad Y_1 = \frac{Z_x + Z^2}{2Z}. \tag{2.14}$$

The involutions on (2.8) and (2.12) given by

$$(Y_1, b, a) \to (-Y_1, -b, a - 6Y_{1,x}), \tag{2.15}$$

$$(Z, \lambda, a) \to (-Z, -\lambda, a - 3Z_x), \tag{2.16}$$

will be seen to define the DT.

3. Method for a One-Family Equation

Let us explain the route to follow, starting from the Painlevé analysis of a PDE with only one family of movable singularities, in order to obtain the BT of this equation. The most recent version of this method can be found in school lecture notes [**8,33**]. A preliminary step consists in performing a local analysis for checking if the equation passes the Painlevé test (a necessary condition of integrability). This analysis also yields the *singular part operator* denoted $\mathcal{D}$. Next,

FIRST STEP. Assume the existence of the transformation

$$u = U + \mathcal{D} \operatorname{Log} \tau, \quad E(u) = 0, \tag{3.1}$$

with u solution of the PDE under consideration, τ some expression related to the ratio of two "entire" functions and U an *a priori* unconstrained function.

SECOND STEP. Choose the order two, then three, then ..., for the unknown scalar Lax pair and represent this Lax pair by an equivalent, multicomponent, Riccati pseudopotential $\mathbf{Y} = (Y_1, \dots)$. A second order scalar Lax pair in canonical form, written here in the case of two independent variables,

$$L_1\psi \equiv \psi_{xx} + \frac{S}{2}\psi = 0, \tag{3.2}$$

$$L_2\psi \equiv \psi_t + C\psi_x - \frac{C_x}{2}\psi = 0, \tag{3.3}$$

$$2[L_1, L_2] \equiv S_t + C_{xxx} + CS_x + 2C_x S = 0, \tag{3.4}$$

is equivalently represented by two Riccati equations in $\mathbf{Y} = \psi_x/\psi = \chi^{-1}$.

A third order scalar Lax pair in canonical form

$$L_1\psi \equiv \psi_{xxx} - a\psi_x - b\psi = 0, \tag{3.5}$$

$$L_2\psi \equiv \psi_t - c\psi_{xx} - d\psi_x - e\psi = 0, \tag{3.6}$$

$$[L_1, L_2] \equiv X_0 + X_1\partial_x + X_2\partial_x^2, \tag{3.7}$$

where X_0, X_1, X_2 depend on a, b, c, d, e and their derivatives is similarly represented by a two-component Riccati pseudopotential $\mathbf{Y} = (Y_1, Y_2)$ of projective type [**1,35,36**], as defined in (2.5)–(2.6) when $K = 0$.

For the Tzitzéica PDE, its invariance under the permutation (Lorentz transformation)

$$\mathcal{P}\colon (\partial_x, \partial_t) \to (\partial_t, \partial_x), \tag{3.8}$$

suggests conserving this invariance in the definition of the Lax pair. Since τ is invariant under $\mathcal{P}$, let us also define the unknown Lax pair as the following invariant one, in the basis (ψ_x, ψ_t, ψ) [**9**],

$$(\partial_x - L)\begin{pmatrix}\psi_x\\ \psi_t\\ \psi\end{pmatrix} = 0, \quad L = \begin{pmatrix} f_1 & f_2 & f_3\\ g_1 & g_2 & g_3\\ 1 & 0 & 0\end{pmatrix}, \tag{3.9}$$

$$(\partial_t - M)\begin{pmatrix}\psi_x\\ \psi_t\\ \psi\end{pmatrix} = 0, \quad M = \begin{pmatrix} g_1 & g_2 & g_3\\ h_1 & h_2 & h_3\\ 0 & 1 & 0\end{pmatrix}, \tag{3.10}$$

with the link

$$\tau = \psi. \tag{3.11}$$

The nine coefficients f_j, g_j, h_j, $j = 1, 2, 3$, are functions to be determined. In the equivalent projective Riccati components (Y_1, Y_2)

$$Y_1 = \frac{\psi_x}{\psi}, \quad Y_2 = \frac{\psi_t}{\psi}, \tag{3.12}$$

with the property $Y_{1,t} = Y_{2,x}$, it is defined as

$$Y_{1,x} = -Y_1^2 + f_1 Y_1 + f_2 Y_2 + f_3, \tag{3.13}$$
$$Y_{2,x} = -Y_1 Y_2 + g_1 Y_1 + g_2 Y_2 + g_3, \tag{3.14}$$
$$Y_{1,t} = -Y_1 Y_2 + g_1 Y_1 + g_2 Y_2 + g_3, \tag{3.15}$$
$$Y_{2,t} = -Y_2^2 + h_1 Y_1 + h_2 Y_2 + h_3. \tag{3.16}$$

One must add to them the six equations $X_j = 0$ defined by the cross-derivative conditions

$$(Y_{1,x})_t - (Y_{1,t})_x = X_0 + X_1 Y_1 + X_2 Y_2 = 0, \tag{3.17}$$
$$(Y_{2,x})_t - (Y_{2,t})_x = X_3 + X_4 Y_1 + X_5 Y_2 = 0, \tag{3.18}$$
$$X_0 = f_{3,t} - g_{3,x} + f_1 g_3 - f_3 g_1 + f_2 h_3 - g_2 g_3 = 0, \tag{3.19}$$
$$X_1 = f_{2,t} - g_{2,x} + f_1 g_2 - f_2 g_1 + f_3 - g_2^2 + f_2 h_2 = 0, \tag{3.20}$$
$$X_2 = f_{1,t} - g_{1,x} + f_2 h_1 - g_1 g_2 - g_3 = 0, \tag{3.21}$$
$$X_3 = h_{3,x} - g_{3,t} + g_3 h_2 - g_2 h_3 + f_3 h_1 - g_1 g_3 = 0, \tag{3.22}$$
$$X_4 = h_{2,x} - g_{2,t} + f_2 h_1 - g_1 g_2 - g_3 = 0, \tag{3.23}$$
$$X_5 = h_{1,x} - g_{1,t} + g_1 h_2 - g_2 h_1 + h_3 - g_1^2 + f_1 h_1 = 0. \tag{3.24}$$

It is worth noticing that there exists no such invariant second order matrix Lax pair.

THIRD STEP. Choose an explicit link

$$\mathcal{D} \operatorname{Log} \tau = f(\psi), \tag{3.25}$$

between the function τ and the solution ψ of a scalar Lax pair. For most PDEs, this is simply (3.11), but, in case of a third order Lax pair, we also have another choice corresponding to the transformation (2.13) which linearizes G25.

FOURTH STEP. With the assumptions (3.1) for a DT, (3.25) for a link between τ and ψ, define the "truncation" for the difference $E(u) - E(U)$

$$E(u) - E(U) = \sum_{j=0}^{-q} E_j(S, C, U) \chi^{j+q} \text{ (for second order)}, \tag{3.26}$$

$$E(u) - E(U) = \sum_{k \ge 0} \sum_{l \ge 0} E_{k,l}(a, b, c, d, e, U) Y_1^k Y_2^l \text{ (for third order)}, \tag{3.27}$$

and solve the set of *determining equations*

$$\forall j \quad E_j(S, C, U) = 0 \quad \text{(for second order)}, \tag{3.28}$$
$$\forall k\, \forall l \quad E_{k,l}(a, b, c, d, e, U) = 0 \quad \text{(for third order)}, \tag{3.29}$$

for the unknown coefficients (S, C) or (a, b, c, d, e) as functions of U.

For the Tzitzéica equation, with the assumptions (3.9)–(3.10) for a Lax pair, (3.25) for a link between τ and ψ, the truncation defined by

$$e^u = \mathcal{D} \operatorname{Log} \tau + e^U, \tag{3.30}$$

$$E(u) \equiv \sum_{k=0}^{3} \sum_{l=0}^{3-k} E_{kl}(f_j, g_j, h_j, U) Y_1^k Y_2^l = 0, \tag{3.31}$$

generates ten determining equations

$$\forall k, l \colon E_{kl}(f_j, g_j, h_j, U) = 0, \tag{3.32}$$

in U and the nine unknown coefficients.

The second, third and fourth steps must be repeated until a success occurs. The process is successful iff

(1) U comes out unconstrained, apart from being a solution of the PDE,
(2) the vanishing of the commutator $[L_1, L_2]$ or $[\partial_x - L, \partial_t - M]$ is equivalent to $E(U) = 0$,
(3) (since this is a $1+1$-dimensional case) the coefficients depend on an arbitrary constant λ, the spectral or Bäcklund parameter.

FIFTH STEP. Obtain the two equations for the BT by eliminating ψ [**5**] between the DT and the scalar Lax pair. This operation may become easier by eliminating the appropriate component of the multicomponent pseudopotential $\mathbf{Y}$ rather than ψ, and this is the only reason for introducing $\mathbf{Y}$ in the second step.

4. From Singularity Analysis to Bäcklund Transformation

Because of their duality [**28, 52**], the SK and KK equations, defined as

$$\mathrm{pSK}(v) \equiv \beta v_t + v_{xxxxx} + \frac{30}{\alpha} v_x v_{xxx} + \frac{60}{\alpha^2} v_x^3 = 0, \tag{4.1}$$

$$\mathrm{pKK}(v) \equiv \beta v_t + v_{xxxxx} + \frac{30}{\alpha} v_x v_{xxx} + \frac{45}{2\alpha} v_{xx}^2 + \frac{60}{\alpha^2} v_x^3 = 0, \tag{4.2}$$

are handled at the same time, with the conservative field u related to the potential one v by $u = v_x$.

Both equations have the PP [**52**]. Each of them has two families of movable singularities, whose leading order $v \sim v_0 \chi^p$ and Fuchs indices are the following [**52**]

$$\mathrm{pSK, F1} \colon p = -1, \quad v_0 = \alpha, \qquad \text{indices } -1, 1, 2, 3, 10, \tag{4.3}$$

$$\mathrm{pSK, F2} \colon p = -1, \quad v_0 = 2\alpha, \qquad \text{indices } -2, -1, 1, 5, 12, \tag{4.4}$$

$$\mathrm{pKK, F1} \colon p = -1, \quad v_0 = \alpha/2, \qquad \text{indices } -1, 1, 3, 5, 7, \tag{4.5}$$

$$\mathrm{pKK, F2} \colon p = -1, \quad v_0 = 4\alpha, \qquad \text{indices } -7, -1, 1, 10, 12. \tag{4.6}$$

The singular part operator $\mathcal{D}$ attached to a given family is $\mathcal{D} = v_0 \partial_x$.

The two families have residues which are not opposite, which makes inapplicable the two-singular manifold method [**37**]. Fortunately, each potential equation possesses in its hierarchy a "minus-one" equation [**55**]

$$\mathrm{pSK}_{-1} \colon v_{xxt} + \frac{6}{\alpha} v_x v_t = 0, \tag{4.7}$$

$$\mathrm{pKK}_{-1} \colon v_t v_{xxt} - \frac{3}{4} v_{xt}^2 + \frac{6}{\alpha} v_x v_t^2 = 0, \tag{4.8}$$

which has only one family. The equation (4.7), initially written by Hirota and Satsuma [**23**], has already been processed successfully [**35, 36**] by the one-family method recalled in Section 3. This is a strong indication that the same method should also succeed for the KK_{-1} and the KK equations, by restricting to their unique common family F1 (4.5).

Eq. (1.3) possesses two families of movable singularities. Denoting $\varphi(x,t)-\varphi_0 = 0$ the equation for the movable singular manifold, and $\mathcal{D}$ the linear singular part operator, the first family has double poles

$$e^u \sim -(2/\alpha)\varphi_x\varphi_t(\varphi-\varphi_0)^{-2}, \qquad \text{indices } (-1,2), \qquad \mathcal{D}=(2/\alpha)\partial_x\partial_t, \tag{4.9}$$

and the second one has simple poles

$$e^{-u} \sim \sqrt{(1/a_2)\varphi_x\varphi_t}(\varphi-\varphi_0)^{-1}, \quad \text{indices } (-1,2). \tag{4.10}$$

Therefore this PDE seems to belong neither to the class of PDEs with only one family, for which the original WTC [**54**] method is well adapted, nor to the class of PDEs with two opposite families [**37, 40**]. Fortunately, it has a one-to-one correspondence [**22**] with another PDE possessing only one family, the Hirota–Satsuma PDE [**23**], defined in potential form as

$$\mathrm{pHS}(w) \equiv w_{xxt} + (6/a)w_x w_t + F(t) = 0, \quad a \neq 0. \tag{4.11}$$

Indeed, in the particular case $F(t) = 0$, the elimination of a_2 in equation (1.3) provides

$$\left(F(t)=0, e^u = \frac{2}{a\alpha}w_t\right) \Longrightarrow \left(e^{-2u}\left(e^{2u}E(u)\right)_x = \left(\frac{\mathrm{pHS}(w)}{w_t}\right)_t\right). \tag{4.12}$$

The single family for the Hirota–Satsuma equation (4.11)

$$w \sim a\varphi_x(\varphi-\varphi_0)^{-1}, \quad \text{indices } (-1,1,6), \qquad \mathcal{D}=a\partial_x, \tag{4.13}$$

together with the link (4.12), proves that the second family (4.10) is irrelevant and must not be considered, thus putting the Tzitzéica PDE in the class of one-family PDEs. This conclusion is confirmed by the negative result of Weiss [**6, 53**] obtained when performing a truncation on e^{-u}.

From the correspondence (4.12) with the Hirota–Satsuma equation, one might think of transposing the well established results for that equation (Lax pair [**26**], DT and BT [**20**], all easy to find by singularity analysis [**35**]) to get the BT of the Tzitzéica equation. Unfortunately, this is impossible [**9**] because the correspondence (4.12) restricts the potential form of the Hirota–Satsuma equation to $F(t) = 0$.

In order to obtain the desired results, one must therefore perform the truncation method on the PDE (1.3).

4.1. Truncation with a second order Lax pair. Such an assumption is equivalent to the original (WTC) version of the singular manifold method.

This one-family truncation for pSK and pKK with the assumption

$$v = v_0\partial_x \operatorname{Log}\psi + V, \tag{4.14}$$

where ψ solution of the second order Lax pair (3.2)–(3.3) and V unconstrained provides in both cases the BT (the x-part of which is the Miura transformation) linking the solutions of pSK and pKK equations:

$$\alpha(w+W/2)_x+(w-W)^2=0, \tag{4.15}$$

$$\begin{aligned}&\left[(w-W)(72W_x^2/\alpha^2+6W_{xxx}/\alpha)-72w_xW_{xx}/\alpha-3W_{xxxx}\right]_x\\&\quad+2\beta(w-W)_t=0,\qquad w=v_{SK},\quad W=v_{KK},\end{aligned} \tag{4.16}$$

a result previously obtained by Hirota [**21**]. As this truncation only determines the link between the solutions of two different PDEs and does not provide any arbitrary constant, the procedure must be repeated with a third order Lax pair.

For the Tzitzéica equation, with the second order Lax pair (3.2)–(3.3), and the identity link (3.11), the results are as follows:

(1) the truncation of e^{-u} has no solution at all [**6, 53**], not even the one-soliton;
(2) the truncation of e^u

$$e^u=\mathcal{D}\operatorname{Log}\tau+e^U,\quad E(u)=0,\quad \mathcal{D}=(2/\alpha)\partial_x\partial_t, \tag{4.17}$$

generates four determining equations, which do admit some particular solutions, although their general solution has not yet been found. Two such particular solutions are a two-wave elliptic solution [**37**], and a particular solution [**44**] linking φ to the pair (τ,τ'), in which τ satisfies the Tzitzéica triplet (1.4)–(1.6) and τ' its adjoint [**27, 47**].

It must be pointed out that, since the Lax order is known (this is three) and is not equal to two, such a WTC-style truncation is incoherent. Indeed, it mixes a second order assumption (the WTC-style truncation) and, in order to succeed, necessarily a third order assumption (such as, for instance, equations like (1.4)–(1.6) for τ). Despite the fact that such an incoherence may still provide the full result for some "robust" equations (Boussinesq, Sawada–Kotera, Hirota–Satsuma), there do exist equations for which it leads to a failure, and the Kaup–Kupershmidt equation is one of them.

4.2. Truncation with a third order Lax pair. With the third order Lax pair (3.5)–(3.6) and the identity link (3.11) corresponding to the linearizing transformation of the G5 ODE, the process is successful for SK but not for KK. It provides the Lax pair [**7, 52**]:

$$L_1=\partial_x^3+6\frac{U}{\alpha}\partial_x-\lambda, \tag{4.18}$$

$$L_2=\beta\partial_t+\left(18\frac{U_x}{\alpha}-9\lambda\right)\partial_x^2+\left(36\frac{U^2}{\alpha^2}-6\frac{U_{xx}}{\alpha}\right)\partial_x-36\lambda\frac{U}{\alpha}, \tag{4.19}$$

$$\beta[L_1,L_2]=(6/\alpha)\mathrm{SK}(U), \tag{4.20}$$

previously obtained in the bilinear formalism by Satsuma and Kaup [**43**].

For KK, one then must choose the G25 link (2.13) to perform the truncation, and the result is [**38**] (notation is $U = V_x$)

$$\lambda'(t) = 0, \quad a = -6U/\alpha, \tag{4.21}$$

$$\frac{\tau_x}{\tau} = \frac{\lambda}{\psi_{xx}/\psi - (1/2)(\psi_x/\psi)^2 + 3(U/\alpha)}, \tag{4.22}$$

$$L_1 = \partial_x^3 + 6\frac{U}{\alpha}\partial_x + 3\frac{U_x}{\alpha} - \lambda, \tag{4.23}$$

$$L_2 = \beta\partial_t - 9\lambda\partial_x^2 + \left(3\frac{U_{xx}}{\alpha} + 36\frac{U^2}{\alpha^2}\right)\partial_x - 3\frac{U_{xxx}}{\alpha} - 72\frac{UU_x}{\alpha^2} - 36\lambda\frac{U}{\alpha}, \tag{4.24}$$

$$\beta[L_1, L_2] = (6/\alpha)\mathrm{KK}(U)\partial_x + (3/\alpha)\mathrm{KK}(U)_x. \tag{4.25}$$

This is the Lax pair given in Ref. [**28**] and the DT given in Ref. [**29**] with:

$$\tau = \psi\psi_{xx} - (1/2)\psi_x^2 + 3(U/\alpha)\psi^2, \quad \tau_x = \lambda\psi^2. \tag{4.26}$$

For the Tzitzéica equation, the determining equations (3.32) and (3.17)–(3.18) have a unique solution [**9**]: the field U must satisfy the Tzitzéica PDE, and the nine coefficients depend on an arbitrary nonzero complex constant λ. Accordingly, one has obtained a Lax pair and a Darboux transformation. This Lax pair is the rewriting in matrix form of the scalar triplet; it reads in its equivalent projective Riccati representation

$$Y_{1,x} = -Y_1^2 + U_xY_1 + \sqrt{a_2}\lambda e^{-U}Y_2, \tag{4.27}$$

$$Y_{2,x} = -Y_1Y_2 - \alpha e^U, \tag{4.28}$$

$$Y_{1,t} = -Y_1Y_2 - \alpha e^U, \tag{4.29}$$

$$Y_{2,t} = -Y_2^2 + U_tY_2 + \sqrt{a_2}\lambda^{-1}e^{-U}Y_1, \tag{4.30}$$

with cross-derivative conditions proportional to the Tzitzéica equation

$$(Y_{1,x})_t - (Y_{1,t})_x = Y_1E(U), \quad (Y_{2,x})_t - (Y_{2,t})_x = Y_2E(U). \tag{4.31}$$

4.3. Bäcklund transformation. The two equations which define the BT result from the elimination [**5**] of the scalar ψ between the DT and the Lax pair.

In order to perform this elimination easily, it is convenient to choose one of the two components of the pseudopotential $\mathbf{Y}$ so as to characterize the DT

$$\mathrm{SK}\colon \frac{v - V}{\alpha} = \frac{\psi_x}{\psi} = Y_1, \tag{4.32}$$

$$\mathrm{KK}\colon \frac{2(v - V)}{\alpha} = \frac{\tau_x}{\tau} = Z. \tag{4.33}$$

The chosen equivalent systems are the canonical projective Riccati system in ($Y_1 = \psi_x/\psi, Y_2 = \psi_{xx}/\psi$) associated with the third order Lax pair (4.18)–(4.19) for SK, and the following system in ($Y_1 = \psi_x/\psi, Z = \tau_x/\tau$) for KK

$$Y_{1,x} = -Y_1^2/2 + \lambda Z^{-1} - 3U/\alpha, \tag{4.34}$$

$$Z_x = 2Y_1Z - Z^2, \tag{4.35}$$

and the corresponding system for the t-derivatives of Y_1 and W explicitly given in [**38**]. The x-BT of SK results from the elimination of Y_2, which provides the equation for $Y_1 = Y$

$$Y_{xx} + 3YY_x + Y^3 + 6(U/\alpha)Y - \lambda = 0, \tag{4.36}$$

followed by the substitution $Y=(v-V)/\alpha$

$$(v-V)_{xx}/\alpha+3(v-V)(v+V)_x/\alpha^2+(v-V)^3/\alpha^3-\lambda=0. \tag{4.37}$$

See again in [**38**] the corresponding expression for the t-part.

The x-BT of KK arises from the elimination of Y_1 between (4.34) and (4.35) which writes for $Z=Y$

$$Y_{xx}-(3/4)Y_x^2/Y+3YY_x/2+Y^3/4+6(U/\alpha)Y-2\lambda=0, \tag{4.38}$$

followed by the substitution $Y=2(v-V)/\alpha$

$$(v-V)_{xx}/\alpha-(3/4)(v-V)_x^2/\big(\alpha(v-V)\big) \\ +3(v-V)(v+V)_x/\alpha^2+(v-V)^3/\alpha^3-\lambda=0. \tag{4.39}$$

An analogous procedure for obtaining the t-part of the BT is described in [**38**].

For the Tzitzéica equation, there exist two different eliminations, depending on which projective component is eliminated, and this provides two BTs, as we now see.

4.3.1. *First elimination*: BT *of the Hirota–Satsuma equation.* Eliminating Y_2 from (4.27), one obtains the two functionally independent equations

$$\text{ODE}\equiv Y_{1,xx}+3Y_1Y_{1,x}+Y_1^3-e^{-U}(e^U)_{xx}Y_1+\alpha\sqrt{a_2}\lambda=0, \tag{4.40}$$

$$\text{PDE}\equiv Y_{1,t}+e^U\big((Y_1Y_{1,x}+Y_1^3)-Y_1^2U_x\big)/(\sqrt{a_2}\lambda)+\alpha e^U=0. \tag{4.41}$$

Their "commutator"

$$[\text{ODE},\text{PDE}]=(Y_{1,xx})_t-(Y_{1,t})_{xx}=Y_1\big(e^{2U}E(U)\big)_x, \tag{4.42}$$

shows the failure to generate the auto-BT of Tzitzéica equation. However, the linearizing transformation $Y_1=\partial_x \operatorname{Log}\psi$, together with the change of variables (4.12), shows that it is, in fact, the BT of the Hirota–Satsuma equation $(\text{pHS}(w))_x=0$, see (4.11). Let us recall that the Painlevé analysis of the Hirota–Satsuma equation [**35**] directly provides this result.

4.3.2. *Second elimination*: BT *of the Tzitzéica equation.* Eliminating Y_1 from (4.28), one similarly obtains

$$\text{ODE}\equiv Y_2Y_{2,xx}-2Y_{2,x}^2-(U_xY_2+3\alpha e^U)Y_{2,x}+\sqrt{a_2}\lambda e^{-U}Y_2^3-\alpha^2e^{2U}=0, \tag{4.43}$$

$$\text{PDE}\equiv Y_2Y_{2,t}+Y_2^3-U_tY_2^2+\sqrt{a_2}\lambda^{-1}(\alpha+e^{-U}Y_{2,x})=0, \tag{4.44}$$

and the vanishing of their "commutator"

$$[\text{ODE},\text{PDE}]=(Y_{2,xx})_t-(Y_{2,t})_{xx}=(3\alpha e^U+U_xY_2+3Y_{2,x}-Y_2\partial_x)E(U), \tag{4.45}$$

is now equivalent to the vanishing of the Tzitzéica equation for U. This elimination therefore generates the auto-BT of Tzitzéica equation, by the substitution

$$Y_2=(\alpha/2)\int(e^u-e^U)\,\mathrm{d}x, \tag{4.46}$$

into (4.43)–(4.44).

The nonlinear ODE (4.43), as well as (4.40), belongs to the equivalence class of the fifth Gambier equation G5, and its linearization

$$Y_2^{-1}=-\alpha^{-1}e^{-U}\partial_x \operatorname{Log}(e^U\psi), \tag{4.47}$$

transforms the BT equations (4.43)–(4.44) into the third order scalar Lax pair of the Gel'fand and Dikii type

$$\mathcal{L}\psi \equiv \psi_{xxx} + (2U_{xx} - U_x^2)\psi_x + \left((2U_{xx} - U_x^2)_x/2 + \sqrt{a_2}\alpha\lambda\right)\psi = 0, \tag{4.48}$$

$$\mathcal{M}\psi \equiv \psi_t + \sqrt{a_2}(\alpha\lambda)^{-1}e^{-2U}(\psi_{xx} + U_x\psi_x + U_{xx}\psi) + \left(U_t + \int (\alpha e^U + a_2 e^{-2U})\,\mathrm{d}x\right)\psi = 0, \tag{4.49}$$

$$[\mathcal{L}, \mathcal{M}] = 3E\partial_x^2 + \left(2(e^U)_x E + E_x\right)\partial_x + (e^U)_x E_x + (3U_{xx} - U_x^2)E, \quad E = E(U). \tag{4.50}$$

The ODE part (4.43) of the BT is equivalently written as

$$\frac{w_{xx}}{w_x} - \frac{W_{xx}}{W_x} - 2\frac{w_x + W_x}{w - W} + \alpha\sqrt{a_2}\lambda\frac{(w-W)^2}{2w_x W_x} = 0, \tag{4.51}$$

with the notation $Y_2 = (\alpha/2)(w - W)$, $e^u = w_x$, $e^U = W_x$.

REMARK. If, instead of the invariant assumption (3.9)–(3.10) with the link (3.11), one had chosen the assumption of a (necessarily noninvariant) Gel'fand and Dikii scalar third order Lax pair, the noninvariant link between τ and ψ would have been the one provided by the linearizing formula (4.47) and the Riccati equation (4.29), i.e. the invertible transformation

$$\tau = e^{-U}(e^U\psi)_x, \quad \psi = -(1/\alpha)e^{-U}\tau_t. \tag{4.52}$$

5. From Bäcklund Transformation to Superposition Principle

5.1. Sawada–Kotera equation. We first derive the SK nonlinear superposition formula previously obtained by Hu and Li [**25**] in the framework of the bilinear Hirota formalism.

Let v_n and $\tilde{v}_n$ be solutions of (4.1) generated by application of the BT to a known solution v_{n-1} (depending of $n-1$ parameters), with, respectively, the Bäcklund parameters λ_n and λ_{n+1}. Assuming that the permutability theorem [**12**] is valid, we denote by v_{n+1} the solution generated by application of the BT to v_n with parameter λ_{n+1}, or by application of the BT to $\tilde{v}_n$ with parameter λ_n. This can be represented by a "Bianchi diagram" and the first step for obtaining the nonlinear superposition formula is to eliminate the four second derivatives of v_{n+1}, v_n, $\tilde{v}_n$, v_{n-1} between the four equations equal to the x-part of the BT (4.37). This yields for v_{n+1} an equation of Riccati type with coefficients depending on v_n, $\tilde{v}_n$ and v_{n-1}:

$$(v_{n+1} - v_{n-1})_x + \frac{1}{\alpha}(v_{n+1}^2 - v_{n-1}^2) + (v_{n+1} - v_{n-1})\left(\frac{v_{n,x} - \tilde{v}_{n,x}}{\tilde{v}_n - v_n} - \frac{1}{\alpha}(v_n + \tilde{v}_n)\right) = 0. \tag{5.1}$$

In the tau-function representation

$$v = \alpha\partial_x \operatorname{Log} f, \tag{5.2}$$

the equation (5.1) linearizes into the second order ODE for f_{n+1}

$$(5.3)\quad (f_{n-1}f_{n+1,xx}-f_{n+1}f_{n-1,xx})(\tilde{f}_n f_{n,x}-\tilde{f}_{n,x}f_n) + (\tilde{f}_n f_{n,xx}-\tilde{f}_{n,xx}f_n)(f_{n-1}f_{n+1,x}-f_{n+1}f_{n-1,x})=0.$$

Introducing the bilinear operator D_x, this last equation is best written as:

$$(5.4)\qquad \partial_x\big(\operatorname{Log} D_x(f_{n+1}\cdot f_{n-1})-\operatorname{Log} D_x(f_n\cdot \tilde{f}_n)\big)=0,$$

which can be integrated once to yield the nonlinear superposition formula:

$$(5.5)\qquad D_x(f_{n+1}\cdot f_{n-1})=K_1(t)D_x(f_n\cdot\tilde{f}_n).$$

Switching to the usual notation

$$(5.6)\qquad f_0\equiv f_{n-1},\quad f_1\equiv f_n,\quad f_2\equiv\tilde{f}_n,\quad f_{12}\equiv f_{n+1},$$

and starting from $f_0=1$, the most general function $f_{12}(\lambda_1,\lambda_2)$ is obtained from the one-parameter functions $f_1(\lambda_1)$, $f_2(\lambda_2)$, $\lambda_1\neq\lambda_2$:

$$(5.7)\qquad f_{12}=K_2(t)+K_1(t)\int^x (f_{1,x}f_2-f_{2,x}f_1)\,\mathrm{d}x,$$

where $K_1(t)$ and $K_2(t)$ are two arbitrary functions.

The construction of the two-soliton solution then follows. Setting in (4.37) $V=0$, and identifying v with v_i, λ with λ_i, for $(i=1,2)$, it results from the DT (4.14) with $\psi\equiv f_i$ solution of the third order Lax pair (4.18)–(4.19) for $U\equiv V_x=0$ that:

$$(5.8)\qquad f_{i,xxx}-\lambda_i f_i=0,\quad \beta f_{i,t}-9\lambda_i f_{i,xx}=0,$$

with general solution

$$(5.9)\quad f_i=A_i e^{p_i x+9p_i^5 t}+B_i e^{r_i x+9r_i^5 t}+C_i e^{s_i x+9s_i^5 t},\qquad A_i,B_i,C_i\quad\text{arbitrary constants},$$

where p_i, r_i, q_i are the three different cubic roots of λ_i subjected to the constraint:

$$(5.10)\qquad p_i^2+r_i^2+p_i r_i=0,$$

as well as two other relations deduced from (5.10) by cyclic permutation of p_i, r_i, s_i. Defining $k_i=r_i-p_i$, one may easily check, taking account of (5.10), that $9(r_i^5-p_i^5)=-k_i^5$. Then, setting $C_i=0$ in (5.9) and choosing $K_1(t)=0$, $K_2(t)=1$ in (5.7), one obtains for the two-soliton solution the well-known expression

$$(5.11)\qquad \begin{aligned} u_{12}&=\alpha\partial_x^2\operatorname{Log}\big[1+e^{\theta_1}+e^{\theta_2}+A_{12}^{\mathrm{SK}}e^{\theta_1+\theta_2}\big],\\ A_{12}^{\mathrm{SK}}&=\frac{(k_1-k_2)^2(k_1^2+k_2^2-k_1k_2)}{(k_1+k_2)^2(k_1^2+k_2^2+k_1k_2)},\end{aligned}$$

with

$$\theta_i=k_i x-k_i^5 t+\delta_i,\quad \delta_i=\operatorname{Log}\frac{B_i(r_i-p_j)(p_i+p_j)}{A_i(r_i+p_j)(p_i-p_j)},\qquad (i,j=1,2),\ i\neq j.$$

5.2. Kaup–Kupershmidt equation. In case of the pKK equation (4.2), the similar elimination of the second derivatives of v_{n+1}, v_n, $\tilde{v}_n$, v_{n-1} between the four equations equal to the x-BT (4.39) yields a differential equation of first order and second degree in v_{n+1}:

$$(5.12)\quad \left(v_{n+1,x} + \frac{2}{\alpha}v_{n+1}^2 - Av_{n+1} + \alpha B\right)^2 = \frac{\alpha^2 C^2 (v_{n+1}^2 - (v_n + \tilde{v}_n)v_{n+1} + v_n\tilde{v}_n)}{(v_n - v_{n-1})(\tilde{v}_n - v_{n-1})},$$

with coefficients

$$(5.13)\quad A = \frac{2}{\alpha}(v_n + \tilde{v}_n) + \frac{\tilde{v}_{n,x} - v_{n,x}}{\tilde{v}_n - v_n}, \quad B = \frac{2}{\alpha^2}v_n\tilde{v}_n + \frac{\tilde{v}_{n,x}v_n - v_{n,x}\tilde{v}_n}{\alpha(\tilde{v}_n - v_n)},$$

$$(5.14)\quad C = B + \frac{1}{\alpha}\left(v_{n-1,x} + \frac{2}{\alpha}v_{n-1}^2 - Av_{n-1}\right).$$

Under the transformation $v_{n+1} = (\alpha/2)\partial_x \operatorname{Log} f_{n+1}$, equation (5.12) becomes

$$(5.15)\quad (f_{n+1,xx} - Af_{n+1,x} + 2Bf_{n+1})^2 = C^2 \frac{\alpha^2 f_{n+1,x}^2 - 2\alpha(\tilde{v}_n + v_n)f_{n+1,x}f_{n+1} + 4v_n\tilde{v}_n f_{n+1}^2}{(v_n - \tilde{v}_{n-1})(\tilde{v}_n - v_{n+1})}.$$

This equation of the Appell type [**2, 10, 11**] has the general solution

$$(5.16)\quad f_{n+1} = c_1^2 f_{n+1}^{(1)} + c_1c_2 f_{n+1}^{(2)} + c_2^2 f_{n+1}^{(3)},$$

in which c_1, c_2 are arbitrary, and $f_{n+1}^{(j)}$, $j = 1, 2, 3$, are three independent solutions of the linear third order equation

$$(5.17)\quad f_{n+1,xxx} + p_1(x)f_{n+1,xx} + p_2(x)f_{n+1,x} + p_3(x)f_{n+1} = 0.$$

The coefficients p_1, p_2, p_3 are functions of f_{n-1}, f_n, $\tilde{f}_n$, with

$$(5.18)\quad v_{n-1} = \frac{\alpha}{2}\partial_x \operatorname{Log} f_{n-1}, \quad v_n = \frac{\alpha}{2}\partial_x \operatorname{Log} f_n, \quad \tilde{v}_n = \frac{\alpha}{2}\partial_x \operatorname{Log} \tilde{f}_n.$$

Switching to the notation (5.6), and starting from $f_0 = 1$, equation (5.17) becomes

$$(5.19)\quad f_{12,xxx} - \left(B\frac{(f_1f_2)_x}{f_{1,x}f_{2,x}} + \frac{f_{1,x}f_{2,xxx} - f_{2,x}f_{1,xxx}}{f_{1,x}f_{2,xx} - f_{2,x}f_{1,xx}}\right)f_{12,xx}$$
$$\times\left(B\frac{f_1f_{2,xx} + f_2f_{1,xx}}{f_{1,x}f_{2,x}} + \frac{f_{1,xx}f_{2,xxx} - f_{2,xx}f_{1,xxx}}{f_{1,x}f_{2,xx} - f_{2,x}f_{1,xx}}\right)f_{12,x} = 0,$$
$$B = (f_{2,xx}f_{1,x} - f_{1,xx}f_{2,x})/\big(2(f_{2,x}f_1 - f_{1,x}f_2)\big).$$

Its general solution in terms of $f_1(\lambda_1)$, $f_2(\lambda_2)$ is

$$(5.20)\quad f_{12} = K_1(t) + K_2(t)\int^x \sqrt{f_{1,x}f_{2,x}} + K_3(t)\left(f_1f_2 - \left(\int^x \sqrt{f_{1,x}f_{2,x}}\right)^2\right).$$

The construction of the two-soliton solution is done as follows. After solving the Lax equations (4.23)–(4.24) with $U = 0$ for the vacuum wave functions $\psi_i(\lambda_i)$ $(i = 1, 2)$, the function f_i is given by

$$(5.21)\quad f_i = \lambda_i \int^x \psi_i^2 \, dx,$$

where ψ_i satisfies the linear superposition (5.9). Choosing $C_i = 0$ in (5.9), the expression (5.21) becomes

$$f_i = \lambda_i \left(\frac{A_i^2}{2p_i} e^{2p_i x + 18p_i^5 t} + \frac{B_i^2}{2r_i} e^{2r_i x + 18r_i^5 t} + \frac{2A_i B_i}{p_i + r_i} e^{(p_i + r_i)x + 9(p_i^5 + r_i^5)t} \right), \tag{5.22}$$

and yields the one-soliton solution

$$\begin{aligned} u_i^{(1)} &= \frac{\alpha}{2} \partial_x^2 \operatorname{Log}(1 + 4e^{(k_i x - k_i^5 t + \delta_i)} + e^{2(k_i x - k_i^5 t + \delta_i)}), \\ \delta_i &= \operatorname{Log}\left(\frac{B_i p_i}{A_i(p_i + r_i)} \right). \end{aligned} \tag{5.23}$$

Then, taking account of (5.21), the expression (5.20) can equivalently be written as

$$K_1(t) = 0, \quad K_2(t) = 0, \quad f_{12} = \lambda_1 \lambda_2 \begin{vmatrix} \int^x \psi_1^2 & \int^x \psi_1 \psi_2 \\ \int^x \psi_1 \psi_2 & \int^x \psi_2^2 \end{vmatrix}. \tag{5.24}$$

Choosing again $C_i = 0$ in (5.21), the two-soliton tau-function becomes

$$\begin{aligned} f_{12} = {} & A_1^2 A_2^2 \frac{(p_1 - p_2)^2}{4p_1 p_2 (p_1 + p_2)^2} e^{2(p_1 + p_2)x + 18(p_1^5 + p_2^5)t} \\ & \times \left(1 + 4(e^{\theta_1} + e^{\theta_2}) + e^{2\theta_1} + e^{2\theta_2} + 8 \frac{2k_1^4 - k_1^2 k_2^2 + 2k_2^4}{(k_1 + k_2)^2 (k_1^2 + k_1 k_2 + k_2^2)} e^{\theta_1 + \theta_2} \right. \\ & \left. + 4A_{12}^{SK} (e^{\theta_1 + 2\theta_2} + e^{\theta_2 + 2\theta_1}) + (A_{12}^{SK})^2 e^{2(\theta_1 + \theta_2)} \right), \end{aligned} \tag{5.25}$$

$$\theta_i = k_i x - k_i^5 t + \Delta_i, \Delta_i = \operatorname{Log}\left(\frac{B_i p_i (r_i - p_j)(p_i + p_j)}{A_i (p_i + r_i)(r_i + p_j)(p_i - p_j)} \right), \; i \neq j, \tag{5.26}$$

and the expression for the two-soliton solution of KK is

$$\begin{aligned} u_{12} = \frac{\alpha}{2} \partial_x^2 \operatorname{Log} \Big(& 1 + 4(e^{\theta_1} + e^{\theta_2}) + e^{2\theta_1} + e^{2\theta_2} + 4A_{12}^{SK} (e^{\theta_1 + 2\theta_2} + e^{\theta_2 + 2\theta_1}) \\ & + 8 \frac{2k_1^4 - k_1^2 k_2^2 + 2k_2^4}{(k_1 + k_2)^2 (k_1^2 + k_1 k_2 + k_2^2)} e^{\theta_1 + \theta_2} + (A_{12}^{SK})^2 e^{2(\theta_1 + \theta_2)} \Big). \end{aligned} \tag{5.27}$$

Now, the expression (5.24) suggests that the N-parameter tau-function may be written as

$$\tau^{(N)} = \det \left[\int^x \psi_i \psi_j \, \mathrm{d}x \right]_{1 \leq i,j \leq N}, \tag{5.28}$$

which indeed coincides with the result obtained in [**30**] by symmetry reduction of the CKP hierarchy.

The N-soliton solution is obtained by computing the second logarithmic derivative of (5.28), setting in the vacuum wave function ψ_i, $C_i = 0$, $1 \leq i \leq N$.

5.3. Tzitzéica equation. Generally, there are two methods to obtain the NLSF by an elimination: either from two copies of the Darboux transformation (as formulated by Darboux, i.e. involving two spectral parameters, not as its degenerate version (1.7)), or from four copies of the Bäcklund transformation.

The Darboux transformation of the Tzitzéica equation, more precisely its Moutard transformation, has been given in Ref. [**48**], which allowed the further derivation of the NLSF [**45**]. In the equivalent tau-function representation defined by

$$e^u = (2/\alpha)\partial_x\partial_t \operatorname{Log} f, \tag{5.29}$$

it reads [**34**]

$$\begin{aligned}(\lambda_2-\lambda_1)(f_{0,xt}f_0 - f_{0,x}f_{0,t})(f_0 f_{12} - f_1 f_2)\\ = -\lambda_2(f_{1,x}f_0 - f_{0,x}f_1)(f_{2,t}f_0 - f_{0,t}f_2) + \lambda_1(f_{2,x}f_0 - f_{0,x}f_2)(f_{1,t}f_0 - f_{0,t}f_1),\end{aligned} \tag{5.30}$$

an expression solvable for f_{12}. The vacuum for building the two-soliton solution is then $\operatorname{Log} f_0 = \alpha/(-a_2/\alpha)^{1/3}xt$.

6. Conclusion

The results of this paper are twofold: the explicit analytic expression for the Bäcklund transformation, and the proof that there indeed exists an application of the truncation method able to yield all the integrability results. This reinforces the power of the singularity approach to PDEs initiated by Weiss *et al.*

References

1. R. Anderson, J. Harnad, and P. Winternitz, *Systems of ordinary differential equations with nonlinear superposition principles*, Phys. D **4** (1982), 164–182.
2. P. Appell, *Sur les équations différentielles algébriques et homogènes par rapport à la fonction inconnue et à ses dérivées*, C. R. Acad. Sc. Paris **104** (1887), 1776–1779.
3. A. V. Bäcklund, *Om ytor med konstant negativ krökning*, Lunds Universitets Arsskrift Avd. 2 **19** (1883).
4. F. J. Bureau, *Differential equations with fixed critical points*, Ann. Mat. Pura Appl. **LXIV** (1964), 229–364.
5. H. H. Chen, *General derivation of Bäcklund transformations from inverse scattering problems*, Phys. Rev. Lett. **33** (1974), 925–928.
6. R. Conte, *Invariant Painlevé analysis of partial differential equations*, Phys. Lett. A **140** (1989), 383–390.
7. ———, *Towards the equivalence between integrability and Painlevé test for partial differential equations*, Nonlinear and Chaotic Phenomena in Plasmas, Solids and Fluids (Edmonton, 1990) (W. Rozmus and J. A. Tuszyński, eds.), World Sci. Publishing, River Edge, NJ, 1991, pp. 94–101.
8. ———, *Exact solutions of nonlinear partial differential equations by singularity analysis*, Direct and Inverse Methods in Nonlinear Evolution Equations (CIME School, Cetraro, 1999) (A. Greco, ed.), Springer, Berlin, 2001, `nlin-SI/0009024`.
9. R. Conte, M. Musette, and A. M. Grundland, *Bäcklund transformation of partial differential equations obtained from the Painlevé–Gambier classification,* II. *Tzitzéica equation*, J. Math. Phys. **40** (1999), 2092–2106.
10. C. M. Cosgrove, *A new formulation of the field equations for the stationary axisymmetric gravitational field* I. *General theory*, J. Phys. A **11** (1978), 2389–2404.
11. C. M. Cosgrove and G. Scoufis, *Painlevé classification of a class of differential equations of the second order and second degree*, Stud. Appl. Math. **88** (1993), 25–87.
12. G. Darboux, *Sur les surfaces dont la courbure totale est constante*, C. R. Acad. Sc. Paris **97** (1883), 848–850, 892–894, 946–949.
13. ———, *Leçons sur la théorie générale des surfaces et les applications géométriques du calcul infinitésimal,* vol. III, Gauthier-Villars, Paris, 1894; Reprinted, *Théorie générale des surfaces*, Chelsea, New York, 1972; Reprinted, Éditions J. Gabay, Paris, 1993.
14. R. K. Dodd and J. D. Gibbon, *The prolongation structure of a higher order Korteweg–de Vries equation*, Proc. Roy. Soc. London A **358** (1977), 287–296.
15. A. P. Fordy, *The Hénon–Heiles system revisited*, Phys. D **52** (1991), 204–210.

16. A. P. Fordy and J. Gibbons, *Some remarkable nonlinear transformations*, Phys. Lett. A **75** (1980), no. 5, 325.
17. B. Gaffet, *A class of* 1d *gas flows soluble by the inverse scattering transform*, Phys. D **26** (1987), 123–139.
18. B. Gambier, *Sur les équations différentielles du second ordre et du premier degré dont l'intégrale générale est à points critiques fixes*, Acta Math. **33** (1910), 1–55.
19. W. Hereman and A. Nuseir, *Symbolic methods to construct exact solutions of nonlinear partial differential equations*, Math. Comp. Simulation **43** (1997), 13–27.
20. R. Hirota, *Direct methods in soliton theory*, Solitons (R. K. Bullough and P. J. Caudrey, eds.), Springer Topics in Current Physics, Springer-Verlag, Berlin, 1980, pp. 157–176.
21. ———, *Fundamental properties of the binary operators in soliton theory and their generalization*, Dynamical Problems in Soliton Systems (S. Takeno, ed.), Springer, Berlin, 1985, pp. 42–49.
22. R. Hirota and A. Ramani, *The Miura transformations of Kaup's equation and of Mikhailov's equation*, Phys. Lett. A **76** (1980), 95–96.
23. R. Hirota and J. Satsuma, *N-soliton solutions of model equations for shallow water waves*, J. Phys. Soc. Japan **40** (1976), 611–612.
24. A. N. W. Hone, *Nonautonomous Hénon–Heiles systems*, Phys. D **118** (1998), 1–16.
25. Xing-biao Hu and Li Yong, *Some resuts on the Caudrey–Dodd–Gibbon–Kotera–Sawada equation*, J. Phys. A **24** (1991), 3205–3212.
26. M. Ito, *An extension of nonlinear evolution equations of the* K-dV (mK-dV) *type to higher orders*, J. Phys. Soc. Japan **49** (1980), 771–778.
27. H. Jonas, *Sopra una classe di trasformazioni asintotiche, applicabili in particolare alle superficie la cui curvatura è proporzionale alla quarta potenza della distanza del piano tangente da uno punto fisso*, Ann. Mat. Pura Appl. **30** (1921), 223–255.
28. D. J. Kaup, *On the inverse scattering problem for cubic eigenvalue problems of the class* $\psi_{xxx} + 6Q\psi_x + 6R\psi = \lambda\psi$, Stud. Appl. Math. **62** (1980), 189–216.
29. D. Levi and O. Ragnisco, *Non-isospectral deformations and Darboux transformations for the third-order spectral problem*, Inverse Problems **4** (1988), 815–828.
30. I. Loris, *On reduced* CKP *equations*, Inverse Problems **15** (1999), 1099–1109.
31. V. B. Matveev and M. A. Salle, *Darboux transformations and solitons*, Springer-Verlag, Berlin, 1991.
32. A. V. Mikhailov, *Integrability of a two-dimensional generalization of the Toda chain*, Pis'ma Zh. Eksp. Teor. Fiz. **30** (1979), 443–448; English transl., Soviet Physics JETP Letters **30** (1979), 414–418.
33. M. Musette, *Painlevé analysis for nonlinear partial differential equations*, The Painlevé Property, One Century Later (R. Conte, ed.), CRM Series in Mathematical Physics, Springer, New York, 1999, pp. 517–572.
34. ———, *Nonlinear superposition formulae of integrable partial differential equations by the singular manifold method*, Direct and Inverse Methods in Nonlinear Evolution Equations (A. Greco, ed.), Springer, Berlin, 2001, about 25 pages.
35. M. Musette and R. Conte, *Algorithmic method for deriving Lax pairs from the invariant Painlevé analysis of nonlinear partial differential equations*, J. Math. Phys. **32** (1991), 1450–1457.
36. ———, *Solitary waves and Lax pairs from polynomial expansions of nonlinear differential equations*, Nonlinear Evolution Equations and Dynamical Systems (M. Boiti, L. Martina, and F. Pempinelli, eds.), World Scientific, Singapore, 1992, 161–170.
37. ———, *The two-singular manifold method,* I. *Modified* KdV *and sine–Gordon equations*, J. Phys. A **27** (1994), 3895–3913.
38. ———, *Bäcklund transformation of partial differential equations from the Painlevé–Gambier classification,* I. *Kaup–Kupershmidt equation*, J. Math. Phys. **39** (1998), 5617–5630.
39. M. Musette and C. Verhoeven, *Nonlinear superposition formula for the Kaup–Kupershmidt partial differential equation*, Phys. D **144** (2000), 211–220.
40. A. Pickering, *The singular manifold method revisited*, J. Math. Phys. **37** (1996), 1894–1927.
41. C. Rogers and S. Carillo, *On reciprocal properties of the Caudrey–Dodd–Gibbon and Kaup–Kupershmidt hierarchies*, Phys. Scripta **36** (1987), 865–869.
42. C. Rogers and W. F. Shadwick, *Bäcklund transformations and their applications*, Academic press, New York, 1982.

43. J. Satsuma and D. J. Kaup, *A Bäcklund transformation for a higher order Korteweg–de Vries equation*, J. Phys. Soc. Japan **43** (1977), 692–697.
44. W. Schief, *The Tzitzéica equation: A Bäcklund transformation interpreted as truncated Painlevé expansion*, J. Phys. A **29** (1996), 5153–5155.
45. ———, *Self-dual Einstein spaces via a permutability theorem for the Tzitzéica equation*, Phys. Lett. A **223** (1996), 55–62.
46. G. Tzitzéica, *Sur une nouvelle classe de surfaces*, C. R. Acad. Sc. Paris **144** (1907), 1257–1259.
47. ———, *Sur une nouvelle classe de surfaces*, Rendiconti del Circolo Matematico di Palermo **25** (1908), 180–187.
48. ———, *Sur une nouvelle classe de surfaces*, C. R. Acad. Sc. Paris **150** (1910), 955–956.
49. ———, *Géométrie différentielle projective des réseaux*, Gauthier-Villars, Paris, 1924.
50. C. Verhoeven, *Licenciaatsverhandeling*, VUB, Brussels, 1999.
51. J. Weiss, *The Painlevé property for partial differential equations.* II: *Bäcklund transformation, Lax pairs, and the Schwarzian derivative*, J. Math. Phys. **24** (1983), 1405–1413.
52. ———, *On classes of integrable systems and the Painlevé property*, J. Math. Phys. **25** (1984), 13–24.
53. ———, *Bäcklund transformation and the Painlevé property*, J. Math. Phys. **27** (1986), 1293–1305.
54. J. Weiss, M. Tabor, and G. Carnevale, *The Painlevé property for partial differential equations*, J. Math. Phys. **24** (1983), 522–526.
55. R. Willox, private communication, 1995.

Dienst Theoretische Natuurkunde, Vrije Universiteit Brussel Pleinlaan 2, B–1050 Brussel, Belgium.

E-mail address, M. Musette: mmusette@vub.ac.be

E-mail address, C. Verhoeven: cverhoev@vub.ac.be

Service de physique de l'état condensé, CEA Saclay F–91191 Gif-sur-Yvette Cedex, France.

E-mail address: conte@amoco.saclay.cea.fr

Centre de Recherches Mathématiques
CRM Proceedings and Lecture Notes
Volume **29**, 2001

Classification of Symmetry-Integrable Evolution Equations

Peter J. Olver, Jan A. Sanders, and Jing Ping Wang

ABSTRACT. This paper describes some recent developments which have made it possible to effectively classify homogeneous systems having infinitely many generalized symmetries, both in the commutative and the noncommutative case. It discusses the program that has to be carried out next to come to an automatic classification mechanism.

1. Introduction

The theory of integrable systems has developed in many directions, and although the interconnections between the different subjects are clearly suggested by the similarity of the results, they are not always so easy to prove or even formulate. The history of the subject experienced two developmental periods. In the first, following the discovery of the Korteweg–de Vries equation, a surprisingly large number of other integrable hierarchies, including mKdV, Sawada–Kotera, Kaup–Kupershmidt, was soon found. However, the second period was more disappointing, as the integrable well quickly dried up, at least in the most basic case of scalar, polynomial evolution equations linear in the highest order derivative. This led to the conjecture that all integrable systems of this particular form had been found. In this paper, we discuss rigorous classification results for both commutative and noncommutative systems, [**SW98b, SW00, OW99**], including a proof of this particular conjecture, and a discussion of the general methods by which such complete classification results are established.

Of the various methods used to characterize integrable differential equations, including existence of infinitely many symmetries and/or conservation laws, soliton solutions, linearization by inverse scattering or differential substitution, Bäcklund transformation, Painlevé property, biHamiltonian structure, recursion operator, formal symmetry of infinite rank, etc., [**Z91, Wan98**], the most fruitful for systematic classification and discovery of new systems has been the characterization integrable systems by the existence a sufficient number of higher order symmetries. The

---

2000 *Mathematics Subject Classification.* Primary: 58G37, 53C50, 53A10; Secondary: 58A15.

P.J.O. supported in part by NSF Grant DMS 98–03154.

J.P.W. supported by a TALENT-stipendium from Netherlands Organization for Scientific Research (NWO).

This is the final form of the paper.

higher order symmetry conditions can be effectively and, in certain cases, completely analyzed by an adaptation of the symbolic method of classical invariant theory, [**Olv99**], after which powerful number-theoretic results on factorizability of polynomials based on Diophantine approximation theory, [**Beu97**], are applied to complete the classification.

In [**SW98b**], it was shown that the integrability of an evolution equation

$$u_t = u_n + f(u, \ldots, u_{n-1}), \quad \text{where } u_n = D_x^n u, \tag{1.1}$$

with f a polynomial starting with terms that are at least quadratic, is determined by the existence of one nontrivial higher order symmetry. This led to the proof of a long-standing conjecture that "in all known cases the existence of one generalized symmetry implies the existence of infinitely many", [**Fok80**], under fairly relaxed conditions. In particular, for homogeneous scalar evolution equations, to prove the integrability of an equation of order 2 we need a symmetry of order 3, for an equation of order 3 we need a symmetry of order 5, for an equation of order 5 we need a symmetry of order 7, and for an equation of order 7 we need a symmetry of order 13, which enable us to give the complete list of integrable homogeneous equations.

The results were quickly generalized to noncommutative polynomial evolution equations of the form (1.1) in which the field variable u takes its values in an associative, noncommutative algebra, [**OW99**]. In this manner, it was rigorously proved that the list of integrable evolution equation in [**OS98**] is complete. These equations can be regarded as quantizing classical integrable systems; see [**FC95**], where the authors treated the Korteweg–de Vries equation.

The classification method consists of several parts: first one has to determine whether a system has an symmetry, or for which parameter values a family of systems has a nontrivial generalized symmetry. Our theorem can then be invoked to show that such system is integrable. The check on the conditions of the theorem involves theoretical results and cannot be done (so far) using only computations. The application of the theorem leads to more calculations, but these involve only lower order terms and can therefore be done effectively for all orders using the symbolic method.

In this paper we will sketch the recent developments of the classification. We will not give the details nor the proofs of the theorems. The classification papers [**SW00, OW99**] may serve as a model for the classification of other types of equations. We will give the lists of (equivalence classes of) integrable equations obtained so far, referring to the relevant papers for the details of how to effectively use these tables.

The method allows not only the classification of integrability with respect to symmetries, but also with respect to other objects like conservation laws. This however turns out to be more complicated than the symmetry analysis. The existence of a nontrivial symmetry is still required, even if one is looking for conservation laws; indeed, one often encounters the equations with only a finite number of conservation laws and no generalized symmetries.

2. Complete Classification Results

In all interesting integrable evolution equations, the right-hand side of equation is a homogeneous differential polynomial under a suitable weighting scheme. The differential equation (1.1) is said to be λ-*homogeneous* of *weight* μ if it admits the one-parameter group of scaling symmetries

$$(x, t, u) \longmapsto (a^{-1}x, a^{-\mu}t, a^{\lambda}u), \quad a \in \mathbb{R}^+.$$

For example, the Korteweg–de Vries equation $u_t = u_{xxx} + uu_x$ is homogeneous of weight 3 for $\lambda = 2$.

Two evolution equations $u_t = K$ and $u_t = Q$ are symmetries of each other if and only if, [**Olv93**],

$$[K, Q] = 0. \tag{2.1}$$

An equation is called *integrable* if it has at infinitely many higher order symmetries.

2.1. Commutative case. In this section, we list all integrable hierarchies which are λ-homogeneous, with $\lambda \geq 0$. The classification theorem states that every λ-homogeneous evolution equation with linear leading term is equivalent, modulo homogeneous transformations in u, to an equation lying in one of the following hierarchies. For $\lambda > 0$ the equivalence transformations are just scalings $u \mapsto \alpha u$, while for $\lambda = 0$ we allow arbitrary change of variables $u \mapsto h(u)$.

• $\lambda = 2$

– Korteweg–de Vries

$$u_t = u_3 + uu_1.$$

– Kaup–Kupershmidt

$$u_t = u_5 + 10uu_3 + 25u_1u_2 + 20u^2u_1.$$

– Sawada–Kotera

$$u_t = u_5 + 10uu_3 + 10u_1u_2 + 20u^2u_1.$$

• $\lambda = 1$

– Burgers'

$$u_t = u_2 + uu_1.$$

– Potential Korteweg–de Vries

$$u_t = u_3 + u_1^2.$$

– Modified Korteweg–de Vries

$$u_t = u_3 + u^2u_1.$$

– Potential Kaup–Kupershmidt

$$u_t = u_5 + 10u_1u_3 + \frac{15}{2}u_2^2 + \frac{20}{3}u_1^3.$$

– Potential Sawada–Kotera

$$u_t = u_5 + 10u_1u_3 + \frac{20}{3}u_1^3.$$

– Kupershmidt equation [**MSS91**, 4.2.6]

$$u_t = u_5 + 5u_1u_3 + 5u_2^2 - 5u^2u_3 - 20uu_1u_2 - 5u_1^3 + 5u^4u_1.$$

• $\lambda = \frac{1}{2}$
– Ibragimov–Shabat [**Cal87**]

$$u_t = u_3 + 3u^2 u_2 + 9uu_1^2 + 3u^4 u_1.$$

• $\lambda = 0$
– Potential Burgers'/Heat equation

$$u_t = u_2 \sim u_t = u_2 + u_1^2.$$

– Potential modified Korteweg–de Vries

$$u_t = u_3 + u_1^3.$$

– Potential Kupershmidt equation

$$u_t = u_5 + 5u_2 u_3 - 5u_1^2 u_3 - 5u_1 u_2^2 + u_1^5.$$

2.2. Noncommutative case. Recently, the analysis of integrable evolution equations in which the field variable u takes its values in an associative, noncommutative algebra, such as matrix, operator, Clifford, and group algebras, has attracted attention. A complete classification for $\lambda > 0$ homogeneous equations with linear leading term was established in [**OW99**]. (The case $\lambda = 0$ poses considerable technical difficulties.) There are only five noncommutative hierarchies, each generalizing one of the preceding commutative hierarchies. Interestingly, whereas the mKdV has two inequivalent noncommutative versions, there is no noncommutative generalization of the Sawada–Kotera, Kaup–Kupershmidt, Kupershmidt, or Ibragimov–Shabat hierarchies.

• $\lambda = 2$
–Korteweg–de Vries

$$u_t = u_3 + uu_1 + u_1 u.$$

• $\lambda = 1$
– Burgers'

$$u_t = u_2 + uu_1 \quad u_t = u_2 + u_1 u.$$

– Potential Korteweg–de Vries

$$u_t = u_3 + u_1^2.$$

– Modified Korteweg–de Vries I

$$u_t = u_3 + u^2 u_1 + u_1 u^2.$$

– Modified Korteweg–de Vries II

$$u_t = u_3 + uu_2 - u_2 u - \tfrac{2}{3} uu_1 u.$$

3. The Symbolic Method

The symbolic method was first introduced by Gel'fand and Dikii, [**GD75**]. It was generalized by Shakiban, [**Sha81, Sha82**], who used it to apply the invariant theory of finite groups to the study of conservation laws of evolution equations, and Ball, Currie, and Olver, [**BCO81, Olv83**], to classify null Lagrangians arising in nonlinear elasticity. In [**Olv83**] the connections with the symbolic method of classical invariant theory were first recognized; see [**Olv99**] for the full details.

The basic idea of the symbolic method is simply to replace u_i, where i is an index—in our case counting the number of derivatives—by ξ^i, where ξ is now a symbol. We see that the basic operation of differentiation, i.e. replacing u_i by

u_{i+1}, is now replaced by multiplication with ξ, as is the case in Fourier transform theory. For higher degree terms with multiple u's, one uses different symbols to denote differentiation; for example, the noncommutative binomial $u_i u_j$ has symbolic form $\xi_1^i \xi_2^j$. In the commutative case, one needs to average over permutations of the differentiation symbols so that $u_i u_j$ and $u_j u_i$ have the same symbolic form. However, in the noncommutative case, this is no longer necessary. In other words, the noncommutative symbolic method works with general polynomials, while in the commutative case one restricts to (multi)-symmetric polynomials.

A differential monomial takes the form $u_I = u_{i_1} u_{i_2} \dots u_{i_k}$. We call k the *degree* of the monomial, $\#I = i_1 + \dots + i_k$ the *index*, and $\max(i_j, j = 1, \dots, k)$ the *order*. We let $\mathcal{U}_n^k$ denote the set of differential polynomials of degree $k+1$ and index n. Let $\mathcal{U}^k = \bigoplus_n \mathcal{U}_n^k$, and $\mathcal{U} = \bigoplus_{n,k\geq 0} \mathcal{U}_n^k$, the algebra of all differential polynomials. The *order* of a differential polynomial is the maximum of the orders of its constituent monomials.

The *transform* or *symbolic form* defines a linear isomorphism between the space $\mathcal{U}^k$ of (non)-commutative differential polynomials of degree $k+1$ and the space $\mathcal{A}^k = \mathbb{R}[\xi_1, \dots, \xi_{k+1}]$ of algebraic polynomials in $k+1$ variables. It is uniquely defined by its action on monomials.

DEFINITION 3.1. The *symbolic form* of a differential monomial is defined as

$$u_{i_1} u_{i_2} \dots u_{i_k} \longmapsto \begin{cases} \xi_1^{i_1} \xi_2^{i_2} \dots \xi_k^{i_k} & (\textit{noncommutative}); \\ \sum_{\pi \in \mathbb{S}^k} \xi_{\pi(1)}^{i_1} \xi_{\pi(2)}^{i_2} \dots \xi_{\pi(k)}^{i_k} & (\textit{commutative}). \end{cases}$$

In general, in analogy with Fourier transforms, we denote the symbolic form of $P \in \mathcal{U}^k$, whether it is commutative or not, by $\widehat{P} \in \mathcal{A}^k$. The transform has two basic properties:

$$\widehat{D_x P}(\xi_1, \dots, \xi_{k+1}) = (\xi_1 + \dots + \xi_{k+1}) \widehat{P}(\xi_1, \dots, \xi_{k+1}),$$

$$\widehat{\frac{\partial P}{\partial u_i}}(\xi_1, \dots, \xi_k) = \frac{1}{i!} \sum_{j=1}^{k+1} \frac{\partial^i \widehat{P}}{(\partial \xi_j)^i}(\xi_1, \dots, \xi_{j-1}, 0, \xi_j, \dots, \xi_k).$$

The following key result is a consequence of these formulae.

PROPOSITION 3.2. Let $K \in \mathcal{U}^m$ and $Q \in \mathcal{U}^n$. Then $D_K(Q) \in \mathcal{U}^{m+n}$, where D_K is the Fréchet derivative of K, and when K and Q are noncommutative,

$$\widehat{D_K[Q]} = \sum_{\tau=1}^{m+1} \widehat{K}\left(\xi_1, \dots, \xi_{\tau-1}, \sum_{\kappa=0}^{n} \xi_{\tau+\kappa}, \xi_{\tau+n+1}, \dots, \xi_{m+n}\right) \widehat{Q}(\xi_\tau, \dots, \xi_{\tau+n});$$

when K and Q are commutative, the right-hand side is needed to be symmetrized.

The following polynomials play a critical role in the analysis.

DEFINITION 3.3. The G-functions are the (commutative) polynomials

$$G_k^{(m)} = \xi_1^k + \dots + \xi_{m+1}^k - (\xi_1 + \dots + \xi_{m+1})^k.$$

The key fact is the following formula for the bracket of a differential polynomial with a linear differential polynomial:

$$\widehat{[u_k, Q]} = G_k^{(m)} \widehat{Q}, \quad \text{whenever } Q \in \mathcal{U}^m. \tag{3.1}$$

This follows directly from Proposition 3.2 and the fact that u_k has symbolic form $\widehat{u_k} = \xi_1^k$. An immediate application is the known result that the space of the symmetries of linear evolution equations $u_t = u_n$ with $n > 1$ is $\mathcal{U}^0$.

The crucial step is the following result, [**Beu97, SW98b**], on the divisibility properties of the G-functions. The proof relies on sophisticated techniques from Diophantine analysis.

PROPOSITION 3.4. We have $G_k^{(m)} = T_k^m H_k^{(m)}$, where $(H_k^{(m)}, H_l^{(m)}) = 1$ for all $2 \le k < l$, and T_k^m is one of the following polynomials:

- $m = 1$:
  - $k = 0 \pmod 2$: $\xi_1\xi_2$.
  - $k = 3 \pmod 6$: $\xi_1\xi_2(\xi_1+\xi_2)$.
  - $k = 5 \pmod 6$: $\xi_1\xi_2(\xi_1+\xi_2)(\xi_1^2+\xi_1\xi_2+\xi_2^2)$.
  - $k = 1 \pmod 6$: $\xi_1\xi_2(\xi_1+\xi_2)(\xi_1^2+\xi_1\xi_2+\xi_2^2)^2$.
- $m = 2$:
  - $k = 0 \pmod 2$: 1.
  - $k = 1 \pmod 2$: $(\xi_1+\xi_2)(\xi_1+\xi_3)(\xi_2+\xi_3)$.
- $m > 2$: 1.

4. Symmetries of λ-Homogeneous Equations

In Section 5 we illustrate the general method given in this section by an elementary example.

Any λ-homogeneous nth order evolution equation can be broken up into its homogeneous components, and so takes the form

$$u_t = K = K^0[u] + K^1[u] + K^2[u] + \dots, \quad \text{where } K^i \in \mathcal{U}^i. \tag{4.1}$$

We assume that $K^0[u] = u_n$ and $0 < \lambda \in \mathbb{Q}$. Note that the index of $K^i[u]$ is $n - i\lambda \ge 0$. When $i\lambda \notin \mathbb{N}$, $K^i[u] = 0$. This reduces the number of relevant λ to a finite set.

Let $S \in \mathcal{U}$ be an mth order symmetry of the evolution equation (4.1). We break up the bracket condition $[S, K] = 0$ into its homogeneous summands, leading to the series of successive symmetry equations

$$\sum_{i+j=r} [S^j, K^i] = 0, \quad \text{for } r = 0, 1, \dots . \tag{4.2}$$

We know that S must have nontrivial linear term, $S^0 \ne 0$, and we can set $S^0 = u_m$ without loss of generality. Clearly we have $[S^0, K^0] = 0$. The next equation to be solved is

$$[u_m, K^1] + [S^1, u_n] = 0, \quad \text{i.e. } G_m^{(1)}\widehat{K^1} = G_n^{(1)}\widehat{S^1}. \tag{4.3}$$

This is trivially satisfied if K has no quadratic terms: $K^1 = 0$. Concentrating on $K^1 \ne 0$, we have the following theorem:

THEOREM 4.1. *Suppose the evolution equation* (4.1) *has a nonzero symmetry S of order $m \ge 2$. Suppose $Q^1 \in \mathcal{U}^1$ is nonzero with the same weight as u_k, where $k \ne m, n$, and k odd if n is odd, that satisfies $[u_n, Q^1] + [K^1, u_k] = 0$. Then there exists a unique symmetry of the form $Q = u_k + Q^1 + Q^2 + \dots$. Moreover, the symmetries Q and S commute.*

We make a very interesting observation. Suppose Q is a nontrivial kth odd order symmetry of (4.1) with odd n, whose quadratic terms have symbolic form:

$$\widehat{Q}^1 = \frac{\widehat{K}^1 G_k^{(1)}}{G_n^{(1)}} = \frac{\widehat{K}^1(\xi_1^2 + \xi_1\xi_2 + \xi_2^2)^{s-s'} H_k^{(1)}}{H_n^{(1)}}.$$

Proposition 3.4 implies that $\lambda \leq 3 + 2\min(s, s')$, where $s' = (n-3)/2 \pmod 3$ and $s = (k-3)/2 \pmod 3$. Then Theorem 4.1 implies that there is a symmetry $Q = u_{2s+3} + Q^1 + \ldots$ of the original equation. The evolution equations defined by Q and K have the same symmetries, so instead of considering K we may consider the equation given by Q, which is of order $q = 2s+3$ for $s = 0$, 1, 2. It follows that we only need to find the symmetries of λ-homogeneous equations (with $\lambda \leq 7$) of order ≤ 7 in order to obtain the complete classification of symmetries of λ-homogeneous scalar polynomial equations starting with linear terms.

A similar observation can be made for even $n > 2$. Suppose we have found a nontrivial symmetry with quadratic term

$$\widehat{Q}^1 = \frac{\widehat{K}^1 G_k^{(1)}}{G_n^{(1)}} = \frac{\widehat{K}^1 G_k^{(1)}}{\xi_1\xi_2 H_n^{(1)}}.$$

This immediately implies $\lambda \leq 2$. Then there is a symmetry $Q = u_2 + Q^1 + \ldots$ of the original equation. Therefore, we only need to find the symmetries of 2nd order equations to get the complete classification of symmetries of λ-homogeneous scalar polynomial equations (with $\lambda \leq 2$) starting with an even linear term.

Finally, we must analyze the case when K has no quadratic terms. Assume that $K^i = 0$ for $i = 1, \ldots, j-1$, and $K^j \neq 0$ for some $j > 1$. In place of (4.3), we now need to solve the leading order equation

$$[u_m, K^j] + [S^j, u_n] = 0.$$

Using (3.1), the symbolic form of this condition is

$$\widehat{S}^j = \frac{\widehat{K}^j G_m^{(j)}}{G_n^{(j)}}. \tag{4.4}$$

Proposition 3.4 implies that this polynomial identity has no solutions when $j \geq 3$, or when $j = 2$ and n is even, since $G_m^{(j)}$ and $G_n^{(j)}$ have no common factors, and the degree of K^j is $n - j\lambda < n$, which is the degree of $G_n^{(j)}$. This implies that there are no symmetries for such equations. When $j = 2$ and n is odd, the equation can only have odd order symmetries, and we have the similar theorem as Theorem 4.1 in this case. Moreover, if the equation (4.4) can be solved for any m, it can also be solved for $m = 3$. By now, we have proved the following:

THEOREM 4.2. *A nontrivial symmetry of a λ-homogeneous equation is part of a hierarchy starting at order* 2, 3, 5 *or* 7.

Only an equation with nonzero quadratic or cubic terms can have a nontrivial symmetry. For such λ, we must find a third order symmetry for a second order equation, a fifth order symmetry for a third order equation, a seventh order symmetry for a fifth order equation with quadratic terms, and the thirteenth order symmetry for a seventh order equation with quadratic terms. It remains to analyze each of these particular cases in detail. The last case can be easily reduced to the case of fifth order equations by determining the quadratic terms of the equation. A straightforward computation, done with the help of MAPLE, completes the proof of our lists. The details of the computation are completed as in the commutative case described in [**SW98a**].

5. Computational Example

To illustrate how the symbolic method works, we give the symbolic calculation for the fifth order symmetry of the Korteweg–de Vries equation. When one computes a symmetry, the natural approach is to do this degree by degree. So for instance, if we have as the equation

$$u_t = K = K^0 + K^1 = u_3 + uu_1 \quad \text{(KdV)},$$

then we compute as a symmetry

$$S = S^0 + S^1 + \cdots = u_5 + a_1 uu_3 + a_2 u_1 u_2 + \ldots .$$

We have to solve $[K^0, S^1] + [K^1, S^0] = 0$, i.e.,

$$\begin{aligned} &D_x^3 S_1 + uD_x S_0 + u_1 S_0 \\ &\qquad = D_x^5 K^1 + a_1 u D_x^3 K^0 + a_1 u_3 K^0 + a_2 u_1 D_x^2 K^0 + a_2 u_2 D_x K^0. \end{aligned}$$

If we translate this to the symbolic method we obtain

$$(\xi_1+\xi_2)^3 \widehat{S}^1 + (\xi_1^5+\xi_2^5)\widehat{K}^1 = (\xi_1+\xi_2)^5 \widehat{K}^1 + (\xi_1^3+\xi_2^3)\widehat{S}^1.$$

Thus we can formally solve

$$\widehat{S}^1 = \frac{(\xi_1+\xi_2)^5 - \xi_1^5 - \xi_2^5}{(\xi_1+\xi_2)^3 - \xi_1^3 - \xi_2^3} \widehat{K}^1,$$

and this is a real solution if $\widehat{S}^1$ turns out to be a polynomial. Thus we have translated our problem into the following question. If we let

$$G_n^{(1)}(\xi_1, \xi_2) = (\xi_1+\xi_2)^n - \xi_1^n - \xi_2^n,$$

then which factors do $G_n^{(1)}$ and $G_m^{(1)}$ have in common? Using the results in Proposition 3.4, we can determine whether the symmetry we are looking for exists or not. In this case, the answer is simple, that is,

$$\widehat{S}^1 = \frac{5}{3}(\xi_1^2 + \xi_1\xi_2 + \xi_2^2), \quad \widehat{K}^1 = \frac{5}{6}(\xi_1^3 + 2\xi_1^2\xi_2 + 2\xi_1\xi_2^2 + \xi_2^3)u^2.$$

Let us compute S^2 by solving $[S^1, K^1] + [S^2, K^0] = 0$. By Proposition 3.2, this leads to

$$\widehat{S}^2 = \frac{5}{6} \frac{(\xi_1+\xi_2)(\xi_2+\xi_3)(\xi_1+\xi_3)(\xi_1+\xi_2+\xi_3)}{(\xi_1+\xi_2+\xi_3)^3 - \xi_1^3 - \xi_2^3 - \xi_3^3} u^3 = \frac{5}{18}(\xi_1+\xi_2+\xi_3)u^3.$$

Note that $[S^2, K^1] = 0$ next degree. Therefore, the fifth order symmetry is

$$S = S^0 + S^1 + S^2 = u_5 + \frac{5}{3}uu_3 + \frac{10}{3}u_1u_2 + \frac{5}{6}u^2u_1.$$

This illustrates both the simplification induced by the symbolic method as well as the role of the G-functions in the whole analysis.

6. Future Developments

The reason that the basic technique works in both commutative and noncommutative situations is simple: in the linear term one does not see any difference between the two, c.f. formula (4.3), and so the same polynomials arise in the symbolic versions. The differences between the two cases only arise during the final, detailed analysis that leads to the complete classification.

There are a number of obvious extensions to the results obtained thus far.

- Systems. There has been a lot of work on classifying certain systems of evolution equations, both in the commutative, [**MSS91**], and noncommutative, [**OS98**], cases. However, these results are far from providing a definitive list of symmetry-integrable systems. Note that the G-functions will depend on the eigenvalues of the linear part of the system, and it is not clear a priori how the results of F. Beukers, [**Beu97**], generalize. Nevertheless, there is some progress in this direction too. In [**BSW98**], it was rigorously proved that an example due Bakirov of a fourth order system of two coupled evolution equations only possesses one nontrivial symmetry, of order six. This shows that, for systems, one higher order symmetry does not necessarily imply infinitely many, and hence integrability. However, Bakirov's example does not violate the more refined version of the conjecture, [**Fok87**], that a system of m evolution equations requires m higher order symmetries in order to be integrable.
- Scalar case with $\lambda < 0$. Some results are known for $\lambda = -1$, which were obtained from the case $\lambda = 0$. Although there seem to be many more integrable systems for $\lambda \leq 0$, this is compensated for by the fact that there are also many more homogeneous transformations. The results for $\lambda = 0$ indicate that one might expect a finite list of symmetry-integrable equations.
- Other algebraic structures. One can think of cyclic commutativity as described in [**Rot99**], where

  $$D_{u^2u_1}[h] = huu_1 + uhu_1 + u^2h_1 \equiv uu_1h + u_1uh + u^2h_1.$$

  This seems to be the right kind of differentiation for the nonpolynomial case, which otherwise seems next to impossible.
- Another direction is to allow coefficients in the equations which do not commute with the field variables.

References

[BCO81] J. M. Ball, J. C. Currie, and P. J. Olver, *Null Lagrangians, weak continuity, and variational problems of arbitrary order*, J. Funct. Anal. **41** (1981), no. 2, 135–174.

[Beu97] F. Beukers, *On a sequence of polynomials*, J. Pure Appl. Algebra **117/118** (1997), 97–103.

[BSW98] F. Beukers, J. A. Sanders, and J. P. Wang, *One symmetry does not imply integrability*, J. Differential Equations **146** (1998), no. 1, 251–260.

[Cal87] F. Calogero, *The evolution partial differential equation* $u_t = u_{xxx} + 3(u_{xx}u^2 + 3u_x^2u) + 3u_xu^4$, J. Math. Phys. **28** (1987), no. 3, 538–555.

[Fok80] A. S. Fokas, *A symmetry approach to exactly solvable evolution equations*, J. Math. Phys. **21** (1980), no. 6, 1318–1325.

[Fok87] ______, *Symmetries and integrability*, Stud. Appl. Math. **77** (1987), no. 3, 253–299.

[FC95] B. Fuchssteiner and A. R. Chowdhury, *A new approach to the quantum* KdV. *Solitons in science and engineering: theory and applications*, Chaos Solitons Fractals **5** (1995), no. 12, 2345–2355.

[GD75] I. M. Gel'fand and L. A. Dikiĭ, *Asymptotic properties of the resolvent of Sturm–Liouville equations, and the algebra of Korteweg–de Vries equations*, Uspehi Mat. Nauk **30** (1975), no. 5 (185), 67–100; English transl., Russian Math. Surveys **30** (1975), no. 5, 77–113.

[MSS91] A. V. Mikhaĭlov, A. B. Shabat, and V. V. Sokolov, *The symmetry approach to classification of integrable equations*, What is Integrability?, Springer Ser. Nonlinear Dynamics, Springer, Berlin, 1991, pp. 115–184.

[Olv83] P. J. Olver, *Hyper-Jacobians, determinantal ideals and weak solutions to variational problems*, Proc. Roy. Soc. Edinburgh Sect. A **95** (1983), nos. 3-4, 317–340.

[Olv93] ———, *Applications of Lie groups to differential equations*, 2nd ed., Grad Texts in Math. **107**, Springer-Verlag, New York, 1993.

[Olv99] ———, *Classical invariant theory*, London Math. Soc. Stud. Texts **44**, Cambridge University Press, Cambridge, 1999.

[OS98] P. J. Olver and V. V. Sokolov, *Integrable evolution equations on associative algebras*, Comm. Math. Phys. **193** (1998), no. 2, 245–268.

[OW99] P. J. Olver and J. P. Wang, *Classification of integrable one-component systems on associative algebras*, Proc. London Math. Soc. **81** (2000), no. 3, 566–586.

[Rot99] G.-C. Rota, *Combinatorial snapshots*, Math. Intelligencer **21** (1999), no. 2, 8–14.

[Sha81] C. Shakiban, *A resolution of the Euler operator*. II, Math. Proc. Cambridge Philos. Soc. **89** (1981), no. 3, 501–510.

[Sha82] ———, *An invariant theoretic characterization of conservation laws*, Amer. J. Math. **104** (1982), no. 6, 1127–1152.

[SW98a] J. A. Sanders and J. P. Wang, *Combining maple and form to decide on integrability questions*, Comput. Phys. Comm. **115** (1998), nos. 2-3, 447–459.

[SW98b] ———, *On the integrability of homogeneous scalar evolution equations*, J. Differential Equations **147** (1998), no. 2, 410–434.

[SW00] ———, *On the integrability of nonpolynomial scalar evolution equations*, J. Differential Equations **166** (2000), no. 1, 132–150.

[Wan98] J. P. Wang, *Symmetries and conservation laws of evolution equations*, Ph.D. thesis, Vrije Universiteit, Amsterdam, 1998.

[Z91] V. E. Zakharov, *What is integrability?*, Springer Ser. Nonlinear Dynam., Springer-Verlag, Berlin, 1991.

School of Mathematics, University of Minnesota, Minneapolis, MN 55455, USA.
E-mail address: `olver@math.umn.edu`
URL: `http://www.math.umn.edu/~olver`

Vrije Universiteit, Faculty of Sciences, Division of Mathematics and Computer Science, Amsterdam, The Netherlands.
E-mail address: `jansa@cs.vu.nl`

School of Mathematics, University of Minnesota, Minneapolis, MN 55455, USA.
Current address: Vrije Universiteit, Faculty of Sciences, Division of Mathematics and Computer Science, Amsterdam, The Netherlands.
E-mail address: `wang@cs.vu.nl`

Centre de Recherches Mathématiques
CRM Proceedings and Lecture Notes
Volume **29**, 2001

Integrability of Evolution Equations and Pseudo-Spherical Surfaces

Enrique G. Reyes

ABSTRACT. It is pointed out that a recent classification of evolution equations

$$u_t = F(x, t, u, u_x, \ldots, u_{x^k}), \tag{0.1}$$

which describe one-parameter families of pseudo-spherical surfaces (*geometric* integrability) provides a systematic procedure to determine a one-parameter family of $\mathrm{sl}(2, \mathbf{R})$-valued linear problems for which (0.1) is the integrability condition (*kinematic* integrability). This observation is used to prove that *every* autonomous second order evolution equation which admits a formal symmetry of infinite rank (*formal* integrability) is kinematically integrable. It is also shown here that this result cannot be extended as proven to third order formally integrable equations. Some work in progress on the converse of the theorem referred to above is also reported.

1. Introduction

When is a partial differential equation integrable? In his talk, Peter Olver calls an evolution equation "integrable" if it possesses generalized symmetries of arbitrarily large order (see also Olver [**14**]), while on the other hand, Fokas, Finkel and Gürses in their respective contributions call an equation (not necessarily of evolutionary type) "integrable" if it is the integrability condition of a one-parameter family of linear problems (see also Fokas, Gel'fand, Finkel, and Liu [**6**]). It is certainly a reasonable endeavour to compare the different notions of integrability one finds in the literature and, if possible, to build "bridges" between them. We will concern ourselves with

(a) kinematic (Faddeev and Takhtajan [**5**]),
(b) formal (Mikhailov, Shabat and Yamilov [**12**]), and
(c) geometric (Chern and Tenenblat [**3**])

2000 *Mathematics Subject Classification.* Primary: 35Q51, 37K10; Secondary: 37K25.

The author thanks Robert Milson for bringing the papers by Terng and Uhlenbeck to his attention. He also thanks the Centre de recherches mathématiques (Université de Montréal) and McGill University for their warm hospitality while this report was being written.

The author holds a post-doctoral fellowship from the Natural Sciences and Engineering Research Council of Canada.

This is the final form of the paper.

integrability.

The main result to be reported here is that formal and kinematic integrability are related, and that it is Chern and Tenenblat's notion what links them. In more precise words, it will be pointed out that *every* autonomous second order evolution equation which is formally integrable is geometrically integrable, that this result cannot be extended as proven to third order formally integrable evolution equations, and that geometric and kinematic integrability are equivalent.

The paper is organized as follows. Kinematic and formal integrability are rigorously introduced in Section 2, following Reyes [**16**]. Geometric integrability is considered in Section 3, and the theorems relating these three notions are discussed in Section 4. Sections 3 and 4 are based on Reyes [**15,16**]. Further remarks, including a short discussion on some work in progress (Reyes [**17**]) appear in Section 5.

The language of jet bundles and prolongations (Krasil'shchik and Verbovetsky [**10**], Olver [**14**]) will be used throughout. Thus, we will fix once and for all a trivial bundle $E \to M$ with base the space M of independent variables, and typical fiber the space of the dependent variable u, and moreover,

(a) a scalar differential equation $\Xi = 0$ in two independent variables x, t, will be often identified with a sub-bundle S^∞ of the infinite jet bundle $J^\infty E$ of E;
(b) the local solutions of $\Xi = 0$ will be identified with local holonomic sections $j^\infty(s)$ of S^∞, in which s is a local section of E; and
(c) if considered as jet coordinates, the function u and its derivatives will be denoted by

$$z_0 := u, \quad z_{0,x^m t^n} := \frac{\partial^{m+n} u}{\partial x^m t^n} \quad \text{and} \quad z_j := \frac{\partial^j u}{\partial x^j} \quad \text{for } j, m, n \geq 1. \tag{1.1}$$

REMARK 1.1. We would like to stress the fact that geometric integrability is particularly interesting from the point of view of this Workshop: Given two equations possessing this structure, a solution of one of them (satisfying a suitably "generic condition", see Section 3 below) determines, under a mild technical condition, a solution of the other (Kamran and Tenenblat [**9**]) so that geometrically integrable equations admit "generalized" Bäcklund transformations!

2. Kinematic and Formal Integrability

We define kinematic integrability first, following Faddeev and Takhtajan [**5**].

DEFINITION 2.1. A differential equation $\Xi = 0$ (not necessarily of evolutionary type) is kinematically integrable if it is the integrability condition of a nontrivial one-parameter family of linear problems

$$\frac{\partial v}{\partial x} = U(\eta)v, \quad \frac{\partial v}{\partial t} = V(\eta)v, \tag{2.1}$$

in which, for each η, $U(\eta)$ and $V(\eta)$ are smooth $\mathrm{sl}(2, \mathbf{R})$-valued functions on $J^\infty E$.

It follows that a differential equation $\Xi = 0$ is kinematically integrable if and only if

(a) the matrix equation

$$\frac{\partial U(\eta)}{\partial t} - \frac{\partial V(\eta)}{\partial x} + [U(\eta), V(\eta)] = 0, \tag{2.2}$$

is identically satisfied whenever $z_0(x,t)$ is a solution of $\Xi = 0$, and

(b) the matrices $U(\eta)$ and $V(\eta)$ satisfy trace $U(\eta) = \text{trace } V(\eta) = 0$.

Equations (2.1) and (2.2) can be interpreted geometrically by means of the *soliton connection* discussed by Crampin, Pirani, and Robinson [**4**], and Faddeev and Takhtajan [**5**].

We will say that a differential equation is *strictly* kinematically integrable if it is kinematically integrable and the diagonal entries of the matrix $U(\eta)$ appearing in (2.1) are η and $-\eta$. For example, equations integrable by the AKNS method are kinematically integrable in the strict sense but, in principle, the cylindrical KdV equation is not (Reyes [**16**]). If an equation is integrable in the strict sense, it is customary to call η the "spectral" parameter, in agreement with the fact that a strictly kinematically integrable equation is the integrability condition of a linear problem of the form

$$\left(\frac{\partial}{\partial x} - U_0\right)v = \eta\,\mathrm{diag}(1,-1)v, \qquad \frac{\partial v}{\partial t} = V(\eta)v,$$

in which $U_0 = U(\eta) - \eta\,\mathrm{diag}(1,-1)$, and that therefore one may hope to solve it by inverse scattering techniques.

We now consider formal integrability. This notion is based on the concept of a "formal symmetry" (Mikhailov, Shabat and Yamilov [**12**]; Mikhailov, Shabat and Sokolov [**11**]) which will not be reviewed here in any generality: only the particular case of a "formal symmetry of infinite rank" is relevant for this paper. Before stating its rigorous definition, we recall the following [**11, 14**]:

(1) The total derivatives with respect to x and t on $J^\infty E$ are the operators

$$D_x = \frac{\partial}{\partial x} + \sum_{m,n} z_{0,x^{m+1}t^n}\frac{\partial}{\partial z_{0,x^m t^n}}, \quad \text{and} \quad D_t = \frac{\partial}{\partial t} + \sum_{m,n} z_{0,x^m t^{n+1}}\frac{\partial}{\partial z_{0,x^m t^n}}. \tag{2.3}$$

(2) If $z_{0,t} = F$ is an evolution equation with equation manifold S^∞, the total derivatives with respect to x and t restricted to S^∞ are the operators

$$D_x = \frac{\partial}{\partial x} + \sum_{k=0}^{\infty} z_{k+1}\frac{\partial}{\partial z_k} \quad \text{and} \quad D_t = \frac{\partial}{\partial t} + \sum_{k=0}^{\infty} D_x^k(F)\frac{\partial}{\partial z_k}. \tag{2.4}$$

(3) If $z_{0,t} = F$ is an evolution equation with equation manifold S^∞ and the function $f(x,t,z_0,\ldots,z_k)$ is the restriction of a smooth function on $J^\infty E$ to S^∞, the *formal linearization* of f is the operator f_* on S^∞ given by

$$f_* = \sum_{i=0}^{k} \frac{\partial f}{\partial z_i} D_x^i. \tag{2.5}$$

Formal symmetries of infinite rank are now introduced thus:

DEFINITION 2.2. Let $z_{0,t} = F(x, z_0, \ldots, z_m)$ be an *autonomous* evolution equation with equation manifold S^∞. A formal symmetry of infinite rank is a formal pseudo-differential operator

$$\Psi = \sum_{-\infty<k\le N} f_k D_x^k, \tag{2.6}$$

on $J^\infty E$ such that the Lax-type equation

$$\Psi_t = [F_*, \Psi], \tag{2.7}$$

in which F_* is the formal linearization of F and Ψ_t is the formal pseudo-differential operator

$$\Psi_t = \sum_{-\infty < k \leq N} D_t(f_k) D_x^k ,$$

holds identically on S^∞.

Equation (2.7) characterizes the important *recursion operators* first introduced by Olver (Olver [**14**] and references therein). It guarantees that if G is a (generalized) symmetry of the equation $z_{0,t} = F$, so is $\Psi(G)$. Of course, one is assuming here that $\Psi(G)$ is a well defined smooth function on $J^\infty E$, a nontrivial hypothesis (that this hypothesis does hold in many interesting cases has been proven recently by Wang [**24**]). In principle, nonetheless, Ψ generates infinite sequences $\Psi^k(G)$, $k \geq 1$, of generalized symmetries of the equation $z_{0,t} = F$.

One can also prove (Mikhailov, Shabat and Yamilov [**12**], Olver [**14**]) the following two important facts:

(a) if Ψ (given by (2.6)) satisfies (2.7), the coefficient f_{-1} is a conserved density of the equation $z_{0,t} = F$, and
(b) if Ψ satisfies (2.7), so does $\Psi^{(m/n)}$ for every m and $n > 0$. Thus, if an autonomous evolution equation possesses a formal symmetry of infinite rank, it has an infinite number of local conservation laws. It does not follow, however, that these conservation laws are nontrivial: Burgers' equation is formally integrable, but it does not have nontrivial local conservation laws other than itself.

These remarks on symmetries and conservation laws of autonomous evolution equations possessing formal symmetries of infinite rank suggest that the existence of a nontrivial (in the sense of not being simply multiplication by a constant) formal pseudo-differential operator Ψ satisfying (2.7) may be certainly advocated as an indicator of integrability. It appears natural then to make the following definition:

DEFINITION 2.3. An autonomous evolution equation $z_{0,t} = F(x, z_0, \ldots, z_m)$ is formally integrable if it possesses a nontrivial formal symmetry of infinite rank.

REMARK 2.4. Formal integrability has been studied in the context of nonautonomous evolution equations by Abellanas and Galindo [**1**], Svinolupov and Sokolov [**19**], and Hernández Heredero, Sokolov and Svinolupov [**8**].

3. Integrability *à la* Chern–Tenenblat

The notion of integrability introduced by Chern and Tenenblat [**3**], what we call *geometric* integrability, is based on the following structure:

DEFINITION 3.1. Let Ξ be a smooth function on the infinite jet bundle $J^\infty E$ equipped with local coordinates determined by (1.1), and consider the differential equation

$$\Xi(x, t, z_0, \ldots, z_{0,x^m t^n}) = 0. \tag{3.1}$$

Equation (3.1) describes pseudo-spherical surfaces if there exist smooth functions $f_{\alpha\beta}$ ($\alpha = 1, 2, 3; \beta = 1, 2$) on $J^\infty E$ for which $f_{11} f_{22} - f_{12} f_{21}$ is not identically zero, such that the pull-back of the one-forms $\omega^\alpha = f_{\alpha 1}\, dx + f_{\alpha 2}\, dt$ by local solutions of (3.1), $\overline{\omega}^\alpha$ say, satisfy the structure equations of a surface of constant Gaussian curvature -1 with metric $(\overline{\omega}^1)^2 + (\overline{\omega}^2)^2$ and connection one-form $\overline{\omega}^3$, namely,

$$d\overline{\omega}^1 = \overline{\omega}^3 \wedge \overline{\omega}^2, \quad d\overline{\omega}^2 = \overline{\omega}^1 \wedge \overline{\omega}^3, \quad \text{and} \quad d\overline{\omega}^3 = \overline{\omega}^1 \wedge \overline{\omega}^2. \tag{3.2}$$

REMARK 3.2. (a) Definition 3.1 can be extended *mutatis mutandis* to the case in which $\Xi = 0$ represents a system of partial differential equations in two independent variables.

(b) The expression "PSS equation" will be sometimes utilized as an abbreviation of the phrase "equation describing pseudo-spherical surfaces".

One may think of the pseudo-spherical surfaces mentioned in the definition as follows. If $s\colon (x,t) \mapsto \big(x, t, z_0(x,t)\big)$ is a local section of E, its domain can be considered as the domain of the inverse of a local chart on the graph Q of the prolonged section $j^\infty(s)$ of $J^\infty E$. If $j^\infty(s)$ is a *generic* solution of an equation $\Xi = 0$ describing pseudo-spherical surfaces with associated one-forms ω^α, that is, if $j^\infty(s)$ is a section of the equation manifold S^∞ of $\Xi = 0$ and (Kamran and Tenenblat [**9**])

$$\big(j^\infty(s)\big)^*(\omega^1 \wedge \omega^2) \neq 0,$$

its graph Q is a submanifold of S^∞ possessing the structure of a pseudo-spherical surfacc, and this structure is determined in (x,t) coordinates by the one-forms $\overline{\omega}^\alpha := (j^\infty(s))^*\omega^\alpha$.

Why is this notion interesting? First, classical Bäcklund transformations, symmetries, and conservation laws of PSS equations may be studied by means of the geometry of pseudo-spherical surfaces (Tenenblat [**21**], Reyes [**16**] and references therein). Second, we will seen in Section 4 that if an equation describes a non-trivial one-parameter family of pseudo-spherical surfaces, it is kinematically integrable and vice versa. Third, as anticipated in the Introduction, if two equations $\Xi(x,t,z_0,\dots) = 0$ and $\overline{\Xi}(\overline{x},\overline{t},\overline{z}_0,\dots) = 0$ describe pseudo-spherical surfaces with associated one-forms $\omega^\alpha = f_{\alpha 1}\,dx + f_{\alpha 2}\,dt$ and $\overline{\omega}^\alpha = \bar{f}_{\alpha 1}\,d\overline{x} + \bar{f}_{\alpha 2}\,d\bar{t}$, and $\bar{f}_{11} = G(\overline{z}_0)$, $G'(\overline{z}_0) \neq 0$, a generic solution of one of them will determine a generic solution of the other. Kamran and Tenenblat [**9**] found this striking generalization of Bäcklund transformations motivated by the classical fact that all surfaces of constant Gaussian curvature -1 are locally indistinguishable.

REMARK 3.3. In their interesting Workshop lectures, Gürses, Fokas, Finkel, Rogers, Schief, and Cieśliński consider an *extrinsic* approach in which (systems of) nonlinear differential equations determine immersed surfaces by means of (generalizations of) Sym's formula (Sym [**20**], Fokas, Gel'fand, Finkel and Liu [**6**]) or appear as the necessary and sufficient condition for the existence of a local embedding of a surface into a three-dimensional space (Sym [**20**], Gürses and Nutku [**7**]). In Chern and Tenenblat's [**3**] approach, one starts with an equation $\Xi = 0$, and then looks for associated one-forms ω^α satisfying (3.2). If one finds them, one obtains pseudo-spherical surfaces described *intrinsically* by the one-forms $\overline{\omega}^\alpha := (j^\infty(s))^*\omega^\alpha$, in which $j^\infty(s)$ is a generic solution of $\Xi = 0$.

We are now ready to introduce geometric integrability:

DEFINITION 3.4. A differential equation (or system of equations) is geometrically integrable (or integrable *à la* Chern–Tenenblat) if it describes a nontrivial one-parameter family of pseudo-spherical surfaces.

Examples of geometrically integrable equations are the equations integrable by two-dimensional AKNS scattering (whenever their associated linear problems are real, see Sasaki [**18**] and Chern and Tenenblat [**3**]) the Burgers equation (see (4.9)–(4.12) below), and all the members of the KdV hierarchy. In particular, the KdV equation

$$z_{0,t} = z_3 + z_0 z_1, \tag{3.3}$$

is geometrically integrable, as it describes pseudo-spherical surfaces with associated one-forms ω^α, $\alpha = 1, 2, 3$, given by

$$\omega^1 = (1 - z_0)\, dx + \left(-z_2 + \eta z_1 - \eta^2 z_0 - \frac{1}{3} z_0^2 + \eta^2 - \frac{2}{9} z_0 + \frac{5}{9}\right) dt, \tag{3.4}$$

$$\omega^2 = \eta\, dx + \left(\eta^3 + \frac{1}{3}\eta z_0 - \frac{1}{3} z_1 + \frac{5}{9}\eta\right) dt, \tag{3.5}$$

$$\omega^3 = \left(\frac{2}{3} - z_0\right) dx + \left(-z_2 + \eta z_1 - \eta^2 z_0 - \frac{1}{3} z_0^2 + \frac{2}{3}\eta^2 - \frac{1}{3} z_0 + \frac{10}{27}\right) dt, \tag{3.6}$$

in which η is a parameter. Further examples appear in the classical paper by Chern and Tenenblat [**3**], Reyes [**15, 16**], the recent monograph by Keti Tenenblat [**21**], and references therein.

4. The Main Result

Let us consider kinematic and geometric integrability first. A differential equation $\Xi = 0$ is kinematically integrable *if and only if* it describes a nontrivial one-parameter family of pseudo-spherical surfaces, that is, if and only if it is geometrically integrable, while strict kinematic integrability amounts to the condition $f_{21} = \eta$. Indeed, one easily sees that Equations (3.2) hold if and only if the linear problem

$$\begin{pmatrix} dv_1 \\ dv_2 \end{pmatrix} = \frac{1}{2} \begin{pmatrix} \omega_2 & \omega_1 - \omega_3 \\ \omega_1 + \omega_3 & -\omega_2 \end{pmatrix} \begin{pmatrix} v_1 \\ v_2 \end{pmatrix}, \tag{4.1}$$

is integrable whenever $z_0(x,t)$ is a solution of $\Xi = 0$, and conversely, any family of linear problems (2.1) satisfying (2.2) whenever $z_0(x,t)$ is a solution of $\Xi = 0$, determines a one-parameter family of one-forms ω^α satisfying (3.2) whenever $z_0(x,t)$ is a solution of $\Xi = 0$, by setting, using an obvious notation, $\omega^1 = (U_{21} + U_{12})\, dx + (V_{21} + V_{12})\, dt$, $\omega^2 = 2U_{11}\, dx + 2V_{11}\, dt$, and $\omega^3 = (U_{21} - U_{12})\, dx + (V_{21} - V_{12})\, dt$.

REMARK 4.1. The one-forms ω^α appearing in Definition 3.1 must satisfy

$$\omega^1 \wedge \omega^2 \not\equiv 0, \tag{4.2}$$

a condition which is not clear it holds for the one-forms ω^α given above. However, one can modify the family of linear problems (2.1)—without changing its integrability condition, of course—in such a way that (4.2) is indeed satisfied. In terms of Crampin, Pirani, and Robinson's [**4**] soliton connection, one just applies an adequate gauge transformation (Reyes [**16**]) to the connection determined by (2.1).

Now we discuss kinematic and formal integrability. Our first theorem is the following (Reyes [**15**]):

THEOREM 4.2. *Every autonomous second order evolution equation*

$$z_{0,t} = F(x, z_0, z_1, z_2),$$

which is formally integrable, is kinematically integrable. Moreover, up to invertible contact transformations of the form

$$x' = \phi(x, z_0, z_1), \quad t' = t, \qquad z_0' = \psi(x, z_0, z_1), \tag{4.3}$$

in which

$$(D_x\phi)\psi_{z_1} = (D_x\psi)\phi_{z_1}, \quad \psi_{z_0}\phi_{z_1} - \phi_{z_0}\psi_{z_1} \neq 0, \tag{4.4}$$

there exists an exhaustive list of representatives of autonomous formally integrable second order equations which are kinematically integrable in the strict sense.

PROOF. (Sketch) Mikhailov, Shabat and Yamilov [**12**, Theorem 5.1] provide an exhaustive list of formally integrable autonomous second order equations. Up to contact transformations of the form (4.3) and (4.4) they are:

$$z_{0,t} = z_2 + a(x)z_0, \tag{4.5}$$

$$z_{0,t} = z_2 + z_0 z_1 + h(x), \tag{4.6}$$

$$z_{0,t} = D_x(z_1 z_0^{-2} + \alpha x z_0 + \beta z_0), \quad \alpha, \beta \in \mathbf{R}, \tag{4.7}$$

$$z_{0,t} = D_x(z_1 z_0^{-2} + x). \tag{4.8}$$

Equations (4.5)–(4.8) *are* geometrically integrable, and their associated one-forms ω^α satisfy $f_{21} = \eta$! These one-forms are found thus. One first *classifies* (Reyes [**15**]) all the equations of the form $z_{0,t} = F(x, t, z_0, \ldots, z_k)$ which describe pseudo-spherical surfaces, under the assumptions that $f_{21} = \eta$, a real parameter, *and* that the equation at hand is *equivalent* to the structure equations (3.2) (thereby generalizing an earlier classification by Chern and Tenenblat [**3**], who considered evolution equations $z_{0,t} = F$ without explicit x, t dependence). One then checks that equations (4.5)–(4.8) appear in that classification. For instance, for the generalized Burgers' equation (4.6), one finds that the η-dependent one-forms

$$\omega^1 = \left(\frac{1}{2}z_0 - \frac{\beta}{\eta}\right) dx + \left(\frac{1}{2}z_1 + \frac{1}{4}z_0^2 + \frac{1}{2}\int h(x)\, dx\right) dt, \tag{4.9}$$

$$\omega^2 = \eta\, dx + \left(\frac{\eta}{2}z_0 + \beta\right) dt, \tag{4.10}$$

$$\omega^3 = -\eta\, dx + \left(-\frac{\eta}{2}z_0 - \beta\right) dt, \tag{4.11}$$

in which β is a solution of the Riccati equation

$$\beta^2 - \eta\beta_x + \frac{\eta^2}{2}\int h(x)\, dx = 0, \tag{4.12}$$

satisfy the structure equations (3.2) whenever $z_0(x, t)$ is a solution of (4.6). □

We pause here to comment on the classification alluded to above. The results one obtains have the following form: An evolution equation $z_{0,t} = F(x, t, z_0, \ldots, z_k)$ describes pseudo-spherical surfaces with associated one-forms $\omega^\alpha = f_{\alpha 1}\, dx + f_{\alpha 2}\, dt$ satisfying $f_{21} = \eta$ if and only if:

(a) the right-hand side of the equation $z_{0,t} = F$ can be written in terms of the functions $f_{\alpha\beta}$ and their derivatives; and

(b) some of the functions $f_{\alpha\beta}$ are described in terms of the others.

For instance, in a generic case, one obtains the following expression for F (Reyes [**15**, Theorem 1]):

$$
\begin{aligned}
F = {} & \frac{1}{L}\sum_{i=1}^{k-1} z_{i+1}(B + f_{22,x})_{z_i} \\
& + \frac{1}{HL}(f_{31}^2 - f_{11}^2)\Big\{\sum_{i=0}^{k-2} z_{i+1}A^i + T - f_{31,z_0}f_{12,x} + f_{11,z_0}f_{32,x}\Big\} \\
& + \frac{f_{11}}{L}(z_1 f_{32,z_0} + f_{32,x}) - \frac{f_{31}}{L}(z_1 f_{12,z_0} + f_{12,x}) + \frac{\eta}{H}(B + f_{22,x}) \\
& - \frac{1}{L}(f_{31,t}f_{11} - f_{11,t}f_{31}),
\end{aligned} \tag{4.13}
$$

while the z_j-derivatives of f_{12} and f_{32}, $1 \le j \le k-1$, are given in terms of f_{11}, f_{22} and f_{31} by the formulas

$$
f_{12,z_j} = -\frac{1}{L}\{f_{11}A^j - f_{11,z_0}(B + f_{22,x})_{z_j}\}, \tag{4.14}
$$

and

$$
f_{32,z_j} = -\frac{1}{L}\{f_{31}A^j - f_{31,z_0}(B + f_{22,x})_{z_j}\}, \tag{4.15}
$$

and further equations for their z_0-derivatives, and the functions f_{12} and f_{32} themselves must be also satisfied. Here, B, H, A^j, T, and L are functions of f_{11}, f_{22}, f_{31} and their derivatives. Their precise formulae are in Reyes [**15**].

REMARK 4.3. The assumption that the PSS equation $\Xi = 0$ with associated one-forms $\omega^\alpha = f_{\alpha 1}\,dx + f_{\alpha 2}\,dt$, is *equivalent* to the structure equations (3.2), implies that one can constrain *a priori* the dependence of the functions $f_{\alpha\beta}$ on the variables z_k, $k \ge 1$ (see Chern and Tenenblat [**3**], Reyes [**15**], and Kamran and Tenenblat [**9**]). These constraints then yield formulae such as (4.13)–(4.15). They are also used to construct Bäcklund transformations for some PSS equations (Chern and Tenenblat [**3**]) and to prove the existence of (generalized, nonlocal) symmetries for equations in this class (Reyes [**16**]). To study equations describing pseudo-spherical surfaces without this technicality, or to show that it is unavoidable, is an important open problem.

Let us go back to the relationship between formal and kinematic integrability. One of course wonders if a theorem analogous to Theorem 4.2 is true for higher order autonomous evolution equations. In actual fact, the proof of the implication "formal integrability $\Rightarrow$ kinematic integrability" sketched above cannot be extended from second to third order equations in complete generality. Indeed, one can check (Reyes [**15**]) that the equation

$$
u_t = (u_x + \delta)^3 u_{xxx}, \quad \delta \in \mathbf{R}, \tag{4.16}
$$

one of the two equations of the form $u_t = f(t, u, u_x)u_{xxx}$ which are formally integrable (Abellanas and Galindo [**1**]), does not appear in the classification mentioned above. Of course, this is also an interesting result by itself! It means that Equation (4.16) is not the integrability condition of an sl$(2, \mathbf{R})$-valued eigenvalue problem depending on at most second derivatives of u. Theorem 4.2 can be extended, however, to a subclass of formally integrable third order equations (Reyes [**15**]):

THEOREM 4.4. *Every autonomous formally integrable third order evolutionary equation of the form*

$$z_{0,t} = z_0^{-3} z_3 + a_2(x, z_0, z_1) z_2^2 + a_1(x, z_0, z_1) z_2 + a_0(x, z_0, z_1), \tag{4.17}$$

is kinematically integrable. Moreover, there exists an exhaustive list of representatives (up to contact transformations of the form (4.3), (4.4)) *of formally integrable third order equations of the form* (4.17) *which are kinematically integrable in the strict sense.*

5. Further Remarks

What about the implication "kinematic integrability $\Rightarrow$ formal integrability"? For instance, what about the converse of Theorem 4.2? Let us consider this problem more closely.

It is often observed (Faddeev and Takhtajan [**5**]) that if an autonomous evolution equation $z_{0,t} = F$, say, is kinematically integrable, it is a member of an infinite hierarchy of kinematically integrable equations (a point also stressed in Ma's Workshop lecture) sharing the x-parts of the associated linear problems, and that these equations generate a sequence of mutually commuting flows. It is also observed (Faddeev and Takhtajan [**5**]) that the existence of a recursion operator (2.6) for $z_{0,t} = F$ then follows. Switching to Chern and Tenenblat's [**3**] point of view, then, one would hope that a given autonomous evolution PSS equation belongs to an infinite hierarchy of PSS equations, and that each member of this hierarchy determines a generalized symmetry of every other member. It is from this perspective that the implication "kinematic integrability $\Rightarrow$ formal integrability" is being currently studied.

The first step, of course, is to construct a hierarchy of PSS equations as above, starting from a given PSS evolution equation $z_{0,t} = F$. It appears that this can indeed be done, the main idea behind such a construction being reminiscent of Mulase's [**13**] observation that the KP hierarchy can be understood as the integrability condition of a linear problem in an infinite-dimensional space. In the present case, the proposed hierarchy of PSS equations appears by imposing the structure equations (3.2) to one-forms defined in an infinite-dimensional space:

Consider the evolutionary autonomous PSS equation $z_{0,t} = F$ with associated one-forms $\omega_0^\alpha = f_{\alpha 1}\, dx + f_{\alpha 2}\, dt$, $\alpha = 1, 2, 3$. Define one-forms ω^α by

$$\omega^\alpha = \omega_0^\alpha + \sum_{i=1}^{\infty} h_{\alpha i}\, d\tau_i, \tag{5.1}$$

in which τ_i, $i \geq 1$, are to be interpreted as extra "flow variables". One has,

THEOREM 5.1. (1) *The one-forms ω^α satisfy the structure equations*

$$d\omega^1 = \omega^3 \wedge \omega^2, \quad d\omega^2 = \omega^1 \wedge \omega^3, \quad d\omega^3 = \omega^1 \wedge \omega^2,$$

in the space of coordinates $(x, t, \tau_1, \tau_2, \dots)$ if and only if so do the triplets
(a) *ω_0^α in (x, t) space,*
(b) *$\omega_i^\alpha := f_{\alpha 1}\, dx + h_{\alpha i}\, d\tau_i$ ($i \geq 1$ fixed) in (x, τ_i) space,*
(c) *$\sigma_{0i}^\alpha := f_{\alpha 2}\, dt + h_{\alpha i}\, d\tau_i$ ($i \geq 1$ fixed) in (t, τ_i) space, and*
(d) *$\sigma_{ij}^\alpha := h_{\alpha i}\, d\tau_i + h_{\alpha j}\, d\tau_j$ ($i \neq j, i, j \geq 1$ fixed) in (τ_i, τ_j) space,*
in which $\alpha = 1, 2, 3$.

(2) *There exist evolution equations* $z_{0,\tau_i} = F_i$, $i \geq 1$, *which describe pseudo-spherical surfaces with associated one-forms* ω_i^α.
(3) *The vector field* $F_j\partial/\partial z_0$ *is a generalized symmetry of the equation* $z_{0,\tau_i} = F_i$, *for all* $i, j = 0, 1, 2, \ldots$; $\tau_0 := t$, $F_0 := F$.

The one-forms (5.1) can be rigorously defined, and the techniques used geometrically integrable equations can be now called upon to study the *hierarchy* of equations $z_{0,\tau_i} = F_i$, $i \geq 0$, appearing in Theorem 5.1. The implication "kinematically integrability $\Rightarrow$ formal integrability" can then be analyzed in great detail. A complete account will be given elsewhere (Reyes [**17**]).

Finally, note that no explicit connection between the theory exposed here and the symplectic and Lie theoretical approaches to integrability (Adams, Harnad and Hurtubise [**2**] and references therein; Terng [**22**], Terng and Uhlenbeck [**23**]) is known. In particular, while it is a classical fact that some PSS equations (KdV, for instance) can be interpreted as Hamiltonian systems on coadjoint orbits of a loop group, it is not known whether one can understand the complete class of geometrically integrable equations in the same terms. This is an important issue which needs to be investigated.

References

1. L. Abellanas and A. Galindo, *A Harry–Dym class of bi-Hamiltonian evolution equations*, Phys. Lett. A **107** (1985), 159–160.
2. M. R. Adams, J. Harnad, and J. Hurtubise, *Darboux coordinates and Liouville–Arnold integration in loop algebras*, Comm. Math. Phys. **155** (1993), 385–413.
3. S. S. Chern and K. Tenenblat, *Pseudo-spherical surfaces and evolution equations*, Stud. Appl. Math. **74** (1986), 55–83.
4. M. Crampin, F. A. E. Pirani, and D. C. Robinson, *The soliton connection*, Lett. Math. Phys. **2** (1977), 15–19.
5. L. D. Faddeev and L. A. Takhtajan, *Hamiltonian methods in the theory of solitons*, Springer Ser. in Soviet Math., Springer-Verlag, Berlin, New York, 1987.
6. A. S. Fokas, I. M. Gel'fand, F. Finkel, and Q. M. Liu, *A formula for constructing infinitely many surfaces on Lie algebras and integrable equations*, Preprint, 1999.
7. M. Gürses and Y. Nutku, *New nonlinear equations from surface theory*, J. Math. Phys. **22** (1981), no. 7, 1393–1398.
8. R. Hernández Heredero, V. V. Sokolov, and S. I. Svinolupov, *Classification of third order integrable evolution equations*, Phys. D **87** (1995), 32–36.
9. N. Kamran and K. Tenenblat, *On differential equations describing pseudospherical surfaces*, J. Differential Equations **115** (1995), no.1, 75–98.
10. J. Krasil'shchik and A. Verbovetsky, *Homological methods in equations of mathematical physics*, Diffiety Institute International Summer School (August 1998), Levoča, Slovakia, Preprint `DIPS 7/98`, `math.DG/9808130`.
11. A. V. Mikhailov, A. B. Shabat, and V. V. Sokolov, *The symmetry approach to classification of integrable equations*, What is integrability? (V. E. Zakharov, ed.), Springer Ser. Nonlinear Dynam., Springer-Verlag, Berlin, 1991, pp. 115–184.
12. A. V. Mikhailov, A. B. Shabat, and R. I. Yamilov, *The symmetry approach to the classification of nonlinear equations*, Complete Lists of Integrable Systems, Russian Math. Surveys **42** (1987), no. 4, 1–63.
13. M. Mulase, *Complete integrability of the Kadomtsev–Petviashvili equation*, Adv. Math. **53** (1984), 57–66.
14. P. Olver, *Applications of Lie groups to differential equations*, 2nd ed., Springer-Verlag, New York, 1993.
15. E. G. Reyes, *Pseudo-spherical surfaces and integrability of evolution equations*, J. Differential Equations **147** (1998), no.1, 195–230; Erratum: J. Differential Equations **153** (1999), no.1, 223–224.

16. ______, *Some geometric aspects of integrability of differential equations in two independent variables*, 1999 (submitted).
17. ______, *Generalized symmetries and hierarchies of geometrically integrable equations*, 1999 (in preparation).
18. R. Sasaki, *Soliton equations and pseudospherical surfaces*, Nuclear Phys. B **154** (1979), 343–357.
19. S. I. Svinolupov and V. V. Sokolov, *Weak nonlocalities in evolution equations*, Mat. Zametki **48** (1990), no. 6, 91–97.
20. A. Sym, *Soliton surfaces and their applications (Soliton geometry from spectral problems)*, Geometric Aspects of the Einstein Equations and Integrable Systems (R. Martini, ed.), Lecture Notes in Phys. **239** (1985), Springer-Verlag, Berlin, pp. 154–231.
21. K. Tenenblat, *Transformations of manifolds and applications to differential equations*, Pitman Monographs Surveys Pure Appl. Math. **93** (1998), Addison Wesley Longman, England.
22. C. L. Terng, *Soliton equations and differential geometry*, J. Differential Geom. **45** (1997), 407–445.
23. C. L. Terng and K. Uhlenbeck, *Poisson actions and scattering theory for integrable systems*, `xxx.lanl.gov e-Print archive, dg-ga/9707004`, 1997.
24. J. P. Wang, *Symmetries and conservation laws of evolution equations*, The Thomas Stieltjes Institute for Mathematics, Amsterdam, 1998.

Department of Mathematics and Statistics, Utah State University, Logan, UT 84322-3900, USA.

Current address: Department of Mathematics, Yale University, P.O. Box 208283, New Haven, CT 06520-8283, USA.

E-mail address: `reyes@math.usu.edu; reyes@math.yale.edu`

Centre de Recherches Mathématiques
CRM Proceedings and Lecture Notes
Volume **29**, 2001

Infinitesimal Bäcklund Transformations of K-Nets. The 2+1-Dimensional Sinh–Gordon System

C. Rogers and W. K. Schief

ABSTRACT. The K-net condition arises, in particular, in the study of isothermic nets. Thus, a line of curvature net which is isothermic is necessarily a K-net. The latter is characterised by the fact that its point equation has equal Laplace invariants and has a canonical form well-adapted to the application of a class of infinitesimal Bäcklund transformations originally introduced by Loewner in 1952 in a gasdynamic context. Here, the background on finite Loewner transformations and their iteration is presented as a preliminary to our discussion of infinitesimal Bäcklund transformations. It is then shown that particular infinitesimal Bäcklund transformations of K-nets lead to the integrable $2+1$-dimensional sinh–Gordon system.

1. The Historical Background to Bäcklund to Loewner Transformations

In the 1940s and 1950s there was considerable interest in the mathematical theory of gasdynamics in high subsonic, trans-sonic and supersonic régimes (Bers [**2**]). In particular, the celebrated Kármán-Tsien approximation to the adiabatic gas law allowed the reduction of the hodograph equations to the Cauchy–Riemann equations in subsonic flow. This led to an extensive search for other multi-parameter model gas laws for which reduction to appropriate canonical forms was available in other flow régimes. In contemporary work, Bergman [**1**] introduced an integral operator expansion method to develop a correction theory based on classical hydrodynamics. This procedure is akin to the Karal–Keller wave-front expansion method developed in inhomogeneous elastodynamics [**5**]. It turns out that termination of Bergman series is linked to the model constitutive approach and, recently, connections have been made with Toda lattice theory [**11**].

In 1950, Loewner [**9**] introduced a class of finite Bäcklund transformations involving two dependent variables viz., relations

$$\mathbb{B}_i(q,\theta,\phi,\psi,\phi_q,\phi_\theta,\psi_q,\psi_\theta;q^*,\theta^*,\phi^*,\psi^*,\phi^*_{q^*},\phi^*_{\theta^*},\psi^*_{q^*},\psi^*_{\theta^*})=0,\quad i=1,\dots,6, \tag{1.1}$$

2000 *Mathematics Subject Classification.* Primary: 35Q51; Secondary: 35Q58.
This is the final form of the paper.

which encode, on application of integrability conditions a link between the hodograph system

$$\begin{pmatrix}\phi\\ \psi\end{pmatrix}_q = \begin{pmatrix}0 & -(1-M^2)/\rho q\\ \rho/q & 0\end{pmatrix}\begin{pmatrix}\phi\\ \psi\end{pmatrix}_\theta, \tag{1.2}$$

and an appropriate canonical form in subsonic, trans-sonic and supersonic régimes, namely

$$\begin{aligned} M<1: &\quad \begin{pmatrix}\phi^*\\ \psi^*\end{pmatrix}_{q^*} = \begin{pmatrix}0 & -1\\ 1 & 0\end{pmatrix}\begin{pmatrix}\phi^*\\ \psi^*\end{pmatrix}_{\theta^*},\\ M\sim 1: &\quad \begin{pmatrix}\phi^*\\ \psi^*\end{pmatrix}_{q^*} = \begin{pmatrix}0 & -q^*\\ 1 & 0\end{pmatrix}\begin{pmatrix}\phi^*\\ \psi^*\end{pmatrix}_{\theta^*},\\ M>1: &\quad \begin{pmatrix}\phi^*\\ \psi^*\end{pmatrix}_{q^*} = \begin{pmatrix}0 & 1\\ 1 & 0\end{pmatrix}\begin{pmatrix}\phi^*\\ \psi^*\end{pmatrix}_{\theta^*}. \end{aligned} \tag{1.3}$$

In the above, ϕ, ψ represent the usual potential and stream functions, ρ and q the gas density and gas speed magnitude and $M(q)$ the local Mach number.

Loewner in [**9**] introduced a class of finite matrix BTs

$$\begin{aligned} \boldsymbol{\Omega}^*_\sigma &= \boldsymbol{A}\boldsymbol{\Omega}_\sigma + \boldsymbol{B}\boldsymbol{\Omega} + \boldsymbol{C}\boldsymbol{\Omega}^*,\\ \boldsymbol{\Omega}^*_\tau &= \widetilde{\boldsymbol{A}}\boldsymbol{\Omega}_\tau + \widetilde{\boldsymbol{B}}\boldsymbol{\Omega} + \widetilde{\boldsymbol{C}}\boldsymbol{\Omega}^*,\\ \boldsymbol{\Omega} &= \begin{pmatrix}\phi\\ \psi\end{pmatrix}, \quad \boldsymbol{\Omega}^* = \begin{pmatrix}\phi^*\\ \psi^*\end{pmatrix}, \end{aligned} \tag{1.4}$$

with a view to mapping the matrix equations

$$\boldsymbol{\Omega}_\sigma = \boldsymbol{S}\boldsymbol{\Omega}_\tau \rightarrow \boldsymbol{\Omega}^*_\sigma = \boldsymbol{S}^*\boldsymbol{\Omega}^*_\tau, \tag{1.5}$$

and preserving the properties[1]

$$\operatorname{tr}\boldsymbol{S} = 0, \quad \det\boldsymbol{S} = -1. \tag{1.6}$$

Integrability conditions applied to the linear BT system (1.4) produce the nonlinear system

$$\begin{aligned} \boldsymbol{B} + \boldsymbol{C}\widetilde{\boldsymbol{A}} - \widetilde{\boldsymbol{A}}_\sigma + (\boldsymbol{A}_\tau - \widetilde{\boldsymbol{B}} - \widetilde{\boldsymbol{C}}\widetilde{\boldsymbol{A}})\boldsymbol{S} &= \boldsymbol{0},\\ \boldsymbol{B}_\tau - \boldsymbol{B}_\sigma + \boldsymbol{C}\widetilde{\boldsymbol{B}} - \widetilde{\boldsymbol{C}}\boldsymbol{B} &= \boldsymbol{0},\\ \boldsymbol{C}_\tau - \widetilde{\boldsymbol{C}}_\sigma + [\boldsymbol{C}, \widetilde{\boldsymbol{C}}] &= \boldsymbol{0}, \end{aligned} \tag{1.7}$$

with

$$\begin{aligned} \boldsymbol{A} &= \widetilde{\boldsymbol{A}},\\ \boldsymbol{B} = \boldsymbol{S}^*\widetilde{\boldsymbol{B}}, &\quad \boldsymbol{C} = \boldsymbol{S}^*\widetilde{\boldsymbol{C}},\\ \boldsymbol{S}^* &= \boldsymbol{A}\boldsymbol{S}\boldsymbol{A}^{-1}. \end{aligned} \tag{1.8}$$

[1]It is readily shown that the hodograph system (1.2) may be written in the form $\boldsymbol{\Omega}_\sigma = \boldsymbol{S}\boldsymbol{\Omega}_\tau$ where $\boldsymbol{S}$ is subject to the conditions (1.6). This is achieved by a change of variable $\sigma = \sigma(q)$, $\tau = \theta$, the nature of which depends on whether $M < 1$, $M \sim 1$ or $M > 1$.

2. Finite Loewner Transformations of Conjugate Nets

The conjugate net equation (Eisenhart [**3**])

$$\boldsymbol{r}_{uv} = \Gamma^1_{12}\boldsymbol{r}_u + \Gamma^2_{12}\boldsymbol{r}_v, \tag{2.1}$$

may be written as

$$\boldsymbol{r}_{uv} = \frac{\partial_v h}{l-h}\boldsymbol{r}_u + \frac{\partial_u l}{h-l}\boldsymbol{r}_v, \tag{2.2}$$

which is a consequence of the matrix equation

$$\begin{pmatrix}\boldsymbol{r}\\ \boldsymbol{r}'\end{pmatrix}_\sigma = \boldsymbol{S}\begin{pmatrix}\boldsymbol{r}\\ \boldsymbol{r}'\end{pmatrix}_\tau, \tag{2.3}$$

where $\sigma = u+v$, $\tau = u-v$ and

$$\boldsymbol{S} = \frac{1}{l-h}\begin{pmatrix} l+h & -2\\ 2hl & -(l+h)\end{pmatrix}. \tag{2.4}$$

In the above,

$$\boldsymbol{r}'_u = h\boldsymbol{r}_u, \quad \boldsymbol{r}'_v = l\boldsymbol{r}_v, \tag{2.5}$$

so that $\boldsymbol{r}'$ determines a parallel net [**3**]. It is noted that the constraints (1.6) on $\boldsymbol{S}$ hold.

An elementary finite Loewner transformation (LT) of the type (1.4) applied to the conjugate net equation (2.3) delivers in a natural manner:

THEOREM 2.1 (The Fundamental Transformation). *The mapping* [**4**]

$$\boldsymbol{r}^* = \boldsymbol{r} - (\theta/\theta')\boldsymbol{r}',$$

where

$$\boldsymbol{r}'_u = h\boldsymbol{r}_u, \quad \boldsymbol{r}'_v = l\boldsymbol{r}_v,$$
$$\theta'_u = h\theta_u, \quad \theta'_v = l\theta_v,$$

takes the conjugate net N *determined by*

$$\boldsymbol{r}_{uv} = \frac{\partial_v h}{l-h}\boldsymbol{r}_u + \frac{\partial_u l}{h-l}\boldsymbol{r}_v,$$

to an associated conjugate net N^* *with*

$$\boldsymbol{r}^*_{uv} = \left(\frac{\partial_v h}{l-h} + (\log\alpha)_v\right)\boldsymbol{r}^*_u + \left(\frac{\partial_u l}{h-l} + (\log\tilde{\alpha})_u\right)\boldsymbol{r}^*_v,$$

where

$$\alpha = 1 - h(\theta/\theta'), \quad \tilde{\alpha} = 1 - l(\theta/\theta').$$

This classical geometric result is equivalent to the binary Darboux transformation of modern soliton theory (Matveev and Salle [**10**]).

Elementary LTs likewise lead to the classical Levy transformations which also map conjugate nets to conjugate nets (Eisenhart [**3**]). These correspond to the basic Darboux transformations of soliton theory.

Loewner in [**9**] restricted his attention to the action of a single finite BT in his study of model gas laws. However, the potential importance of the iteration of these BTs in that context is alluded to in his seminal paper of 1950 in the following manner:

"It is of great theoretical and practical interest to determine all systems of partial differential equations whose form can be brought arbitrarily close to one of these canonical forms by such composite transformations".

The application of iterated LTs in the context of model constitutive laws in nonlinear elastodynamics has been discussed by Rogers [**13**]. The iteration of finite LTs in general was dealt with recently in [**14**]. Therein it was shown that the general class of finite LTs may be completely parametrised in terms of "eigenfunctions" and "adjoint eigenfunctions" of $\boldsymbol{S}$, that is, $\boldsymbol{\Omega}_i$ and $\boldsymbol{\Psi}_i$ such that, respectively

$$\boldsymbol{\Omega}_{i\sigma} = \boldsymbol{S}\boldsymbol{\Omega}_{i\tau}, \quad \boldsymbol{\Psi}_{i\sigma} = \boldsymbol{S}^T\boldsymbol{\Psi}_{i\psi}. \tag{2.6}$$

A geometric interpretation in terms of conjugate nets of a result presented in [**14**] yields:

THEOREM 2.2. *Conjugate nets determined by* (2.3)–(2.4) *are mapped*

$$\boldsymbol{\Omega}_\sigma = \boldsymbol{S}\boldsymbol{\Omega}_\tau \to \boldsymbol{\Omega}^*_\sigma = \boldsymbol{S}^*\boldsymbol{\Omega}^*_\tau,$$

under

$$\boldsymbol{\Omega}^* = \boldsymbol{\Omega}^*_1\boldsymbol{N}_{10} + \boldsymbol{\Omega}^*_2,$$

where

$$\begin{aligned}\boldsymbol{\Omega}^*_1 &= -(\boldsymbol{\Omega}_3 - \boldsymbol{\Omega}_1\boldsymbol{N}_{11}^{-1}\boldsymbol{N}_{13}^{-1})^{-1}\boldsymbol{\Omega}_1\boldsymbol{N}_{11}^{-1},\\ \boldsymbol{\Omega}^*_2 &= (\boldsymbol{\Omega}_3 - \boldsymbol{\Omega}_1\boldsymbol{N}_{11}^{-1}\boldsymbol{N}_{13})^{-1},\end{aligned}$$

with

$$\boldsymbol{S}^* = (\boldsymbol{\Omega}^*_2 + \boldsymbol{\Omega}^*_1\boldsymbol{\Psi}_1^T)\boldsymbol{S}(\boldsymbol{\Omega}^*_2 + \boldsymbol{\Omega}^*_1\boldsymbol{\Psi}_1^T)^{-1},$$

where the binary potential $\boldsymbol{N}_{ij}$ associated with the pair of (*adjoint*) *eigenfunctions* $(\boldsymbol{\Psi}_i, \boldsymbol{\Omega}_j)$ *is introduced via*

$$\boldsymbol{N}_{ij,\sigma} = \boldsymbol{\Psi}_i^T\boldsymbol{\Omega}_{j\sigma}, \quad \boldsymbol{N}_{ij,\tau} = \boldsymbol{\Psi}_i^T\boldsymbol{\Omega}_{j\tau}.$$

The generic LT corresponds to the iterated action of binary Darboux transformations (boosted by gauge transformations) acting on the conjugate net equations. The iterated action of classical Darboux (Levy) transformations is also shown in [**14**] to be conveniently accommodated by a matrix LT.

In summary, a compact representation for iterated finite LTs may be obtained in terms of eigenfunctions. The important observation was made in [**14**] that the multi-parameter model laws obtained via iteration of the LTs are associated with multi-soliton solutions evaluated at constant time. The importance of multi-parameter nonlinear constitutive laws as bench marks in electromagnetic theory has recently been alluded to by Sjöberg [**15**].

3. Koenigs Nets

The conjugate net equation (2.1) admits Laplace invariants

$$H = -(\Gamma^1_{12})_u + \Gamma^1_{12}\Gamma^2_{12}, \tag{3.1}$$

$$K = -(\Gamma^2_{12})_v + \Gamma^1_{12}\Gamma^2_{12}, \tag{3.2}$$

that is, H, K are invariant under a scaling $\boldsymbol{r}' = \xi(u,v)\boldsymbol{r}$.

If $H = K$ then the net is said to be a Koenigs net or K-net (Eisenhart [**3**]) and

$$(\Gamma^1_{12})_u = (\Gamma^2_{12})_v. \tag{3.3}$$

On introduction of the potential $\log\sqrt{\rho}$ such that

$$\Gamma^1_{12} = (\log\sqrt{\rho})_v, \quad \Gamma^2_{12} = (\log\sqrt{\rho})_u, \tag{3.4}$$

the conjugate net equation reduces to the K-net equation

$$\boldsymbol{r}_{uv} = \partial_v(\log\sqrt{\rho})\boldsymbol{r}_u + \partial_u(\log\sqrt{\rho})\boldsymbol{r}_v. \tag{3.5}$$

In terms of the independent variables σ, τ, (3.5) becomes

$$\begin{gathered}(1/\sqrt{\rho})\Box(\boldsymbol{r}/\sqrt{\rho}) - (\boldsymbol{r}/\sqrt{\rho})\Box(1/\sqrt{\rho}) = \boldsymbol{0},\\ (\Box = \partial^2/\partial\sigma^2 - \partial^2/\partial\tau^2),\end{gathered} \tag{3.6}$$

or

$$\frac{\partial}{\partial\tau}(\rho^{-1}\boldsymbol{r}_\tau) - \frac{\partial}{\partial\sigma}(\rho^{-1}\boldsymbol{r}_\sigma) = \boldsymbol{0}. \tag{3.7}$$

The latter allows the introduction of a vector potential $\bar{\boldsymbol{r}}$ such that

$$\bar{\boldsymbol{r}}_\sigma = \rho^{-1}\boldsymbol{r}_\tau, \quad \bar{\boldsymbol{r}}_\tau = \rho^{-1}\boldsymbol{r}_\sigma, \tag{3.8}$$

whence we obtain the matrix equation

$$\begin{pmatrix}\boldsymbol{r}\\ \bar{\boldsymbol{r}}\end{pmatrix}_\sigma = \begin{pmatrix}0 & \rho\\ \rho^{-1} & 0\end{pmatrix}\begin{pmatrix}\boldsymbol{r}\\ \bar{\boldsymbol{r}}\end{pmatrix}_\tau. \tag{3.9}$$

The latter is convenient for the application of Bäcklund transformations corresponding to Loewner-type infinitesimal deformation of K-nets.

4. Loewner's Infinitesimal BTs. The Soliton Connection

Loewner, in 1952, again in a gasdynamics context, introduced the notion of an infinitesimal matrix BT [**9**]. Therein, the increment $\boldsymbol{\Omega}^* - \boldsymbol{\Omega} = \Delta\boldsymbol{\Omega}$ is entered instead of $\boldsymbol{\Omega}^*$ and all matrices are assumed to involve a continuous parameter ε. There results a class of infinitesimal BTs

$$\begin{aligned}\boldsymbol{\Omega}_{\sigma\varepsilon} &= \boldsymbol{\Lambda}\boldsymbol{\Omega}_\sigma + \boldsymbol{A}\boldsymbol{\Omega} + \boldsymbol{C}\boldsymbol{\Omega}_\varepsilon,\\ \boldsymbol{\Omega}_{\tau\varepsilon} &= \boldsymbol{\Lambda}\boldsymbol{\Omega}_\tau + \boldsymbol{B}\boldsymbol{\Omega} + \boldsymbol{D}\boldsymbol{\Omega}_\varepsilon,\end{aligned} \tag{4.1}$$

acting on a matrix equation

$$\boldsymbol{\Omega}_\sigma = \boldsymbol{S}\boldsymbol{\Omega}_\tau. \tag{4.2}$$

In the original gasdynamics setting, Loewner proposed the study of hodograph systems (4.2) asymptotically reducible via the infinitesimal BTs as $\varepsilon \to \infty$ to the appropriate canonical form of (1.3) in subsonic, trans-sonic and supersonic régimes.

It is readily shown that compatibility of the system (4.1)–(4.2) imposes six constraints on the six matrices $\boldsymbol{A}$, $\boldsymbol{B}$, $\boldsymbol{C}$, $\boldsymbol{D}$, $\boldsymbol{\Lambda}$ and $\boldsymbol{S}$, namely

$$\begin{gathered}\boldsymbol{A} = \boldsymbol{S}\boldsymbol{B}, \quad \boldsymbol{C} = \boldsymbol{S}\boldsymbol{D},\\ \boldsymbol{B}_\sigma - \boldsymbol{A}_\tau + \boldsymbol{D}\boldsymbol{A} - \boldsymbol{C}\boldsymbol{B} = \boldsymbol{0},\\ \boldsymbol{D}_\sigma - \boldsymbol{C}_\tau + [\boldsymbol{D}, \boldsymbol{C}] = \boldsymbol{0},\\ (-\boldsymbol{\Lambda}_\tau + \boldsymbol{B} + \boldsymbol{D}\boldsymbol{\Lambda})\boldsymbol{S} + (\boldsymbol{\Lambda}_\sigma - \boldsymbol{A} - \boldsymbol{C}\boldsymbol{\Lambda}) = \boldsymbol{0},\end{gathered} \tag{4.3}$$

together with the "evolution"

$$\boldsymbol{S}_\varepsilon = [\boldsymbol{\Lambda}, \boldsymbol{S}]. \tag{4.4}$$

In analogy with the finite LT case, the above conditions (4.3) can be completely parametrised in terms of eigenfunctions and adjoint eigenfunctions.

Thus, the zero-curvature condition $(4.3)_4$ admits the general solution

$$C = F_\sigma F^{-1}, \quad D = F_\tau F^{-1}, \tag{4.5}$$

whence the condition $(4.3)_2$ implies that

$$F_\sigma = SF_\tau, \tag{4.6}$$

while the constraint $(4.3)_3$ reduces to

$$(F^{-1}A)_\tau = (F^{-1}B)_\sigma. \tag{4.7}$$

The latter allows the introduction of a matrix potential G via

$$A = FG_\sigma, \quad B = FG_\tau, \tag{4.8}$$

and insertion of these relations into (4.3), shows that

$$(FG)_\sigma = S(FG)_\tau. \tag{4.9}$$

Finally, the condition $(4.3)_5$ reduces to

$$R_\sigma = R_\tau S, \tag{4.10}$$

where

$$R = F^{-1}\Lambda - G, \tag{4.11}$$

and we obtain the following result:

SUMMARY. *The Loewner class of infinitesimal BTs adopts the form*

$$\begin{aligned} \partial_\sigma(F^{-1}\Omega_\varepsilon) &= \partial_\sigma(G\Omega) + R\Omega_\sigma, \\ \partial_\tau(F^{-1}\Omega_\varepsilon) &= \partial_\tau(G\Omega) + R\Omega_\tau, \end{aligned} \tag{4.12}$$

with integrability condition

$$\partial_\sigma(R\Omega_\sigma) = \partial_\tau(R\Omega_\tau), \tag{4.13}$$

where

$$\begin{aligned} F_\sigma = SF_\tau, \quad (FG)_\sigma = S(FG)_\tau, \\ (R^T)_\sigma = S^T(R^T), \end{aligned} \tag{4.14}$$

while the evolution of S is given by

$$S_\varepsilon = [\Lambda, S], \tag{4.15}$$

where

$$\Lambda = F(G + R). \tag{4.16}$$

Thus, F and FG represent "eigenfunctions" of S while R^T is an "adjoint eigenfunction".

It is noted that the integrability condition (4.13) implies the existence of a bilinear potential N associated with the pair (R, Ω) such that

$$dN = Rd\Omega. \tag{4.17}$$

5. The 2+1-Dimensional Sinh–Gordon System

If we set

$$\boldsymbol{F} = \begin{pmatrix} \alpha_1 & \alpha_2 \\ \beta_1 & \beta_2 \end{pmatrix}, \quad \boldsymbol{FG} = \begin{pmatrix} \alpha_3 & \alpha_4 \\ \beta_3 & \beta_4 \end{pmatrix}, \tag{5.1}$$

then the "eigenfunction" relations $(4.14)_{1,2}$ show that each pair $\{\alpha_i, \beta_i\}$, $i = 1, \ldots, 4$ satisfies

$$\alpha_{i\sigma} = \rho\beta_{i\tau}, \quad \beta_{i\sigma} = \rho^{-1}\alpha_{i\tau}, \tag{5.2}$$

while, if

$$\boldsymbol{R} = \begin{pmatrix} \beta_5 & \alpha_5 \\ \beta_6 & \alpha_6 \end{pmatrix}, \tag{5.3}$$

it is seen that $\{\alpha_i, \beta_i\}$, $i = 5, 6$ again satisfy pairs of equations of the type (4.12).

If we now turn to the evolution (4.15) and set $\boldsymbol{\Lambda} = [\lambda_{ij}]$ then

$$\boldsymbol{S}_\varepsilon = \begin{pmatrix} \lambda_{12}\rho^{-1} - \lambda_{21}\rho & (\lambda_{11} - \lambda_{22})\rho \\ (\lambda_{22} - \lambda_{11})\rho^{-1} & \lambda_{21}\rho - \lambda_{12}\rho^{-1} \end{pmatrix}. \tag{5.4}$$

Since $\boldsymbol{S}_\varepsilon$ must have vanishing diagonal elements if the K-net is to be continuously deformed to a K-net then

$$\lambda_{12} - \lambda_{21}\rho^2 = 0, \tag{5.5}$$

where

$$(\log \rho)_\varepsilon = \lambda_{11} - \lambda_{22}. \tag{5.6}$$

In the sequel, we set $\lambda_{12} = \lambda_{21} = 0$ so that

$$\begin{gathered} \alpha_4 + \alpha_1\alpha_5 + \alpha_2\alpha_6 = 0, \quad \beta_3 + \beta_1\beta_5 + \beta_2\beta_6 = 0, \\ (\log \rho)_\varepsilon = \alpha_3 - \beta_4 + \alpha_1\beta_5 + \alpha_2\beta_6 - \beta_1\alpha_5 - \beta_2\alpha_6. \end{gathered} \tag{5.7}$$

If we now make the specialisation $\boldsymbol{F} = \boldsymbol{\sigma}_1$ where $\boldsymbol{\sigma}_1$ is the Pauli matrix then $(4.17)_{1,2}$ show that

$$\alpha_4 + \alpha_6 = 0, \quad \beta_3 + \beta_5 = 0, \tag{5.8}$$

whence

$$\beta_4 + \beta_6 = \text{const} = c_1, \quad \alpha_3 + \alpha_5 = \text{const} = c_2, \tag{5.9}$$

so that

$$(\log \sqrt{\rho})_\varepsilon = \alpha_3 - \beta_4 + (c_1 - c_2)/2. \tag{5.10}$$

The constant $(c_1 - c_2)/2$ may be taken to be zero without loss of generality.

Thus, the compatibility conditions for the particular class of infinitesimal deformations of K-nets under consideration reduce to

$$\begin{gathered} (\log \sqrt{\rho})_\varepsilon = \alpha_3 - \beta_4, \\ \frac{\partial}{\partial\sigma}\left(\frac{\partial\alpha_3}{\partial\sigma}/\rho\right) - \frac{\partial}{\partial\tau}\left(\frac{\partial\alpha_3}{\partial\tau}/\rho\right) = 0, \\ \frac{\partial}{\partial\sigma}\left(\rho\frac{\partial\beta_4}{\partial\sigma}\right) - \frac{\partial}{\partial\tau}\left(\rho\frac{\partial\beta_4}{\partial\tau}\right) = 0. \end{gathered} \tag{5.11}$$

Introduction of the potentials α, β via

$$\alpha_3 = \partial\alpha/\partial\varepsilon, \quad \beta_4 = \partial\beta/\partial\varepsilon. \tag{5.12}$$

Now produces the integrable 2 + 1-dimensional sinh–Gordon system [**6**]

$$
\begin{aligned}
\frac{\partial}{\partial\sigma}\left(e^{2(\alpha-\beta)}\beta_{\sigma\varepsilon}\right) - \frac{\partial}{\partial\tau}\left(e^{2(\alpha-\beta)}\beta_{\tau\varepsilon}\right) &= 0, \\
\frac{\partial}{\partial\sigma}\left(e^{-2(\alpha-\beta)}\alpha_{\sigma\varepsilon}\right) - \frac{\partial}{\partial\tau}\left(e^{-2(\alpha-\beta)}\alpha_{\tau\varepsilon}\right) &= 0.
\end{aligned} \tag{5.13}
$$

In fact, more generally, a reinterpretation and generalisation of the infinitesimal Loewner transformations (4.1)–(4.2) has been shown to provide a master 2 + 1-dimensional integrable system [**6,7**]. Indeed, it has been demonstrated by Oevel and Schief in [**12**] that a hierarchy of Loewner-type systems is readily constructed which is compatible with the general multi-component Kadomtsev–Petviashvili (KP) and modified Kadomtsev–Petriashvili (mkP) hierarchies of Sato theory. The Loewner-type systems possess the advantage that they admit a natural parametrisation with canonical subclasses. Moreover, they allow geometric interpretation in terms of the infinitesimal deformation of conjugate nets.

References

1. S. Bergman, *Integral operators in the theory of partial differential equations*, Springer-Verlag, Berlin and New York, 1968.
2. L. Bers, *Mathematical aspects of subsonic and transonic gasdynamics*, John Wiley & Sons Inc., New York, 1958.
3. L. Eisenhart, *Transformations of surfaces*, Princeton University Press, Princeton, 1923.
4. H. Jonas, *Über die Transformation der konjugierten Systeme und über den gemeinsamen Ursprung der Bianchischen Permutabilitätstheoreme*, Sitzungsber. d. Berlin. Math. Ges. **14** (1915), 96–118.
5. F. C. Karal Jr. and J. B. Keller, *Elastic wave propagation in homogeneous and inhomogeneous media*, J. Acoust. Soc. Amer. **31** (1959), 694–705.
6. B. G. Konopelchenko and C. Rogers, *On generalised Loewner systems: Novel integrable equations in* 2 + 1*-dimensions*, J. Math. Phys. **34** (1993), no. 1, 214–242.
7. ———, *On* 2 + 1*-dimensional nonlinear systems of Loewner-type*, Phys. Lett A **158** (1991), no. 8, 391–397.
9. C. Loewner, *A transformation theory of partial differential equations of gas dynamics*, Tech. Notes Nat. Adv. Comm. Aeronaut. **2065** (1950), 1–56.
8. ———, *Generation of solutions of systems of partial differential equations by composition of infinitesimal Bäcklund transformations*, J. Anal. Math. **2** (1953), 219–242.
10. V. B. Matveev and M. A. Salle, *Darboux transformations and solitons*, Springer Series in Nonlinear Dynamics, Springer-Verlag, Berlin, 1991.
11. J. J. C. Nimmo, W. K. Schief, and C. Rogers, *Termination of Bergman series. Connection to the B_n Toda system* (M. J. Ablowitz and P. A. Clarkson, eds.), J. Engrg. Math. **36** (1999), 137–148 (Special Issue on Applications of Solitons and Symmetries).
12. W. Oevel and W. K. Schief, *Squared eigenfunctions of the (modified)* KP *hierarchy and scattering problems of Loewner type*, Rev. Math. Phys. **6** (1994), no. 6, 1301–1338.
13. C. Rogers, *Iterated Bäcklund-type transformations and the propagation of disturbances in nonlinear elastic materials*, J. Math. Anal. Appl. **49** (1975), 638–648.
14. W. K. Schief and C. Rogers, *Loewner transformations. Adjoint and binary Darboux connections*, Stud. Appl. Math. **100** (1998), no. 4, 391–422.
15. D. Sjöberg, *Reconstruction of nonlinear material properties for homogeneous, isotropic slabs using electromagnetic waves*, Inverse Problems **15** (1999), no. 2, 431–444.

Department of Applied Mathematics, University of New South Wales, Sydney 2052, Australia.

E-mail address, C. Rogers: `c.rogers@maths.unsw.edu.au`

E-mail address, W. K. Schief: `schief@maths.unsw.edu.au`

Centre de Recherches Mathématiques
CRM Proceedings and Lecture Notes
Volume **29**, 2001

Isothermic Surfaces and the Calapso Equation: The Full Monty

W. K. Schief

Abstract. A vector analogue of the classical Calapso equation governing isothermic surfaces in $\mathbb{R}^{n+2}$ is introduced. It is shown that this vector Calapso system admits a (nonlocal) scalar Lax pair based on the classical Moutard equation. The analogue of Darboux's Bäcklund transformation for isothermic surfaces in $\mathbb{R}^3$ is derived and formulated in terms of the classical Moutard transformation acting on the scalar Lax pair. A permutability theorem for isothermic surfaces is set down which manifests itself in an explicit superposition principle for the vector Calapso system. This superposition principle in vectorial form is shown to constitute an integrable discretization of the vector Calapso system and therefore defines discrete isothermic surfaces in $\mathbb{R}^{n+2}$.

1. Introduction

The classical theory of isothermic surfaces in Euclidean space $\mathbb{R}^3$ was developed in the 19th century by various geometers such as Lamé and Darboux. A milestone was set by Darboux [**16**] in 1899 who derived a Bäcklund transformation for isothermic surfaces in terms of a linear representation containing an arbitrary parameter. This remarkable development established the integrable nature of isothermic surfaces. Darboux's transformation was discussed "in extenso" by Bianchi [**3**] who proved the existence of an associated permutability theorem. Later, Demoulin [**19**] demonstrated that Darboux's transformation possesses the important constant cross-ratio property. A fourth-order differential equation which encapsulates isothermic surfaces was set down by Calapso in 1903 [**9**].

Recently, isothermic surfaces in $\mathbb{R}^3$ and $\mathbb{R}^4$ have attracted considerable interest in connection with their interpretation in terms of integrability. This research was prompted by Cieśliński, Goldstein and Sym [**14**] who reformulated isothermic surfaces in $\mathbb{R}^3$ as "soliton surfaces" [**36**]. Darboux's classical Bäcklund transformation for isothermic surfaces in $\mathbb{R}^3$ has been related to a matrix Darboux transformation [**12**] by means of a suitable extension of Sym's formula [**36**] and re-interpreted in the context of curved flats in symmetric spaces [**8**]. The discrete analogue of this matrix Darboux transformation has been discussed in [**13**] based on the integrable discretization of isothermic surfaces proposed by Bobenko and Pinkall [**5**].

2000 *Mathematics Subject Classification.* Primary: 53A05; Secondary: 58F07.
This is the final form of the paper.

The discretization of isothermic surfaces in $\mathbb{R}^4$ along with an associated Bäcklund transformation have been the subject of [**23**]. Isothermic surfaces in spaces of arbitrary dimension have been studied earlier in [**32**] in the context of isometric deformations of surfaces.

Here, we give an overwiew of novel results obtained in connection with the study of isothermic surfaces in $\mathbb{R}^{n+2}$. For brevity, we only state the main theorems and refer to [**34**] for the details. Thus, the key observation is that an alternative formulation of the Gauß–Weingarten equations for isothermic surfaces in $\mathbb{R}^{n+2}$ provides both an explicit superposition principle for solutions of the Gauß–Mainardi–Codazzi equations associated with isothermic surfaces and a "tractable" algebraic formulation of discrete isothermic surfaces. Remarkably, the classical Moutard equation [**28**], which is shown to form part of an integrable vector analogue of the Calapso equation, turns out to be key to our investigations. We make contact with Clifford algebras, the discrete Korteweg–de Vries equation, discrete holomorphic functions and the nonlinear σ-model.

2. The Geometry of Isothermic Surfaces

A surface $\Sigma \subset \mathbb{R}^{n+2}$ embedded in Euclidean space $\mathbb{R}^{n+2}$ is termed *isothermic* if it admits a parametrization

$$\boldsymbol{r}\colon \mathbb{R}^2 \to \mathbb{R}^{n+2}, \quad (x,y) \mapsto \boldsymbol{r}(x,y), \tag{2.1}$$

which is *conjugate*, *orthogonal* and *conformal* [**20**]. In analytic terms, this means that the position vector $\boldsymbol{r}$ obeys the three conditions

$$\begin{aligned} \text{CONJUGACY}: &\quad \boldsymbol{r}_{xy} = A\boldsymbol{r}_x + B\boldsymbol{r}_y,\\ \text{ORTHOGONALITY}: &\quad \boldsymbol{r}_x \cdot \boldsymbol{r}_y = 0,\\ \text{CONFORMALITY}: &\quad \boldsymbol{r}_x^2 = \boldsymbol{r}_y^2, \end{aligned} \tag{2.2}$$

where A and B are some functions of x and y. If

$$\underline{\boldsymbol{N}} = (\boldsymbol{N}^1, \dots, \boldsymbol{N}^n), \tag{2.3}$$

constitutes an orthonormal basis of the normal bundle, that is

$$\underline{\boldsymbol{N}} \cdot \boldsymbol{r}_x = \underline{\boldsymbol{N}} \cdot \boldsymbol{r}_y = 0, \quad \boldsymbol{N}^i \cdot \boldsymbol{N}^k = \delta^{ik}, \tag{2.4}$$

then the *first fundamental form* I and the set of n *second fundamental forms* $\underline{\mathrm{II}} = (\mathrm{II}^1, \dots, \mathrm{II}^n)$ may be parametrized according to

$$\begin{aligned} \mathrm{I} &= d\boldsymbol{r}\cdot d\boldsymbol{r} = e^{2\theta}(dx^2 + dy^2),\\ \underline{\mathrm{II}} &= -d\underline{\boldsymbol{N}}\cdot d\boldsymbol{r} = -e^{\theta}(\underline{h}_1 dx^2 + \underline{h}_2 dy^2). \end{aligned} \tag{2.5}$$

Since the coordinate system is orthogonal, one can always find an orthonormal basis $\underline{\boldsymbol{N}}$ satisfying $\boldsymbol{N}_x^i \cdot \boldsymbol{N}^k = \boldsymbol{N}_y^i \cdot \boldsymbol{N}^k = 0$ [**21**] and hence the *Gauß–Weingarten equations* associated with isothermic surfaces in $\mathbb{R}^{n+2}$ assume the form

$$\begin{aligned} \begin{pmatrix} \boldsymbol{X}\\ \boldsymbol{Y}\\ \underline{\boldsymbol{N}} \end{pmatrix}_x &= \begin{pmatrix} 0 & -\theta_y & -\underline{h}_1^{\mathrm{T}}\\ \theta_y & 0 & 0\\ \underline{h}_1 & 0 & 0 \end{pmatrix} \begin{pmatrix} \boldsymbol{X}\\ \boldsymbol{Y}\\ \underline{\boldsymbol{N}} \end{pmatrix},\\ \begin{pmatrix} \boldsymbol{X}\\ \boldsymbol{Y}\\ \underline{\boldsymbol{N}} \end{pmatrix}_y &= \begin{pmatrix} 0 & \theta_x & 0\\ -\theta_x & 0 & -\underline{h}_2^{\mathrm{T}}\\ 0 & \underline{h}_2 & 0 \end{pmatrix} \begin{pmatrix} \boldsymbol{X}\\ \boldsymbol{Y}\\ \underline{\boldsymbol{N}} \end{pmatrix}, \end{aligned} \tag{2.6}$$

where the unit tangent vectors $\boldsymbol{X}$ and $\boldsymbol{Y}$ are defined by

$$\boldsymbol{X} = e^{-\theta}\boldsymbol{r}_x, \quad \boldsymbol{Y} = e^{-\theta}\boldsymbol{r}_y. \tag{2.7}$$

The integrability conditions for the linear system (2.6), which constitute the *Gauß–Mainardi–Codazzi equations* associated with the fundamental forms (2.5), then read

$$\underline{h}_{1y} = \theta_y \underline{h}_2, \quad \underline{h}_{2x} = \theta_x \underline{h}_1, \quad \Delta\theta + \underline{h}_1^{\mathrm{T}}\underline{h}_2 = 0, \tag{2.8}$$

with the Laplacian $\Delta = \partial_x^2 + \partial_y^2$. The latter is a vector version of the classical *isothermic system* [**20**]. Accordingly, we here term it *vector isothermic system.* It is implicitly contained in [**32**].

Classical constant mean curvature, minimal, Bonnet and Enneper surfaces and Dupin cyclides are isothermic as are quadrics and surfaces of revolution [**20**]. Moreover, isothermic surfaces may be confined to spheres $\mathbb{S}^{n+1} \subset \mathbb{R}^{n+2}$ or, in other words, they may also be defined in Riemannian spaces of constant curvature.

3. Generation and Discretization of Isothermic Surfaces

The transformation theory for conjugate coordinates is well-established [**21**]. In particular, the *Fundamental transformation* [**24**] constitutes a mapping between surfaces which preserves conjugate coordinates, that is *conjugate nets* are transformed into conjugate nets. The basic ingredient in the Fundamental transformation is a *parallel conjugate net.* Thus, if $\boldsymbol{r}$ parametrizes a conjugate net on a surface Σ, that is $\boldsymbol{r}$ obeys a linear hyperbolic equation of the form $(2.2)_1$, then a conjugate net $\boldsymbol{r}'$ on a surface Σ' is said to be parallel to the conjugate net on Σ if the tangent vectors to the coordinate lines $x = \text{const}$ and $y = \text{const}$ are parallel at corresponding points on Σ and Σ'. This implies that there exist functions $\mathcal{X}$ and $\mathcal{Y}$ such that

$$\boldsymbol{r}'_x = \mathcal{X}\boldsymbol{r}_x, \quad \boldsymbol{r}'_y = \mathcal{Y}\boldsymbol{r}_y. \tag{3.1}$$

Conversely, any solution $(\mathcal{X}, \mathcal{Y})$ of the compatibility condition of (3.1), viz

$$\mathcal{X}_y = A(\mathcal{Y} - \mathcal{X}), \quad \mathcal{Y}_x = B(\mathcal{X} - \mathcal{Y}), \tag{3.2}$$

gives rise to a conjugate net $\boldsymbol{r}'$ which is parallel to $\boldsymbol{r}$. Now, if r is a scalar solution of $(2.2)_1$ and r' is a corresponding solution of the scalar version of (3.1), that is

$$r'_x = \mathcal{X}r_x, \quad r'_y = \mathcal{Y}r_y, \tag{3.3}$$

then it is readily shown that $(2.2)_1$ is form-invariant under the Fundamental transformation

$$\mathbb{B}\colon \boldsymbol{r} \to \tilde{\boldsymbol{r}} = \boldsymbol{r} - \frac{r}{r'}\boldsymbol{r}', \tag{3.4}$$

with some functions $\tilde{A}$ and $\tilde{B}$.

It turns out that the functions $\mathcal{X}$ and $\mathcal{Y}$ may be constrained in such a way that the Fundamental transformation preserves the orthogonality condition $(2.2)_2$. In this case, the Fundamental transformation has been termed *generalized Ribaucour transformation* [**21**] since in $\mathbb{R}^3$ one retrieves a transformation which was set down by Ribaucour in connection with the correspondence of lines of curvature on the envelopes of sphere congruences.

Similarly, it can be shown [**34**] that an admissible additional constraint guarantees that the generalized Ribaucour transformation leaves invariant the conformality property $(2.2)_3$. The scalar function r then forms part of the $n+4$-dimensional linear system

$$\begin{pmatrix} X \\ Y \\ \underline{N} \\ r \\ r^* \end{pmatrix}_x = \begin{pmatrix} 0 & -\theta_y & -\underline{h}_1^{\mathrm{T}} & me^{-\theta} & me^{\theta} \\ \theta_y & 0 & 0 & 0 & 0 \\ \underline{h}_1 & 0 & 0 & 0 & 0 \\ e^{\theta} & 0 & 0 & 0 & 0 \\ e^{-\theta} & 0 & 0 & 0 & 0 \end{pmatrix} \begin{pmatrix} X \\ Y \\ \underline{N} \\ r \\ r^* \end{pmatrix},$$

(3.5)

$$\begin{pmatrix} X \\ Y \\ \underline{N} \\ r \\ r^* \end{pmatrix}_y = \begin{pmatrix} 0 & \theta_x & 0 & 0 & 0 \\ -\theta_x & 0 & -\underline{h}_2^{\mathrm{T}} & -me^{-\theta} & me^{\theta} \\ 0 & \underline{h}_2 & 0 & 0 & 0 \\ 0 & e^{\theta} & 0 & 0 & 0 \\ 0 & -e^{-\theta} & 0 & 0 & 0 \end{pmatrix} \begin{pmatrix} X \\ Y \\ \underline{N} \\ r \\ r^* \end{pmatrix}$$

which is compatible modulo the vector isothermic system (2.8). In $\mathbb{R}^3$, the above-mentioned particular Ribaucour transformation is identical with Darboux's classical Bäcklund transformation for isothermic surfaces [**16**]. In fact, the linear representation (3.5) containing the arbitrary constant parameter m was introduced in this context. Remarkably, for $m = 0$, the linear system (3.5) is nothing but a scalar version of the Gauß–Weingarten equations (2.6). This observation is particularly important in connection with a *purely algebraic* iteration of the Bäcklund transformation for isothermic surfaces [**34**].

Bianchi [**3**] observed that Darboux's transformations $\mathbb{B}_i$ for distinct parameters m_i *commute.* Thus, if Σ_1 and Σ_2 are two Bäcklund transforms of an isothermic surface Σ then the $\mathbb{B}_2$-transform Σ_{12} of Σ_1 and the $\mathbb{B}_1$-transform Σ_{21} of Σ_2 coincide. The property

$$\mathbb{B}_2 \circ \mathbb{B}_1 = \mathbb{B}_1 \circ \mathbb{B}_2, \tag{3.6}$$

is also known as *closure of the Bianchi diagram.* Accordingly, any point $\boldsymbol{r}$ on the seed isothermic surface Σ and its Bäcklund transforms $\boldsymbol{r}_1$, $\boldsymbol{r}_2$ and $\boldsymbol{r}_{12}$ on the surfaces Σ_1, Σ_2 and Σ_{12} respectively constitute the corners of a quadrilateral as illustrated in Figure 1. In the sequel, we refer to this quadrilateral as a *Bianchi quadrilateral.*

The Bianchi quadrilateral admits three important geometric properties. Firstly, the Bianchi quadrilateral associated with the generic Fundamental transformation is readily shown to be *planar* [**21**]. Secondly, Bianchi [**4**] pointed out that if the Fundamental transformation is restricted to the Ribaucour transformation then the Bianchi quadrilateral is *cyclic*, that is the Bianchi quadrilateral is inscribed in a circle. Thirdly, Demoulin [**19**] deduced that the *cross-ratio* of the Bianchi quadrilateral associated with Darboux's Bäcklund transformation for isothermic surfaces is *independent* of x and y. In fact, the cross-ratio reads

$$\frac{|\boldsymbol{r}_1 - \boldsymbol{r}||\boldsymbol{r}_{12} - \boldsymbol{r}_2|}{|\boldsymbol{r}_2 - \boldsymbol{r}||\boldsymbol{r}_{12} - \boldsymbol{r}_1|} = \left|\frac{m_2}{m_1}\right|, \tag{3.7}$$

where m_1 and m_2 are the Bäcklund parameters corresponding to $\mathbb{B}_1$ and $\mathbb{B}_2$. The geometry of the Bianchi quadrilateral is displayed in Figure 1. It is emphasized that these classical results are equally valid in Euclidean spaces of arbitrary dimension [**34**].

It is remarkable that planarity, cyclicity and constant cross-ratio of quadrilaterals may be regarded as the discrete analogues of conjugacy, orthogonality and

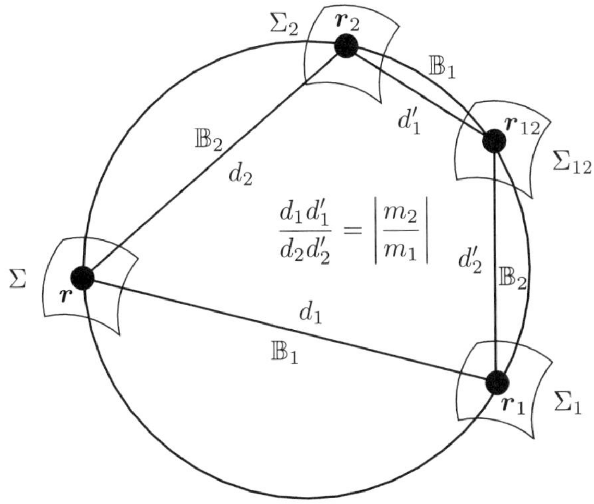

FIGURE 1. Planarity, cyclicity and constant cross-ratio of the Bianchi quadrilateral.

conformality. Indeed, Bobenko and Pinkall [**5**] proposed the following definition: A lattice (*discrete surface*)

$$\boldsymbol{r}\colon \mathbb{Z}^2 \to \mathbb{R}^{n+2}, \quad (n_1, n_2) \mapsto \boldsymbol{r}(n_1, n_2), \tag{3.8}$$

is termed *discrete isothermic* if the quadrilaterals are cyclic (and therefore planar) and the cross-ratio is independent of n_1, n_2. It is not difficult to show that this definition of discrete isothermic surfaces reduces to the classical one in the natural continuum limit.

The integrability of discrete isothermic surfaces has been shown explicitly in the cases $n = 1$ [**5**] and $n = 2$ [**23**]. It is evident that iterative application of two infinite sequences of Darboux's Bäcklund transformations generates a two-parameter (x, y) family of discrete isothermic surfaces. Moreover, in $\mathbb{R}^3$, application of an additional suite of Bäcklund transformations gives rise to triply "orthogonal" systems of discrete isothermic surfaces [**13**].

4. The Vector Calapso System

In principle, it is possible to derive an explicit nonlinear superposition principle for solutions of the vector isothermic system (2.8) which are generated by the Bäcklund transformation for isothermic surfaces. However, this *permutability theorem* at the nonlinear level appears to be rather complicated. In contrast, if one reformulates the Gauß–Mainardi–Codazzi equations in the manner proposed by Calapso then the permutability theorem assumes a remarkably compact vectorial form. Thus, it is readily shown that if $(\theta, \underline{h}_1, \underline{h}_2)$ is a solution of the vector isothermic system then the functions

$$\underline{z} = \frac{1}{\sqrt{2}}(\underline{h}_1 + \underline{h}_2), \quad \rho = e^{-\theta}(e^{\theta})_{xy}, \tag{4.1}$$

satisfy the *vector Calapso system*

$$\underline{z}_{xy} = \rho \underline{z}, \quad \Delta\rho + (\underline{z}^2)_{xy} = 0. \tag{4.2}$$

Conversely, it has been demonstrated that any solution of the vector Calapso system may be mapped to a class of solutions of the vector isothermic system [**34**].

In the classical case $n = 1$, one may eliminate the function ρ to obtain the *Calapso equation* [**9**]

$$\Delta\left(\frac{z_{xy}}{z}\right) + (z^2)_{xy} = 0. \tag{4.3}$$

For $n = 2$, the vector Calapso system turns out to coincide with the stationary reduction, i.e. $u_t = 0$, $\underline{z} = (\Re(u), \Im(u))$, of the *(Benney–Roskes-) Davey–Stewartson II equation* [**1,17**]

$$\mathrm{i}u_{\mathrm{t}} = u_{\mathrm{xy}} - \rho u, \quad \Delta\rho + (|u|^2)_{\mathrm{xy}} = 0. \tag{4.4}$$

This observation is due to Ferapontov [**22**] in the case of the classical Calapso equation which corresponds to the *real* reduction of the Davey–Stewartson II equation.

In Minkowski space $\mathbb{M}^{n+2}$ with space-like normal bundle, the vector Calapso system takes the form

$$\underline{z}_{xy} = \rho\underline{z}, \quad \Box\rho + (\underline{z}^2)_{xy} = 0, \tag{4.5}$$

where $\Box = \partial_x^2 - \partial_y^2$ designates the usual d'Alembert operator. In the scalar case, we therefore obtain the *Zoomeron equation* [**10, 11**]

$$\Box\left(\frac{z_{xy}}{z}\right) + (z^2)_{xy} = 0. \tag{4.6}$$

The two-component case corresponds to the *Boomeron equation* [**10, 11**] as discussed in a forthcoming paper [**18**]. As in the Euclidean case, the Boomeron equation may be regarded as the stationary reduction of the *Davey–Stewartson III equation* [**7**]

$$\mathrm{i}u_{\mathrm{t}} = u_{\mathrm{xy}} - \rho u, \quad \Box\rho + (|u|^2)_{\mathrm{xy}} = 0. \tag{4.7}$$

5. A Scalar Lax Pair Based on the Moutard Equation

Motivated by the construction of *generalized Loewner systems* [**31**], for instance the *vector Ernst equation* of general relativity [**26**], a Lax pair for the vector Calapso system has been sought in [**34**] which contains the *Moutard equation* $(4.2)_1$ as the "scattering problem". The latter has been used extensively in differential geometry and soliton theory (see [**34**] and references therein). Thus, the scalar (nonlocal) Lax pair

$$\psi_{xy} = \rho\psi, \quad \Box\psi + (f + k)\psi = \underline{z}^{\mathrm{T}}\underline{S}(\underline{z}, \psi), \tag{5.1}$$

where k is an arbitrary constant and the *bilinear potential* $\underline{S}(\underline{z}, \psi)$ is defined by

$$\underline{S}_x = \underline{z}_x\psi - \underline{z}\psi_x, \quad \underline{S}_y = \underline{z}\psi_y - \underline{z}_y\psi, \tag{5.2}$$

is consistent if and only if $(\underline{z}, \rho)$ is a solution of the vector Calapso system (4.2). More precisely, the bilinear potential $\underline{S}$ is well-defined if and only if $\underline{z}$ is a solution of the Moutard equation $(4.2)_1$. Then, the compatibility condition of (5.1) leads to the Frobenius system

$$f_x = 2\rho_y + (\underline{z}^2)_x, \quad f_y = -2\rho_x - (\underline{z}^2)_y, \tag{5.3}$$

which, in turn, is compatible modulo the remaining scalar equation $(4.2)_2$ of the vector Calapso system. Since f is only defined up to a constant of integration, the trivial invariance $f \to f + k$ explains the existence of the constant parameter k.

The Moutard equation was discussed by Moutard [**28**] with regard to its transformation into canonical forms which may be integrated. It turns out that this *Moutard transformation* indeed leaves invariant the vector Calapso system. In fact, if ψ^1 is a solution of the Lax pair (5.1) subject to the admissible constraint

$$2\psi^1 \Delta \psi^1 - 2(\nabla \psi^1)^2 + \underline{z}^2 \psi^1 = \underline{S}^2(\underline{z}, \psi^1), \tag{5.4}$$

then

$$\mathbb{B}_1 : \underline{z}_1 = \frac{\underline{S}(\underline{z}, \psi^1)}{\psi^1}, \quad \rho_1 = \rho - 2(\ln \psi^1)_{xy}, \tag{5.5}$$

is another solution of the vector Calapso system. It is interesting to note that this particular Moutard transformation induces at the surface level a transformation which is identical with the Bäcklund transformation for isothermic surfaces formulated in Section 3. This is consistent with the fact that the Lax pair (5.1) may be retrieved from the linear system (3.5).

6. An Explicit Permutability Theorem for the Vector Calapso System

For the derivation of an explicit permutability theorem, it is now necessary to bring the vector Calapso system into the form

$$\underline{z}_{xy} = a_y \underline{z}, \quad a_y = \varepsilon b_x, \quad a_x + \varepsilon b_y + \underline{z}^2 = 0, \tag{6.1}$$

where $\varepsilon^2 = 1$ and $\rho = a_y = \varepsilon b_x$, $f = -a_x + \varepsilon b_y$. The action of the Moutard transformation on a, b and ε is then readily shown to be

$$\mathbb{B}_1 : a_1 = a - 2(\ln \psi^1)_x, \quad b_1 = -b + 2\varepsilon(\ln \psi^1)_y, \quad \varepsilon_1 = -\varepsilon. \tag{6.2}$$

The constant $\varepsilon = \pm 1$ has been introduced for purely aesthetic reasons, that is in order to be able to cast the permutability theorem into a compact vectorial form. To this end, one expresses the ratios ψ^1_x/ψ^1, ψ^1_y/ψ^1 and $\underline{S}(\underline{z}, \psi^1)/\psi^1$ in terms of $\underline{z}, \underline{z}_1, a, a_1, b, b_1$ and eliminates them from the Lax pair and the defining relations for $\underline{S}$ by means of the transformation formulae $(5.5)_1$, (6.2). This process leads to three differential relations and nine additional relations associated with the (compound) Bäcklund transformations $\mathbb{B}_2$, $\mathbb{B}_2 \circ \mathbb{B}_1$, $\mathbb{B}_1 \circ \mathbb{B}_2$. If we take into account that the Bianchi diagram commutes then one can extract three *purely algebraic relations* from these twelve differential relations. It is noted that this elimination procedure is analogous to that for the sine–Gordon equation based on the classical Bäcklund transformation [**25**]. Thus, we have the following result [**34**]:

If $\underline{u}_1$ and $\underline{u}_2$ are two solutions of the vector Calapso system related to a seed solution $\underline{u} = (a, b, \sqrt{2}\underline{z})$ by the Bäcklund transformations $\mathbb{B}_1$ and $\mathbb{B}_2$ then $\underline{u}_{12}$ as given by

$$\underline{u}_{12} - \underline{u} = \frac{4(k_2 - k_1)}{(\underline{u}_1 - \underline{u}_2)^2}(\underline{u}_1 - \underline{u}_2), \tag{6.3}$$

is a fourth solution of the vector Calapso system and constitutes the Bäcklund transform of both $\underline{u}_1$ and $\underline{u}_2$, that is $\underline{u}_{12} = \mathbb{B}_2(\underline{u}_1) = \mathbb{B}_1(\underline{u}_2)$.

It is interesting to note that in the case $n = 2$ and $b \to ib$ corresponding to isothermic surfaces in Minkowski space the permutability theorem (6.3) constitutes a simplified version of that set down for the Boomeron equation in [**11**]. The latter contains scalar and vector components and has been derived in connection with a matrix version of the classical Darboux transformation [**15**] applied to a matrix Schrödinger equation.

7. The Discrete Calapso Equation

Permutability theorems for integrable equations may be interpreted naturally as integrable difference equations. The prototypical example is the classical permutability theorem for the sine–Gordon equation. By construction, the permutability theorem may also be regarded as a Bäcklund transformation for itself due to the commutativity of the associated Bianchi diagram (see [**34**] and references therein). In the present case, we may regard the permutablity theorem (6.3), that is

$$\underline{u}_{12} - \underline{u} = \frac{\lambda_1 - \lambda_2}{(\underline{u}_1 - \underline{u}_2)^2}(\underline{u}_1 - \underline{u}_2), \tag{7.1}$$

where $\lambda_i = -4k_i$, as a difference equation with the vector-valued function $\underline{u} = \underline{u}(n_1, n_2)$ and $\underline{u}_1 = \underline{u}(n_1{+}1, n_2)$, $\underline{u}_2 = \underline{u}(n_1, n_2{+}1)$, $\underline{u}_{12} = \underline{u}(n_1{+}1, n_2{+}1)$ depending on the discrete variables n_1, n_2. By construction, this difference equation represents a discretization of the vector Calapso system since the Bäcklund transformation $\mathbb{B}$ discretizes isothermic surfaces as discussed in Section 3. Indeed, it has been shown in [**34**] that the *discrete (vector) Calapso equation* (7.1) reduces to its continuous counterpart (6.1) in the natural continuum limit.

The discrete Calapso equation admits a matrix Lax pair which takes the form of the Lax pair for the *discrete Korteweg–de Vries* (KdV) *equation* [**30**] with *Clifford algebra*-valued entries [**34**]. Indeed, in the case $N = n + 2 = 1$, the discrete Calapso equation reduces to the discrete KdV equation

$$(u_{12} - u)(u_1 - u_2) = \lambda_1 - \lambda_2. \tag{7.2}$$

For $N = 2$, the two-component discrete Calapso equation may be identified with the complexified KdV equation and, in turn, mapped via the *discrete Miura transformation* [**29**] to the constant cross-ratio equation

$$\frac{(g_1 - g)(g_{12} - g_2)}{(g_2 - g)(g_{12} - g_1)} = \frac{\lambda_2}{\lambda_1}, \tag{7.3}$$

governing *discrete holomorphic functions* [**5**]. The case $N = 3$ corresponds to a discretization of the classical Calapso equation. However, if we interpret the inner product of two vectors $\underline{f}$ and $\underline{g}$ as $< \underline{f}, \underline{g} >= \underline{f}^{\mathrm{T}} \mathcal{S} \underline{g}$, where $\mathcal{S}$ denotes a suitable constant symmetric matrix, then (7.1) may also be viewed as a discrete Calapso equation in Minkowski space. Accordingly, discrete Zoomeron and Boomeron equations are obtained in the cases $N = 3$ and $N = 4$ respectively.

A Bäcklund transformation for the discrete Calapso equation is readily obtained by considering another Bäcklund transformation $\mathbb{B}_\mu$ associated with a Bäcklund parameter μ. If we introduce the notation $\tilde{\underline{u}} = \mathbb{B}_\mu(\underline{u})$ then the two permutability theorems corresponding to $[\mathbb{B}_1, \mathbb{B}_\mu] = [\mathbb{B}_2, \mathbb{B}_\mu] = 0$ read

$$\tilde{\underline{u}}_1 - \underline{u} = \frac{\mu - \lambda_1}{(\tilde{\underline{u}} - \underline{u}_1)^2}(\tilde{\underline{u}} - \underline{u}_1), \quad \tilde{\underline{u}}_2 - \underline{u} = \frac{\mu - \lambda_2}{(\tilde{\underline{u}} - \underline{u}_2)^2}(\tilde{\underline{u}} - \underline{u}_2). \tag{7.4}$$

These may be interpreted as difference equations and therefore constitute *Bäcklund equations* for the discrete Calapso equation. Thus, if $\underline{u}$ is a solution of the discrete Calapso equation (7.1) then the system (7.4) is compatible and $\tilde{\underline{u}}$ is another solution of the discrete Calapso equation [**34**].

8. The Discrete $O(N)$ Nonlinear σ-Model

In the last section of this paper, we record a canonical reduction of the discrete Calapso equation which delivers an important connection with nonlinear σ-models and, in particular, (discrete) pseudospherical surfaces. Thus, it is natural to impose the constraint $\underline{u}^2 = 1$ on the discrete Calapso equation written in the form

$$\underline{u}_{12} + \underline{u} = \frac{\lambda_1 + \lambda_2}{(\underline{u}_1 + \underline{u}_2)^2}(\underline{u}_1 + \underline{u}_2). \tag{8.1}$$

As shown in [**34**], this constraint leads to

$$\underline{u}_{12} + \underline{u} = \frac{\underline{u}^{\mathrm{T}}(\underline{u}_1 + \underline{u}_2)}{1 + \underline{u}_1^{\mathrm{T}}\underline{u}_2}(\underline{u}_1 + \underline{u}_2), \quad \underline{u}^2 = 1, \tag{8.2}$$

together with the admissible *first integrals*

$$\lambda_1 = 2\underline{u}^{\mathrm{T}}\underline{u}_1, \quad \lambda_2 = 2\underline{u}^{\mathrm{T}}\underline{u}_2. \tag{8.3}$$

In the natural continuum limit, we therefore obtain the integrable $O(N)$ *nonlinear σ-model* relevant in Lagrangian field theory [**33**]. The (discrete) $O(3)$ nonlinear σ-model is known to constitute the (discrete) Gauß map of (discrete) pseudospherical surfaces [**6**]. In the context of Bäcklund transformations, this observation implies that Bianchi's classical permutability theorem for pseudospherical surfaces is encapsulated in the permutability theorem (6.3) for isothermic surfaces. Moreover, the local $O(4) \cong SU(2) \times SU(2)$ isomorphism has been used to show that the $O(4)$ nonlinear σ-model is equivalent to the Pohlmeyer–Regge–Lund system [**27, 33**] which, in turn, has been linked to the Maxwell–Bloch and self-induced transparency (SIT) equations as well as the system governing stimulated Raman scattering (SRS) [**35**]. The integrable discrete counterpart of these physically significant systems is therefore represented by the discrete $O(4)$ nonlinear σ-model $(8.2)_{N=4}$.

As in the continuous case, the discrete Calapso equation admits a scalar Lax pair which assumes a remarkably simple form in the context of the discrete $O(N)$ nonlinear σ-model [**34**]. Application of the natural continuum limit provides the compact Lax pair

$$\psi_{xy} + (\underline{u}_x^{\mathrm{T}}\underline{u}_y)\psi = 0, \quad \lambda\psi = \underline{u}^{\mathrm{T}}\underline{S}(\underline{u}, \psi), \tag{8.4}$$

for the $O(N)$ nonlinear σ-model, where the bilinear potential $\underline{S}$ is defined by

$$\underline{S}_x = \underline{u}_x\psi - \underline{u}\psi_x, \quad \underline{S}_y = \underline{u}\psi_y - \underline{u}_y\psi. \tag{8.5}$$

The Bäcklund transformation (7.4) then reduces to

$$\begin{gathered}(\tilde{\underline{u}} - \underline{u})_x = -\frac{\underline{u}_x^{\mathrm{T}}\tilde{\underline{u}}}{\mu + 1}(\tilde{\underline{u}} + \underline{u}), \quad (\tilde{\underline{u}} + \underline{u})_y = -\frac{\underline{u}_y^{\mathrm{T}}\tilde{\underline{u}}}{\mu - 1}(\tilde{\underline{u}} - \underline{u}), \\ \tilde{\underline{u}}^{\mathrm{T}}\underline{u} = \mu, \quad \tilde{\underline{u}}^2 = 1.\end{gathered} \tag{8.6}$$

It is interesting to note that the "spectral parameter" μ has a distinct geometric meaning. Thus, the property $(8.6)_3$ implies that the angle between the unit vectors $\underline{u}$ and $\tilde{\underline{u}}$ is constant. This is the direct analogue of the requirement which led to Bäcklund's generalization of the Bianchi transformation for pseudospherical surfaces [**2, 20**], namely that the angle between the normals $\boldsymbol{N}$ and $\widetilde{\boldsymbol{N}}$ to a pseudospherical surface Σ and its Bäcklund transform $\widetilde{\Sigma}$ respectively be constant. The Bianchi transformation corresponds to $\mu = 0$, in which case $\underline{u}$ and $\tilde{\underline{u}}$ are orthogonal. It is emphasized that the relation $(8.6)_3$ is equally valid in the discrete setting.

References

1. D. J. Benney and G. J. Roskes, *Wave instabilities*, Stud. Appl. Math. **48** (1969), 377–385.
2. L. Bianchi, *Ricerche sulle superficie a curvatura costante e sulle elicoidi*, Ann. Scuola Norm. Sup. Pisa **2** (1879), 285–340.
3. ———, *Ricerche sulle superficie isoterme e sulla deformazione delle quadriche*, Ann. di Matem. **11** (1905), 93–157.
4. ———, *Lezioni di geometria differenziale* **1-4**, Zanichelli, Bologna, 1923–1927.
5. A. Bobenko and U. Pinkall, *Discrete isothermic surfaces*, J. Reine Angew. Math. **475** (1996), 187–208.
6. ———, *Discrete surfaces with constant negative Gaussian curvature and the Hirota equation*, J. Differential Geom. **43** (1996), no. 3, 527–611.
7. M. Boiti, F. Pempinelli, and P. C. Sabatier, *Nonlinear evolution equations from an inverse spectral problem*, Applications of Analytic and Geometric Methods to Nonlinear Differential Equations (P. A. Clarkson, ed.), Kluwer, Dordrecht, 1993, pp. 89–108.
8. F. Burstall, U. Hertrich-Jeromin, F. Pedit, and U. Pinkall, *Curved flats and isothermic surfaces*, Math. Z. **225** (1997), no. 2, 199–209.
9. P. Calapso, *Sulla superficie a linee di curvatura isoterme*, Rend. Circ. Mat. Palermo **17** (1903), 275–286.
10. F. Calogero and A. Degasperis, *Coupled nonlinear evolution equations solvable via the inverse spectral transform, and solitons that come back: The boomeron*, Lett. Nuovo Cimento **16** (1976), no. 14, 425–433.
11. ———, *Bäcklund transformations, nonlinear superposition principle, multisoliton solutions and conserved quantities for the "boomeron" nonlinear evolution equation*, Lett. Nuovo Cimento **16** (1976), no. 14, 434–438.
12. J. Cieśliński, *The Darboux–Bianchi transformation for isothermic surfaces. Classical results versus the soliton approach*, Differential Geom. Appl. **7** (1997), 1–28.
13. ———, *The Bäcklund transformation for discrete isothermic surfaces*, Symmetries and Integrability of Difference Equations (P. Clarkson and F. Nijhoff, eds.), London Math. Soc. Lecture Note Ser., vol. 255, Cambridge University Press, 1999, pp. 109–121.
14. J. Cieśliński, P. Goldstein, and A. Sym, *Isothermic surfaces in* $\mathbf{E}^3$ *as soliton surfaces*, Phys. Lett. A **205** (1995), 37–43.
15. G. Darboux, *Sur une proposition relative des équations linéaires*, C.R. Acad. Sci. Paris **96** (1882), 1456–1459.
16. ———, *Sur les surfaces isothermiques*, C. R. Acad. Sci. Paris **128** (1899), 1299–1305.
17. A. Davey and K. Stewartson, *On three-dimensional packets of surface waves*, Proc. Roy. Soc. London Ser. A **338** (1974), 101–110.
18. A. Degasperis, C. Rogers, and W.K. Schief, *Isothermic surfaces generated via Darboux–Bäcklund transformations. Boomeron and Zoomeron connections*, (in preparation).
19. A. Demoulin, *Sur les systèmes et les congruences* K, C. R. Acad. Sci. Paris **150** (1910), 156–159.
20. L. P. Eisenhart, *A treatise on the differential geometry of curves and surfaces*, Dover Publications, New York, 1960.
21. ———, *Transformation of surfaces*, Chelsea, New York, 1962.
22. E. V. Ferapontov, *Surfaces in Lie sphere geometry and the stationary Davey–Stewartson hierarchy*, SFB 288 Preprint **287**, Technische Universität Berlin, 1997.
23. U. Hertrich-Jeromin, T. Hoffmann, and U. Pinkall, *A discrete version of the Darboux transform for isothermic surfaces*, Discrete Integrable Geometry and Physics (A. I. Bobenko and R. Seiler, eds.), Clarendon Press, Oxford, 1999, pp. 59–81.
24. H. Jonas, *Über die Transformation der konjugierten Systeme und über den gemeinsamen Ursprung der Bianchischen Permutabilitätstheoreme*, Sitzungsber. Berl. Math. Ges. **14** (1915), 96–118.
25. G. L. Lamb Jr, *Bäcklund transformations at the turn of the century*, Bäcklund Transformations (R. M. Miura, ed.), Lecture Notes in Math. **515**, Springer-Verlag, New York, 1976, pp. 69–79.
26. B. Léauté and G. Marcilhacy, *On a particular solution of the equation of Ernst with n fields, parametrized by an arbitrary harmonic function*, J. Math. Phys. **26** (1985), no. 8, 1938–1941.

27. F. Lund and T. Regge, *Unified approach to strings and vortices with soliton solutions*, Phys. Rev. D **14** (1976), no. 6, 1524–1535.
28. T. Moutard, *Sur la construction des équations de la forme* $1/z\ \partial^2 z/\partial x\partial y = \lambda(x,y)$, *qui admettent une intégrale générale explicite*, J. École Polytechnique **45** (1878), 1–11.
29. F. Nijhoff and H. Capel, *The discrete Korteweg–de Vries equation*, Acta Appl. Math. **39** (1995), 133–158.
30. F. W. Nijhoff, G. R. W. Quispel, and H. W. Capel, *Direct linearization of nonlinear difference-difference equations*, Phys. Lett. A **97** (1983), no. 4, 125–128.
31. W. Oevel and W. K. Schief, *Squared eigenfunctions of the* (*modified*) KP *hierarchy and scattering problems of Loewner type*, Rev. Math. Phys. **6** (1994), no. 6, 1301–1338.
32. B. Palmer, *Isothermic surfaces and the Gauss map*, Proc. Amer. Math. Soc. **104** (1988), no. 3, 876–884.
33. K. Pohlmeyer, *Integrable Hamiltonian systems and interactions through quadratic constraints*, Comm. Math. Phys. **46** (1976), no. 3, 207–221.
34. W. K. Schief, *Isothermic surfaces in spaces of arbitrary dimension: Integrability, discretization and Bäcklund transformations. A discrete Calapso equation*, Stud. Appl. Math. **106** (2001), 85–137.
35. H. Steudel, *Space-time symmetry of self-induced transparency and of stimulated Raman scattering*, Phys. Lett. A **156** (1991), 491–492.
36. A. Sym, *Soliton surfaces and their application. Soliton geometry from spectral problems*, Geometric Aspects of the Einstein Equations and Integrable Systems (R. Martini, ed.), Lecture Notes in Phys. **239**, Springer-Verlag, Berlin (1985), 154–231.

School of Mathematics, The University of New South Wales, Sydney, NSW 2052, Australia.

E-mail address: `schief@maths.unsw.edu.au`

Centre de Recherches Mathématiques
CRM Proceedings and Lecture Notes
Volume **29**, 2001

Bäcklund Transformations Induced by Symmetries Application: Discrete mKdV

R. Schmid

ABSTRACT. We give a general method for the derivation of Bäcklund transformations induced by symmetries. We demonstrate this method on two easy examples, the KdV equation and the discrete mKdV equation $\partial q_n/\partial t = (1+q_n^2)(q_{n+1}-q_{n-1})$ with simple point symmetries. In connection with symplectic numerical algorithms we are particularly interested in discrete equations.

1. Outline of the Method

We consider a nonlinear partial differential equation in the form

$$u_t = F(u, u_x, u_{xx}, \dots), \quad u = u(x,t), \tag{1.1}$$

or in discrete form

$$\frac{\partial u_n}{\partial t} = F(u_n, u_{n+i}, \dots), \quad n, i \in \mathbf{Z}. \tag{1.2}$$

We assume that this partial differential equation is associated with a **linear** integrable system

$$\begin{cases} \Phi_x = M\Phi, \\ \Phi_t = N\Phi, \end{cases} \tag{1.3}$$

which means that the compatibility condition $\Phi_{xt} = \Phi_{tx}$ is equivalent of u being a solution of the partial differential equation (1.1), where $\Phi = \binom{\varphi_1}{\varphi_2}$ and M, N are 2×2 matrices with entries functions of $u, u_x, \dots$, and a parameter λ.

More in general, the integrable equation (1.1) can be obtained by assuming the existence of a compatible nonlinear system of degree k and order n, depending on k potential functions $\varphi_1, \dots, \varphi_k$ i.e. the matrices M and N in (1.3) are replaced by nonlinear functions $\Omega(\varphi_1, \dots, \varphi_k, u, u_x, \dots, \partial_x^n u)$ and $\Theta(\varphi_1, \dots, \varphi_k, u, u_x, \dots, \partial_x^n u)$ respectively. For example, the system

$$\begin{cases} \varphi_x = \Omega(\varphi, u) = -\lambda + \varphi^2 + u, \\ \varphi_t = \Theta(\varphi, u, u_x, u_{xx}) = 4\lambda(-\lambda + \varphi^2) + 2(\lambda + \varphi^2)u + 2u^2 + 2\varphi u_x + u_{xx}, \end{cases}$$

2000 *Mathematics Subject Classification.* Primary: 37K35; Secondary: 35Q53.
We thank X. Yang for many fruitful discussions and suggestions.
This is the final form of the paper.

is a nonlinear representation of an integrable system of degree 1 and order 2 associated with the KdV equation. For $k = 1$ these were called pseudopotentials in [**12**], [**18**]. For this more general case we refer to [**19**], [**17**]. Here we restrict ourself to the linear (matrix) case.

We derive the Riccati form of the equation by letting $q := \varphi_1/\varphi_2$ and obtain

$$\begin{cases} q_x = q_x\big(M(u,q,\lambda)\big), \\ q_t = q_t\big(N(u,q,\lambda)\big). \end{cases} \tag{1.4}$$

Solving the first equation in (1.4) for u we get a Miura transform as $u = f(\lambda, q, q_x, \dots)$, and by plugging u into the second equation of (1.4) we obtain the **modified** partial differential equation (mpde) of (1.1) as

$$q_t = K(\lambda, q, q_x, \dots). \tag{1.5}$$

THEOREM 1.1. *If the pde* (1.1), *or* (1.2) *has two different linear systems* (*or two different Riccati equations*) *but the corresponding modified equations are identical, then we can define a Bäcklund transformation for this pde.*

PROOF. Let

$$\begin{cases} \Phi'_x = M'\Phi', \\ \Phi'_t = N'\Phi', \end{cases} \tag{1.6}$$

be a second linear system for (1.1). As before we get the Riccati equations for q'_x and q'_t , the Miura transform as $u' = g(\lambda', q', q'_x, \dots)$ and the corresponding modified equation $q'_t = L(\lambda', q', q'_x, \dots)$. Since by assumption $L = K$, we get another Miura transform $u' = g(\lambda, u, u_x, \dots)$. Now we can define the Bäcklund transformation by $u' = g(f^{-1}(u))$. □

REMARK. In order to define this Bäcklund transformation we must be able to solve the equation $u = f(\lambda, q, q_x, \dots)$, which we can in all examples considered, like KdV, sine–Gordon, nonlinear Schrödinger equations etc. including their discrete versions [**16**], [**17**].

2. Example: KdV

We illustrate this method on the simplest example, the KdV equation $u_t = 6uu_x + u_{xxx}$.

A first linear system for the KdV equation is [**3**]

$$\Phi_x = \begin{Bmatrix} \lambda & u \\ -1 & -\lambda \end{Bmatrix} \Phi, \tag{2.1}$$

$$\Phi_t = \begin{Bmatrix} (4\lambda^3 + 2\lambda u + u_x) & (u_{xx} + 2\lambda u_x + 4\lambda^2 u + 2u^2) \\ (-4\lambda^2 - 2u) & (-4\lambda^3 - 2\lambda u - u_x) \end{Bmatrix} \Phi. \tag{2.2}$$

With $q = \varphi_1/\varphi_2$ we get the Riccati equations

$$\begin{cases} q_x = 2\lambda q + q^2 + u, \\ q_t = 2(4\lambda^3 + 2\lambda u + u_x)q + (4\lambda^2 + 2u)q^2 + (u_{xx} + 2\lambda u_x + 4\lambda^2 u + 2u^2). \end{cases} \tag{2.3}$$

Now for the Miura transform we obtain $u = q_x - 2\lambda - q^2$ which leads to the modified equation

$$q_t = -6q^2 q_x + q_{xxx} - 12\lambda q q_x. \tag{2.4}$$

Note that for $\lambda = 0$ this is the regular mKdV equation.

A second linear system fo KdV is given by

$$\Phi'_x = \begin{Bmatrix} -\lambda' & -u' \\ 1 & \lambda' \end{Bmatrix} \Phi', \tag{2.5}$$

$$\Phi'_t = \begin{Bmatrix} (-4\lambda'^3 - 2\lambda' u' + u'_x) & (-u'_{xx} + 2\lambda' u'_x - 4\lambda'^2 u' - 2u'^2) \\ (4\lambda'^2 + 2u') & (4\lambda'^3 + 2\lambda' u' - u'_x) \end{Bmatrix} \Phi'. \tag{2.6}$$

This leads to the Riccati equations

$$\begin{cases} q'_x = & -2\lambda' q' - q'^2 - u', \\ q'_t = & -2(4\lambda'^3 + 2\lambda' u' - u'_x) q' - (4\lambda'^2 + 2u') q'^2 \\ & \qquad -(u'_{xx} - 2\lambda' u'_x + 4\lambda'^2 u' + 2u'^2). \end{cases} \tag{2.7}$$

For the Miura transform we obtain $u' = -q'_x - 2\lambda' - q'^2$ which leads to the modified equation

$$q'_t = -6q'^2 q'_x + q'_{xxx} - 12\lambda' q' q'_x. \tag{2.8}$$

Notice that we have the same modified equations (2.4) and (2.8) for both linear systems (2.1)–(2.2) and (2.5)–(2.6). This leads to the Bäcklund transformation $u' = g(f^{-1}(u))$ where $u = f(q) = q_x - 2\lambda q - q^2$ and $u' = g(q) = -q_x - 2\lambda q - q^2$; [**16**]. This Bäcklund transformation is the same as obtained in [**3**] using different methods. For the discrete equations though we will obtain new Bäcklund transformation.

3. Symmetries

A second linear system often occurs form **symmetries**. We have the following theorem [**16**]:

THEOREM 3.1. *If the partial differential equation is invariant under a point transformation* $T : (x,t) \mapsto (x', t')$ *but the linear system* Φ *is not invariant under* T, *then* $\Phi' := T(\Phi)$ *defines a second linear system for this partial differential equation.*

Our example above was constructed from the point symmetry of the KdV equation $T(x,t) = (\pm x, \pm t)$. More complicated symmetries will be investigated in a future paper.

4. Discrete Equations: mKdV

In this section we apply the same method to discrete equations and illustrate it with the discrete mKdV equation [**1**]

$$\frac{\partial q_n}{\partial t} = (1 + q_n^2)(q_{n+1} - q_{n-1}), \quad n \in \mathbf{Z}. \tag{4.1}$$

A linear integrable system system for (4.1) is given by [**1**]

$$V_{n+1} = M_n V_n, \quad \frac{\partial V_n}{\partial t} = N_n V_n, \tag{4.2}$$

where

$$V_n = \begin{pmatrix} V_{n1} \\ V_{n2} \end{pmatrix}, \quad M_n = \begin{pmatrix} \eta & q_n \\ -q_n & \frac{1}{\eta} \end{pmatrix}, \quad N_n = \begin{pmatrix} A_n & B_n \\ C_n & D_n \end{pmatrix},$$

$$A_n = \eta^2 + q_{n-1}q_n, \quad B_n = q_n\eta + q_{n-1}\frac{1}{\eta},$$

$$C_n = -q_{n-1}\eta - q_n\frac{1}{\eta}, \quad D_n = \frac{1}{\eta^2} + q_{n-1}q_n.$$

The Riccati form with $u_n := V_{n1}/V_{n2}$ is given by

$$u_{n+1} = \frac{u_n\eta + q_n}{1/\eta - q_n u_n}, \tag{4.3}$$

$$\frac{\partial u_n}{\partial t} = (A_n - D_n)u_n + B_n - C_n u_n^2. \tag{4.4}$$

Solving (4.3) for q_n we obtain the Miura transform as in the continuous case and plugging it into (4.4) we get the modified equation as

$$\frac{\partial u_n}{\partial t} = \frac{u_{n+1} - u_n(\eta^2 + 1/\eta^2) - u_n^3}{1 + u_n u_{n+1}} + \frac{u_n^3 + u_n(\eta^2 + 1/\eta^2) - u_{n-1}}{1 + u_n u_{n-1}}. \tag{4.5}$$

We can solve for q_n and get the Miura transformation $q_n = \phi(\eta, u, u_n, \dots)$ as

$$q_n = \frac{u_{n+1}1/\eta - u_n\eta}{1 + u_n u_{n+1}}. \tag{4.6}$$

The Miura transform $q = \phi(u)$ maps a solution u of the modified equation (4.5) into a solution q of the original equation (4.1).

A simple symmetry of the discrete mKdV equation (4.1) is given by

$$T : n \longmapsto m = -n, \quad t \longmapsto s = -t. \tag{4.7}$$

However, the linear system (4.2) is **not** invariant under this symmetry T. Hence by Theorem 1.1, we obtain another linear system for the mKdV equation (4.1) by applying the transformation T to the linear system (4.2) and we get

$$V'_{n-1} = \begin{pmatrix} \eta' & q'_n \\ -q'_n & \frac{1}{\eta'} \end{pmatrix} V'_n, \quad \frac{\partial V'_n}{\partial t} = \begin{pmatrix} A'_n & B'_n \\ C'_n & D'_n \end{pmatrix} V'_n, \tag{4.8}$$

where

$$A'_n = -(\eta'^2 + q'_{n-1}q'_n), \qquad B'_n = -\left(q'_n\eta' + q'_{n+1}\frac{1}{\eta'}\right),$$

$$C'_n = -\left(-q'_{n+1}\eta' - q'_n\frac{1}{\eta'}\right), \quad D'_n = -\left(\frac{1}{\eta'^2} + q'_{n+1}q'_n\right).$$

As before we obtain the Riccati form with $v_n := V'_{n1}/V'_{n2}$ as

$$v_{n-1} = \frac{v_n\eta' + q'_n}{1/\eta' - q'_n v_n}, \tag{4.9}$$

$$\frac{\partial v_n}{\partial t} = (A'_n - D'_n)v_n + B'_n + C'_n v_n^2. \tag{4.10}$$

Solving (4.9) for q'_n and plugging it into (4.10) we obtain the modified equation:

$$\frac{\partial v_n}{\partial t} = \frac{v_{n+1} - v_n(\eta'^2 + 1/\eta'^2) - v_n^3}{1 + v_n v_{n+1}} + \frac{v_n^3 + v_n(\eta'^2 + 1/\eta'^2) - v_{n-1}}{1 + v_n v_{n-1}}. \tag{4.11}$$

Note that this second modified equation (4.11) is identical with the first modified equation (4.5). So we can define a Bäcklund transformation following our Theorem 1.1. First we get a Miura transformation ψ from solving (4.9) as

$$q'_n = \psi(\eta', v, v_n, \dots) = \frac{v_{n-1} 1/\eta' - v_n \eta'}{1 + v_n v_{n-1}}. \tag{4.12}$$

The Miura transform $q' = \psi(v)$ maps a solution v of the modified equation (4.11) into a solution q' of the original equation (4.1). Now we can define the Bäcklund transformation according to Theorem 1.1.

5. Bäcklund Transformation for the Discrete mKdV Equation

Two solutions q and q' of the mKdV equation (4.1) are obtained from the same solution u of the modified equation using the different Miura transformations ϕ:

$$q_n = \phi(\eta, u, u_n, \dots) = \frac{u_{n+1} 1/\eta - u_n \eta}{1 + u_n u_{n+1}}. \tag{5.1}$$

Therefore, by Theorem 1.1 we have the **Bäcklund** transformation

$$q' = \psi\big(\phi^{-1}(q)\big), \tag{5.2}$$

explicitly

$$q'_{n+1} = \frac{u_n(1/\eta^2 + \eta^2) - q_n(u_n^2 1/\eta + \eta)}{\eta u_n^2 + 1/\eta}. \tag{5.3}$$

As promised, we can indeed solve the equation and for $q_n = 0$ we can easily solve for u_n and get a nontrivial solution.

6. A Nontrivial Solution to the Discrete mKdV Equation

Using this Bäcklund transformation (5.3) we can easily construct nontrivial solutions for the discrete mKdV equation (4.1) from the trivial solution $q = 0$.

Indeed for $q = 0$ we get

$$V_{n1} = c_1 \eta^n \exp(\eta^2 t), \quad V_{n2} = c_2 \eta^{-n} \exp(\eta^{-2} t). \tag{6.1}$$

Hence, the corresponding solution to the modified equation is

$$u_n = \frac{V_{n1}}{V_{n2}} = c\eta^{2n} \exp\big((\eta^2 - \eta^{-2})t\big). \tag{6.2}$$

Using the Bäcklund transformation (5.3) we get a **nontrivial solution** of the discrete mKdV equation (4.1) as:

$$q'_{n+1} = \frac{1}{2}(\eta^{-2} - \eta^2)\,\mathrm{sech}\big(c_0 + (2n+1)\ln(\eta) + (\eta^2 - \eta^{-2})t\big), \tag{6.3}$$

where $c_0 = \ln(c)$ is a constant and η is the parameter.

We obtained similar results for the discrete Schrödinger equation, the discrete Sine–Gordon equation and the discrete KdV equation [**16**].

In a next paper we will investigate Bäcklund transformations induced by more complicated symmetries as described e.g. in [**5**], [**7**], [**8**], [**13**],[**14**] as well as Bäcklund transformations induced by second linear systems which are not induced by symmetries [**9**].

References

1. M. J. Ablowitz and H. Segur, *Solitons and the inverse scattering transform*, SIAM, Philadelphia, 1981.
2. R. L. Anderson and N. J. Ibragimov, *Lie–Bäcklund transformations in applications*, SIAM, Philadelphia, 1979.
3. H.-H. Chen, *General derivation of Bäcklund transformation from inverse scattering problems*, Phys. Rev. Lett. **33** (1974), 925–928.
4. A. R. Chowdhury, *A Darboux–Bäcklund transformation associated with a discrete nonlinear Schrödinger equation*, Lett. Math. Phys. **7** (1983), 313–317.
5. R. Floreanini and L. Vinet, *Lie symmetries of finite difference equations*, J. Math. Phys. **36** (1995), no. 12, 7024–7042.
6. A. S. Fokas, *Symmetries and integrability*, Stud. Appl. Math. **77** (1987), no. 3, 253–299.
7. R. Hernandez Heredero, D. Levi, and P. Winternitz, *Symmetries of the discrete Burgers equation*, J. Phys. A **32** (1999), no.14, 2685–2695.
8. D. Levi and P. Winternitz, *Symmetries of discrete dynamical systems*, J. Math. Phys. **37** (1996), no. 11, 5551–5576.
9. W.-X. Ma, *Darboux transformations for Lax integrable systems in* $2n$ *dimensions*, Lett. Math. Phys. **39** (1997), no. 1, 33–49.
10. V. B. Matveev and M. A. Salle, *Darboux transforms and solitons*, Springer-Verlag, Berlin, 1991.
11. A. V. Mikhailov, A. B. Shabat, and V. V. Sokolov, *The symmetry approach to classification of integrable systems*, What is Integrability? (V. E. Zakharov, ed.), Springer-Verlag, Berlin, 1991, pp. 115–184.
12. M. C. Nucci, *Riccati-type pseudopotentials and their applications*, Nonlinear Equations in Applied Sciences (W. Ames & C. Rogers, eds.), Academic Press, San Diego, 1992, pp. 399–436.
13. P. J. Olver and V. V. Sokolov, *Integrable evolution equations on associative algebras*, Comm. Math. Phys. **193** (1998), no. 2, 245–268.
14. P. J. Olver and V. V. Sokolov, *Non-Abelian integrable systems of derivative nonlinear Schrödinger type*, Inverse Problems **14** (1998), no. 6, L5–L8.
15. C. Rogers and W. F. Shadwick, *Bäcklund transformations and their applications*, Academic Press, New York, 1982.
16. R. Schmid and X. Yang, *Bäcklund transformations induced by symmetries*, Phys. Lett. A **195** (1994), no. 1, 63–73.
17. ———, *On Bäcklund transformations for soliton equations*, Preprint, 1999.
18. H. D. Wahlguist and F. B. Estabrook, *Bäcklund transformations for solutions of the Korteweg–de Vries equation*, Phys. Rev. Lett. **31** (1973), 1386–1390.
19. X. Yang, *On Bäcklund transformations for nonlinear differential equations*, Ph.D. Thesis, Emory University, 1994.

Department of Mathematics, Emory University, Atlanta, GA 30322, USA
E-mail address: rudolf@mathcs.emory.edu

Centre de Recherches Mathématiques
CRM Proceedings and Lecture Notes
Volume **29**, 2001

Darboux Transformation for a Spectral Problem Quadratic in the Spectral Parameter

H. Steudel

Abstract. Darboux transformations for a spectral problem of second order in the spectral parameter are established by a remarkably simple approach. N-fold Darboux transforms, both over vacuum and over finite background, then can be written in terms of Vandermonde-type determinants. Several nonlinear evolution equations, e.g., the Derivative Nonlinear Schrödinger equation, are connected with such a scattering problem. Using structural properties of Vandermonde-type determinants we are able to generate computer pictures for the collision process of 8 solitons.

1. Introduction

We have in mind a spectral problem defined by

$$\phi_x = \zeta(J\zeta + Q)\phi \equiv U\phi, \quad J = \begin{pmatrix} -\mathrm{i} & 0 \\ 0 & \mathrm{i} \end{pmatrix}, \quad Q = \begin{pmatrix} 0 & q(x) \\ r(x) & 0 \end{pmatrix}, \tag{1.1}$$

where ϕ is a two-component column, $\phi = (\varphi_1, \varphi_2)^T$, and ζ denotes the spectral parameter. It is also called *quadratic bundle*, and there is some theory developed in the literature [**5, 6**] around such problems concerning Darboux/Bäcklund transformations, soliton solutions, Hamiltonian structure etc. Our rather elementary approach to Darboux transformations for (1.1), however, seems to be new. Note that, in some sense, the Darboux transformation is even simpler as that for the AKNS (Ablowitz, Kaup, Newell and Segur [**1**]) system, c.f. [**13, 16**], because—in the language of [**13**]—no *matrix spectral parameter* is required.

As well-known, iterated Darboux transformations with respect to the Sturm–Liouville problem are given in terms of Wronski determinants [**3**] which might be understood as Vandermonde determinants with powers replaced by multiple derivatives. On the other hand, AKNS problems as well as some other 2×2 scattering problems lead to *Vandermonde-type determinants* [**13, 16, 19**]. Due to their structural properties [**19**], the systematic use of such determinants does not only provide

2000 *Mathematics Subject Classification.* Primary: 35Q51, 35Q58; Secondary: 35Q55.

The author would like to thank K. Baumgärtel, V. S. Gerdzhikov, A. M. Kamchatnov, R. Meinel, and A. A. Zabolotskii for stimulating discussions, and he is gratefull to the referee for useful comments, in particular for pointing at the "shuffle product". The research was supported by the Max Planck Society.

This is the final form of the paper.

a convenient shorthand notation. This approach leads to particularly compact and transparent N-soliton formulas which are well-suited for their numerical evaluation.

A theorem giving the one-step Darboux transformation is formulated in Section 2. In Section 3 we establish the N-step Darboux transformation. The spectral problem (1.1) defines some class of integrable partial differential equations in quite an analogous way to the well-known AKNS class. In Section 4 we give four examples as representatives of this class including the Derivative Nonlinear Schrödinger (DNLS) equation. At the example of the DNLS equation we present some figures depicting N-soliton solutions. In the Appendix the reader may find the definition of Vandermonde-type determinants as well as a crucial reduction formula according to [**19**].

2. Darboux Transformations

From (1.1) it is trivial to derive a Riccati equation for the quotient $\beta \equiv \varphi_2/\varphi_1$,

$$\beta_x = \zeta(r + 2i\zeta\beta - q\beta^2). \tag{2.1}$$

A Darboux transformation now can be defined by the subsequent theorem.

THEOREM. *Given one particular solution* $\{\phi_1(x), \zeta_1, q(x), r(x)\}$ *to* (1.1) *or* $\{\beta_1(x), \zeta_1, q(x), r(x)\}$ *to* (2.1) *we define the matrix*

$$M(\zeta) = \begin{pmatrix} -\beta_1\zeta/\zeta_1 & 1 \\ 1 & -\alpha_1\zeta/\zeta_1 \end{pmatrix}, \quad \alpha_1 \equiv 1/\beta_1. \tag{2.2}$$

Then from any solution $\{\phi, \zeta, q, r\}$ *to* (1.1) *a new solution* $\{\tilde{\phi}, \zeta, \tilde{q}, \tilde{r}\}$ *is found by the Darboux transformation*

$$\tilde{\phi} = M\phi, \quad \tilde{q} = \beta_1(\beta_1 q - 2i\zeta_1), \quad \tilde{r} = \alpha_1(\alpha_1 r + 2i\zeta_1). \tag{2.3}$$

The proof is easy and straightforward and might be omitted here.

For later use it is of importance to notice that the matrix function M alternatively could be characterized by the properties:

$$1)\ M(\zeta_1)\phi_1 \equiv 0, \quad 2)\ M(0) = \begin{pmatrix} 0 & 1 \\ 1 & 0 \end{pmatrix}. \tag{2.4}$$

Principally, there is some arbitrariness in the choice of the normalization 2).

Reduction. If $r = q^*$ (with the star denoting the complex conjugate value), ζ_1 real, and β_1 is chosen of modulus 1 then it follows $\alpha_1 = \beta_1^*$, $\tilde{r} = \tilde{q}^*$. I.e., the symmetry $r = q^*$ is conserved.

Commutativity. It is of importance to notice and is not difficult to check that Darboux transformations of the type just defined commute.

3. The N-Fold Darboux Transform

Now we return to the spectral problem $\phi_x = U\phi$ with U given by (1.1) (with no reduction as far). Given N solutions $\{q, r, \beta_k, \zeta_k\}$, $k = 1, \ldots N$ to equation (2.1) then for the N-fold Darboux transform of any solution $\{q, r, \phi, \zeta\}$ the wave function is a polynomial of Nth order in ζ,

$$\phi^{[N]} = \sum_{k=0}^{N} P_k \zeta^{N-k} \phi. \tag{3.1}$$

From (2.2) together with commutativity now it follows

$$\sum_{k=0}^{N} P_k \zeta_j^{N-k} \phi_j = 0, \quad j = 1 \dots N. \tag{3.2}$$

From the iteration of (2.2) the coefficients P_k get the structure

$$P_{2l-1} = \begin{pmatrix} 0 & p_{2l-1} \\ s_{2l-1} & 0 \end{pmatrix}, \quad P_{2l} = \begin{pmatrix} p_{2l} & 0 \\ 0 & s_{2l} \end{pmatrix}, \quad p_N = s_N = 1. \tag{3.3}$$

Equation (3.2) splits into two systems of linear equations for the coefficients p_k and s_k respectively. We have to distinguish whether N is odd or even. As will be seen below, only the coefficients p_0, p_1, s_0, s_1 are required. We will write down p_0, p_1 explicitly in terms of Vandermonde-type determinants as introduced in [**19**], c.f. the Appendix of the present paper.

- $N = 2n + 1$

$$\sum_{l=1}^{n} p_{2l-1} \zeta_j^{2(n-l+1)} + \sum_{l=0}^{n} p_{2l} \zeta_j^{2n-2l+1} \alpha_j = -1. \tag{3.4}$$

$$\sum_{l=1}^{n} s_{2l-1} \zeta_j^{2(n-l+1)} + \sum_{l=0}^{n} s_{2l} \zeta_j^{2n-2l+1} \beta_j = -1. \tag{3.5}$$

$$p_0 = -\frac{\mathcal{V}_{n+1,n}(1; \zeta_j \alpha_j | \zeta_j^2)}{\mathcal{V}_{n,n+1}(\zeta_j^2; \zeta_j \alpha_j | \zeta_j^2)}. \tag{3.6}$$

$$p_1 = (-1)^n \frac{\mathcal{V}_{n,n+1}(1; \zeta_j \alpha_j | \zeta_j^2)}{\mathcal{V}_{n,n+1}(\zeta_j^2; \zeta_j \alpha_j | \zeta_j^2)}. \tag{3.7}$$

- $N = 2n$

$$\sum_{l=0}^{n-1} p_{2l} \zeta_j^{2(n-l)} + \sum_{l=1}^{n} p_{2l-1} \zeta_j^{2(n-l)+1} \beta_j = -1. \tag{3.8}$$

$$\sum_{l=0}^{n-1} s_{2l} \zeta_j^{2(n-l)} + \sum_{l=1}^{n} s_{2l-1} \zeta_j^{2(n-l)+1} \alpha_j = -1. \tag{3.9}$$

$$p_0 = (-1)^n \frac{\mathcal{V}_{nn}(1; \zeta_j \beta_j | \zeta_j^2)}{\mathcal{V}_{nn}(\zeta_j^2; \zeta_j \beta_j | \zeta_j^2)}. \tag{3.10}$$

$$p_1 = \frac{\mathcal{V}_{n+1,n-1}(1; \zeta_j \beta_j | \zeta_j^2)}{\mathcal{V}_{nn}(\zeta_j^2; \zeta_j \beta_j | \zeta_j^2)}. \tag{3.11}$$

In both cases—N odd or even—s_0 and s_1 are easily found from the respective formulas for p_0 and p_1 because it holds generally

$$s_k = p_k(\alpha \leftrightarrow \beta). \tag{3.12}$$

Let us write the transformed spectral problem in the form

$$\phi_x^{[N]} = \zeta(J\zeta + Q^{[N]})\phi, \quad Q^{[N]} = \begin{pmatrix} 0 & q^{[N]}(x) \\ r^{[N]}(x) & 0 \end{pmatrix}. \tag{3.13}$$

Substitution of (3.1) into (1.1), (3.13) and comparison of powers in ζ leads to

$$
\begin{aligned}
&[P_0, J] = 0,\\
&[P_1, J] + P_0 Q - Q^{[N]} P_0 = 0,\\
&P_{k-1,x} + [P_{k+1}, J] + P_k Q - Q^{[N]} P_k = 0, \quad k = 1 \dots N-1,\\
&P_{N-1,x} + P_N Q - Q^{[N]} P_N = 0,
\end{aligned}
\tag{3.14}
$$

and from the second equation of this system we get

$$
q^{[N]} = \frac{p_0 q + 2\mathrm{i} p_1}{s_0}, \quad r^{[N]} = \frac{s_0 r - 2\mathrm{i} s_1}{p_0}. \tag{3.15}
$$

In case of the reduction $r = q^*$ one has to take the ζ_j as real or pairs of complex conjugate values and gets $s_k = p_k^*$.

4. Four Examples from a Class of Integrable Systems

We wish to establish pairs of matrix functions $U(x,t,\zeta)$, $V(x,t,\zeta)$ with U having the structure given by (1.1) such that the simultaneous system

$$
\phi_x = U\phi, \quad \phi_t = V\phi \equiv \begin{pmatrix} A & B \\ C & -A \end{pmatrix}, \tag{4.1}
$$

is compatible. The compatibility condition $U_t - V_x + [U,V] = 0$ leads to

$$
\begin{aligned}
A_x &= \zeta(Cq - Br),\\
B_x + 2\mathrm{i}\zeta^2 B &= \zeta(q_t - 2Aq),\\
C_x - 2\mathrm{i}\zeta^2 C &= \zeta(r_t + 2Ar).
\end{aligned}
\tag{4.2}
$$

There is also a Riccati form of the system (4.1), with $\beta \equiv \varphi_2/\varphi_1$,

$$
\beta_x = 2\mathrm{i}\zeta^2\beta + \zeta(r - q\beta^2), \quad \beta_t = -2A\beta + C - B\beta^2. \tag{4.3}
$$

EXAMPLE 1. The choice

$$
q = r^* = q_1 + \mathrm{i} q_2, \tag{4.4}
$$

$$
A = -\mathrm{i}p, \quad B = C = \delta/\zeta, \tag{4.5}
$$

with q_1, q_2 and p being real functions of x and t and $\delta = \text{const}$ leads to the integrable system

$$
p_x = -2\delta q_2, \quad q_{1t} = p q_2, \quad q_{2t} = \delta - p q_1. \tag{4.6}
$$

EXAMPLE 2.

$$
A = \frac{\mathrm{i}}{2}\left(bc + \frac{1}{2\zeta^2}\right), \quad B = -\frac{\mathrm{i}b(x,t)}{2\zeta}, \quad C = \frac{\mathrm{i}c(x,t)}{2\zeta^2}. \tag{4.7}
$$

With this specification we arrive at

$$
\begin{aligned}
b_x &= q, \quad q_t = b + \mathrm{i}bcq\\
c_x &= r, \quad r_t = c - \mathrm{i}bcr.
\end{aligned}
\tag{4.8}
$$

Possible reduction: $r = \pm q^*$, $c = \pm b^*$.

Both Examples 1 and 2 presumably can be found in the literature though, at present, we are unable to give quotations.

EXAMPLE 3. (Due to Zabolotskii [**21**])

$$A = \frac{-\mathrm{i}\zeta^2 bc}{\zeta^2+1}, \quad B = \frac{-\mathrm{i}\zeta b^2(x,t)}{\zeta^2+1}, \quad C = \frac{-\mathrm{i}\zeta c^2(x,t)}{\zeta^2+1}, \tag{4.9}$$

leads to the system

$$\begin{aligned} b_x &= \mathrm{i}(b - \mathrm{i}cq) \ , & q_t &= 2b(b - \mathrm{i}cq) \\ c_x &= \mathrm{i}(-c + \mathrm{i}br), & r_t &= 2c(-c + \mathrm{i}br). \end{aligned} \tag{4.10}$$

Reduction: $r = -q^*$, $c = +b^*$.

EXAMPLE 4. The derivative nonlinear Schrödinger equation [**11**]

$$A = -\mathrm{i}(2\zeta^4 + qr\zeta^2),\ B = 2q\zeta^3 + (\mathrm{i}q_x + q^2 r)\zeta,\ C = 2r\zeta^3 + (-\mathrm{i}r_x + r^2 q)\zeta, \tag{4.11}$$

results in

$$\mathrm{i}q_t + q_{xx} - \mathrm{i}(q^2 r)_x = 0, \quad -\mathrm{i}r_t + r_{xx} + \mathrm{i}(r^2 q)_x = 0. \tag{4.12}$$

Reduction: $r = \pm q^*$.

Now it is easy to establish N-soliton solutions for all four examples. Starting from some *seed solution* we should solve the simultaneous Riccati system (4.3) for N values of the spectral parameter. Then we have to substitute the respective solutions β_j into the formulas given in Section 3. If, in particular, the seed solution is such that the Riccati equations are explicitly solvable we arrive at explicit analytical N-soliton formulas. In the next section we will present some results concerning Example 4.

5. N-Soliton Solutions for the DNLS Equation

Let us consider the DNLS equation (4.12) with the reduction $r = q^*$. There is a rich literature [**2, 4, 7–12, 14, 15, 20**] on this equation. If we start from vacuum $q = 0$ the Riccati equations (4.3) become linear and are easily solved by

$$\beta = C \exp\bigl[2\mathrm{i}\zeta^2(x + 2\zeta^2 t)\bigr]. \tag{5.1}$$

If we take $\zeta = \zeta_1$ real the one-step Bäcklund transformation applied to vacuum does not lead to a localized solution but to a monochromatic wave. Taking a pair ζ_1, $\zeta_2 = \zeta_1^*$ and choosing $\beta_2 = 1/\beta_1^*$ we recover the one-soliton solution given already by Kaup [**11**], and for N such pairs we get the N-soliton solution in terms of Vandermonde-type determinants.

Instead we may start from the wave

$$q = a e^{\mathrm{i}(kx - lt)}, \quad r = q^*, \quad l = k(k - a^2), \tag{5.2}$$

with real a, k as the seed solution. Then putting

$$\beta = -\mathrm{i}\gamma \exp\bigl[\mathrm{i}k\bigl(-x + (k - a^2)t\bigr)\bigr], \tag{5.3}$$

we get a Riccati system with constant coefficients,

$$\gamma_x = -\frac{1}{v}\gamma_t = \mathrm{i}\left[2m\gamma + \zeta a(1 + \gamma^2)\right], \quad v = k - a^2 - 2\zeta^2, \quad m = \zeta^2 - k/2. \tag{5.4}$$

Its general solution is

$$\gamma = \frac{-(m + Kf) + \mathrm{i}Kd\coth[d(x - vt)]}{(Km + f) + \mathrm{i}d\coth[d(x - vt)]}, \tag{5.5}$$

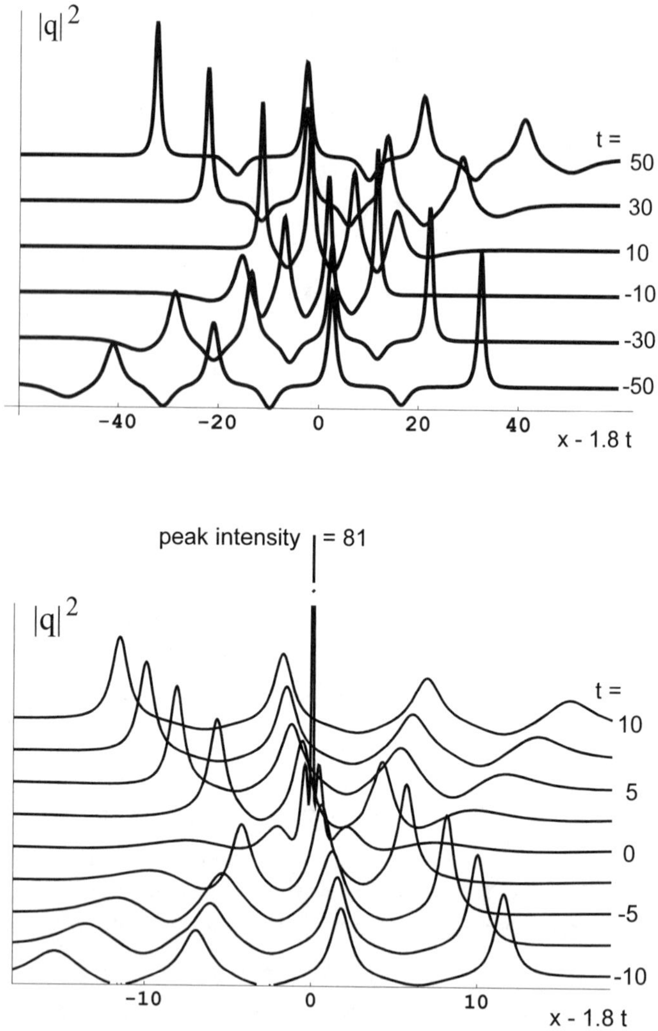

FIGURE 1. Collision of four bright and four dark DNLS solitons over background intensity 1. Upper and lower parts differ in their scales only. There are 8 equidistant positive eigenvalues with $\zeta_1 = 0.189$ being the lowest and $\zeta_8 = 0.811$ the highest.

where K is a constant and $d = \sqrt{a^2\zeta^2 - m^2}$. a, k, ζ should be chosen such that d is real. Note that there is a rather simple asymptotic behavior of γ because the space and time variables appear only in the argument of the hyperbolic cotangent. As an example we depict the collision of eight solitons, four bright and four dark ones. Both parts of Fig. 1 show the same example differing in the scale only. We also generated figures of a 10-soliton collision. However, then a numerical breakdown occured at the collision center.

Another type of solutions is achieved if we take pairs of complex conjugate eigenvalues and start from the same seed solution. A pulsating solution generated from a two-fold Bäcklund transformation with one such pair is depicted in Fig. 2.

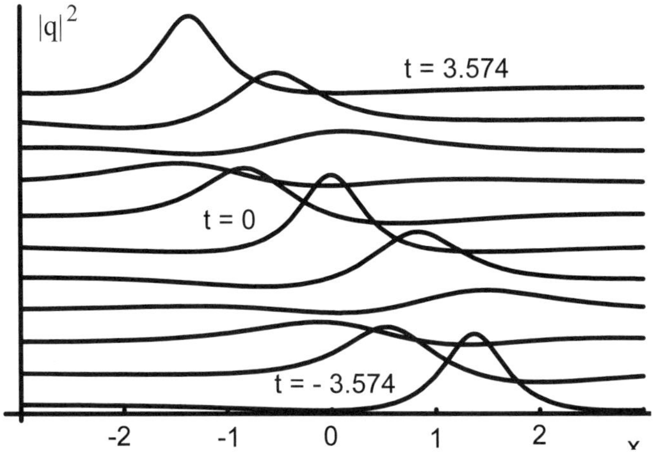

FIGURE 2. Pulsating DNLS-soliton generated from a two-fold Bäcklund transformation with $\zeta_1 = (1+i)/2$, $\zeta_2 = \zeta_1^*$ with background intensity 1 as in Figure 1.

Of course our formulas might be applied as well for a combination of N real and M pairs of conjugate complex eigenvalues.

Appendix A. Vandermonde-Like Determinants

Genuine Vandermonde-determinants are defined as follows and can be expressed as products of differences,

$$\mathcal{V}_N(x_1,\dots,x_N) := \begin{vmatrix} 1 & x_1 & x_1^2 & \dots & x_1^{N-1} \\ 1 & x_2 & x_2^2 & \dots & x_2^{N-1} \\ \dots & \dots & \dots & \dots & \dots \\ 1 & x_N & x_N^2 & \dots & x_N^{N-1} \end{vmatrix} = \prod_{i>j}(x_i - x_j). \tag{A.1}$$

Vandermonde-like determinants are defined by the formula [**18, 19**]

$$\mathcal{V}_{MN}(a_r;b_r|x_r) \tag{A.2}$$
$$:= \begin{vmatrix} a_1 & a_1x_1 & \dots & a_1x_1^{M-1} & b_1 & b_1x_1 & \dots & b_1x_1^{N-1} \\ a_2 & a_2x_2 & \dots & a_2x_2^{M-1} & b_2 & b_2x_2 & \dots & b_2x_2^{N-1} \\ \dots & \dots & \dots & \dots & \dots & \dots & \dots & \dots \\ a_{M+N} & a_{M+N}x_{M+N} & \dots & a_{M+N}x_{M+N}^{M-1} & b_{M+N} & b_{M+N}x_{M+N} & \dots & b_{M+N}x_{M+N}^{N-1} \end{vmatrix},$$

where $r = 1, 2 \dots M+N$.

The most important property is given by the *reduction formula*

$$\mathcal{V}_{MN}(a_r;b_r|x_r) \tag{A.3}$$
$$= \sum_P \varepsilon_P \prod_{j=1}^{M} a_{s(j)} \prod_{k=M+1}^{M+N} b_{s(k)} \mathcal{V}_M\big(x_{s(1)}..x_{s(M)}\big)\mathcal{V}_N\big(x_{s(N+1)}..x_{s(M+N)}\big).$$

The sum goes over all permutations $P = \big(s(1),\dots,s(M+N)\big)$ of $(1,2,\dots,M+N)$ such that $s(i) < s(j)$ for $i < j \leq N$ as well as for $N < i < j$. $\varepsilon_P = +1$ for P even or -1 for P odd. By this formula any Vandermonde-like determinant generally is expressed as a sum over binary products of genuine Vandermonde determinants. Thus the $(N+M)!$ terms of the determinant are replaced by $\binom{M+N}{M}$ terms. E.g., for $M=N=5$, 3628800 terms are reduced to 252 terms.

It was the referee who noticed that sums of just the same structure as in (A.3) appear in the definition of "shuffle products" in certain Lie algebras [**17**]. Maybe there is some deeper connection hidden behind this observation.

References

1. M. J. Ablowitz, D. J. Kaup, A. C. Newell, and H. Segur, *The inverse scattering transform-Fourier analysis for nonlinear problems*, Stud. Appl. Math. **53** (1974), 249–315.
2. K. Baumgärtel, *Soliton approach to magnetic holes*, J. Geophys. Res. **104** (1999), 28295–28308.
3. M. M. Crum, *Associated Sturm–Liouville problems*, Quart. J. Oxford Ser. **6** (1955), 121–127.
4. H. Eichhorn, *Bäcklund transformation and N-soliton solution for the generalized nonlinear Schrödinger equation*, Ann. Physik **44** (1987), 261–264.
5. V. S. Gerdzhikov and M. I. Ivanov, *The quadratic bundle of a general form and the nonlinear evolution equations* I, II, Bulgar. J. Phys. **10** (1983), 13–26; 130–142 (Russian).
6. V. S. Gerdzhikov, M. I. Ivanov, and P. P. Kulish, *Quadratic bundle and nonlinear equations*, Teoret. Mat. Fiz. **44** (1980), 342–357 (Russian); English transl., Theoret. Math. Phys. **44** (1980), 784–795.
7. Y. Ichikawa, K. Konno, M. Wadati, and H. Sanuki, *Spiky soliton in circularly polarized Alfvén waves*, J. Phys. Soc. Japan **48** (1980), 279–286.
8. A. M. Kamchatnov, *Propagation of ultrashort periodic pulses in nonlinear fiber light guides*, Zh. Exp. Teor. Fiz. **97** (1990), 144–153 (Russian); Engl. transl., Sov. Phys. JETP **70** (1990), 80–84.
9. ———, *New approach to periodic solutions of integrable equations and nonlinear theory of modulational instability*, Phys. Rep. **286** (1997), 199–270.
10. A. M. Kamchatnov, S. A. Darmanyan, and F. Lederer, *Formation of solitons on the sharp front of the pulse in an optical fiber*, Phys. Lett. A **245** (1998), 259–264.
11. D. J. Kaup and A. C. Newell, *An exact solution for a derivative nonlinear Schrödinger equation*, J. Math. Phys. **19** (1978), 798–801.
12. T. Kawata and H. Inoue, *Exact solutions of the derivative nonlinear Schrödinger equation under the nonvanishing conditions*, J. Phys. Soc. Japan **44** (1978), 1968–1976.
13. V. B. Matveev and M. A. Salle, *Darboux transformations and solitons*, Springer-Verlag, Berlin, 1991.
14. E. Mjølhus, *Nonlinear Alfvén waves and the* DNLS *equation: Oblique aspects*, Phys. Scripta **40** (1989), 227–237.
15. E. Mjølhus and T. Hada, *Soliton theory of quasi-parallel* MHD *waves*, Nonlinear Waves and Chaos in Plasma (T. Hada and H. Matsumoto, eds.), Terrapub, Tokyo, 1997, pp. 121–168.
16. G. Neugebauer and R. Meinel, *General N-soliton solution of the* AKNS *class on arbitrary background*, Phys. Lett. A **100** (1984), 467–470.
17. R. Ree, *Lie elements and an algebra associated with shuffles*, Ann. of Math. **68** (1958), 210–220.
18. H. Steudel, R. Meinel, and D. J. Kaup, *Solution of degenerate two-photon propagation from Bäcklund transformations*, J. Modern Opt. **44** (1997), 287–303.
19. H. Steudel, R. Meinel, and G. Neugebauer, *Vandermonde-like determinants and N-fold Darboux/Bäcklund transformations*, J. Math. Phys. **38** (1997), 4692–4695.
20. M. Wadati, K. Konno, and Y. Ichikawa, *A generalization of inverse scattering method*, J. Phys. Soc. Japan **46** (1979), 1965–1966.
21. A. A. Zabolotskii, private communication.

Institut für Physik der Humboldt-Universität, Arbeitsgruppe "Nichtklassische, Strahlung", Invalidenstr 110, D-10115 Berlin, Germany.

E-mail address: `steudel@physik.hu-berlin.de`

Centre de Recherches Mathématiques
CRM Proceedings and Lecture Notes
Volume **29**, 2001

Separation of Variables and Darboux Transformations

Z. Thomova and P. Winternitz

ABSTRACT. A classification of maximal Abelian subalgebras (MASAs) of the pseudoeuclidean Lie algebras is presented. It is then used to generate coordinate systems in the pseudo-Euclidean space $M(p,q)$ in which the Laplace–Beltrami equation allows the separation of variables. The MASAs induce ignorable variables; i.e. variables that do not figure explicitly in the metric tensor. Darboux transformations are then used to introduce solvable potentials depending on these ignorable variables.

1. Introduction

The purpose of this article is to show how classical Darboux transformations [**1**] can be applied to partial differential equations. The context will be that of the separation of variables in Laplace and wave equations, or more generally, in Laplace–Beltrami equations on homogeneous spaces. We shall use separable coordinates with a maximal number of ignorable variables, i.e. coordinates that do not figure in the metric tensor and hence in the Laplace–Beltrami operator. The ignorable variables are induced by the action of maximal Abelian subgroups of the isometry group of the corresponding space. A classification of the maximal Abelian subalgebras (MASAs) of the isometry algebra will provide a classification of coordinate systems with a maximal number of ignorable variables.

In Euclidean spaces only two types of ignorable variables exist. Cartesian ones correspond to translation generators P_i in a MASA, polar ones to rotation generators M_{jk}. In pseudoeuclidean spaces $E(p,q)$, $p \geq q \geq 1$ the situation is much richer, since MASAs of the isometry algebra can involve nilpotent elements in $o(p,q)$. This is already true for Minkowski space ($q = 1$), as we shall see below.

Darboux transformations have gained new importance, new applications and new geometric applications with the advent of soliton theory (see Ref. [**2, 3**] and the references therein).

2000 *Mathematics Subject Classification.* Primary: 22E60; Secondary: 35J05.

The authors thank D. Levi for useful discussions. One of the authors (Z.T.) thanks the Centre de recherches mathématiques for hospitality during her visits. The research of Z.T. was partially supported by a PDQWL Individual Award from UUP SUNY-Utica. The research of P.W. was partly supported by research grants from NSERC of Canada, FCAR du Québec and NATO.

This is the final form of the paper.

They were first introduced by Darboux in 1882 in a study of linear ordinary differential equations [**1**], in particular the Sturm–Liouville problem

$$\psi''(x) - u(x)\psi(x) = -E\psi(x). \tag{1.1}$$

Indeed, let us assume that we know two distinct solutions ψ and ψ_0 of the above equation corresponding to the eigenvalues E and E_0, respectively. The Darboux transformation

$$\tilde{\psi} = \left(\frac{d}{dx} - \frac{\psi_0'}{\psi_0}\right)\psi, \tag{1.2}$$

provides us with a solution of a related equation, namely

$$\tilde{\psi}''(x) - \tilde{u}(x)\tilde{\psi}(x) = -E\tilde{\psi}(x), \tag{1.3}$$

$$\tilde{u}(x) = u - 2\left(\frac{\psi_0'}{\psi_0}\right)'. \tag{1.4}$$

Eq. (1.2) gives us solutions of new Sturm–Liouville problems in terms of known ones. Notice that the general eigenvalue E is the same in the original equation (1.1) and in the new equation (1.2). The "specific" eigenvalue E_0 only figures via the eigenfunction ψ_0 in the new potential $\tilde{u}(x)$ of Eq. (1.4) and the new solution $\tilde{\psi}(x)$ in Eq. (1.2).

In Section 2 we sum up some previously obtained results [**5, 6**] on the MASAs of the pseudoeuclidean algebras $e(p,q)$ with an emphasis on $e(p,0)$ and $e(p,1)$. Section 3 is devoted to separable coordinate systems in Euclidean and Minkowski spaces related to these MASAs. Finally, in Section 4 we use Darboux transformations to introduce multidimensional solvable potentials into Laplace and wave equations.

2. MASAs of $e(p,q)$

2.1. General formulation. The pseudoeuclidean Lie algebra $e(p,q)$ is the semidirect sum of the pseudoorthogonal Lie algebra $o(p,q)$ and an Abelian algebra $T(n)$ of translations

$$e(p,q) = o(p,q) \ni T(n), \quad n = p + q. \tag{2.1}$$

We will make use of the following matrix representation of the Lie algebra $e(p,q)$ and the corresponding Lie group $E(p,q)$. We introduce an "extended metric"

$$K_e = \begin{pmatrix} K & 0 \\ 0 & 0_1 \end{pmatrix}, \tag{2.2}$$

where K satisfies

$$K = K^T \in \mathbb{R}^{n\times n}, \qquad n = p + q, \quad \det K \neq 0, \tag{2.3}$$

$$\operatorname{sgn} K = (p,q), \quad p \geq q \geq 0. \tag{2.4}$$

Here $\operatorname{sgn} K$ denotes the signature of K, where p and q are the numbers of positive and negative eigenvalues, respectively. Then $X_e \in e(p,q)$ and $H \in E(p,q)$ are represented as

$$X_e(X,\alpha) \equiv X_e = \begin{pmatrix} X & \alpha^T \\ 0 & 0 \end{pmatrix}, \qquad X \in \mathbb{R}^{n\times n}, \quad \alpha \in \mathbb{R}^{1\times n}, \tag{2.5}$$

$$H = \begin{pmatrix} G & a^T \\ 0 & 1 \end{pmatrix}, \qquad G \in \mathbb{R}^{n\times n}, \quad a \in \mathbb{R}^{1\times n}, \tag{2.6}$$

$$XK + KX^T = 0, \quad GKG^TK, \quad X_eK_e + K_eX_e^T = 0. \tag{2.7}$$

The vector $\alpha \in \mathbb{R}^{1\times n}$ represents the translations. We say that the translations are of positive, negative or zero length if

$$\alpha K \alpha^T > 0, \quad \alpha K \alpha^T < 0, \quad \alpha K \alpha^T = 0, \tag{2.8}$$

respectively. Zero length vectors are called isotropic.

We will be classifying maximal Abelian subalgebras of the pseudoeuclidean Lie algebra $e(p,q)$ into conjugacy classes under the action of the pseudoeuclidean Lie group $E(p,q)$.

2.2. Classification strategy. The classification of MASAs of $e(p,q)$ is based on the fact that $e(p,q)$ is the semidirect sum of the Lie algebra $o(p,q)$ and an Abelian ideal $T(n)$ (the translations). We proceed in five steps.

1. Classify subalgebras $T(k_+,k_-,k_0)$ of $T(n)$. They are characterized by a triplet (k_+,k_-,k_0), where k_+,k_- and k_0 are the number of positive length, negative length and isotropic vectors, respectively. They are represented by matrices of type (2.5) with $X=0$.
2. Find the centralizer $C(k_+,k_-,k_0)$ of $T(k_+,k_-,k_0)$ in $o(p,q)$

$$C(k_+,k_-,k_0) = \{X \in o(p,q) \mid [X, T(k_+,k_-,k_0)] = 0\}. \tag{2.9}$$

3. Construct all MASAs $M(k_+,k_-,k_0)$ of $C(k_+,k_-,k_0)$ and classify them under the action of the normalizer $N \equiv \mathrm{Nor}[T(k_+,k_-,k_0), G]$ of $T(k_+,k_-,k_0)$ in the group $G \sim E(p,q)$.
4. Obtain a representative list of all splitting MASAs of $e(p,q)$ as direct sums

$$M(k_+,k_-,k_0) \oplus T(k_+,k_-,k_0), \tag{2.10}$$

and keep only those amongst them that are indeed maximal (and mutually inequivalent).
5. Construct all nonsplitting MASAs from the splitting ones and classify them under the action of the group $N \ni T(n)$. A nonsplitting MASA is any MASA of $e(p,q)$ not conjugate to one of the form (2.10). This involves some elementary cohomology theory and the procedure is described elsewhere [**6**].

2.3. MASAs of $e(p,0)$ and $e(p,1)$. In this section we shall list all splitting MASAs of $e(p,0)$ and $e(p,1)$. The results are given as theorems. For the proofs we refer to the original articles [**5, 6**].

Let us first consider the real Euclidean algebra $e(n,0)$ with a basis $\{P_i, L_{jk}, i = 1,\dots,n, 1 \le k \le n\}$ consisting of infinitesimal translations and rotations. The structure of the MASAs is very simple and we have:

THEOREM 2.1. *Every* MASA *of* $e(p,0)$ *splits into the direct sum* $M(k) = F(k) \oplus T(k)$ *and is* $E(p,0)$ *conjugate to precisely one subalgebra with*

$$F(k) = \{L_{12}, L_{34}, \dots, L_{2l-1,2l}\}, \quad T(k) = \{P_{2l+1}, \dots, P_p\},$$

where k is such that $p-k$ is even ($p-k=2l$).

Now let us consider the Lie algebra $e(n,1)$. Every MASA of $e(n,1)$ can be represented by a matrix pair $\{M, K_e\}$, satisfying

$$M(k_+,k_-,k_0) \equiv M = \begin{pmatrix} M_0 & & & & & & \gamma^T \\ & M_1 & & & & & 0 \\ & & \ddots & & & & \vdots \\ & & & M_l & & & 0 \\ & & & & 0_{k_+} & & x^T \\ & & & & & & 0_1 \end{pmatrix}, \tag{2.11}$$

$$K_e = \begin{pmatrix} K_0 & & & \\ & I_{2l} & & \\ & & I_{k_+} & \\ & & & 0_1 \end{pmatrix}, \quad \operatorname{sgn} K_0 = (n-k_+-2l, 1),$$

where $M_i = \left(\begin{smallmatrix} 0 & a_i \\ -a_i & 0 \end{smallmatrix}\right)$, $a_i \in \mathbf{R}$, $x \in \mathbf{R}^{1\times k_+}$, $M_0 \in \mathbf{R}^{(n-k_+-2l+1)\times(n-k_+-2l+1)}$, $M_0K_0 + K_0M_0^T = 0$. The conditions on M_0, γ and x are given in two theorems.

THEOREM 2.2. *Three different kinds of splitting* MASA*s of* $e(p,1)$ *exist. They are characterized by the triplet* (k_+,k_-,k_0):

A) $M(k_+,1,0)$, $0 \le k_+ \le p$

$$M_0 = 0 \in \mathbf{R}, \quad \gamma^T = z \in \mathbf{R} \quad \text{and} \quad K_0 = -1, \tag{2.12}$$

$p-k_+$ *is even*, $0 \le l \le p-k_+/2$, $d = \dim M(k_+,1,0) = 1+l+k_+$, $[p+3/2] \le d \le p+1$.

B) $M(k_+,0,0)$, $0 \le k_+ \le p-1$

$$M_0 = \begin{pmatrix} c & 0 \\ 0 & -c \end{pmatrix}, \quad \gamma^T \begin{pmatrix} 0 \\ 0 \end{pmatrix}, \quad K_0 = \begin{pmatrix} 0 & 1 \\ 1 & 0 \end{pmatrix}, \tag{2.13}$$

where $p-k_+$ *is odd*, $0 \le l \le p-k_+-1/2$, $d = \dim M(k_+,0,0) = 1+l+k_+$, $[p+2/2] \le d \le p$.

C) $M(k_+,0,1)$, $0 \le k_+ \le p-2$

$$M_0 = \begin{pmatrix} 0 & \alpha & 0 \\ 0 & 0 & -\alpha^T \\ 0 & 0 & 0 \end{pmatrix}, \quad \gamma^T = \begin{pmatrix} z \\ 0_\mu \\ 0 \end{pmatrix}, \quad K_0 = \begin{pmatrix} & & 1 \\ & I_\mu & \\ 1 & & \end{pmatrix}, \tag{2.14}$$

where $1 \le \mu \le p-1$ *and* $0 \le l \le p-k_+-2/2$, $z \in \mathbf{R}$, $\alpha \in \mathbf{R}^{1\times\mu}$, $d = \dim M(k_+,0,1) = \mu+l+k_++1$, $[p+3/2] \le d \le p$.

All entries a_i, x, z, α *and* c *are free.*

THEOREM 2.3. *Nonsplitting* MASA*'s of* $e(p,1)$ *are obtained from splitting ones of type* C *in Theorem* 2.2 *and are conjugate to precisely one* MASA *of the form*

(i) *for* $\mu \ge 2$

$$M_0 = \begin{pmatrix} 0 & \alpha & 0 \\ 0 & 0 & -\alpha^T \\ 0 & 0 & 0 \end{pmatrix}, \quad \gamma^T = \begin{pmatrix} z \\ A\alpha^T \\ 0 \end{pmatrix}, \tag{2.15}$$

where A *is a diagonal matrix with* $a_1 = 1 \ge |a_2| \ge \cdots \ge |a_\mu| \ge 0$ *and* $\operatorname{Tr} A = 0$, K_0 *is as in* (2.14).

(ii) *for* $\mu = 1$ *we have a special case for which the nonsplitting* MASA *has the following form*

$$M_0 = \begin{pmatrix} 0 & a & 0 \\ 0 & 0 & -a \\ 0 & 0 & 0 \end{pmatrix}, \quad \gamma^T = \begin{pmatrix} z \\ 0 \\ a \end{pmatrix}, \quad K_0 = \begin{pmatrix} 0 & 0 & 1 \\ 0 & 1 & 0 \\ 1 & 0 & 0 \end{pmatrix}. \tag{2.16}$$

No other nonsplitting MASAs *of* $e(p,1)$ *exist.*

3. Separation of Variables in the Laplace–Beltrami Operator

Specifically for the spaces $M(p,q)$ of this article, we generate the coordinates as follows. We use the realization (2.6) of the group $E(p,q)$, but restrict H to be a maximal Abelian subgroup of $E(p,q)$. We have $G = \langle \exp X \rangle$, where X is one of the MASAs we have constructed. We then write

$$\begin{pmatrix} x \\ 1 \end{pmatrix} = e^X \begin{pmatrix} s \\ 1 \end{pmatrix}, \quad s \in \mathbb{R}^{p+q}, \tag{3.1}$$

where s represents a vector in a subspace of $M(p,q)$ parameterized by nonignorable variables $(s_1, \dots, s_k)$, and X is a MASA of $e(p,q)$, parametrized by a set of ignorable variables.

Euclidean space $M(p)$. In Cartesian coordinates we have

$$\Delta = \frac{\partial^2}{\partial x_1^2} + \cdots + \frac{\partial^2}{\partial x_n^2}. \tag{3.2}$$

Space $M(p)$ is split into a direct sum of one and two-dimensional spaces. In each $M(1)$ we have a Cartesian coordinate x_i, corresponding to the translation P_i. In each subspace $M(2)$ we have polar coordinates, e.g. $M_{12} = \partial/\partial\alpha_1$ corresponds to

$$x_1 = s_1 \cos\alpha_1, \quad x_2 = s_1 \sin\alpha_1, \tag{3.3}$$

with α_1 ignorable.

Minkowski space $M(p,1)$. The Laplace–Beltrami operator in Cartesian coordinates is written as

$$\Box_{p,1}\Psi = E\Psi, \quad \Delta_{LB} \equiv \Box_{p,1} = \frac{\partial^2}{\partial x_1^2} + \cdots + \frac{\partial^2}{\partial x_p^2} - \frac{\partial^2}{\partial x_0^2}. \tag{3.4}$$

We introduce a separable system of coordinates in each indecomposable subspace of $M(p,1)$. Space $M(1,0)$ corresponds to a Cartesian coordinate, $M(2,0)$ to polar coordinate as in Eq. (3.3). Now let us consider the coordinates corresponding to $M(k,1)$.

M(0,1): x_0 (Cartesian coordinate).
M(1,1): $x_0 = s\cosh\alpha$, $x_1 = s\sinh\alpha$ (hyperbolic coordinates),
$x_0 = s\sinh\alpha$, $x_2 = s\cosh\alpha$
for $x_0^2 - x_1^2 = \pm s^2$, (respectively).
M(2,1): There are two ignorable variables, z and a and we have

$$x_0 + x_1 = r\sqrt{2} + 2a, \quad x_0 - x_1 = ra^2\sqrt{2} + \frac{2}{3}a^3 - z\sqrt{2}, \\ x_2 = -a^2 - ar\sqrt{2}. \tag{3.5}$$

And the operator in this $M(2,1)$ subspace of $M(p,1)$ is:

$$\Box_{2,1} = \sqrt{2}\frac{\partial^2}{\partial r \partial z} + \frac{1}{2}\frac{1}{r^2}\frac{\partial^2}{\partial a^2} + \frac{1}{r^2}\frac{\partial}{\partial r^2} - \frac{\sqrt{2}}{r^2}\frac{\partial^2}{\partial r \partial a} + \frac{1}{\sqrt{2}}\frac{1}{r}\frac{\partial}{\partial z} - \frac{1}{r^3}\frac{\partial}{\partial r} + \frac{1}{\sqrt{2}}\frac{1}{r^3}\frac{\partial}{\partial a}. \tag{3.6}$$

Let us consider the space $M(k,1)$ with $k \geq 2$ and the splitting MASA (2.14). Applying Eq. (3.1) with

$$X = \begin{pmatrix} 0 & \alpha & 0 & z \\ 0 & 0 & -\alpha^T & 0 \\ 0 & 0 & 0 & 0 \\ 0 & 0 & 0 & 0 \end{pmatrix}, \quad s = \begin{pmatrix} 0 \\ \vdots \\ 0 \\ r \end{pmatrix}, \quad r \in \mathbb{R}, \tag{3.7}$$

we obtain the coordinates

$$\begin{aligned} x_k + x_0 &= r\sqrt{2}, \quad x_k - x_0 = -r\alpha\alpha^T \frac{1}{\sqrt{2}} + z\sqrt{2} \\ x_1 &= -r\alpha_1 \\ &\vdots \\ x_{k-1} &= -r\alpha_{k-1}. \end{aligned} \tag{3.8}$$

The wave operator in these coordinates is

$$\Box_{k,1} = 2\frac{\partial^2}{\partial z \partial r} + \frac{k-1}{r}\frac{\partial}{\partial z} + \frac{1}{r^2}\sum_{i=1}^{k-1}\frac{\partial^2}{\partial \alpha_i^2}. \tag{3.9}$$

The variables z and α_i are ignorable (only r figures in Eq. (3.9)) and solution of the wave equation then separates

$$\psi = R(r)e^{mz}\prod_{i=1}^{k-1} e^{b_i \alpha_i}, \tag{3.10}$$

with $R(r)$

$$R(r) = r^{1-k/2}\exp\left(\frac{1}{r}\frac{\sum_{i=1}^{k-1} b_i^2}{2m}\right)\exp\left(\frac{Er}{2m}\right). \tag{3.11}$$

Now consider the space $M(k,1)$ for $k \geq 3$ and the nonsplitting MASA (2.15). The coordinates we obtain are

$$\begin{aligned} x_k + x_0 &= r\sqrt{2} \\ x_k - x_0 &= \frac{1}{\sqrt{2}}(2z - r\alpha\alpha^T + \alpha A \alpha^T) \\ x_1 &= (q_1 - r)\alpha_1 \\ &\vdots \\ x_{k-1} &= (q_{k-1} - r)\alpha_{k-1}. \end{aligned} \tag{3.12}$$

The wave operator is

$$\Box_{k,1} = 2\frac{\partial^2}{\partial z \partial r} - \left(\sum_{i=1}^{k-1}\frac{1}{(q_i - r)}\right)\frac{\partial}{\partial z} + \sum_{i=1}^{k-1}\frac{1}{(q_i - r)^2}\left(\frac{\partial^2}{\partial \alpha_i^2}\right). \tag{3.13}$$

We see that α_k and z are ignorable variables. The solution of the wave equation then separates and we have

$$\Psi = R(r)e^{mz}\prod_{i=1}^{k-1} e^{b_i\alpha_i}, \tag{3.14}$$

with $R(r)$ equal to

$$R(r) = \prod_{i=1}^{k-1}(q_i - r)^{-1/2}\exp\left(-\frac{1}{2m}\sum_{i=1}^{k-1}\frac{b_i^2}{q_i - r}\right)\exp\left(\frac{Er}{2m}\right). \tag{3.15}$$

4. Darboux Transformations

4.1. Examples of Darboux transformations in 1 dimension. As stated in the introduction the Darboux transformations (1.2) relates solutions of two different one-dimensional Schrödinger equation (1.1) and (1.3) with potentials $u(x)$ and $\tilde{u}(x)$, respectively.

Let us give several elementary examples to be used below. We always start from a solvable system for which all solutions are known.

1. The free particle: $u(x) = 0$.

   For $E_0 = 0$ in Eq. (1.1) we have $\psi_0 = ax + b$. With no loss of generality we can shift b to $b = 0$, since we assume $a \neq 0$. Eq. (1.4) then gives an inverse square potential

$$\tilde{u}(x) = \frac{2}{x^2}. \tag{4.1}$$

   For $E_0 = \omega^2 > 0$ the solution of Eq. (1.1) is $\psi_0 = A\sin(\omega x + \alpha)$. Again with no loss of generality we can shift α to $\alpha = 0$ and obtain the Poeschl–Teller potential [**4**]

$$\tilde{u}(x) = \frac{2\omega^2}{\sin^2 \omega x}. \tag{4.2}$$

   For $E_0 = -\omega^2 < 0$ we take $\psi_0 = A\sinh\omega x$ and obtain a hyperbolic version of the potential (4.2), namely

$$\tilde{u}(x) = \frac{2\omega^2}{\sinh^2 \omega x}. \tag{4.3}$$

2. The harmonic oscillator: $u(x) = \omega^2 x^2$.

   Let us choose the solution $\psi_0 = xe^{(\varepsilon\omega/2)x^2}$, $\varepsilon = \pm 1$, corresponding to $E_0 = -3\varepsilon\omega$. We obtain

$$\tilde{u}(x) = \omega^2 x^2 + \frac{2}{x^2} - 2\varepsilon\omega. \tag{4.4}$$

3. The Coulomb potential: $u(x) = 2\alpha/x$.

   Let us choose the solution $\psi_0 = xe^{\alpha x}$ corresponding to the energy $E_0 = -\alpha^2$. The new potential is

$$\tilde{u}(x) = \frac{2\alpha}{x} + \frac{2}{x^2}. \tag{4.5}$$

4.2. Darboux transformations in Laplace–Beltrami equations. One interpretation of the separation variables in a Laplace–Beltrami equation is that we require the separated solutions $\psi(z_1, \dots z_n)$ to be the eigenfunctions of a complete set of n second order commuting operators in the enveloping algebra of the isometry algebra:

$$\triangle_{LB}\psi = -E\psi, \tag{4.6}$$

$$Y_i\psi = \lambda_i\psi, \quad i = 1, \dots, n-1, \tag{4.7}$$

$$[Y_i, Y_k] = 0, \quad [\triangle_{LB}, Y_i] = 0,$$

$$\psi = \prod_{i=1}^{n} \psi_i(z_i). \tag{4.8}$$

The specific feature of ignorable variables α_i is that for them the corresponding operator Y_i is actually the square of an element of the isometry algebra $Y_i = X^2$. The function $\psi_i(\alpha_i)$ then satisfies

$$X\psi_i = \frac{\partial}{\partial \alpha_i} = \lambda_i\psi_i, \tag{4.9}$$

and hence we have $\psi_i = e^{\lambda_i \alpha_i}$. The only way that ignorable variables figure in the Laplace–Beltrami operator is via the derivative operator $\partial/\partial\alpha_i$. If the coordinates are orthogonal, only second derivatives are present. Such is always the case in real Euclidean spaces and often also in Minkowski ones. We can then replace Eq. (4.9) by

$$\frac{\partial^2}{\partial \alpha_i^2}\psi_i = \mu_i\psi_i. \tag{4.10}$$

A Darboux transformation can then be applied to Eq. (4.10) to introduce a solvable potential into Eq. (4.10) and also into the original Eq. (4.6). If the separable coordinates are not orthogonal, then terms of the type $\partial^2/\partial z \partial\alpha$ may occur, where α is ignorable, z not. Then the potential introduced into (4.10) will destroy separability.

Let us now look at specific coordinate systems in Minkowski space and consider the Laplace–Beltrami equation with potential

$$(\Box_{k,1} - u)\psi = -E\psi. \tag{4.11}$$

We start with the coordinate system (3.8). To preserve separation of variables the potential u must have the form

$$u = f(r) + \frac{1}{r^2}\sum_{1}^{k-1} v_i(\alpha_i). \tag{4.12}$$

The ignorable variable z is not orthogonal (see Eq. (3.9)), however the variables α_i do enter in the correct form. Writing $\psi(x) = R(r)Z(z)\prod_{i=1}^{k-1} A_i(\alpha_i)$, we obtain the following equations for each A_i

$$\left(\frac{\partial^2}{\partial\alpha_i^2} + v_i(\alpha_i)\right)A_i = -\mu_i A_i, \quad i = 1, \dots, k-1, \tag{4.13}$$

and, for each α_i separately, we can perform a Darboux transformation. Starting with the potential $v_i(\alpha_i) = 0$ we obtain a potential of the form (4.1), (4.2), or (4.3) with x replaced by α_i. Similarly, starting with a harmonic oscillator, or Coulomb potential we obtain (4.4), or (4.5), respectively, again with $x = \alpha_i$. Returning to Cartesian coordinates, we have

$$\alpha_i = -\frac{x_i\sqrt{2}}{x_k + x_0}, \quad i = 1, \dots k-1, \tag{4.14}$$

in all potentials.

The situation in the coordinates (3.12) is quite similar. A separable potential has the form (4.12), the equations for $A_i(\alpha_i)$ have the form (4.13). Darboux transformations can be used to induce all the potentials (4.1)–(4.5) with

$$x = \alpha_i = \frac{x_i\sqrt{2}}{\sqrt{2}q_i - x_k - x_0}, \quad i = 1, \dots, k-1. \tag{4.15}$$

The Darboux transformation can be made for each variable α_i separately and a different one can be chosen in each case.

5. Conclusions

The examples considered in this contribution are typical for a more general situation. Instead of pseudo-euclidean spaces we could consider the separation of variables in other homogeneous spaces. The corresponding isometry or conformal groups will provide us with tools for separating variables in the Laplace–Beltrami operator. Darboux transformations can then be used to generate integrable systems with nontrivial interactions in these spaces. This can be done for ignorable variables that are orthogonal. In the examples of Section 4 these are the ignorable variables α_i, not howerer the variable z, that is ignorable but not orthogonal to the variable r. A crucial step in this procedure and in the construction of coordinate systems with ignorable variables is classification of MASAs of the corresponding isometry or conformal Lie algebra.

References

1. G. Darboux, *Sur une proposition relative aux équations linéaires*, C. R. Acad. Sci. Paris **94** (1882), 1456–1459.
2. D. Levi, O. Ragnisco, and A. Sym, *Dressing method vs. classical Darboux transformation*, Nuovo Cimento B **83** (1984), 34–42.
3. V. B. Matveev and M. A. Salle, *Darboux transformations and solitons*, Springer-Verlag, Berlin, 1991.
4. G. Poeschl and E. Teller, *Bemerkungen zur Quantenmechanik des anharmonischen Oszillators*, Z. Physik **83** (1933), 143–151.
5. Z. Thomova and P. Winternitz, *Maximal Abelian subgroups of the isometry and conformal groups of Euclidean and Minkowski spaces*, J. Phys. A **31** (1998), 1831–1858.
6. ———, *Maximal Abelian subalgebras of $e(p,q)$ algebras*, Linear Algebra Appl. **291** (1999), 245–274.

SUNY Institute of Technology, School of Arts and Sciences, P.O. Box 3050, Utica, NY 13504-3050, USA.

E-mail address, Z. Thomova: `thomovz@sunyit.edu`

Centre de recherches mathématiques, Université de Montréal, C.P. 6128, Succ. Centre-ville, Montréal, Québec, H3C 3J7, Canada.

E-mail address: `wintern@crm.umontreal.ca`

Centre de Recherches Mathématiques
CRM Proceedings and Lecture Notes
Volume **29**, 2001

Bäcklund Transformations as Nonlinear Ordinary Differential, or Difference Equations with Superposition Formulas

P. Winternitz

ABSTRACT. A review is presented of the theory of nonlinear ordinary differential equations, the general solution of which can be expressed as a function of a finite number of particular solutions. Such equations include matrix Riccati equations and occur as Bäcklund transformations.

1. Introduction

Let us consider the well known Bäcklund transformation for the sine–Gordon equation

$$z_x - \tilde{z}_x = 2a \sin\frac{z+\tilde{z}}{2}, \quad z_y + \tilde{z}_y = \frac{2}{a}\sin\frac{z-\tilde{z}}{2}, \tag{1.1}$$

relating two solutions, z and $\tilde{z}$, of the sine–Gordon equation, $z_{xy} = \sin z$. The sine–Gordon equation satisfied by z and $\tilde{z}$, can be rationalized by putting $u = \tan z/4$, $\tilde{u} = \tan\tilde{z}/4$. We then obtain

$$\begin{aligned} u_x &= \frac{1}{1+\tilde{u}^2}\left[a\tilde{u} + \tilde{u}_x + a(1-\tilde{u}^2)u + (\tilde{u}_x - a\tilde{u})u^2\right], \\ u_y &= \frac{1}{1+\tilde{u}^2}\left[-\frac{1}{a}\tilde{u} - \tilde{u}_y + \frac{1}{a}(1-\tilde{u}^2)u + \left(\frac{\tilde{u}}{a} - \tilde{u}_y\right)u^2\right], \end{aligned} \tag{1.2}$$

where both u and $\tilde{u}$ satisfy

$$u_{xy} = \frac{u}{1+u^2}(2u_xu_y + 1 - u^2). \tag{1.3}$$

Thus, the Bäcklund transformation (1.2) for the "disguised" sine–Gordon equation (1.3) has the form of two Riccati equations for $u(x,y)$, once $\tilde{u}(x,y)$ is given.

The Riccati equation allows a "nonlinear superposition formula", i.e. its general solution can be written as a function of any 3 distinct particular solutions y_1, y_2

2000 *Mathematics Subject Classification.* Primary: 34A26; Secondary: 22E60.

The research reported here was partly supported by research grants from NSERC of Canada, and FCAR du Québec and NATO.

This is the final form of the paper.

and y_3 as

$$y(t) = \frac{y_2(y_1 - y_3)C - y_3(y_1 - y_2)}{(y_1 - y_3)C - (y_1 - y_2)}, \tag{1.4}$$

where C is an arbitrary constant.

The purpose of this lecture is to review the theory of nonlinear ordinary differential equations with superposition formulas. By this we mean systems of first order ODEs for which the general solution can be expressed as a function of a finite number of particular ones. Such equations occur as Bäcklund transformations for integrable nonlinear partial differential equations and their classification and analysis should lead to new types of integrable equations.

2. Indecomposable and Decomposable Systems of ODEs with Superposition Formulas

The study of nonlinear ODEs with superposition formulas was reduced to a problem in Lie group theory by S. Lie himself [**16**]. His result can be summed up as follows.

THEOREM 2.1 (S. Lie). *The general solution of* ODEs

$$\dot{\vec{y}} = \vec{\eta}(\vec{y}, t), \quad \vec{y}, \vec{\eta} \in F^n, \quad t \in F, \quad (F \sim \mathbb{R} \text{ or } \mathbb{C}), \tag{2.1}$$

can be written in the form

$$\vec{y}(t) = \vec{F}(\vec{y}_1, \dots, \vec{y}_m, c_1, \dots, c_n), \tag{2.2}$$

where $\vec{y}_1, \dots, \vec{y}_m$ are particular solutions of Eq. (2.1) *and $c_1, \dots, c_n$ are arbitrary constants, if and only if*

1. *The equation has the form*

$$\dot{\vec{y}} = \sum_{k=1}^{r} Z_k(t)\vec{\eta}_k(\vec{y}). \tag{2.3}$$

2. *The vector fields*

$$X_k = \sum_{\mu=1}^{n} \eta_k^{\mu}(\vec{y})\partial_{y^{\mu}}, \tag{2.4}$$

generate a finite dimensional Lie algebra L

$$[X_k, X_l] = \sum_{m=1}^{d} f_{kl}^{m} X_m, \quad d < \infty. \tag{2.5}$$

The number of fundamental solutions m, the number of equations n and the dimension d of the Lie algebra L satisfy

$$mn \geq d. \tag{2.6}$$

The classification of all ODEs with superposition formulas thus reduces to a classification of all finite-dimensional subalgebras of $\mathrm{diff}(n, F)$, the infinite dimensional Lie algebra of local diffeomorphisms of an n-dimensional complex, or real space. This turns out to be a very difficult task that has been completely solved only for $n = 1$ and $n = 2$ [**17**]. For $n = 1$ the result is very simple: the only finite dimensional subalgebras of diff$(1, F)$ are sl$(2, F)$ and its subalgebras. Up to local diffeomorphisms, there is just one realization, namely

$$X_1 = \frac{d}{dy}, \quad X_2 = y\frac{d}{dy}, \quad X_3 = y^2\frac{d}{dy}. \tag{2.7}$$

For $n = 2$ there exist infinitely many distinct Lie algebras and their dimension d can be arbitrary [**17**].

To continue to $n \geq 3$ and to obtain results valid for all values of n, it is necessary to proceed differently, making use of geometric ideas. First of all, we introduce the concept of indecomposability of the system of ODEs (2.3).

Let the variables $\{y_1, \ldots, y_n\}$ be local coordinates on some manifold M. Let G be a Lie group that acts transitively and effectively on M. We can then identify M as a quotient space $M \sim G/G_0$, where $G_0 \subset G$ is the isotropy group of the origin in M. The Lie algebra L of Lie's theorem is realized by the vector fields (2.4) and corresponds to the group G. The subgroup G_0 gives rise to the Lie algebra $L_0 \subset L$ of vector fields that vanish at the origin.

Thus, a system of ODEs satisfying the criteria of Lie's theorem is actually associated with a Lie algebra pair $\{L_0, L\}$. We shall call the system of ODEs (2.3) decomposable if the coordinates y_i can be split into two subsets $(y_1, \ldots, y_n) \sim (u_1, \ldots, u_{n_1}, v_1, \ldots, v_{n_2})$, $n = n_1 + n_2$, $n_1 \geq 1$, $n_2 \geq 1$ in such a manner that the variables u_i themselves obey a set of equations, satisfying Lie's theorem and not involving the variables v_j. This will occur if all vector fields (2.4) simultaneously have the form

$$X_k = \sum_{i=1}^{n_1} \xi^i(\vec{u})\partial_{u_i} + \sum_{j=1}^{n_2} \eta^j(\vec{u}, \vec{v})\partial v_j. \tag{2.8}$$

The crucial point is that the coefficients ξ^i depend only on the variables $(u_1, \ldots, u_{n_1})$. If no such splitting of the coordinates exists, the system of ODEs is indecomposable.

Globally, the existence of such coordinates means that there exists a G-invariant foliation of the manifold M. Otherwise, we say that the action of G on M is primitive.

Locally, the absence of an invariant foliation of M, and hence the indecomposability of the system ODEs, implies that the pair of Lie algebras L_0 and L form a *transitive primitive effective Lie algebra*. This means:

1. L_0 is a maximal subalgebra of L.
2. L_0 does not contain an ideal of L.

A classification of transitive primitive Lie algebras exists [**5, 6, 9, 14, 15, 19**] and it was used to classify the indecomposable systems of ODEs with superposition formulas [**18, 21, 22**]. The classification can be summed up as a theorem.

THEOREM 2.2. *Five types of transitive primitive Lie algebras exist. They are*

1. *L simple, L_0 a maximal parabolic irreducible subalgebra.*
2. *L simple, L_0 a maximal parabolic reducible subalgebra.*
3. *L simple, L_0 a maximal reductive subalgebra.*
4. *$L \sim K \oplus K$, K simple, $L_0 \sim K_D$ ($K_D \sim K$ is the "diagonal" subalgebra of $K \oplus K$).*
5. *L an affine Lie algebra, L_0 a reductive subalgebra.*

Here we shall consider only one example, belonging to the first class listed in Theorem 2.2. Thus, L will be a simple Lie algebra, $L \sim \mathrm{sl}(N, F)$ and L_0 a maximal parabolic irreducible subalgebra that we shall denote $p(r, s)$. We put

$$N = r + s, \quad r \geq 1, \quad s \geq 1, \quad n = r \cdot s. \tag{2.9}$$

Elements ξ of $\mathrm{sl}(N, F)$ and ξ_0 of $p(r, s)$ will be represented by block matrices

$$\begin{gathered} \xi = \begin{pmatrix} C & A \\ -D & -B \end{pmatrix}, \quad C \in F^{r \times r}, \quad B \in F^{s \times s}, \\ \operatorname{Tr} C = \operatorname{Tr} B, \quad A \in F^{r \times s}, \quad D \in F^{s \times r}, \end{gathered} \tag{2.10}$$

and ξ_0 will be as in Eq. (2.10) with $A \equiv 0$. The manifold

$$M \sim G/G_0 \sim \mathrm{SL}(N, F)/P(r, s),$$

is in this case constructed as a Grassmannian of r planes in F^N. We first introduce homogeneous coordinates

$$U = \begin{pmatrix} X \\ Y \end{pmatrix} \sim \begin{pmatrix} XH \\ YH \end{pmatrix}, \quad X \in F^{r \times s}, \quad Y \in F^{s \times s}, \quad H \in \mathrm{GL}(s, F). \tag{2.11}$$

In the neighbourhood of the origin $X = 0$, $Y = I$ we introduce affine coordinates $W = XY^{-1}$, thus getting rid of the redundancy inherent in the affine coordinates. The coordinates are the matrix elements of the matrices U and W, respectively.

In homogeneous coordinates we write a system of linear equations

$$\begin{pmatrix} \dot{X} \\ \dot{Y} \end{pmatrix} = \begin{pmatrix} C & A \\ -D & -B \end{pmatrix} \begin{pmatrix} X \\ Y \end{pmatrix}, \quad \operatorname{Tr} C = \operatorname{Tr} B. \tag{2.12}$$

The affine coordinates $W = XY^{-1}$ then satisfy a system of matrix Riccati equations

$$\dot{W} = A + WB + CW + WDW. \tag{2.13}$$

The number of equations is $n = r \cdot s$ and we have $N = r + s$. The vector fields (2.4) of Lie's theorem are

$$\begin{gathered} \hat{A}_{\alpha a} = \partial_{w_{\alpha a}}, \quad \widehat{B}_{ba} = w_{\alpha b} \partial w_{\alpha a}, \quad \widehat{C}_{\alpha\beta} = w_{\beta a} \partial_{w_{\alpha a}}, \\ \widehat{D}_{b\beta} = w_{\alpha b} w_{\beta a} \partial_{w_{\alpha a}}, \end{gathered} \tag{2.14}$$

and by construction, they generate $\mathrm{sl}(N, F)$ (greek letters vary in the range $1 \leq \alpha \leq r$, latin ones in $1 \leq a \leq s$).

The superposition formula is given by the group action, corresponding to the vector fields (2.14), i.e.

$$W(t) = \big(G_{11}(t) W_0 + G_{12}(t)\big)\big(G_{21}(t) W_0 + G_{22}(t)\big)^{-1}, \tag{2.15}$$

where the matrix functions $G_{ik}(t)$ are expressed in terms of known solutions of Eq. (2.13) and W_0 is a constant matrix, related to the initial condition.

For $r > s = 1$ Eq. (2.13) is a projective Riccati equation. The number of equations in $n = r$, the number of solutions needed is $m = n + 2$ [**1, 2**]. For $r = s$ we have square matrix Riccati equations. The number of solutions needed for $r = s \geq 2$ is $m = 5$ [**11**].

Projective Riccati equations occur as Bäcklund transformations e.g. for the Boussinesq equation [**4**], or the multidimensional sine–Gordon equations [**3**]. Matrix Riccati equations occur in the same role e.g. for σ models [**10, 20**].

Decomposable systems of ODEs with superposition formulas can also be constructed systematically. The equations and superposition formulas are generated by imprimitive group actions. The case that has been considered so far is $M \sim \mathrm{SL}(N,F)/P_0$, where P_0 is a nonmaximal parabolic subgroup [**12, 13**]. This leads to equations with polynomial nonlinearities; the polynomials being of order 3 or higher. Such equations have so far, to our knowledge, not appeared as Bäcklund transformations. The space M in this case has the structure of a flag manifold. Integrable systems on flag manifolds have been studied [**24**] and we expect the corresponding Bäcklund transformations to have the form induced by imprimitive group actions.

3. Difference Equations with Superposition Formulas

A problem that has been recently addressed [**8**] is that of discretizing nonlinear ODEs with superposition formulas, while preserving their essential properties. These are:

1. The superposition formula itself.
2. The fact that these equations are linearizable (by an embedding into a higher dimensional space).
3. The equations have the Painlevé property.

To preserve these properties, the discretization must preserve the underlying group structure.

The ODEs (2.3) can be symbolically written as

$$\dot{y}^{\mu}(t) = \hat{l}(t)y^{\mu}(t) \quad 1 \leq \mu \leq n, \tag{3.1}$$

$$\hat{l}(t) = \sum_{k=1}^{r} Z_k(t) \sum_{\alpha=1}^{n} \eta_k^{\alpha}(\vec{y}) \frac{\partial}{\partial y^{\alpha}}. \tag{3.2}$$

We see that the operator $\hat{l}(t)$ is a time dependent vector field following some path in the Lie algebra L as t varies. For t fixed we can integrate the vector fields representing L and obtain a group element $g \in G$. As t varies we obtain a path $g(t)$ in the Lie group G, corresponding to the given path in the Lie algebra.

The discretized equation (2.3) will have the form

$$\vec{y}(t+\sigma) = g(t)\vec{y}(t), \tag{3.3}$$

where σ is the lattice step and $g(t)$ is an operator representing the group action.

This discretization has been carried out for all homogeneous spaces $M \sim G/G_0$ where G is simple and G_o maximal parabolic [**8**].

Here we shall just present one example, namely the discrete matrix Riccati equation. To obtain it we make use of the linearization (2.12) of the system (2.13). The linear equations (2.12) are discretized by putting:

$$\begin{pmatrix} X(t+\sigma) \\ y(t+\sigma) \end{pmatrix} = \begin{pmatrix} G_{11}(t) & G_{12}(t) \\ G_{21}(t) & G_{22}(t) \end{pmatrix} \begin{pmatrix} X(t) \\ Y(t) \end{pmatrix}. \tag{3.4}$$

Viewing (X, Y) as homogeneous coordinates and passing to affine ones $W = XY^{-1}$, we obtain the discrete matrix Riccati equation as

$$W(t+\sigma) = \big(G_{11}(t)W(t) + G_{12}(t)\big)\big(G_{21}(t)W(t) + G_{22}(t)\big)^{-1}. \tag{3.5}$$

The continuous limit is obtained by taking the group element $g(t)$ close to the origin

$$G_{11} = I + \varepsilon C, \quad G_{12} = \varepsilon A, \quad G_{21} = -\varepsilon D, \quad G_{22} = I - \varepsilon B,$$

and then identifying the group parameter ε with the step σ

$$\lim_{\sigma\to 0}\frac{W(t+h)-W(t)}{\sigma}=A+WB+CW+WDW. \tag{3.6}$$

Thus, we have a method of associating a difference equation with every ODE with a superposition formula. We simply replace the Lie algebra action by the corresponding group action.

The superposition formula has the same form as for the ODE. The group elements must now be expressed in terms of m solutions of the difference equation. The difference equations are linearizable by construction. Movable singularities of solutions can only be produced by the projection from the space where the group acts linearly to the space $M\sim G/G_0$. Hence the Painlevé property for ODEs and its discrete analog, singularity confinement [**7**].

The described method of discretizing ODEs with superposition formulas has an additional benefit, at least for ODEs with constant coefficients. Namely, the ODEs and their discretizations have identical solution sets [**23**].

Here we present the result for the matrix Riccati equation.

THEOREM 3.1. *Consider the matrix Riccati equation* (2.13) *with* A, B, C *and* D *constant matrices and its discretization* (3.5), *where*

$$\begin{pmatrix} G_{11} & G_{12}\\ G_{21} & G_{22}\end{pmatrix}=\exp\sigma Z,\quad Z\equiv\begin{pmatrix} C & A\\ -D & -B\end{pmatrix}. \tag{3.7}$$

The solutions of Eq. (2.13) *and* (3.5) *coincide exactly.*

PROOF. Let us rewrite the discrete linear system (3.4) in the form

$$e^{\sigma(d/dt)}\xi(t)=e^{\sigma Z}\xi(t),\quad \xi(t)\equiv\begin{pmatrix} X(t)\\ Y(t)\end{pmatrix}. \tag{3.8}$$

Expanding both sides of Eq. (3.8) into a power series and comparing different powers of σ, we have

$$\frac{d^n}{dt^n}\xi=Z^n\xi. \tag{3.9}$$

For $n=1$ we obtain the original equation (2.13) which is satisfied by assumption. For $n=2$ we have

$$\frac{d^2}{dt^2}\xi=\frac{d}{dt}(Z\xi)=Z\frac{d}{dt}\xi=Z^2\xi, \tag{3.10}$$

so (3.9) with $n=2$ is satisfied (by virtue of the fact that $dZ/dt=0$). Using induction it is easy to show that relation (3.9) for all n holds as a consequence of it holding for $n=1$ (for Z a constant matrix). Thus, every solution of Eq. (2.13) is also a solution of Eq. (3.5). Vice versa, let $\xi(t)$ be the general solution of the discrete equation (3.8). It depends on $(r+s)\cdot s$ constants, as does the general solution of the continuous equation (3.9) with $n=1$. Hence the two solution sets coincide.

As an example, consider the real constant coefficient Riccati equation itself. The equation can be transformed into

$$\dot{y}=\omega(y^2+\kappa),\quad \kappa=\begin{cases} 1\\ -1\\ 0\end{cases}. \tag{3.11}$$

Let us consider the case $\kappa = 1$. The discretization is

$$y(t+\sigma) = \frac{y(t)\cos\sigma\omega + \sin\sigma\omega}{-y(t)\sin\sigma\omega + \cos\sigma\omega}. \tag{3.12}$$

It is easy to check that the general solution of both these equations is

$$y(t) = \tan h\omega(t - t_0). \quad \square \tag{3.13}$$

4. Conclusion

A systematic method for constructing Bäcklund transformations for a given partial differential equation exists [**25, 26**]. It makes use of exterior differential forms but can also be formulated in terms of an overdetermined system of differential equations. It involves the construction of a closed ideal of differential forms. This closure is achieved by Lie theoretical requirements that force the Bäcklund transformations, when written as a system of first order ODEs, to have the form (2.3)–(2.5).

Thus the theory of ODEs with superposition formulas provides a setting for classifying Bäcklund transformations.

The Bäcklund transformations readily available in the literature correspond to just one of the classes of transitive primitive Lie algebras listed in Theorem 2.2. The case is that of L simple and L_0 maximal parabolic irreducible. The nonlinearities in the ODEs (and in the Bäcklund transformations) are quadratic (as in matrix Riccati equations).

We expect the other classes of ODEs and also their discretizations to occur in a similar role for other types of integrable systems.

References

1. R. L. Anderson, *A nonlinear superposition principle admitted by coupled Riccati equations of the projective type*, Lett. Math. Phys. **4** (1980), 1–7.
2. R. L. Anderson, J. Harnad, and P. Winternitz, *Group theoretical approach to superposition rules for systems of Riccati equations*, Lett. Math. Phys. **5** (1981), 143–148.
3. R. Beals and K. Tenenblat, *Inverse scattering and the Bäcklund transformations for the generalized wave and generalized Sine–Gordon equations*, Stud. Appl. Math. **78** (1988), 227–256.
4. H. H. Chen, *General derivation of Bäcklund transformations from inverse scattering problems*, Phys. Rev. Lett. **33** (1974), 925–928.
5. M. Demazure, *Classification des algèbres de Lie filtrées*, Séminaire Bourbaki, vol. 1966-1977, 19e année, W. A. Benjamin, New York, 1968, pp. 326.01–326.11.
6. M. Golubitsky, *Primitive actions and maximal subgroups of Lie groups*, J. Differential Geom. **7** (1972), 175–191.
7. B. Grammaticos, A. Ramani, and K. M. Tamizhmani, *Non-proliferation of pre-images in integrable mappings*, J. Phys. A **27** (1994), 559–566.
8. G. Grammaticos, A. Ramani, and P. Winternitz, *Discretizing families of linearizable equations*, Phys. Lett. A **245** (1998), 382–388.
9. V. W. Guillemin and S. Sternberg, *An algebraic model of transitive differential geometry*, Bull. Amer. Math. Soc. **70** (1964), 16–47.
10. J. Harnad, Y. Saint-Aubin, and S. Shnider, *The soliton correlation matrix and the reduction problem for integrable systems*, Comm. Math. Phys. **93** (1984), 33–56.
11. J. Harnad, P. Winternitz, and R. L. Anderson, *Superposition principles for matrix Riccati equations*, J. Math. Phys. **24** (1983), 1062–1072.
12. M. Havliček, S. Pošta, and P. Winternitz, *Nonlinear superposition formulas based on imprimitive group action*, J. Math. Phys. **40** (1999), 3104–3122.
13. ———, *Superposition formulas based on nonprimitive group action*, these Proceedings.

14. S. Kobayashi and T. Nagano, *On filtered Lie algebras and geometric structures* I, J. Math. Mech. **13** (1964), 875–907.
15. ———, *On filtered Lie algebras and geometric structures* II, J. Math. Mech. **14** (1965), 513–521.
16. S. Lie, *Vorlesungen über continuirliche Gruppen mit geometrischen und anderem Anwendungen*, Bearbeitet und herausgegeben von Dr. G. Scheffers, Teubner, Leipzig, 1893.
17. ———, *Theorie der Transformationgruppen*, Dritter Abschnitt, Abteilun. I., Unter Mitwirkung von Pr. F. Engel, Teubner, Leipzig, 1893.
18. L. Michel and P. Winternitz, *Families of transitive primitive maximal simple Lie subalgebras of* diff n, Adv. Math. Sci., CRM's 25 years, vol. 11, CRM Proceedings and Lecture Notes, AMS, Providence, R.I., 1997, pp. 451–479.
19. T. Ochiai, *Classification of the finite nonlinear primitive Lie algebras*, Trans. Amer. Math. Soc. **124** (1966), 313–322.
20. A. T. Ogielski, M. K. Prasad, A. Sinha, and L. L. Chau Wang, *Bäcklund transformations and local conservation laws for principle chiral fields*, Phys. Lett. B **91** (1980), 387–391.
21. S. Schnider and P. Winternitz, *Nonlinear equations with superposition principles and the theory of transitive primitive Lie algebras*, Lett. Math. Phys. **8** (1984), 69–78.
22. ———, *Classification of systems of nonlinear ordinary differential equations with superposition principles*, J. Math. Phys. **25** (1984), 3155–3165.
23. A. Turbiner and P. Winternitz, *Solutions of nonlinear differential and difference equations with superposition formulas*, Lett. Math. Phys. **50** (1999), 189–201.
24. K. Ueno and K. Takasaki, *Toda lattice hierarchy*, Adv. Stud. Pure Math. **4**, North Holland, Amsterdam, 1984.
25. H. D. Wahlquist and F. B. Estabrook, *Prolongation structures and nonlinear evolution equations* I, II, J. Math. Phys. **16** (1995), 1–7.
26. ———, *Prolongation structures and nonlinear evolution equations* I, II, J. Math. Phys. **17** (1976), 1403–1411.

Centre de recherches mathématiques, Université de Montréal, C.P. 6128, Succ. Centre-ville, Montréal, Québec, H3C 3J7, Canada.

E-mail address: `wintern@crm.umontreal.ca`